This is an updated and expanded second edition of a successful and well-reviewed text presenting a detailed exposition of the modern theory of supermanifolds, including a rigorous account of the super-analogs of all the basic structures of ordinary manifold theory.

The exposition opens with the theory of analysis over supernumbers (Grassman variables), Berezin integration, supervector spaces and the superdeterminant. This basic material is then applied to the theory of supermanifolds, with an account of super-analogs of Lie derivatives, connections, metric, curvature, geodesics, Killing flows, conformal groups, etc. The book goes on to discuss the theory of super Lie groups, super Lie algebras, and invariant geometrical structures on coset spaces. Complete descriptions are given of all the simple super Lie groups. The book then turns to applications. Chapter 5 contains an account of the Peierls bracket for superclassical dynamical systems, super Hilbert spaces, path integration for fermionic quantum systems, and simple models of Bose–Fermi supersymmetry. The sixth and final chapter, which is new in this revised edition, examines dynamical systems for which the topology of the configuration supermanifold is important. A concise but complete account is given of the path-integral derivation of the Chern–Gauss–Bonnet formula for the Euler–Poincaré characteristic of an ordinary manifold, which is based on a simple extension of a point particle moving freely in this manifold to a supersymmetric dynamical system moving in an associated supermanifold. Many exercises are included to complement the text.

Review comment on the first edition

'*Supermanifolds* is destined to become the standard work for all serious study of super-symmetric theories of physics.' *Nature*

CAMBRIDGE MONOGRAPHS ON
MATHEMATICAL PHYSICS

General Editors: P. V. Landshoff, D. R. Nelson, D. W. Sciama, S. Weinberg

SUPERMANIFOLDS

SECOND EDITION

Cambridge Monographs on Mathematical Physics

A. M. Anile *Relativistic Fluids and Magneto-Fluids*
J. Bernstein *Kinetic Theory in the Early Universe*
N. D. Birrell and P. C. W. Davies *Quantum Fields in Curved Space*†
D. M. Brink *Semiclassical Methods in Nucleus–Nucleus Scattering*
J. C. Collins *Renormalization*†
P. D. B. Collins *An Introduction to Regge Theory and High Energy Physics*
M. Creutz *Quarks, Gluons and Lattices*†
F. de Felice and C. J. S. Clarke *Relativity on Curved Manifolds*
B. DeWitt *Supermanifolds, second edition*†
P. G. O. Freund *Introduction to Supersymmetry*†
F. G. Friedlander *The Wave Equation on a Curved Space-Time*
J. A. H. Futterman, F. A. Handler and R. A. Matzner *Scattering from Black Holes*
M. Göckeler and T. Schücker *Differential Geometry, Gauge Theories and Gravity*†
M. B. Green, J. H. Schwarz and E. Witten *Superstring Theory, volume 1: Introduction*†
M. B. Green, J. H. Schwarz and E. Witten *Superstring Theory, volume 2: Loop Amplitudes, Anomalies and Phenomenology**
S. W. Hawking and G. F. R. Ellis *The Large-Scale Structure of Space-Time*†
F. Iachello and A. Arima *The Interacting Boson Model*
F. Iachello and P. van Isacker *The Interacting Boson–Fermion Model*
C. Itzykson and J.-M. Drouffe *Statistical Field Theory, volume 1: From Brownian Motion to Renormalization and Lattice Gauge Theory*†
C. Itzykson and J.-M. Drouffe *Statistical Field Theory, volume 2: Strong Coupling, Monte Carlo Methods, Conformal Field Theory, and Random Systems*†
J. I. Kapusta *Finite-Temperature Field Theory*
D. Kramer, H. Stephani, M. A. H. MacCallum and E. Herlt *Exact solutions of Einstein's Field Equations*
N. H. March *Liquid Metals: Concepts and Theory*
L. O'Raifeartaigh *Group Structure of Gauge Theories*†
A. Ozorio de Almeida *Hamiltonian systems: Chaos and Quantization*†
R. Penrose and W. Rindler *Spinors and Space-time, volume 1: Two-Spinor Calculus and Relativistic Fields*†
R. Penrose and W. Rindler *Spinors and Space-time, volume 2: Spinor and Twistor Methods in Space-Time Geometry*
S. Pokorski *Gauge Field Theories*†
V. N. Popov *Functional Integrals and Collective Excitations*†
R. Rivers *Path Integral Methods in Quantum Field Theory*†
R. G. Roberts *The Structure of the Proton*
W. C. Saslaw *Gravitational Physics of Stellar and Galactic Systems*†
J. M. Stewart *Advanced General Relativity*
R. S. Ward and R. O. Wells Jr *Twistor Geometry and Field Theories*†

† Issued as a Paperback

SUPERMANIFOLDS

SECOND EDITION

BRYCE DEWITT

Jane and Roland Blumberg Professor of Physics, University of Texas at Austin

CAMBRIDGE UNIVERSITY PRESS

CAMBRIDGE

NEW YORK PORT CHESTER

MELBOURNE SYDNEY

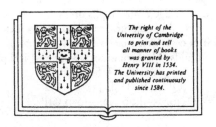

CAMBRIDGE UNIVERSITY PRESS
Cambridge, New York, Melbourne, Madrid, Cape Town, Singapore,
São Paulo, Delhi, Dubai, Tokyo, Mexico City

Cambridge University Press
The Edinburgh Building, Cambridge CB2 8RU, UK

Published in the United States of America by
Cambridge University Press, New York

www.cambridge.org
Information on this title: www.cambridge.org/9780521423779

First published 1984
First paperback edition 1985
Reprinted with corrections 1987
Second edition 1992

A catalogue record for this publication is available from the British Library

Library of Congress Cataloguing in Publication Data

DeWitt, Bryce S. (Bryce Seligman), 1923–
Supermanifolds/Bryce DeWitt. – 2nd edn.
p. cm. – (Cambridge monographs on mathematical physics)
Includes bibliographical references and index.
ISBN 0 521 41320 6.– 0 521 42377 5 (pbk.)
1. Supermanifolds (Mathematics) 2. Mathematical physics.
I. Title. II. Series.
QC20.7.M24D47 1992
530.1′5–dc20 91–11070 CIP

ISBN 978-0-521-41320-6 Hardback
ISBN 978-0-521-42377-9 Paperback

Contents

Preface to first edition

This book is an outgrowth of a book on quantum gravity that the author started to write nine years ago in collaboration with Christopher Isham. It began as an Appendix to the quantum gravity book, but subsequent developments modified the original plan. Firstly, new results in quantum gravity, particularly in supergravity and in the applications of topology to quantum field theory, appeared so rapidly that the timing of the collaborative volume became inopportune. Secondly, the theory of supermanifolds had so many loose ends which needed to be dealt with that the original Appendix grew beyond reasonable size limits and turned into a book in its own right.

A previous generation of theoretical physicists could function adequately with a knowledge of the theory of ordinary manifolds and ordinary Lie groups. With the discovery of Bose–Fermi supersymmetry all this changed. Nowadays the theorist must know about supermanifolds and super Lie groups. The purpose of the present volume is to provide him with an easily accessible account of these mathematical structures. Mathematicians will find much of this book incomplete and expressed in language that they have nowadays passed beyond, but it is probably pitched about right for the average physicist. It still has something of the character of an Appendix in its lack of any account of how it relates to supergravity and other locally supersymmetric theories. For a time it was to have appeared as Volume I of a two-volume work on supermanifolds and supersymmetry written in collaboration with Peter van Nieuwenhuizen and Peter West. However, delays caused by new developments in supergravity theory, particularly in higher-dimensional Kaluza–Klein versions of supergravity theory, rendered this linkage impractical. The second volume will ultimately appear (in the same Cambridge University Press series), but rather than delay the first volume further, the decision was made to publish the two as separate books. While waiting for the second book to appear, the reader of the present volume who wishes to establish linkages to physics will have to content himself with studying the elementary applications of supermanifold theory selected in chapter 5 of this volume and with reading the already vast literature on supersymmetric theories.

The author wishes to express his gratitude to the following for their support at various times during the writing of this book: The Warden and Fellows of All Souls College, Oxford, where the book was begun, the John Simon Guggenheim Foundation, the United States National Science Foundation, the North Atlantic Treaty Organization and the University of Texas.

Preface to the second edition

At the end of the fifth and last chapter of the first edition the author wrote that if the book were ever to be revised it would include an account of the beautiful work of E. Witten and of L. Alvarez-Gaumé on supersymmetry, Morse theory and the Atiyah–Singer index theorem. Chapter 6 of this revised edition is a partial fulfilment of the promise. The aim of chapter 6, like that of chapter 5, is almost exclusively pedagogical. Unlike chapter 5, however, chapter 6 deals with nontrivial supermanifolds, and the author discovered that there are numerous fine points in the theory of the Feynman functional integral for such supermanifolds that are not adequately covered in the literature, even on a formal level. To be pedagogically helpful the book *has* to deal with these issues, given the fact that, despite the essential role it plays in chapter 6, the functional integral is used in a formal way rather than as a rigorous tool. This has meant that, in order to keep the reader's confidence, the author has had to expend a large part of his effort on the functional integral itself and hence could include only a little of the flavor of the index theorem, as it touches the Euler–Poincaré characteristic. The effort to display the internal consistency of the functional integral formalism has nevertheless been useful in that it presents a challenge to the student to attempt what must surely be possible, namely, to establish the functional integral at last on a fully rigorous basis for both bosonic and fermionic systems.

1
Analysis over supernumbers

1.1 Supernumbers and superanalytic functions

Grassmann algebras

Let ζ^a, $a = 1, \ldots N$, be a set of generators for an algebra, which anticommute:

$$\zeta^a \zeta^b = -\zeta^b \zeta^a, \qquad (\zeta^a)^2 = 0, \quad \text{for all } a, b. \tag{1.1.1}$$

The algebra is called a *Grassmann algebra* and will be denoted by Λ_N. We shall usually, though not always, deal with the formal limit $N \to \infty$. The corresponding algebra will be denoted by Λ_∞.

The elements $1, \zeta^a, \zeta^a\zeta^b, \ldots,$ where the indices in each product are all different, form an infinite basis for Λ_∞. When N is finite the sequence terminates at $\zeta^1 \ldots \zeta^N$ and there are only 2^N distinct basis elements. Under addition as well as multiplication by a complex number, the elements of Λ_N form a linear vector space of 2^N dimensions; the elements of Λ_∞ form an infinite-dimensional vector space. As algebras over the complex numbers (which is the only field we shall consider) Λ_N and Λ_∞ are associative but not commutative (excluding the trivial cases $N = 0, 1$).

Supernumbers

The elements of Λ_∞ will be called *supernumbers*. Every supernumber can be expressed in the form

$$z = z_B + z_S, \tag{1.1.2}$$

where z_B is an ordinary complex number and

$$z_S = \sum_{n=1}^{\infty} \frac{1}{n!} c_{a_1 \ldots a_n} \zeta^{a_n} \ldots \zeta^{a_1}, \tag{1.1.3}$$

the c's also being complex numbers. The c's are completely antisymmetric in their indices, and summation over repeated indices is to be understood unless otherwise stated. The number z_B will be called the *body* and the remainder z_S will be called the *soul* of the supernumber z. If Λ_∞ is replaced by Λ_N (N finite) then the soul of a supernumber is always nilpotent:

$$z_S^{N+1} = 0. \tag{1.1.4}$$

When N is infinite the soul need not be nilpotent.

When N is finite the condition $\zeta^a z = 0$ for all a implies that z has the form $z = c\zeta^1 \ldots \zeta^N$ for some complex number c. When N is infinite the condition $\zeta^a z = 0$ for all a implies $z = 0$.

A supernumber has an inverse if and only if its body is nonvanishing. The inverse, which is unique, is given by the formula

$$z^{-1} = z_B^{-1} \sum_{n=0}^{\infty} (-z_B^{-1} z_S)^n. \tag{1.1.5}$$

Series of this kind may be introduced to extend any analytic function f on the complex numbers to a supernumber-valued function on Λ_∞:

$$f(z) = \sum_{n=0}^{\infty} \frac{1}{n!} f^{(n)}(z_B) z_S^n. \tag{1.1.6}$$

Here $f^{(n)}(z_B)$ denotes the nth derivative of f at the point z_B in the complex plane, and the definition is valid for all z_B that are not singular points of f. Because of (1.1.4) the 'Taylor series' (1.1.6) terminates when N is finite. By substituting (1.1.3) into (1.1.6) one can obtain an expansion of $f(z)$ in terms of the basis elements of Λ_∞. When N is infinite expressions (1.1.5) and (1.1.6), as well as their expansions in terms of basis elements, are formal infinite series. The coefficient of each term is unique and finite.

One may consider matrices whose elements are supernumbers. The body of a matrix is then defined as the ordinary matrix obtained by replacing each element with its body. The soul of a matrix is the remainder. A square matrix has an inverse, and is said to be *nonsingular*, if and only if its body is nonsingular. The inverse is unique and is given by a formula analogous to (1.1.5).

c-numbers and a-numbers

Any supernumber may be split into its even and odd parts:

$$z = u + v, \tag{1.1.7}$$

$$u = z_B + \sum_{n=1}^{\infty} \frac{1}{(2n)!} c_{a_1 \ldots a_{2n}} \zeta^{a_{2n}} \ldots \zeta^{a_1}, \tag{1.1.8}$$

$$v = \sum_{n=0}^{\infty} \frac{1}{(2n+1)!} c_{a_1 \ldots a_{2n+1}} \zeta^{a_{2n+1}} \ldots \zeta^{a_1}. \tag{1.1.9}$$

We shall often confine our attention to supernumbers that are either purely odd or purely even. Odd supernumbers anticommute among themselves and will be called *a-numbers*. Even supernumbers commute with everything and will be called *c-numbers*. A-numbers possess no body and hence are not invertible. The set of all c-numbers is a commutative subalgebra of Λ_∞, which will be denoted by $\mathbf{C}_c$. The set of all a-numbers will be denoted by $\mathbf{C}_a$; it is *not* a subalgebra. The product of two c-numbers, or of two a-numbers, is a c-number. The product of an a-number and a c-number is an a-number. The square of every a-number vanishes.

If Λ_∞ is replaced by Λ_N, $\mathbf{C}_c$ and $\mathbf{C}_a$ become 2^{N-1}-dimensional vector spaces. In the formal limit $N \to \infty$ they may continue to be regarded as vector spaces, but we shall not give them a norm or even a topology.

Superanalytic functions of supernumbers

Just as ordinary analysis can be constructed as the theory of analytic mappings of the complex plane into itself, so can an analytic theory of functions of c-numbers and a-numbers be built up by studying mappings from $\mathbf{C}_c$ or $\mathbf{C}_a$ to Λ_∞.

Consider first $\mathbf{C}_a$. Let Λ_∞ for the moment be replaced by Λ_N and let f be a mapping from $\mathbf{C}_a$ into Λ_N. Since, when N is finite, $\mathbf{C}_a$ and Λ_N are finite-dimensional vector spaces over the complex numbers, one has a body of conventional theory on which to draw in order to define, for example, the condition that f be a differentiable mapping at a point v of $\mathbf{C}_a$. Mere differentiability, however, does not involve the algebraic structure of $\mathbf{C}_a$ and Λ_N. What is more interesting is to pass immediately to the formal limit $N \to \infty$ and to demand that f be *superanalytic* at that point. By this is meant the following: let v be given an arbitrary infinitesimal a-number displacement $\mathrm{d}v$. Then its image $f(v)$ in Λ_∞ must suffer a displacement which, for all $\mathrm{d}v$. takes the form

$$\mathrm{d}f(v) = \mathrm{d}v\left[\frac{\overrightarrow{\mathrm{d}}}{\mathrm{d}v}f(v)\right] = \left[f(v)\frac{\overleftarrow{\mathrm{d}}}{\mathrm{d}v}\right]\mathrm{d}v, \tag{1.1.10}$$

where the coefficients $\frac{\overrightarrow{\mathrm{d}}}{\mathrm{d}v}f(v)$ and $f(v)\frac{\overleftarrow{\mathrm{d}}}{\mathrm{d}v}$ are independent of $\mathrm{d}v$ and depend (at most) only on v. These coefficients are called, respectively, the *left* and *right derivatives* of f with respect to v.

It can be shown (see exercise **1.1**) that the general solution of eq. (1.1.10) is

$$f(v) = a + bv, \tag{1.1.11}$$

where a and b are arbitrary constant elements of Λ_∞. That is, a superanalytic function of an a-number variable is simply a linear function! It is therefore superanalytic everywhere in $\mathbf{C}_a$ (no singularities). If the coefficient b is separated into its even and odd parts,

$$b = b_e + b_o, \tag{1.1.12}$$

then one may write

$$f(v)\frac{\overleftarrow{\mathrm{d}}}{\mathrm{d}v} = b_e + b_o, \quad \frac{\overrightarrow{\mathrm{d}}}{\mathrm{d}v} f(v) = b_e - b_o, \tag{1.1.13}$$

and one sees that the left and right derivatives of f are, in fact, constants, independent of v, a fact that may be expressed in the form

$$f(v)\frac{\overleftarrow{\mathrm{d}}\,\overleftarrow{\mathrm{d}}}{\mathrm{d}v\,\mathrm{d}v} = \frac{\overrightarrow{\mathrm{d}}}{\mathrm{d}v} f(v)\frac{\overleftarrow{\mathrm{d}}}{\mathrm{d}v} = \frac{\overrightarrow{\mathrm{d}}\,\overrightarrow{\mathrm{d}}}{\mathrm{d}v\,\mathrm{d}v} f(v) = 0. \tag{1.1.14}$$

If the range of f is contained in $\mathbf{C}_c$ then a is even, b is odd, and $\frac{\overrightarrow{\mathrm{d}}}{\mathrm{d}v} f(v) = -f(v)\frac{\overleftarrow{\mathrm{d}}}{\mathrm{d}v}$. If, on the other hand, the range of f is contained in $\mathbf{C}_a$ then a is odd, b is even and $\frac{\overrightarrow{\mathrm{d}}}{\mathrm{d}v} f(v) = f(v)\frac{\overleftarrow{\mathrm{d}}}{\mathrm{d}v}$.

Superanalytic mappings f from $\mathbf{C}_c$ to Λ_∞ are defined similarly:

$$\mathrm{d}f(u) = \mathrm{d}u\left[\frac{\overrightarrow{\mathrm{d}}}{\mathrm{d}u} f(u)\right] = \left[f(u)\frac{\overleftarrow{\mathrm{d}}}{\mathrm{d}u}\right]\mathrm{d}u. \tag{1.1.15}$$

But here the similarity ends. Because $\mathrm{d}u$ is a c-number it follows that

$$\frac{\overrightarrow{\mathrm{d}}}{\mathrm{d}u} f(u) = f(u)\frac{\overleftarrow{\mathrm{d}}}{\mathrm{d}u}, \tag{1.1.16}$$

so there is no need to distinguish between right and left derivatives. Moreover, the class of superanalytic functions of a c-number variable is infinitely richer than the class represented by eq. (1.1.11). For example, for every ordinary analytic function f on the complex numbers there is a superanalytic function over $\mathbf{C}_c$ analogous to (1.1.6):

$$f(u) \overset{\text{def}}{=} \sum_{n=0}^{\infty} \frac{1}{n!} f^{(n)}(u_{\mathrm{B}}) u_{\mathrm{S}}^n, \tag{1.1.17}$$

where u_{B} and u_{S} are respectively the body and soul of u. The general solution of eq. (1.1.15) has the form

$$f(u) = \sum_{n=0}^{\infty} \frac{1}{n!} f_{a_1 \ldots a_n}(u) \zeta^{a_n} \ldots \zeta^{a_1}, \tag{1.1.18}$$

where the $f_{a_1 \ldots a_n}(u)$ are functions like (1.1.17). If the range of f is contained in $\mathbf{C}_c$ then the $f_{a_1 \ldots a_n}$, with n odd, vanish. If the range is contained in $\mathbf{C}_a$ then the $f_{a_1 \ldots a_n}$, with n even, vanish.

Integration of superanalytic functions of supernumbers

The theory of integration too may be generalized from the ordinary complex plane $\mathbf{C}$ to the spaces $\mathbf{C}_c$ and $\mathbf{C}_a$. Consider first $\mathbf{C}_c$. The singularities (poles, branch points, etc.) of the functions (1.1.17) and (1.1.18) are located at specific values of u_B, independent of u_S. Therefore, in speaking of the location of these singularities relative to a given curve in $\mathbf{C}_c$, one may always imagine the curve to be projected onto the u_B plane and use such conventional terms as 'left', 'right', 'above', 'below', 'inside', 'on', etc. Any line integral of the form $\int_{u_1}^{u_2} f(u)\,du$ depends only on the endpoints and on the homotopic relation of the curve to the various singularities of f, and not on the specific curve. If f is superanalytic on and inside a *closed* curve in $\mathbf{C}_c$ then the curve may be continuously deformed to a point without crossing any singularity and

$$\oint f(u)\,du = 0. \tag{1.1.19}$$

More generally, if f is superanalytic on the curve and superanalytic inside except at a finite number of poles, then

$$\oint f(u)\,du = 2\pi\mathrm{i} \times (\text{sum of residues at the poles}). \tag{1.1.20}$$

Note that if f has the general form (1.1.18) the residues may be arbitrary supernumbers.

Line integrals of superanalytic functions over $\mathbf{C}_a$ do not behave analogously. Integrals $\oint f(v)\,dv$ over closed curves in $\mathbf{C}_a$ do *not* vanish unless $f(v)$ is itself the derivative of a superanalytic function, i.e., a constant (see eqs. (1.1.13) and (1.1.14)). In particular the integral $\oint v\,dv$ depends in a continuous fashion on the contour. We shall presently wish to attach an alternative meaning to the symbol $\int v\,dv$, for which there is no such ambiguity. But before doing so we need to take a look at functions of a real variable.

1.2 Real supernumbers. Differentiable functions of real *c*-numbers, and their integrals

Complex conjugation

In order to define *real* supernumbers we have to make some rules about complex conjugation (denoted here by an asterisk*). The laws of complex conjugation of sums and products of supernumbers will be taken in the form

$$(z + z')^* = z^* + z'^*, \qquad (zz')^* = z'^* z^*, \qquad \text{for all } z, z' \text{ in } \Lambda_\infty. \quad (1.2.1)$$

The complex conjugate of the body of a supernumber will be taken to be its ordinary complex conjugate, and the generators of Λ_∞ will be assumed to be 'real':

$$\zeta^{a*} = \zeta^a, \quad \text{for all } a. \qquad (1.2.2)$$

Evidently

$$(\zeta^a \zeta^b \ldots \zeta^c)^* = \zeta^c \ldots \zeta^b \zeta^a, \qquad (1.2.3)$$

and from this, together with the anticommutation law (1.1.1), one may infer that the basis element $\zeta^{a_1} \ldots \zeta^{a_n}$ is real when $\frac{1}{2}n(n-1)$ is even and imaginary when $\frac{1}{2}n(n-1)$ is odd. (As for ordinary numbers, a supernumber z is said to be *real* if $z^* = z$ and *imaginary* if $z^* = -z$.) A general element of Λ_∞ is real if and only if both its body and soul are real. The soul will be real if and only if the coefficients $c_{a_1 \ldots a_n}$ in the expansion (1.1.3) are real when $\frac{1}{2}n(n-1)$ is even and imaginary when $\frac{1}{2}n(n-1)$ is odd.

We shall denote by $\mathbf{R}_c$ the subset of all real elements of $\mathbf{C}_c$ and by $\mathbf{R}_a$ the subset of all real elements of $\mathbf{C}_a$. The set $\mathbf{R}_c$ is a subalgebra of $\mathbf{C}_c$. The product of two real c-numbers is a real c-number. The product of a real c-number and a real a-number is a real a-number. The product of two real a-numbers is an *imaginary* c-number.

The symbol 'x' will generally be used to denote a real variable, whether over $\mathbf{R}_c$ or over $\mathbf{R}_a$. When it is necessary to emphasize which of the two domain spaces is relevant we shall sometimes revert to using the symbols 'u' and 'v', with the understanding that their values are restricted to be real.

Functions, distributions and integrals over $\mathbf{R}_c$

A function from $\mathbf{R}_c$ to Λ_∞ need *not* be the restriction to $\mathbf{R}_c$ of a superanalytic function over $\mathbf{C}_c$ in order to be differentiable in the sense of eq. (1.1.15) *with* du *now restricted to* $\mathbf{R}_c$. Let f be any C^∞ function of an ordinary real variable. It may be generalized to a differentiable function over $\mathbf{R}_c$ in complete analogy with eq. (1.1.17):

$$f(x) = \sum_{n=0}^{\infty} \frac{1}{n!} f^{(n)}(x_{\mathrm{B}}) x_{\mathrm{S}}^n. \qquad (1.2.4)$$

If Λ_∞ is replaced by Λ_N the series (1.2.4) terminates at $n = [N/2]$, and f need be only $C^{[N/2]+1}$. Note that in the limit $N \to \infty$ there is no convergence problem, for (1.2.4) is a formal series.

Instead of starting with a C^∞ function one may equally well start with an arbitrary *distribution*. Since the notion of the derivative of a distribution is well defined, eq. (1.2.4) may then be regarded as defining a *distribution* over $\mathbf{R}_c$. The presence of a soul in the independent variable evidently has little practical effect on the variety of functions with which one may work in applications of the theory. In this respect $\mathbf{R}_c$ is a harmless generalization of its own subspace $\mathbf{R}$, the real line.

Consider now a contour that is restricted to lie wholly within $\mathbf{R}_c$. The value of the integral, over this contour, of a function f defined over $\mathbf{R}_c$, will depend only on the endpoints of the contour, provided merely that f is differentiable in the sense of eq. (1.1.15) with $\mathrm{d}u$ restricted to $\mathbf{R}_c$. That is, *f does not have to be analytic.* A schematic proof of this fact may be constructed along the lines of the suggested proof for exercise **1.2**, in the case in which f, when restricted to $\mathbf{R}$, is C^∞. More generally, suppose that f is a distribution possessing only a discrete set of singular points (describable, for example, in terms of delta functions and their derivatives). Let $F(x)$ be the corresponding generalization to $\mathbf{R}_c$ of the indefinite integral $\int f(x)\,\mathrm{d}x$. Then taking, for simplicity, the case of a contour between two points a and b in $\mathbf{R}_c$ for which x_S is a smooth single-valued function of x_B, one may write

$$\begin{aligned}
\int_a^b f(x)\,\mathrm{d}x &= \sum_{n=0}^{\infty} \frac{1}{n!}\int_{a_\mathrm{B}}^{b_\mathrm{B}} f^{(n)}(x_\mathrm{B})x_\mathrm{S}^n(x_\mathrm{B})[1 + x_\mathrm{S}'(x_\mathrm{B})]\,\mathrm{d}x_\mathrm{B}\\
&= \sum_{n=0}^{\infty} \int_{a_\mathrm{B}}^{b_\mathrm{B}} f^{(n)}(x_\mathrm{B})\left[\frac{1}{n!}x_\mathrm{S}^n(x_\mathrm{B}) + \frac{1}{(n+1)!}\frac{\mathrm{d}}{\mathrm{d}x_\mathrm{B}}x_\mathrm{S}^{n+1}(x_\mathrm{B})\right]\mathrm{d}x_\mathrm{B}\\
&= \int_{a_\mathrm{B}}^{b_\mathrm{B}} f[1 + x_\mathrm{S}'(x_\mathrm{B})]\,\mathrm{d}x_\mathrm{B}\\
&\quad + \sum_{n=1}^{\infty} \frac{1}{n!}[f^{(n-1)}(b_\mathrm{B})b_\mathrm{S}^n - f^{(n-1)}(a_\mathrm{B})a_\mathrm{S}^n]\\
&\quad + \sum_{n=1}^{\infty} \int_{a_\mathrm{B}}^{b_\mathrm{B}} \left[-\frac{1}{n!}f^{(n-1)}(x_\mathrm{B})\frac{\mathrm{d}}{\mathrm{d}x_\mathrm{B}}x_\mathrm{S}^n(x_\mathrm{B})\right.\\
&\qquad \left. + \frac{1}{(n+1)!}f^{(n)}(x_\mathrm{B})\frac{\mathrm{d}}{\mathrm{d}x_\mathrm{B}}x_\mathrm{S}^{n+1}(x_\mathrm{B})\right]\mathrm{d}x_\mathrm{B}\\
&= F(b) - F(a),
\end{aligned} \tag{1.2.5}$$

provided neither a_B nor b_B lies on a singular point of any of the $f^{(n)}$. The derivation may easily be generalized to include contours for which the soul is a multi-valued function of the body.

Fourier transforms over $\mathbf{R}_c$

The contour independence of the above result implies that in working with integrals over $\mathbf{R}_c$ one may for many purposes proceed as if one were working over $\mathbf{R}$. A striking illustration of this is provided by the theory of Fourier transforms, which remains totally unchanged in form under the generalization from $\mathbf{R}$ to $\mathbf{R}_c$. The essence of this theory is summed up in the formula

$$\delta(x) = (2\pi)^{-1}\int_{\mathbf{R}_c} \mathrm{e}^{ipx}\,\mathrm{d}p \overset{\text{def}}{=} \lim_{\varepsilon\to +0}(2\pi)^{-1}\int_{\mathbf{R}_c} \mathrm{e}^{ipx-\varepsilon p^2}\,\mathrm{d}p, \qquad (1.2.6)$$

where the limit is understood to be taken after all integrations involving $\delta(x)$ have been performed. The symbol '$\int_{\mathbf{R}_c}$' means 'integrate over any contour in $\mathbf{R}_c$ the bodies of whose endpoints tend to $-\infty$ and $+\infty$ respectively'. How the soul behaves along the contour is completely irrelevant. Because the integrand is an entire function that vanishes at the endpoints independently of their souls, the contour may be displaced until it coincides with $\mathbf{R}$, without affecting the value of the integral. All the usual theory thereupon applies. It applies, in fact, even if x itself possesses a soul (!), for (1.2.6) then implies

$$\delta(x) = \sum_{n=0}^{\infty}\frac{1}{n!}\,\delta^{(n)}(x_{\mathrm{B}})x_{\mathrm{S}}^n, \qquad (1.2.7)$$

and if f is any function of the form (1.2.4) where none of the $f^{(n)}$ are singular at $x_{\mathrm{B}} = 0$, then its product with $\delta(x)$ is a function of the form to which eq. (1.2.5) applies, which means that $\int_C f(x)\delta(x)\,\mathrm{d}x$ depends only on the endpoints of the contour C. If the bodies of the endpoints are $-\infty$ and $+\infty$ respectively, then the integrand vanishes at these endpoints, and the contour may be displaced until it coincides with $\mathbf{R}$. This implies

$$\int_{\mathbf{R}_c} f(x)\delta(x)\,\mathrm{d}x = f(0), \qquad (1.2.8)$$

even if the contour does not pass through the point $x = 0$! From now on we shall often omit the subscript $\mathbf{R}_c$ on the symbol '$\int_{\mathbf{R}_c}$', in analogy with the frequent custom of omitting the $\pm\infty$ on '$\int_{-\infty}^{\infty}$'.

1.3 Functions and integrals over $\mathbf{R}_a$

Basic definitions

To develop an integration theory for functions on $\mathbf{R}_a$, one must proceed rather differently, indeed in a manner that appears, at first sight, bizarre.

By so doing one can obtain a theory that displays remarkable analogies to integration theory over $\mathbf{R}_c$. The first thing that one must do is give up the idea that a definite integral is associated with a family of paths all having the same, or related, endpoints. We have seen, in any case, that such integrals are path independent only in the trivial case in which the integrand is a constant.

We shall confine our attention to functions differentiable in the sense of eq. (1.10) with $\mathrm{d}v$ now restricted to $\mathbf{R}_a$ and v replaced by x. All such functions have the form (cf. eq. (1.1.11))

$$f(x) = a + bx, \tag{1.3.1}$$

where a and b are constant supernumbers.† Because (1.3.1) is a linear form in x, in order to give a meaning to the symbol '$\int f(x)\mathrm{d}x$', one has only to decide what meaning to give to the symbols '$\int \mathrm{d}x$' and '$\int x\,\mathrm{d}x$'. The integrals of all other functions will then be determined by the rules

$$\int [f(x) + g(x)]\,\mathrm{d}x = \int f(x)\,\mathrm{d}x + \int g(x)\,\mathrm{d}x, \quad \text{for all } f \text{ and } g \text{ on } \mathbf{R}_a{}^{\ddagger} \tag{1.3.2}$$

$$\int af(x)\,\mathrm{d}x = a\int f(x)\,\mathrm{d}x, \quad \text{for all } a \text{ in } \Lambda_\infty \text{ and all } f \text{ on } \mathbf{R}_a. \tag{1.3.3}$$

In choosing the basic integrals we shall be guided by analogy with the equation

$$\int \left[\frac{\mathrm{d}}{\mathrm{d}x} f(x)\right]\mathrm{d}x = 0, \tag{1.3.4}$$

which holds for differentiable functions (or distributions) $f(x)$ on $\mathbf{R}_c$ satisfying $f(x) \xrightarrow[x_B \to \pm\infty]{} 0$. If we require eq. (1.3.4) to hold *also* on $\mathbf{R}_a$ then we must necessarily have

$$\int \mathrm{d}x \stackrel{\text{def}}{=} 0, \tag{1.3.5}$$

$$\int x\,\mathrm{d}x \stackrel{\text{def}}{=} Z, \tag{1.3.6}$$

where Z is some constant supernumber.

In order to accept the definitions (1.3.5) and (1.3.6) the reader must give

† If f takes its values in $\mathbf{R}_a$ then a is a real a-number and b a real c-number. If f takes its values in $\mathbf{R}_c$ then a is a real c-number and b is an imaginary a-number.

‡ Here, and in what follows, by the phrase 'all f on $\mathbf{R}_a$' is meant 'all f of the form (1.3.1) with x in $\mathbf{R}_a$'.

up conventional prejudices. Measure-theoretical notions play no role here. Integration over $\mathbf{R}_a$ becomes a purely formal procedure, the utility of which rests ultimately on the naturalness with which it can be used to encode certain algebraic information. One may note already that eqs. (1.3.2), (1.3.3), (1.3.5) and (1.3.6) together imply the law of shifting the integration variable and the law of integration by parts:

$$\int f(x+a)\,\mathrm{d}x = \int f(x)\,\mathrm{d}x, \tag{1.3.7}$$

$$\int f(x)\frac{\overleftarrow{\mathrm{d}}}{\mathrm{d}x}g(x)\,\mathrm{d}x = \int f(x)\frac{\overrightarrow{\mathrm{d}}}{\mathrm{d}x}g(x)\,\mathrm{d}x, \tag{1.3.8}$$

for all f and g on $\mathbf{R}_a$ and all a in $\mathbf{R}_a$. The proofs of eqs. (1.3.7) and (1.3.8) are easy exercises. If f takes its values in $\mathbf{R}_c$ then (1.3.8) may be rewritten in the more familiar form

$$\int f(x)\left[\frac{\overrightarrow{\mathrm{d}}}{\mathrm{d}x}g(x)\right]\mathrm{d}x = -\int\left[\frac{\overrightarrow{\mathrm{d}}}{\mathrm{d}x}f(x)\right]g(x)\,\mathrm{d}x.$$

Equation (1.3.6) will be supplemented by the convention

$$\int x\,\mathrm{d}x = -\int \mathrm{d}x\,x. \tag{1.3.9}$$

That is, the symbol 'dx' will be treated formally as if dx were an a-number. It is not, however, to be imagined as being, like x, a real a-number; nor is the formal bodilessness of dx to be regarded as implying that the constant Z in eq. (1.3.6) has vanishing body. The only condition that the anticommutativity of dx imposes is that Z be a c-number.

Fourier transforms over $\mathbf{R}_a$

We shall fix Z by drawing on another analogy (in addition to eq. (1.3.4)) with integration theory over $\mathbf{R}_c$. First note that if Z has nonvanishing body then an analog for the delta function exists for integrals over $\mathbf{R}_a$, namely

$$\delta(x) = Z^{-1}x. \tag{1.3.10}$$

As may be readily verified this function satisfies

$$\int f(x)\delta(x)\,\mathrm{d}x = f(0), \quad \text{for all } f \text{ on } \mathbf{R}_a, \tag{1.3.11}$$

the order of factors in the integrand being important. It is a remarkable fact that if Z is chosen appropriately this delta function may be expressed as a 'Fourier integral', in complete analogy with (1.2.6):

$$\delta(x) = (2\pi)^{-1} \int e^{ipx}\,dp. \tag{1.3.12}$$

Here p is an a-number and the integrand is to be regarded as a function on $\mathbf{R}_a \times \mathbf{R}_a$.[†] Since the series for the exponential terminates, the integral is readily evaluated:

$$\begin{aligned}(2\pi)^{-1} \int e^{ipx}\,dp &= (2\pi)^{-1} \int (1 + ipx)\,dp = -(2\pi)^{-1} ix \int p\,dp \\ &= (2\pi i)^{-1} Zx.\end{aligned} \tag{1.3.13}$$

Equating expressions (1.3.10) and (1.3.13) one infers

$$Z^2 = 2\pi i. \tag{1.3.14}$$

The phase of Z will be fixed by the convention

$$Z = (2\pi i)^{\frac{1}{2}} = (2\pi)^{\frac{1}{2}} e^{\pi i/4}. \tag{1.3.15}$$

It will be noted that, contrary to the situation on $\mathbf{R}_c$, the delta function on $\mathbf{R}_a$ is an odd function of its argument: $\delta(-x) = -\delta(x)$. A related fact is that the order of p and x in the exponent of (1.3.12) is important, and a choice of order has to be made. It is perfectly possible to develop the theory with the opposite ordering, but then the sign in front of i in many of the previous equations must be changed.

With the delta function expressible in the form (1.3.12) a theory of Fourier transforms may be developed. The Fourier transform of a function f on $\mathbf{R}_a$ will be defined by

$$\tilde{f}(p) \overset{\text{def}}{=} (2\pi)^{-\frac{1}{2}} \int f(x) e^{ixp}\,dx, \tag{1.3.16}$$

the order of factors being, as usual, important. If f has the form (1.3.1) then its Fourier transform is readily computed to be

$$\tilde{f}(p) = i^{\frac{1}{2}} b + i^{-\frac{1}{2}} ap, \tag{1.3.17}$$

from which one immediately sees that the original function is regained by taking the Fourier transform twice.[‡] Another proof of this is as follows:

$$\begin{aligned}\tilde{\tilde{f}}(x) &= (2\pi)^{-\frac{1}{2}} \int \tilde{f}(p) e^{ipx}\,dp \\ &= (2\pi)^{-1} \iint f(x') e^{ix'p}\,dx' e^{ipx}\,dp\end{aligned}$$

[†] Since px is imaginary the integrand takes its values in $\mathbf{R}_c$.

[‡] Note that no intervening operation of reflection in the origin ($x \to -x$) is required as in the case of ordinary Fourier transforms on $\mathbf{R}_c$.

$$= -(2\pi)^{-1}\iint f(x')e^{-ip(x'-x)}\,dp\,dx'$$

$$= \int f(x')\delta(x'-x)\,dx' = f(x). \tag{1.3.18}$$

Integrals over $\mathbf{R}_a^n$

The steps of the above proof illustrate how multiple integrals are built up by composition of single integrals. The general theory of integrals over $\mathbf{R}_a^n (\overset{\text{def}}{=} \mathbf{R}_a \times \ldots \times \mathbf{R}_a$, n factors) is based on such composition. The 'volume element' in $\mathbf{R}_a^n$ is defined to be

$$d^n x \overset{\text{def}}{=} i^{n(n-1)/2}\,dx^1 \ldots dx^n, \tag{1.3.19}$$

where $(x^1, \ldots, x^n)$ denotes an arbitrary point of $\mathbf{R}_a^n$. The n-dimensional delta function is then given by

$$\delta(x) = (2\pi i)^{-n/2} i^{n(n-1)/2} x^1 \ldots x^n, \tag{1.3.20}$$

where 'x' is an abbreviation for '$(x^1, \ldots, x^n)$'. The proof is straightforward:

$$\int f(x)\delta(x)\,d^n x = (2\pi i)^{-n/2}(-1)^{n(n-1)/2}\int f(x)x^1 \ldots x^n\,dx^1 \ldots dx^n$$

$$= (2\pi i)^{-n/2} f(0)\int x^1 \ldots x^n\,dx^n \ldots dx^1$$

$$= (2\pi i)^{-n/2} f(0)\int x^1\,dx^1 \ldots \int x^n\,dx^n = f(0).$$

Here $f(x)$ is any differentiable function on $\mathbf{R}_a^n$, namely any function having the general form

$$f(x) = \sum_{r=0}^{n} \frac{1}{r!} a_{\alpha_1 \ldots \alpha_r} x^{\alpha_r} \ldots x^{\alpha_1}, \tag{1.3.21}$$

where the indices α_i range from 1 to n and the a's are arbitrary supernumbers completely antisymmetric in these indices.

The delta function may also be expressed as a Fourier integral:

$$\delta(x) = (2\pi)^{-n}\int e^{ip_\alpha x^\alpha}\,d^n p$$

$$= (2\pi)^{-n} i^{n(n-1)/2}\int e^{ip_1 x^1} \ldots e^{ip_n x^n}\,dp_1 \ldots dp_n$$

$$= (2\pi)^{-n} \mathrm{i}^{n(n-1)/2} \int \mathrm{e}^{\mathrm{i}p_1 x^1} \mathrm{d}p_1 \ldots \mathrm{e}^{\mathrm{i}p_n x^n} \mathrm{d}p_n$$

$$= \mathrm{i}^{n(n-1)/2} \delta(x^1) \ldots \delta(x^n). \tag{1.3.22}$$

An alternative derivation is the following:

$$(2\pi)^{-n} \int \mathrm{e}^{\mathrm{i}p_\alpha x^\alpha} \mathrm{d}^n p = (2\pi)^{-n} \mathrm{i}^{n(n-1)/2} \int \sum_{r=0}^{n} \frac{(-1)^r}{r!} x^{\alpha_1} p_{\alpha_1} \ldots x^{\alpha_r} p_{\alpha_r} \mathrm{d}p_1 \ldots \mathrm{d}p_n$$

$$= (2\pi \mathrm{i})^{-n} \mathrm{i}^{n(n-1)/2} \frac{1}{n!} \int x^{\alpha_1} \ldots x^{\alpha_n} p_{\alpha_n} \ldots p_{\alpha_1} \mathrm{d}p_1 \ldots \mathrm{d}p_n$$

$$= (2\pi \mathrm{i})^{-n} \mathrm{i}^{n(n-1)/2} x^1 \ldots x^n \int p_n \ldots p_1 \mathrm{d}p_1 \ldots \mathrm{d}p_n$$

$$= (2\pi \mathrm{i})^{-n/2} \mathrm{i}^{n(n-1)/2} x^1 \ldots x^n.$$

The n-dimensional Fourier transform is defined by

$$\tilde{f}(p) = (2\pi)^{-n/2} \int f(x) \mathrm{e}^{\mathrm{i}x^\alpha p_\alpha} \mathrm{d}^n x. \tag{1.3.23}$$

When $f(x)$ is expressed in the general form (1.3.21) its Fourier transform takes the form

$$\tilde{f}(p) = \sum_{r=0}^{n} \frac{1}{r!} \tilde{a}^{\alpha_1 \ldots \alpha_r} p_{\alpha_r} \ldots p_{\alpha_1}, \tag{1.3.24}$$

where

$$\tilde{a}^{\alpha_1 \ldots \alpha_r} = \mathrm{i}^{n/2} (-\mathrm{i})^{n(n-1)/2} \frac{(-1)^{nr} \mathrm{i}^r}{(n-r)!} a_{\beta_1 \ldots \beta_{n-r}} \varepsilon^{\beta_{n-r} \ldots \beta_1 \alpha_1 \ldots \alpha_r}, \tag{1.3.25}$$

ε being the antisymmetric permutation symbol with n indices. Using the fact that

$$\delta(-x) = (-1)^n \delta(x), \tag{1.3.26}$$

where '$-x$' is an abbreviation for '$(-x^1, \ldots, -x^n)$', one may show that $\tilde{\tilde{f}} = f$ by a proof patterned on (1.3.18).

It should be obvious to the reader that integrals over $\mathbf{R}_a$ can be combined with integrals over $\mathbf{R}_c$ to produce multiple integrals over $\mathbf{R}_c^m \times \mathbf{R}_a^n$. Such integrals play an important role in supermanifold theory and we shall devote considerable attention to them. One of the important problems is to determine the rules for changing variables in such integrals. Both linear and nonlinear transformations of variables will be studied. But before undertaking this task we shall find it convenient to introduce the concept of a *supervector space*.

1.4 Supervector spaces

Definition

A *supervector space* is a set $\mathfrak{S}$ of elements called *supervectors*, together with a collection of mappings, having special properties as follows:

(1) There exists a binary-operation mapping $+ : \mathfrak{S} \times \mathfrak{S} \to \mathfrak{S}$, called *addition*, which is commutative and associative. In conventional notation one writes

$$+(\mathbf{X}, \mathbf{Y}) \stackrel{\text{def}}{=} \mathbf{X} + \mathbf{Y}, \quad \text{for all } \mathbf{X}, \mathbf{Y} \text{ in } \mathfrak{S}, \tag{1.4.1}$$

and hence

$$\mathbf{X} + \mathbf{Y} = \mathbf{Y} + \mathbf{X}, \tag{1.4.2}$$

$$\mathbf{X} + (\mathbf{Y} + \mathbf{Z}) = (\mathbf{X} + \mathbf{Y}) + \mathbf{Z} \stackrel{\text{def}}{=} \mathbf{X} + \mathbf{Y} + \mathbf{Z}, \quad \text{for all } \mathbf{X}, \mathbf{Y}, \mathbf{Z} \text{ in } \mathfrak{S}. \tag{1.4.3}$$

(2) There exists an element $\mathbf{0}$ of $\mathfrak{S}$ such that

$$\mathbf{X} + \mathbf{0} = \mathbf{X}, \quad \text{for all } \mathbf{X} \text{ in } \mathfrak{S}. \tag{1.4.4}$$

(3) For every supervector $\mathbf{X}$ in $\mathfrak{S}$ there exists another supervector in $\mathfrak{S}$ denoted by $-\mathbf{X}$ such that

$$-\mathbf{X} + \mathbf{X} = \mathbf{0}. \tag{1.4.5}$$

It is easy to verify that $\mathbf{0}$ is unique and that, for every $\mathbf{X}$ in $\mathfrak{S}$, $-\mathbf{X}$ is unique. The supervector $\mathbf{0}$ is called the *zero supervector* and $-\mathbf{X}$ is called the *negative* of $\mathbf{X}$. It is conventional to define a binary operation called *subtraction* by

$$\mathbf{X} - \mathbf{Y} \stackrel{\text{def}}{=} \mathbf{X} + (-\mathbf{Y}), \quad \text{for all } \mathbf{X}, \mathbf{Y} \text{ in } \mathfrak{S}. \tag{1.4.6}$$

(4) For every supernumber α there exist two mappings, $\alpha_L : \mathfrak{S} \to \mathfrak{S}$ and $\alpha_R : \mathfrak{S} \to \mathfrak{S}$, called *left multiplication* and *right multiplication* respectively and conventionally expressed by the notation

$$\alpha_L(\mathbf{X}) \stackrel{\text{def}}{=} \alpha\mathbf{X}, \quad \alpha_R(\mathbf{X}) \stackrel{\text{def}}{=} \mathbf{X}\alpha, \quad \text{for all } \mathbf{X} \text{ in } \mathfrak{S}. \tag{1.4.7}$$

These mappings satisfy the linear laws

$$(\alpha + \beta)\mathbf{X} = \alpha\mathbf{X} + \beta\mathbf{X}, \qquad \mathbf{X}(\alpha + \beta) = \mathbf{X}\alpha + \mathbf{X}\beta, \tag{1.4.8}$$

$$\alpha(\mathbf{X} + \mathbf{Y}) = \alpha\mathbf{X} + \alpha\mathbf{Y}, \qquad (\mathbf{X} + \mathbf{Y})\alpha = \mathbf{X}\alpha + \mathbf{Y}\alpha, \tag{1.4.9}$$

$$(\alpha\beta)\mathbf{X} = \alpha(\beta\mathbf{X}) \stackrel{\text{def}}{=} \alpha\beta\mathbf{X}, \qquad \mathbf{X}(\alpha\beta) = (\mathbf{X}\alpha)\beta \stackrel{\text{def}}{=} \mathbf{X}\alpha\beta, \tag{1.4.10}$$

$$1\mathbf{X} = \mathbf{X}, \qquad \mathbf{X}1 = \mathbf{X}, \tag{1.4.11}$$

for all α, β in Λ_∞ and all $\mathbf{X}$, $\mathbf{Y}$ in $\mathfrak{S}$. In (1.4.11) 1 is the ordinary number one. It is straightforward to verify that

$$0\mathbf{X} = \mathbf{X}0 = \mathbf{0}, \tag{1.4.12}$$

$$\alpha\mathbf{0} = \mathbf{0}\alpha = \mathbf{0}, \tag{1.4.13}$$

for all $\mathbf{X}$ in $\mathfrak{S}$ and all α in Λ_∞. In (1.4.12) 0 is the ordinary number zero.

(5) Left and right multiplication are related. Firstly,

$$(\alpha\mathbf{X})\beta = \alpha(\mathbf{X}\beta) \overset{\text{def}}{=} \alpha\mathbf{X}\beta, \tag{1.4.14}$$

for all α, β in Λ_∞ and all $\mathbf{X}$ in $\mathfrak{S}$. Secondly, if α is a c-number then it commutes with all supervectors. That is

$$\alpha\mathbf{X} = \mathbf{X}\alpha, \quad \text{for all } \alpha \text{ in } \mathbf{C}_c \text{ and all } \mathbf{X} \text{ in } \mathfrak{S}. \tag{1.4.15}$$

Thirdly, for every $\mathbf{X}$ in $\mathfrak{S}$ there exist *unique* supervectors $\mathbf{U}$ and $\mathbf{V}$ in $\mathfrak{S}$ such that

$$\mathbf{X} = \mathbf{U} + \mathbf{V}, \tag{1.4.16}$$

$$\alpha\mathbf{U} = \mathbf{U}\alpha, \qquad \alpha\mathbf{V} = -\mathbf{V}\alpha, \quad \text{for all } \alpha \text{ in } \mathbf{C}_a. \tag{1.4.17}$$

The supervectors $\mathbf{U}$ and $\mathbf{V}$ are called respectively the *even* and *odd* parts of $\mathbf{X}$.

Properties (1.4.16) and (1.4.17) are analogs of properties that hold for supernumbers themselves. Thus for every supernumber z there exist unique supernumbers u and v such that $z = u + v$ and $\alpha u = u\alpha$, $\alpha v = -v\alpha$, for all a-numbers α. Note that u and v are *not* unique if Λ_∞ is replaced by Λ_N. In that case u and v are determined only *modulo* a multiple of $\zeta^1 \ldots \zeta^N$.

If the odd part of a supervector vanishes (i.e., equals the zero supervector) the supervector is said to be of *type c*. If its even part vanishes it is said to be of *type a*. The zero supervector is the only supervector that is simultaneously c-type and a-type. A supervector that has a definite type will be called *pure*. Similarly, a supernumber that is either a c-number or an a-number will be called *pure*. For pure supernumbers and supervectors eqs. (1.4.15)–(1.4.17) may be summarized in the formula

$$\alpha\mathbf{X} = (-1)^{\alpha\mathbf{X}}\mathbf{X}\alpha, \tag{1.4.18}$$

where a notational convention has been introduced that will be used frequently in the rest of this book: each symbol appearing in an exponent of (-1) is to be understood as assuming the value 0 or 1 according as the corresponding quantity is c-type or a-type. We shall sometimes write equations like (1.4.18) even when it has not been stated that α and $\mathbf{X}$ are pure. In that case the equation is to be understood as describing a property that strictly holds as written only when the quantities involved *are* pure but which extends by linearity in an unambiguous fashion to the case in which the quantities are impure.

(6) There exists a mapping $*: \mathfrak{S} \to \mathfrak{S}$ called *complex conjugation*, conventionally written in the form

$$*(\mathbf{X}) \stackrel{\text{def}}{=} \mathbf{X}^*, \quad \text{for all } \mathbf{X} \text{ in } \mathfrak{S}, \tag{1.4.19}$$

which satisfies

$$\mathbf{X}^{**} = \mathbf{X}, \tag{1.4.20}$$

$$(\mathbf{X} + \mathbf{Y})^* = \mathbf{X}^* + \mathbf{Y}^*, \tag{1.4.21}$$

$$(\alpha\mathbf{X})^* = \mathbf{X}^*\alpha^*, \quad (\mathbf{X}\alpha)^* = \alpha^*\mathbf{X}^*, \tag{1.4.22}$$

for all $\mathbf{X}$, $\mathbf{Y}$ in $\mathfrak{S}$ and all α in Λ_∞. It is easy to verify that the complex conjugate of a pure supervector is pure, the type remaining unaffected by the complex conjugation mapping. A supervector $\mathbf{Z}$ will be said to be *real* if $\mathbf{Z}^* = \mathbf{Z}$, *imaginary* if $\mathbf{Z}^* = -\mathbf{Z}$, and *complex* otherwise. A complex supervector can always be decomposed into its real and imaginary parts:

$$\mathbf{Z} = \mathbf{X} + \mathrm{i}\mathbf{Y}, \tag{1.4.23}$$

$$\mathbf{X} = \tfrac{1}{2}(\mathbf{Z} + \mathbf{Z}^*) = \mathbf{X}^*, \qquad \mathbf{Y} = -\tfrac{1}{2}\mathrm{i}(\mathbf{Z} - \mathbf{Z}^*) = \mathbf{Y}^*. \tag{1.4.24}$$

The zero supervector is the only supervector that is simultaneously real and imaginary. Note that the product of a real *c*-number and a real supervector is a real supervector. The product of a real *a*-number and a real *c*-type supervector is a real *a*-type supervector, but the product of a real *a*-number and a real *a*-type supervector is an *imaginary* *c*-type supervector.

This completes the list of properties satisfied by the elements of a supervector space and its associated mappings. We now consider specific structures that may be introduced into supervector spaces.

Bases

Let $\{{}_i\mathbf{e}\}$ be a discrete set of supervectors in a supervector space $\mathfrak{S}$. The ${}_i\mathbf{e}$ are said to be *linearly independent* if and only if no supervector of form $c^i{}_i\mathbf{e}$ with the c's in Λ_∞ vanishes unless all the c's vanish. A linearly independent set $\{{}_i\mathbf{e}\}$ is called a *complete linearly independent set* or a *basis* if every supervector $\mathbf{X}$ in $\mathfrak{S}$ can be expressed in the form

$$\mathbf{X} = X^i{}_i\mathbf{e}, \quad \text{for some } X^i \text{ in } \Lambda_\infty. \tag{1.4.25}$$

Because of the linear independence of the basis the X^i are unique. They are called the components of $\mathbf{X}$ with respect to the basis $\{{}_i\mathbf{e}\}$.

A comment should be made about the unusual convention that has been adopted here of writing the indices on the basis supervectors as prefixes rather than suffixes. Prefixed indices will occur repeatedly throughout this book. They will be employed to maintain the following convention: index pairs will be contracted in a straightforward way only if the indices in each pair are adjacent, with one being a *right* index and the other a *left* index. (Otherwise index-dependent powers of (-1) must generally be inserted.) The common convention that one of the indices in each contracted pair be an upper one and the other a lower one will also be maintained.

A supervector space may be nontrivial and yet have no basis. An example is the subspace obtained by multiplying all the elements of a given supervector space by a fixed *a*-number or a fixed nilpotent *c*-number. If a supervector space *has* a basis, consisting of d supervectors, it is said to have *total dimension d*. The total dimension, which may be finite or infinite, is an invariant of the supervector space. This may be seen as follows. Let $\{{}_i\mathbf{e}\}$ and $\{{}_i\bar{\mathbf{e}}\}$ be two distinct bases. Since both are bases one may write

$$ {}_i\mathbf{e} = {}_iK^j\,{}_j\bar{\mathbf{e}}, \qquad {}_i\bar{\mathbf{e}} = {}_i\bar{K}^j\,{}_j\mathbf{e}, \tag{1.4.26}$$

where the ${}_iK^j$, ${}_i\bar{K}^j$ are unique supernumbers. Furthermore

$$ K\bar{K} = 1_d, \qquad \bar{K}K = 1_{\bar{d}}, \tag{1.4.27}$$

where d is the number of $\mathbf{e}$'s, $\bar{d}$ is the number of $\bar{e}$'s, 1_d and $1_{\bar{d}}$ are unit matrices of dimensionality d and $\bar{d}$ respectively, and K and $\bar{K}$ are the matrices formed out of the ${}_iK^j$ and ${}_i\bar{K}^j$. Equations (1.4.27) imply

$$ K_{\mathrm{B}}\bar{K}_{\mathrm{B}} = 1_d, \qquad \bar{K}_{\mathrm{B}}K_{\mathrm{B}} = 1_{\bar{d}}. \tag{1.4.28}$$

But since the bodies K_{B} and $\bar{K}_{\mathrm{B}}$ are ordinary matrices these equations can hold simultaneously only if $d = \bar{d}$, qed. Evidently K and $\bar{K}$ must be nonsingular square matrices with $K^{-1} = \bar{K}$, $K_{\mathrm{B}}^{-1} = \bar{K}_{\mathrm{B}}$.

Pure bases

Let $\{{}_i\mathbf{e}\}$ be any basis of a supervector space having finite total dimension d. Let the $\mathbf{e}$'s be decomposed into their even and odd parts:

$$ {}_i\mathbf{e} = {}_i\mathbf{f} + {}_i\mathbf{g}, \quad \text{for all } i, \tag{1.4.29}$$

where the $\mathbf{f}$'s are all *c*-type and the $\mathbf{g}$'s are all *a*-type.

We may write

$$ {}_i\mathbf{f} = {}_iM^j\,{}_j\mathbf{e}, \qquad {}_i\mathbf{g} = {}_iN^j\,{}_j\mathbf{e}, \tag{1.4.30}$$

where the ${}_iM^j$, ${}_iN^j$ are unique supernumbers. Equations (1.4.29) and (1.4.30) together imply

$$(1_d 1_d)\begin{pmatrix} M \\ N \end{pmatrix} = M + N = 1_d, \tag{1.4.31a}$$

where M and N are the square matrices formed out of the ${}_iM^j$ and ${}_iN^j$. Equation (1.4.31a) in turn implies

$$(1_d 1_d)\begin{pmatrix} M_{\mathrm{B}} \\ N_{\mathrm{B}} \end{pmatrix} = M_{\mathrm{B}} + N_{\mathrm{B}} = 1_d. \tag{1.4.31b}$$

Since the rank of $(1_d 1_d)$ is d and the rank 1_d is d, the rank of the $2d \times d$ matrix $\begin{pmatrix} M_{\mathrm{B}} \\ N_{\mathrm{B}} \end{pmatrix}$ must also be d. Let the ranks of the bodies M_{B} and N_{B} be r and s respectively. We cannot have $r + s < d$, for otherwise the rank of $\begin{pmatrix} M_{\mathrm{B}} \\ N_{\mathrm{B}} \end{pmatrix}$ would be less than d. We also cannot have $r + s > d$, as may be shown by the following arguments.

First note that eqs. (1.4.29) and (1.4.30) imply

$${}_i\mathbf{f} = {}_iM^j({}_j\mathbf{f} + {}_j\mathbf{g}), \qquad {}_i\mathbf{g} = {}_iN^j({}_j\mathbf{f} + {}_j\mathbf{g}). \tag{1.4.32}$$

Now decompose the supernumbers ${}_iM^j$, ${}_iN^j$ into their even and odd parts,

$${}_iM^j = {}_i\overset{\mathrm{e}}{M}{}^j + {}_i\overset{\mathrm{o}}{M}{}^j, \qquad {}_iN^j = {}_i\overset{\mathrm{e}}{N}{}^j + {}_i\overset{\mathrm{o}}{N}{}^j \tag{1.4.33}$$

and note that multiplication of a pure supervector by an a-number changes its type whereas multiplication by a c-number leaves its type unaffected. From this infer

$${}_i\overset{\mathrm{o}}{M}{}^j\,{}_j\mathbf{f} + {}_i\overset{\mathrm{e}}{M}{}^j\,{}_j\mathbf{g} = \mathbf{0}, \qquad {}_i\overset{\mathrm{e}}{N}{}^j\,{}_j\mathbf{f} + {}_i\overset{\mathrm{o}}{N}{}^j\,{}_j\mathbf{g} = 0, \tag{1.4.34}$$

or, suppressing indices,

$$\begin{pmatrix} \overset{\mathrm{e}}{N} & \overset{\mathrm{o}}{N} \\ \overset{\mathrm{o}}{M} & \overset{\mathrm{e}}{M} \end{pmatrix}\begin{pmatrix} \mathbf{f} \\ \mathbf{g} \end{pmatrix} = \begin{pmatrix} \mathbf{0} \\ \vdots \\ \mathbf{0} \end{pmatrix}. \tag{1.4.35}$$

Next observe that $\overset{\mathrm{e}}{M}_{\mathrm{B}} = M_{\mathrm{B}}$ and $\overset{\mathrm{e}}{N}_{\mathrm{B}} = N_{\mathrm{B}}$. Since the ranks of M_{B} and N_{B} are r and s respectively, one may introduce ordinary matrices A, C, E and G, of dimensions $r \times d$, $d \times r$, $s \times d$ and $d \times s$ respectively, such that the square matrices $AM_{\mathrm{B}}C$ and $EN_{\mathrm{B}}G$ are nonsingular. One may also introduce ordinary matrices B, D, F and H, of dimensions $n \times d$, $d \times n$, $m \times d$ and $d \times m$ respectively, where

$$n = d - r, \qquad m = d - s, \tag{1.4.36}$$

such that the $d \times d$ matrices $\begin{pmatrix} A \\ B \end{pmatrix}$, (CD), $\begin{pmatrix} E \\ F \end{pmatrix}$ and (GH) are all nonsingular. Let the inverses of (CD) and (GH) be denoted by $\begin{pmatrix}\bar{C}\\ \bar{D}\end{pmatrix}$ and $\begin{pmatrix}\bar{G}\\ \bar{H}\end{pmatrix}$ respectively, where $\bar{C}$, $\bar{D}$, $\bar{G}$ and $\bar{H}$ are ordinary matrices of dimensions $r \times d$, $n \times d$, $s \times d$ and $m \times d$ respectively. Then eq. (1.4.35) may be rewritten in the equivalent form

$$\begin{pmatrix}0\\ \vdots\\ 0\end{pmatrix} = \begin{pmatrix}\begin{pmatrix}E\\ F\end{pmatrix} & 0\\ 0 & \begin{pmatrix}A\\ B\end{pmatrix}\end{pmatrix}\begin{pmatrix}\overset{e}{N} & \overset{o}{N}\\ \overset{o}{M} & \overset{e}{M}\end{pmatrix}\begin{pmatrix}(GH) & 0\\ 0 & (CD)\end{pmatrix}\begin{pmatrix}\begin{pmatrix}\bar{G}\\ \bar{H}\end{pmatrix} & 0\\ 0 & \begin{pmatrix}\bar{C}\\ \bar{D}\end{pmatrix}\end{pmatrix}\begin{pmatrix}\mathbf{f}\\ \mathbf{g}\end{pmatrix}$$

$$= \begin{pmatrix} E\overset{e}{N}G & E\overset{e}{N}H & E\overset{o}{N}C & E\overset{o}{N}D\\ F\overset{e}{N}G & F\overset{e}{N}H & F\overset{o}{N}C & F\overset{o}{N}D\\ A\overset{o}{M}G & A\overset{o}{M}H & A\overset{e}{M}C & A\overset{e}{M}D\\ B\overset{o}{M}G & B\overset{o}{M}H & B\overset{e}{M}C & B\overset{e}{M}D\end{pmatrix}\begin{pmatrix}\bar{G}\mathbf{f}\\ \bar{H}\mathbf{f}\\ \bar{C}\mathbf{g}\\ \bar{D}\mathbf{g}\end{pmatrix}. \tag{1.4.37}$$

The first and third rows of eq. (1.4.37) imply

$$\begin{pmatrix} E\overset{e}{N}G & E\overset{o}{N}C\\ A\overset{o}{M}G & A\overset{e}{M}C\end{pmatrix}\begin{pmatrix}\bar{G}\mathbf{f}\\ \bar{C}\mathbf{g}\end{pmatrix} = -\begin{pmatrix} E\overset{e}{N}H & E\overset{o}{N}D\\ A\overset{o}{M}H & A\overset{e}{M}D\end{pmatrix}\begin{pmatrix}\bar{H}\mathbf{f}\\ \bar{D}\mathbf{g}\end{pmatrix}. \tag{1.4.38}$$

The body of the matrix on the left of this equation is $\mathrm{diag}(EN_B G, AM_B C)$ which, as we have seen, is nonsingular. The matrix itself is therefore nonsingular and it follows that the $s + r$ supervectors $\bar{G}\mathbf{f}$, $\bar{C}\mathbf{g}$ can be expressed as linear combinations of the m c-type supervectors $\bar{H}\mathbf{f}$ and the n a-type supervectors $\bar{D}\mathbf{g}$. But

$$\mathbf{e} = (G\bar{G} + H\bar{H})\mathbf{f} + (C\bar{C} + D\bar{D})\mathbf{g}, \tag{1.4.39}$$

whence it follows that the original basis supervectors ${}_i\mathbf{e}$ can themselves be expressed as linear combinations of the $\bar{H}\mathbf{f}$ and the $\bar{D}\mathbf{g}$. This in turn implies that $m + n \geq d$ or $r + s \leq d$, qed.

We have already seen that the inequality cannot hold. Therefore $m + n = d$. The $\bar{H}\mathbf{f}$ and $\bar{D}\mathbf{g}$ evidently constitute a *pure basis*, consisting of m c-type supervectors and n a-type supervectors. We have thus proved the

Theorem

Every supervector space having finite total dimension has a pure basis

The ordered pair (m, n) is called the *dimension* of the supervector space. The dimension, like the total dimension, is an invariant of the supervector space. This may be seen as follows. Let $\{{}_i\mathbf{e}\}$ and $\{{}_i\bar{\mathbf{e}}\}$ be two distinct pure bases, related as in eqs. (1.4.26). Let the e's and ē's be ordered so that the c-type basis vectors come before the a-type basis vectors. Then the matrix K has the block structure

$$K = \begin{pmatrix} A & C \\ D & B \end{pmatrix}, \tag{1.4.40}$$

where the elements of the submatrices A and B are c-numbers, while the elements of the submatrices C and D are a-numbers. We have seen earlier that the body of K, which now has the form

$$K_{\mathrm{B}} = \begin{pmatrix} A_{\mathrm{B}} & 0 \\ 0 & B_{\mathrm{B}} \end{pmatrix}, \tag{1.4.41}$$

must be nonsingular. But this can be the case only if A_{B} and B_{B} are square matrices. From this the invariance of the dimension (m, n) follows.

In the case of infinite-dimensional supervector spaces the existence of a basis does not automatically imply the existence of a pure basis. However, if one pure basis exists then an infinity of others exists.

Pure real bases

Let $\{{}_i\mathbf{e}\}$ be a pure basis of a supervector space of finite dimension (m, n). Let the e's be decomposed into their real and imaginary parts:

$${}_i\mathbf{e} = {}_i\mathbf{f} + \mathrm{i}\,{}_i\mathbf{g}, \quad {}_i\mathbf{f}^* = {}_i\mathbf{f}, \quad {}_i\mathbf{g}^* = {}_i\mathbf{g}, \quad \text{for all } i. \tag{1.4.42}$$

We may write

$${}_i\mathbf{f} = {}_iF^j\,{}_j\mathbf{e}, \quad {}_i\mathbf{g} = {}_iG^j\,{}_j\mathbf{e}, \tag{1.4.43}$$

where the ${}_iF^j$, ${}_iG^j$ are unique pure supernumbers. If the index i is called c-type or a-type according as ${}_i\mathbf{e}$ is c-type or a-type, then the ${}_iF^j$, ${}_iG^j$ are c-numbers when i and j have the same type and are a-numbers when i and j have opposite type. Let F and G be the matrices formed out of the ${}_iF^j$ and ${}_iG^j$. If the basis supervectors ${}_i\mathbf{e}$ are ordered so that the c-type e's come first, followed by the a-type ones, then F and G have the block structures

$$F = \begin{pmatrix} A & C \\ D & B \end{pmatrix}, \quad G = \begin{pmatrix} P & R \\ S & Q \end{pmatrix}, \tag{1.4.44}$$

where the elements of A, B, P and Q are c-numbers, while the elements of C, D, R and S are a-numbers. A and P are $m \times m$ matrices, B and Q are $n \times n$ matrices, C and R are $m \times n$ matrices; and D and S are $n \times m$ matrices.

Equations (1.4.42) and (1.4.43) together imply

$$(1_d \; \mathrm{i}1_d)\begin{pmatrix} F \\ G \end{pmatrix} = F + \mathrm{i}G = 1_d, \qquad d = m + n, \tag{1.4.45}$$

and hence

$$\begin{pmatrix} A_{\mathrm{B}} + \mathrm{i}P_{\mathrm{B}} & 0 \\ 0 & B_{\mathrm{B}} + \mathrm{i}Q_{\mathrm{B}} \end{pmatrix} = F_{\mathrm{B}} + \mathrm{i}G_{\mathrm{B}} = 1_d = \begin{pmatrix} 1_m & 0 \\ 0 & 1_n \end{pmatrix},$$

or

$$(1_m \; \mathrm{i}1_m)\begin{pmatrix} A_{\mathrm{B}} \\ P_{\mathrm{B}} \end{pmatrix} = 1_m, \qquad (1_n \; \mathrm{i}1_n)\begin{pmatrix} B_{\mathrm{B}} \\ Q_{\mathrm{B}} \end{pmatrix} = 1_n. \tag{1.4.46}$$

The ranks of the matrices $\begin{pmatrix} A_{\mathrm{B}} \\ P_{\mathrm{B}} \end{pmatrix}$, $\begin{pmatrix} B_{\mathrm{B}} \\ Q_{\mathrm{B}} \end{pmatrix}$, $\begin{pmatrix} F_{\mathrm{B}} \\ G_{\mathrm{B}} \end{pmatrix}$ are evidently m, n and d respectively.

Equations (1.4.42) and (1.4.43) also imply

$${}_i\mathbf{f} = {}_iF^j\,{}_j\mathbf{f} + \mathrm{i}\,{}_iF^j\,{}_j\mathbf{g}, \qquad {}_i\mathbf{g} = {}_iG^j\,{}_j\mathbf{f} + \mathrm{i}\,{}_iG^j\,{}_j\mathbf{g}, \tag{1.4.47}$$

whence

$$\begin{aligned} \begin{pmatrix} \mathbf{0} \\ \mathbf{0} \end{pmatrix} &= \begin{pmatrix} {}_i\mathbf{f} - {}_i\mathbf{f}^* \\ {}_i\mathbf{g} - {}_i\mathbf{g}^* \end{pmatrix} \\ &= \begin{pmatrix} {}_iF^j\,{}_j\mathbf{f} + \mathrm{i}\,{}_iF^j\,{}_j\mathbf{g} - {}_j\mathbf{f}\,{}_iF^{j*} + \mathrm{i}\,{}_j\mathbf{g}\,{}_iF^{j*} \\ {}_iG^j\,{}_j\mathbf{f} + \mathrm{i}\,{}_iG^j\,{}_j\mathbf{g} - {}_j\mathbf{f}\,{}_iG^{j*} + \mathrm{i}\,{}_j\mathbf{g}\,{}_iG^{j*} \end{pmatrix} \\ &= \begin{pmatrix} {}_iF^j - (-1)^{j(i+j)}\,{}_iF^{j*} & \mathrm{i}[{}_iF^j + (-1)^{j(i+j)}\,{}_iF^{j*}] \\ {}_iG^j - (-1)^{j(1+j)}\,{}_iG^{j*} & \mathrm{i}[{}_iG^j + (-1)^{j(1+j)}\,{}_iG^{j*}] \end{pmatrix} \begin{pmatrix} {}_j\mathbf{f} \\ {}_j\mathbf{g} \end{pmatrix}. \end{aligned} \tag{1.4.48}$$

Here an index appearing in an exponent of (-1) is to be understood as assuming the value 0 or 1 according as it is c-type or a-type. Note that the summation convention for repeated indices does not apply to indices appearing in exponents of (-1). Such indices may participate in summations induced by their appearance twice elsewhere in an expression, but they themselves may not induce summations.

Now let $\mathbf{f}_c, \mathbf{g}_c, \mathbf{f}_a, \mathbf{g}_a$ denote the 1-column matrices whose elements are respectively the c-type $\mathbf{f}$'s, the c-type $\mathbf{g}$'s, the a-type $\mathbf{f}$'s and the a-type $\mathbf{g}$'s. Then eq. (1.4.48) may be rewriten in the equivalent form

$$\begin{pmatrix} A - A^* & \mathrm{i}(A + A^*) & C + C^* & \mathrm{i}(C - C^*) \\ P - P^* & \mathrm{i}(P + P^*) & R + R^* & \mathrm{i}(R - R^*) \\ D - D^* & \mathrm{i}(D + D^*) & B - B^* & \mathrm{i}(B + B^*) \\ S - S^* & \mathrm{i}(S + S^*) & Q - Q^* & \mathrm{i}(Q + Q^*) \end{pmatrix} \begin{pmatrix} \mathbf{f}_c \\ \mathbf{g}_c \\ \mathbf{f}_a \\ \mathbf{g}_a \end{pmatrix} = \begin{pmatrix} \mathbf{0} \\ \vdots \\ \\ \mathbf{0} \end{pmatrix}. \tag{1.4.49}$$

Suppose $(X \;\; Y)$ is a $1 \times 2m$ real matrix such that[†]

$$(X \;\; Y)\begin{pmatrix} A_{\mathrm{B}} - A_{\mathrm{B}}^* & \mathrm{i}(A_{\mathrm{B}} + A_{\mathrm{B}}^*) \\ P_{\mathrm{B}} - P_{\mathrm{B}}^* & \mathrm{i}(P_{\mathrm{B}} + P_{\mathrm{B}}^*) \end{pmatrix} = (0 \ldots 0). \tag{1.4.50}$$

[†] $(X \;\; Y)$ may be assumed to be real because the matrix to which it is applied is pure imaginary.

Then also

$$(X\,Y)\begin{pmatrix} A_{\mathrm{B}} \\ P_{\mathrm{B}} \end{pmatrix} = (0\ldots 0), \quad (X\,Y)\begin{pmatrix} A_{\mathrm{B}}^* \\ P_{\mathrm{B}}^* \end{pmatrix} = (0\ldots 0). \tag{1.4.51}$$

Evidently the rank of the pure imaginary matrix

$$\begin{pmatrix} A_{\mathrm{B}} - A_{\mathrm{B}}^* & \mathrm{i}(A_{\mathrm{B}} + A_{\mathrm{B}}^*) \\ P_{\mathrm{B}} - P_{\mathrm{B}}^* & \mathrm{i}(P_{\mathrm{B}} + P_{\mathrm{B}}^*) \end{pmatrix} \tag{1.4.52}$$

cannot be less than m, for otherwise the rank of the matrix $\begin{pmatrix} A_{\mathrm{B}} \\ P_{\mathrm{B}} \end{pmatrix}$ would be less than m. Similarly, the rank of the pure imaginary matrix

$$\begin{pmatrix} B_{\mathrm{B}} - B_{\mathrm{B}}^* & \mathrm{i}(B_{\mathrm{B}} + B_{\mathrm{B}}^*) \\ Q_{\mathrm{B}} - Q_{\mathrm{B}}^* & \mathrm{i}(Q_{\mathrm{B}} + Q_{\mathrm{B}}^*) \end{pmatrix} \tag{1.4.53}$$

cannot be less than n, for otherwise the rank of the matrix $\begin{pmatrix} B_{\mathrm{B}} \\ Q_{\mathrm{B}} \end{pmatrix}$ would be less than n.

Let the ranks of the matrices (1.4.52) and (1.4.53) be $m'(\geq m)$ and $n'(\geq n)$ respectively. Then one may introduce ordinary real matrices $(K\,K')$, $\begin{pmatrix} M \\ M' \end{pmatrix}$, $(W\,W')$ and $\begin{pmatrix} Y \\ Y' \end{pmatrix}$, of dimensions $m' \times 2m$, $2m \times m'$, $n' \times 2n$ and $2n \times n'$ respectively, such that the pure imaginary square matrices

$$(K\,K')\begin{pmatrix} A_{\mathrm{B}} - A_{\mathrm{B}}^* & \mathrm{i}(A_{\mathrm{B}} + A_{\mathrm{B}}^*) \\ P_{\mathrm{B}} - P_{\mathrm{B}}^* & \mathrm{i}(P_{\mathrm{B}} + P_{\mathrm{B}}^*) \end{pmatrix}\begin{pmatrix} M \\ M' \end{pmatrix} \tag{1.4.54}$$

and

$$(W\,W')\begin{pmatrix} B_{\mathrm{B}} - B_{\mathrm{B}}^* & \mathrm{i}(B_{\mathrm{B}} + B_{\mathrm{B}}^*) \\ Q_{\mathrm{B}} - Q_{\mathrm{B}}^* & \mathrm{i}(Q_{\mathrm{B}} + Q_{\mathrm{B}}^*) \end{pmatrix}\begin{pmatrix} Y \\ Y' \end{pmatrix} \tag{1.4.55}$$

are nonsingular. One may also introduce ordinary real matrices $(L\,L')$, $\begin{pmatrix} N \\ N' \end{pmatrix}$, $(X\,X)$ and $\begin{pmatrix} Z \\ Z' \end{pmatrix}$, of dimensions $(2m - m') \times 2m$, $2m \times (2m - m')$, $(2n - n') \times 2n$ and $2n \times (2n - n')$ respectively, such that the square matrices

$$\begin{pmatrix} K & K' \\ L & L' \end{pmatrix}, \quad \begin{pmatrix} M & N \\ M' & N' \end{pmatrix}, \quad \begin{pmatrix} W & W' \\ X & X' \end{pmatrix} \quad \text{and} \quad \begin{pmatrix} Y & Z \\ Y' & Z' \end{pmatrix} \tag{1.4.56}$$

are all nonsingular.

Let the inverses of the second and fourth of the matrices (1.4.56) be denoted by

$$\begin{pmatrix} \bar{M} & \bar{M}' \\ \bar{N} & \bar{N}' \end{pmatrix} \quad \text{and} \quad \begin{pmatrix} \bar{Y} & \bar{Y}' \\ \bar{Z} & \bar{Z}' \end{pmatrix} \tag{1.4.57}$$

respectively. Then eq. (1.4.49) may be rewritten in the equivalent form

$$
\begin{pmatrix} \mathbf{0} \\ \vdots \\ \\ \mathbf{0} \end{pmatrix} = \begin{pmatrix} K & K' & 0 & 0 \\ L & L' & 0 & 0 \\ 0 & 0 & W & W' \\ 0 & 0 & X & X' \end{pmatrix}
$$

$$
\times \begin{pmatrix} A+A^* & \mathrm{i}(A+A^*) & C+C^* & \mathrm{i}(C-C^*) \\ P-P^* & \mathrm{i}(P+P^*) & R+R^* & \mathrm{i}(R-R^*) \\ D-D^* & \mathrm{i}(D+D^*) & B-B^* & \mathrm{i}(B+B^*) \\ S-S^* & \mathrm{i}(S+S^*) & Q-Q^* & \mathrm{i}(Q+Q^*) \end{pmatrix}
$$

$$
\times \begin{pmatrix} M & N & 0 & 0 \\ M' & N' & 0 & 0 \\ 0 & 0 & Y & Z \\ 0 & 0 & Y' & Z' \end{pmatrix} \begin{pmatrix} \bar{M} & \bar{M}' & 0 & 0 \\ \bar{N} & \bar{N}' & 0 & 0 \\ 0 & 0 & \bar{Y} & \bar{Y}' \\ 0 & 0 & \bar{Z} & \bar{Z}' \end{pmatrix} \begin{pmatrix} \mathbf{f}_c \\ \mathbf{g}_c \\ \mathbf{f}_a \\ \mathbf{g}_a \end{pmatrix} \qquad (1.4.58)
$$

The first and third rows of this equation imply

$$
\begin{pmatrix} (K\ K') \begin{pmatrix} A-A^* & \mathrm{i}(A+A^*) \\ P-P^* & \mathrm{i}(P+P^*) \end{pmatrix} \begin{pmatrix} M \\ M' \end{pmatrix} & (K\ K') \begin{pmatrix} C+C^* & \mathrm{i}(C-C^*) \\ R+R^* & \mathrm{i}(R-R^*) \end{pmatrix} \begin{pmatrix} Y \\ Y' \end{pmatrix} \\ (W\ W') \begin{pmatrix} D-D^* & \mathrm{i}(D+D^*) \\ S-S^* & \mathrm{i}(S+S^*) \end{pmatrix} \begin{pmatrix} M \\ M' \end{pmatrix} & (W\ W') \begin{pmatrix} B-B^* & \mathrm{i}(B+B^*) \\ Q-Q^* & \mathrm{i}(Q+Q^*) \end{pmatrix} \begin{pmatrix} Y \\ Y' \end{pmatrix} \end{pmatrix} \begin{pmatrix} M\mathbf{f}_c + \bar{M}'\mathbf{g}_c \\ \\ \bar{Y}\mathbf{f}_a + \bar{Y}'\mathbf{g}_a \end{pmatrix}
$$

$$
= - \begin{pmatrix} (K\ K') \begin{pmatrix} A-A^* & \mathrm{i}(A+A^*) \\ P-P^* & \mathrm{i}(P+P^*) \end{pmatrix} \begin{pmatrix} N \\ N' \end{pmatrix} & (K\ K') \begin{pmatrix} C+C^* & \mathrm{i}(C-C^*) \\ R+R^* & \mathrm{i}(R-R^*) \end{pmatrix} \begin{pmatrix} Z \\ Z' \end{pmatrix} \\ (W\ W') \begin{pmatrix} D-D^* & \mathrm{i}(D+D^*) \\ S-S^* & \mathrm{i}(S+S^*) \end{pmatrix} \begin{pmatrix} N \\ N' \end{pmatrix} & (W\ W') \begin{pmatrix} B-B^* & \mathrm{i}(B+B^*) \\ Q-Q^* & \mathrm{i}(Q+Q^*) \end{pmatrix} \begin{pmatrix} Z \\ Z' \end{pmatrix} \end{pmatrix} \begin{pmatrix} \bar{N}\mathbf{f}_c + \bar{N}'\mathbf{g}_c \\ \\ \bar{Z}\mathbf{f}_a + \bar{Z}'\mathbf{g}_a \end{pmatrix}
$$

(1.4.59)

Because of the nonsingularity of the matrices (1.4.54) and (1.4.55) the body of the matrix appearing on the left of eq. (1.4.59) is nonsingular. Therefore the m' real c-type supervectors $\bar{M}\mathbf{f}_c + \bar{M}'\mathbf{g}_c$ and the n' real a-type supervectors $\bar{Y}\mathbf{f}_a + \bar{Y}'\mathbf{g}_a$ can be expressed as linear combinations of the $2m - m'$ real c-type supervectors $\bar{N}\mathbf{f}_c + \bar{N}'\mathbf{g}_c$ and the $2n - n'$ real a-type supervectors $\bar{Z}\mathbf{f}_a + \bar{Z}'\mathbf{g}_a$. But we also, have, in an obvious notation,

$$
\begin{pmatrix} \mathbf{e}_c \\ \mathbf{e}_a \end{pmatrix} = \begin{pmatrix} \mathbf{f}_c + \mathrm{i}\mathbf{g}_c \\ \mathbf{f}_a + \mathrm{i}\mathbf{g}_a \end{pmatrix} = \begin{pmatrix} M + \mathrm{i}M' & N + \mathrm{i}N' & 0 & 0 \\ 0 & 0 & Y + \mathrm{i}Y' & Z + \mathrm{i}Z' \end{pmatrix} \begin{pmatrix} \bar{M}\mathbf{f}_c + M'\mathbf{g}_c \\ \bar{N}\mathbf{f}_c + \bar{N}'\mathbf{g}_c \\ \bar{Y}\mathbf{f}_a + \bar{Y}'\mathbf{g}_a \\ \bar{Z}\mathbf{f}_a + \bar{Z}'\mathbf{g}_a \end{pmatrix},
$$

(1.4.60)

which implies that the original basis supervectors can themselves be expressed as linear combinations of the $\bar{N}\mathbf{f}_c + \bar{N}'\mathbf{g}_c$ and the $\bar{Z}\mathbf{f}_a + \bar{Z}'\mathbf{g}_a$.

Evidently we must have $2m - m' \geq m$ and $2n - n' \geq n$, or $m' \leq m$ and $n' \leq n$. We have already seen that the inequalities cannot hold. Therefore $m = m'$, $n = n'$, and we have proved the

Theorem

Every supervector space having finite total dimension has a basis that is both pure and real

Standard bases

In practice it is convenient to work with pure bases for which the *c*-type basis supervectors are real and the *a*-type supervectors are pure imaginary. Such bases, which will be called *standard bases*, can be constructed from pure real bases by multiplying all the *a*-type basis supervectors by i. A standard basis is characterized by

$$ {}_i\mathbf{e}^* = (-1)^i\,{}_i\mathbf{e}. \qquad (1.4.61)$$

Let $\mathbf{X}$ be a real supervector having components X^i with respect to a standard basis. Then

$$X^i\,{}_i\mathbf{e} = \mathbf{X} = \mathbf{X}^* = {}_i\mathbf{e}^* X^{i*} = (-1)^i\,{}_i\mathbf{e} X^{i*}. \qquad (1.4.62)$$

If $\mathbf{X}$ is *c*-type then X^i is *c*-type or *a*-type according as the index i is *c*-type or *a*-type. If $\mathbf{X}$ is *a*-type the type association is reversed. Therefore

$$X^i\,{}_i\mathbf{e} = (-1)^{i+i(\mathrm{X}+i)}\, X^{i*}\,{}_i\mathbf{e} = (-1)^{i\mathrm{X}}\, X^{i*}\,{}_i\mathbf{e}, \qquad (1.4.63)$$

whence it follows that

$$X^{i*} = (-1)^{i\mathrm{X}} X^i \quad (\mathbf{X}\text{ real}). \qquad (1.4.64)$$

If $\mathbf{X}$ is a real *c*-type supervector then all its components with respect to a standard basis are seen to be real.

1.5 Linear transformations, supertransposition and dual supervector spaces

Change of basis

Let $\{{}_i\mathbf{e}\}$ and $\{{}_i\bar{\mathbf{e}}\}$ be two bases of a supervector space $\mathfrak{S}$. They must be related by equations of the form (1.4.26) with $\bar{K} = K^{-1}$. If they are pure bases then the matrix K has the block form (1.4.40) where the elements of the submatrices A and B are *c*-numbers, while the elements of the submatrices C and D are *a*-numbers. If both bases are standard bases then the elements of A, B and C are real while the elements of D are pure imaginary. We shall confine our attention to pure bases, occasionally imposing the additional requirement that the bases be standard.

Let X^i and $\bar{X}^i$ be the components of a supervector $\mathbf{X}$ relative to the bases $\{{}_i\mathbf{e}\}$ and $\{{}_i\bar{\mathbf{e}}\}$ respectively. Then

$$X^i{}_iK^j{}_j\bar{\mathbf{e}} = X^i{}_i\mathbf{e} = \mathbf{X} = \bar{X}^i{}_i\bar{\mathbf{e}}, \tag{1.5.1}$$

whence follows the transformation law for components:

$$\bar{X}^i = X^j{}_jK^i. \tag{1.5.2}$$

By taking the body of both sides of this equation it is easy to see that if the components of two supervectors have identical bodies relative to one pure basis then they have identical bodies relative to every other. Supervectors may therefore be divided, in a basis-independent way, into equivalence classes of supervectors whose components have identical bodies. A supervector is said to have *vanishing body* if it is in the same equivalence class as the zero supervector. Otherwise it is said to have nonvanishing body.

Let $\{{}_i\bar{\bar{\mathbf{e}}}\}$ be a third pure basis, with

$${}_i\mathbf{e} = {}_iK^j{}_j\bar{\mathbf{e}}, \quad {}_i\bar{\mathbf{e}} = {}_iL^j{}_j\bar{\bar{\mathbf{e}}}. \tag{1.5.3}$$

Then $\{{}_i\mathbf{e}\}$ and $\{{}_i\bar{\bar{\mathbf{e}}}\}$ are related by

$${}_i\mathbf{e} = {}_i(KL)^j{}_j\bar{\bar{\mathbf{e}}}, \tag{1.5.4}$$

where KL is the matrix product of K and L:

$${}_i(KL)^j = {}_iK^k{}_kL^j. \tag{1.5.5}$$

Note that when the matrix elements are supernumbers, as here, the order of the symbols 'K' and 'L' specifies not only which elements get multiplied together and added but also the order of the elements themselves in each term of each sum.

Shifting indices. The supertranspose

It turns out to be a convenience to be able to shift some of the indices that are encountered, from right to left and vice versa. The simplest of all shifting conventions is

$${}^iX \overset{\text{def}}{=} X^i, \quad \mathbf{X}\ c\text{-type}. \tag{1.5.6}$$

That is, if $\mathbf{X}$ is a c-type supervector then the indices on its components relative to a pure basis may be placed either on the right, as heretofore, or on the left. Equation (1.5.2) may now be written in the alternative form

$${}^i\bar{X} = {}^jX{}_jK^i = (-1)^{j(i+j)}{}_jK^{i\,j}X = {}^iK^{\sim}{}_j{}^jX, \tag{1.5.7}$$

where

$$ {}^iK^{\sim}_j \stackrel{\text{def}}{=} (-1)^{j(i+j)}\,{}_jK^i \tag{1.5.8}$$

The matrix $K^{\sim}$ formed out of the ${}^iK^{\sim}_j$ is called the *supertranspose* of the matrix K. Expressing K in the block form (1.4.40) one finds

$$K^{\sim} = \begin{pmatrix} A^{\sim} & -D^{\sim} \\ C^{\sim} & B^{\sim} \end{pmatrix}, \tag{1.5.9}$$

where $A^{\sim}$, $B^{\sim}$, $C^{\sim}$, $D^{\sim}$ are the ordinary transposes of the submatrices A, B, C, D respectively. It is straightforward to show that the supertranspose of the product of two matrices, such as the product (1.5.5), obeys the familiar rule

$$(KL)^{\sim} = L^{\sim}K^{\sim}. \tag{1.5.10}$$

Proof

$$\begin{aligned} {}^i(KL)^{\sim}_j &= (-1)^{j(i+1)}\,{}_j(KL)^i = (-1)^{j(i+1)}\,{}_jK^k\,{}_kL^i \\ &= (-1)^{j(i+1)+j(k+1)+k(i+1)+(j+k)(k+i)}\,{}^iL^{\sim}_k\,{}^kK^{\sim}_j. \end{aligned}$$

It is easy to verify that the exponent in the final expression is always an even number, qed.

It is also easy to show that supertransposition commutes with inversion:

$$K^{\sim-1} = K^{-1\sim}. \tag{1.5.11}$$

Proof

$$\begin{aligned} {}^i(K^{-1\sim}K^{\sim})_j &= {}^iK^{-1\sim}{}_k\,{}^kK^{\sim}_j \\ &= (-1)^{k(i+1)+j(k+1)+(i+k)(k+j)}\,{}_jK^k\,{}_kK^{-1i} \\ &= (-1)^{j(i+1)}\,{}_j\delta^i = {}^i\delta_j, \end{aligned}$$

where ${}_j\delta^i$ and ${}^i\delta_j$ are alternative symbols for the conventional Kronecker delta. Evidently

$$1_{(m,n)}{}^{\sim} = 1_{(m,n)}, \tag{1.5.12}$$

where $1_{(m,n)}$ is the unit matrix of dimension (m, n).

Suppose now that the supervector $\mathbf{X}$ is pure but not necessarily of type c. Then

$$\bar{X}^i = X^j\,{}_jK^i = (-1)^{(\mathbf{X}+j)(i+j)}\,{}_jK^iX^j = (-1)^{\mathbf{X}(i+j)}\,{}^iK^{\sim}_j\,X^j, \tag{1.5.13}$$

which shows that if the definition (1.5.6) is generalized to

$${}^iX \stackrel{\text{def}}{=} (-1)^{\mathbf{X}i}X^i, \tag{1.5.14}$$

then once again (cf. eq. (1.5.7))

$$ {}^i\bar{X} = {}^iK^{\sim}_j\,{}^jX. \tag{1.5.15}$$

Equation (1.5.14) will be chosen as the general law for shifting indices on the components of pure supervectors. For impure supervectors the law follows from (1.5.14) by linearity upon decomposing **X** into its even and odd parts.

Extensions of the supertransposition rules

Equation (1.5.8) defines the supertranspose of a matrix whose elements have their left indices in the lower position and their right indices in the upper position. It is also useful to define the supertransposes of matrices for which the index positions are reversed, for example the matrix $K^{\sim}$. Let L be such a matrix. Then

$$ {}_iL^{\sim\,j} \overset{\text{def}}{=} (-1)^{i(i+j)}\,{}^jL_i. \tag{1.5.16}$$

With this convention supertransposition satisfies

$$ K^{\sim\sim} = K. \tag{1.5.17}$$

Furthermore the rules (1.5.10) and (1.5.11) remain intact.

Supertransposes can profitably be defined even for matrices whose elements have both indices in the lower position or both indices in the upper position. Matrices of this type are encountered in the theory of quadratic forms, for example $X^i\,{}_iM_j\,{}^jY$ where **X** and **Y** are supervectors. Suppose **X** and **Y** are both c-type. Then

$$ X^i\,{}_iM_j\,{}^jY = (-1)^{i+j+ij}\,{}^jY\,{}_iM_j\,X^i = Y^j\,{}_jM^{\sim}_i\,{}^iX,$$

where

$$ {}_iM^{\sim}_j \overset{\text{def}}{=} (-1)^{i+j+ij}\,{}_jM_i. \tag{1.5.18}$$

More generally, if **X** and **Y** are pure but not necessarily c-type, one has

$$\begin{aligned} X^i\,{}_iM_j\,{}^jY &= (-1)^{\mathbf{X}i+\mathbf{Y}j+(\mathbf{X}+i)(\mathbf{Y}+i)+(i+j)(\mathbf{Y}+j)}\,Y^j\,{}_iM_j\,{}^iX \\ &= (-1)^{\mathbf{XY}}\,Y^j\,{}_jM^{\sim}_i\,{}^iX. \end{aligned} \tag{1.5.19}$$

Note that the definition (1.5.18) yields $M^{\sim\sim} = M$.

If M is a nonsingular matrix its inverse is a matrix whose elements have both indices in the upper position. Let N be such a matrix. In this case define

$$ {}^iN^{\sim j} \overset{\text{def}}{=} (-1)^{ij}\,{}^jN^i, \tag{1.5.20}$$

which satisfies $N^{\sim\sim} = N$. With this definition the rule (1.5.11) continues to hold. Proof:

$$\begin{aligned} {}^iM^{-1\sim k}\,{}_kM^{\sim}_j &= (-1)^{ik+k+j+jk+(i+k)(j+k)}\,{}_jM_k\,{}^kM^{-1i} \\ &= (-1)^{j(i+1)}\,{}_j\delta^i = {}^i\delta_j. \end{aligned}$$

Moreover, with the definitions (1.5.8), (1.5.16), (1.5.18) and (1.5.20), the rule (1.5.10) holds no matter what positions the indices on the elements of K and L have, provided only that the adjacent indices that are contracted in the matrix multiplication have opposite positions, one up and one down. The proof of this will be left as an exercise for the reader.

It should be noted that all of these rules apply *only* to matrices whose elements are c-type when both indices have the same type and a-type when the two indices are of opposite type. Occasionally matrices are encountered whose elements have a different type structure. The rules must then be modified accordingly. (See, for example, exercises **1.9** and **1.10** at the end of the chapter.)

If M is a matrix whose elements have both indices in the upper position or both indices in the lower position then M is said to be *supersymmetric* if $M^{\sim} = M$ and *antisupersymmetric* if $M^{\sim} = -M$. The inverse of a nonsingular supersymmetric matrix is supersymmetric, and the inverse of a nonsingular antisupersymmetric matrix is antisupersymmetric.

Dual supervector spaces

Let $\mathfrak{S}$ be a supervector space. The supervector space *dual* to $\mathfrak{S}$, denoted by $\mathfrak{S}^*$, is defined to be the set of all mappings $\omega: \mathfrak{S} \to \Lambda_\infty$, conventionally expressed by the notation

$$\omega(\mathbf{X}) \stackrel{\text{def}}{=} \mathbf{X}\cdot\omega, \quad \text{for all } \mathbf{X} \text{ in } \mathfrak{S}, \tag{1.5.21}$$

which satisfy the linear laws

$$(\alpha\mathbf{X})\cdot\omega = \alpha(\mathbf{X}\cdot\omega) \stackrel{\text{def}}{=} \alpha\mathbf{X}\cdot\omega, \tag{1.5.22}$$

$$(\mathbf{X}+\mathbf{Y})\cdot\omega = \mathbf{X}\cdot\omega + \mathbf{Y}\cdot\omega, \tag{1.5.23}$$

for all α in Λ_∞ and all $\mathbf{X}, \mathbf{Y}$ in $\mathfrak{S}$. $\mathbf{X}\cdot\omega$ is called the *inner product* of $\mathbf{X}$ and ω.

The set $\mathfrak{S}^*$ is given the structure of a supervector space by defining

$$\mathbf{X}\cdot(\omega+\sigma) \stackrel{\text{def}}{=} \mathbf{X}\cdot\omega + \mathbf{X}\cdot\sigma, \tag{1.5.24}$$

$$\mathbf{X}\cdot(\alpha\omega) \overset{\text{def}}{=} (\mathbf{X}\alpha)\cdot\omega, \tag{1.5.25}$$

$$\mathbf{X}\cdot(\omega\alpha) \overset{\text{def}}{=} (\mathbf{X}\cdot\omega)\alpha \overset{\text{def}}{=} \mathbf{X}\cdot\omega\alpha, \tag{1.5.26}$$

for all $\mathbf{X}$ in $\mathfrak{S}$, all α in Λ_∞, and all ω, σ in $\mathfrak{S}^*$. It is easy to verify that $\mathfrak{S}^*$ has all the properties (1)–(4) of section 1.4. Proofs of the analogs of eqs. (1.4.2), (1.4.3), (1.4.8), (1.4.9), (1.4.10), (1.4.11), (1.4.14) and (1.4.15) are straightforward. The zero supervector in $\mathfrak{S}^*$ is the mapping that sends every $\mathbf{X}$ in $\mathfrak{S}$ into the ordinary number zero, whence the analogs of eqs. (1.4.12) and (1.4.13) immediately follow. The negative of an element ω of $\mathfrak{S}^*$ is defined by $\mathbf{X}\cdot(-\omega) = -(\mathbf{X}\cdot\omega)$ for all $\mathbf{X}$ in $\mathfrak{S}$.

The unique even and odd parts of ω are defined by

$$\omega = \mathbf{u} + \mathbf{v}, \tag{1.5.27}$$

$$\mathbf{X}\cdot\mathbf{u} = (\mathbf{U}\cdot\omega)_e + (\mathbf{V}\cdot\omega)_o, \tag{1.5.28}$$

$$\mathbf{X}\cdot\mathbf{v} = (\mathbf{U}\cdot\omega)_o + (\mathbf{V}\cdot\omega)_e, \tag{1.5.29}$$

for all $\mathbf{X}$ in $\mathfrak{S}$, where $\mathbf{U}$ and $\mathbf{V}$ are the even and odd parts of $\mathbf{X}$ respectively. The c-type elements of $\mathfrak{S}^*$ map c-type elements of $\mathfrak{S}$ into c-numbers and a-type elements of $\mathfrak{S}$ into a-numbers. With a-type elements of $\mathfrak{S}^*$ the association is reversed. From these facts follows the analog of eq. (1.4.18):

$$\alpha\omega = (-1)^{\alpha\omega}\omega\alpha. \tag{1.5.30}$$

Finally, the analogs of eqs. (1.4.20), (1.4.21) and (1.4.22) follow from the definition

$$\mathbf{X}\cdot\omega^* \overset{\text{def}}{=} (-1)^{\mathbf{X}\omega}(\mathbf{X}^*\cdot\omega)^*, \tag{1.5.31}$$

for all $\mathbf{X}$ in $\mathfrak{S}$ and all ω in $\mathfrak{S}^*$. In particular,

$$\begin{aligned} \mathbf{X}\cdot(\alpha\omega)^* &= (-1)^{\mathbf{X}(\alpha+\omega)}(\mathbf{X}^*\cdot(\alpha\omega))^* = (-1)^{\mathbf{X}(\alpha+\omega)}((\alpha^*\mathbf{X})^*\cdot\omega)^* \\ &= (-1)^{\mathbf{X}(\alpha+\omega)+(\alpha+\mathbf{X})\omega}(\alpha^*\mathbf{X})\cdot\omega^* = (-1)^{\alpha(\mathbf{X}+\omega)}\alpha^*(\mathbf{X}\cdot\omega^*) \\ &= (\mathbf{X}\cdot\omega^*)\alpha^* = \mathbf{X}\cdot(\omega^*\alpha^*), \end{aligned} \tag{1.5.32}$$

$$\begin{aligned} \mathbf{X}\cdot(\omega\alpha)^* &= (-1)^{\mathbf{X}(\alpha+\omega)}(\mathbf{X}^*\cdot(\omega\alpha))^* = (-1)^{\mathbf{X}(\alpha+\omega)}\alpha^*(\mathbf{X}^*\cdot\omega)^* \\ &= (-1)^{\alpha\mathbf{X}}\alpha^*(\mathbf{X}\cdot\omega^*) = (\mathbf{X}\alpha^*)\cdot\omega^* = \mathbf{X}\cdot(\alpha^*\omega^*). \end{aligned} \tag{1.5.33}$$

It is not difficult to verify that real c-type elements of $\mathfrak{S}^*$ map real elements of $\mathfrak{S}$ into real supernumbers. Real a-type elements of $\mathfrak{S}^*$ map real c-type elements of $\mathfrak{S}$ into real a-numbers, but they map real a-type elements of $\mathfrak{S}$ into *imaginary* c-numbers.

Dual bases

If $\mathfrak{S}$ has a basis $\{{}_i\mathbf{e}\}$ (which will be assumed to be pure), then because of the linear laws (1.5.22) and (1.5.23), an element ω of $\mathfrak{S}^*$ is uniquely defined by its action on the basis supervectors. Evidently, for all $\mathbf{X}$ in $\mathfrak{S}$ and all ω in $\mathfrak{S}^*$,

$$\mathbf{X}\cdot\omega = X^i\,{}_i\omega, \tag{1.5.34}$$

where

$${}_i\omega \overset{\text{def}}{=} {}_i\mathbf{e}\cdot\omega. \tag{1.5.35}$$

The ${}_i\omega$ are called the *components of* ω *with respect to the basis* $\{{}_i\mathbf{e}\}$. Under the change of basis (1.4.26), with $\bar{K} = K^{-1}$, they transform according to the law

$${}_i\bar{\omega} = {}_iK^{-1\,j}\,{}_j\omega. \tag{1.5.36}$$

One may also write

$$\omega = \mathbf{e}^i\,{}_i\omega, \tag{1.5.37}$$

where $\{\mathbf{e}^i\}$ is a set of elements of $\mathfrak{S}^*$ defined by

$${}_i\mathbf{e}\cdot\mathbf{e}^j = {}_i\delta^j, \quad \text{for} \quad \text{all} \quad i, j. \tag{1.5.38}$$

Under the change of basis (1.4.26) the $\mathbf{e}^i$ suffer the change

$$\bar{\mathbf{e}}^i = \mathbf{e}^j\,{}_jK^i. \tag{1.5.39}$$

The $\mathbf{e}^i$ are linearly independent because if ω is the zero supervector in $\mathfrak{S}^*$ then, by virtue of eq. (1.5.35), all the ${}_i\omega$ vanish. Equation (1.5.37) implies that $\{\mathbf{e}^i\}$ is a complete set. Therefore $\{\mathbf{e}^i\}$ is a basis for $\mathfrak{S}^*$. It is known as the basis *dual* to $\{{}_i\mathbf{e}\}$. Note that

$$X^i = X^j\,{}_j\delta^i = X^j\,{}_j\mathbf{e}\cdot\mathbf{e}^i = \mathbf{X}\cdot\mathbf{e}^i. \tag{1.5.40}$$

Evidently every element of $\mathfrak{S}$ is uniquely defined by the actions of the elements of the dual basis upon it.

It is easy to verify that the dual to a pure basis is also pure. Moreover, if $\mathfrak{S}$ has finite dimension (m, n) then $\mathfrak{S}^*$ also has dimension (m, n). If $\{{}_i\mathbf{e}\}$ is a standard basis in $\mathfrak{S}$ then its dual $\{\mathbf{e}^i\}$ is said to be a standard basis in $\mathfrak{S}^*$. Standard bases in $\mathfrak{S}^*$ are characterized by

$$\mathbf{e}^{i*} = \mathbf{e}^i. \tag{1.5.41}$$

That is, *all* the basis supervectors are real, which is to be contrasted with (1.4.61). This has the consequence that if ω is real then its components with respect to a standard basis satisfy

$${}_i\omega^* = (-1)^{i(\omega+1)}\,{}_i\omega \quad (\omega \text{ real}), \tag{1.5.42}$$

which is to be contrasted with (1.4.64).

Further index-shifting conventions

Just as it is convenient to shift indices on the components of supervectors $\mathbf{X}$ in $\mathfrak{S}$, so is it convenient also to shift indices on the components of dual supervectors ω in $\mathfrak{S}^*$. Begin by writing

$$X^i{}_{,i}\omega = (-1)^{\mathrm{X}i+(\mathrm{X}+i)(\omega+i)}{}_i\omega{}^iX = (-1)^{\mathrm{X}\omega+i(\omega+1)}{}_i\omega{}^iX. \qquad (1.5.43)$$

This relation suggests that one adopt the convention

$$\omega_i \overset{\text{def}}{=} (-1)^{i(\omega+1)}{}_i\omega, \qquad (1.5.44)$$

which yields

$$X^i{}_{,i}\omega = (-1)^{\mathrm{X}\omega}\omega_i{}^iX. \qquad (1.5.45)$$

By analogy with eq. (1.5.34) it is natural to define an alternative inner product

$$\omega\cdot\mathbf{X} \overset{\text{def}}{=} \omega_i{}^iX = (-1)^{\mathrm{X}\omega}\mathbf{X}\cdot\omega, \qquad (1.5.46)$$

and hence an alternative set of mappings from $\mathfrak{S}$ to Λ_∞. It is easy to verify that these mappings satisfy analogs of the laws (1.5.22), (1.5.25) and (1.5.26):

$$\omega\cdot(\mathbf{X}\alpha) = (-1)^{\omega(\mathrm{X}+\alpha)}(\mathbf{X}\alpha)\cdot\omega = (-1)^{\omega\mathrm{X}}\mathbf{X}\cdot\omega\alpha = (\omega\cdot\mathbf{X})\alpha \overset{\text{def}}{=} \omega\cdot\mathbf{X}\alpha, \qquad (1.5.47)$$

$$(\omega\alpha)\cdot\mathbf{X} = (-1)^{\mathrm{X}(\omega+\alpha)}\mathbf{X}\cdot\omega\alpha = (-1)^{\omega(\mathrm{X}+\alpha)}\alpha\mathbf{X}\cdot\omega = \omega\cdot(\alpha\mathbf{X}), \qquad (1.5.48)$$

$$(\alpha\omega)\cdot\mathbf{X} = (-1)^{\mathrm{X}(\omega+\alpha)}\mathbf{X}\cdot(\alpha\omega) = (-1)^{\mathrm{X}\omega}\alpha\mathbf{X}\cdot\omega = \alpha(\omega\cdot\mathbf{X}) \overset{\text{def}}{=} \alpha\omega\cdot\mathbf{X}. \qquad (1.5.49)$$

Therefore if the definition (1.5.46) is extended to the impure case by imposition of the linear laws

$$\omega\cdot(\mathbf{X}+\mathbf{Y}) \overset{\text{def}}{=} \omega\cdot\mathbf{X} + \omega\cdot\mathbf{Y}, \qquad (1.5.50)$$

$$(\omega+\sigma)\cdot\mathbf{X} \overset{\text{def}}{=} \omega\cdot\mathbf{X} + \sigma\cdot\mathbf{X}, \qquad (1.5.51)$$

it follows that the alternative mappings themselves constitute a supervector space.

Neither set of mappings has any intrinsic property that gives it a preferred status over the other. It is therefore convenient to abuse logic slightly by regarding the symbol 'ω' as representing *both* the mapping $\mathbf{X} \mapsto \mathbf{X}\cdot\omega$ and the mapping $\mathbf{X} \mapsto \omega\cdot\mathbf{X}$. (The two are identical when ω is *c*-type.) Which mapping is actually involved in a given instance is always clear from the equations. The only error that one might be led to make in adopting this viewpoint concerns the basis supervectors. The basis $\{\mathbf{e}^i\}$ in $\mathfrak{S}^*$ is *not* dual to the basis $\{{}_i\mathbf{e}\}$ in $\mathfrak{S}$ under both kinds of mappings. One must introduce alternative basis supervectors defined by

$$\mathbf{e}_i \stackrel{\text{def}}{=} (-1)^i\, {}_i\mathbf{e}, \quad {}^i\mathbf{e} \stackrel{\text{def}}{=} \mathbf{e}^i \tag{1.5.52}$$

(assuming the bases to be pure). Only then can one write analogs for eqs. (1.4.25), (1.5.35), (1.5.37), (1.5.38) and (1.5.40):

$$\begin{aligned} {}^i\mathbf{e}\cdot\mathbf{e}_j &= (-1)^{ij}\,\mathbf{e}_j\cdot{}^i\mathbf{e} = (-1)^{j(i+1)}\,{}_j\mathbf{e}\cdot\mathbf{e}^i \\ &= (-1)^{j(i+1)}\,{}_j\delta^i = {}^i\delta_j, \end{aligned} \tag{1.5.53}$$

$$\omega\cdot\mathbf{e}_i = (-1)^{i+\omega i}\,{}_i\mathbf{e}\cdot\omega = (-1)^{i(\omega+1)}\,{}_i\omega = \omega_i, \tag{1.5.54}$$

$${}^i\mathbf{e}\cdot\mathbf{X} = (-1)^{i\mathbf{X}}\,\mathbf{X}\cdot\mathbf{e}^i = (-1)^{i\mathbf{X}}X^i = {}^iX, \tag{1.5.55}$$

$$\mathbf{X} = X^i\,{}_i\mathbf{e} = (-1)^{\mathbf{X}i+i+i(\mathbf{X}+i)}\,\mathbf{e}_i\,{}^iX = \mathbf{e}_i\,{}^iX, \tag{1.5.56}$$

$$\omega = \mathbf{e}^i\,{}_i\omega = (-1)^{i(\omega+1)+i(i+\omega)}\,\omega_i\,{}^i\mathbf{e} = \omega_i\,{}^i\mathbf{e}, \tag{1.5.57}$$

We record here also the analogs of the transformation laws (1.5.3), (1.5.36) and (1.5.39):

$$\mathbf{e}_i = \bar{\mathbf{e}}_j\,{}^jK^{\sim}{}_i, \tag{1.5.58}$$

$$\bar{\omega}_i = \omega_j\,{}^jK^{-1\sim}{}_i, \tag{1.5.59}$$

$${}^i\bar{\mathbf{e}} = {}^iK^{\sim}{}_j\,{}^j\mathbf{e}. \tag{1.5.60}$$

We mention finally a few additional index-shifting conventions that are convenient from time to time. These involve the elements of matrices like those discussed at the beginning of this section. We shall define†

$$K_i{}^j \stackrel{\text{def}}{=} (-1)^i\,{}_iK^j, \tag{1.5.61}$$

$$L^i{}_j \stackrel{\text{def}}{=} {}^iL_j, \tag{1.5.62}$$

$$M_{ij} \stackrel{\text{def}}{=} (-1)^i\,{}_iM_j, \tag{1.5.63}$$

$$N^{ij} \stackrel{\text{def}}{=} {}^iN^j. \tag{1.5.64}$$

These rules yield

$$K^{\sim i}{}_j = (-1)^{ij}\,K_j{}^i, \tag{1.5.65}$$

$$L^{\sim}{}_i{}^j = (-1)^{ij}\,L^j{}_i, \tag{1.5.66}$$

$$M^{\sim}{}_{ij} = (-1)^{ij}\,M_{ji}, \tag{1.5.67}$$

$$N^{\sim ij} = (-1)^{ij}\,N^{ji}. \tag{1.5.68}$$

† These conventions are identical with those for c-type tensor fields. (See eqs. (1.11.34), Chapter 2.)

Note also that with these rules $\delta^i{}_j$ is an ordinary Kronecker delta but $\delta_i{}^j$ is not.

1.6 The supertrace and the superdeterminant

The supertrace

Let K be a matrix whose elements ${}_iK^j$ have the left index in the lower position and the right index in the upper position. Let these elements fall into type classes according to the block structure depicted in eq. (1.4.40). We define the *supertrace of* K as follows:

$$\operatorname{str} K \overset{\text{def}}{=} (-1)^i\, {}_iK^i = K_i{}^i. \tag{1.6.1}$$

Alternatively, let L be a matrix with similar block structure but whose elements have the left index in the upper position and the right index in the lower position. Again we define a supertrace:

$$\operatorname{str} L \overset{\text{def}}{=} (-1)^i\, {}^iL_i = (-1)^i\, L^i{}_i. \tag{1.6.2}$$

When K or L is expressed in the block form (1.4.40) the supertrace becomes

$$\operatorname{str} K = \operatorname{tr} A - \operatorname{tr} B, \tag{1.6.3}$$

where 'tr' denotes the ordinary trace. Note that

$$\operatorname{str} 1_{(m,n)} = \operatorname{tr} 1_m - \operatorname{tr} 1_n = m - n. \tag{1.6.4}$$

Definitions (1.6.1) and (1.6.2) together guarantee that the supertrace is invariant under supertransposition:

$$\operatorname{str} K^{\sim} = (-1)^i\, {}^iK^{\sim}{}_i = (-1)^{i+i(i+i)}\, {}_iK^i = (-1)^i\, {}_iK^i = \operatorname{str} K. \tag{1.6.5}$$

They also have the more important property of yielding the cyclic invariance law.[†] Let M and N be any two matrices (having the usual block structure) the positions of the indices on the elements of which allow M and N to be multiplied together yielding a product whose elements have one upper index and one lower index. Examples are ${}_iM_j$ and ${}^iN^j$, ${}_iM^j$ and ${}_iN^j$, and ${}^iM_j$ and ${}^iN_j$. Choosing the first example, we have

$$\begin{aligned}\operatorname{str}(MN) &= (-1)^i\, {}_iM_j\, {}^jN^i = (-1)^{i+(i+j)^2}\, {}^jN^i\, {}_iM_j \\ &= (-1)^j\, {}^jN^i\, {}_iM_j = \operatorname{str}(NM).\end{aligned} \tag{1.6.6}$$

The same result holds also with the other examples.

[†] The only other definitions that would yield this law are $\operatorname{str} K = -(-1)^i\, {}_iK^i$ and $\operatorname{str} L = -(-1)^i\, {}^iL_i$. Definitions (1.6.1) and (1.6.2) reduce to the ordinary trace when there are no *a*-type indices.

The superdeterminant

Ordinary determinants and ordinary traces are related through the classical variational law

$$\delta \ln \det A = \mathrm{tr}(A^{-1}\delta A). \tag{1.6.7}$$

One may define a *superdeterminant* by integrating an analogous variational law involving the supertrace,[†]

$$\delta \ln \mathrm{sdet}\, M \overset{\mathrm{def}}{=} \mathrm{str}(M^{-1}\,\delta M), \tag{1.6.8}$$

starting from the boundary condition

$$\mathrm{sdet}\, 1_{(m,n)} \overset{\mathrm{def}}{=} 1. \tag{1.6.9}$$

The supertrace is defined only for matrices whose elements have one upper and one lower index. The product $M^{-1}\,\delta M$ is always in this class regardless of the position of the indices on the elements of the matrix M. Therefore the superdeterminant is defined (subject to certain nonsingularity conditions that will appear) for matrices belonging to any of the four types K, L, M and N considered in section 1.5.

Using the cyclic invariance of the supertrace one may write

$$\begin{aligned} \delta \ln \mathrm{sdet}(KM) &= \mathrm{str}[(KM)^{-1}\delta(KM)] = \mathrm{str}[M^{-1}K^{-1}(\delta K M + K\delta M)] \\ &= \mathrm{str}(K^{-1}\delta K) - \mathrm{str}(M^{-1}\delta M) \\ &= \delta \ln \mathrm{sdet}\, K + \delta \ln \mathrm{sdet}\, M \\ &= \delta \ln[(\mathrm{sdet}\, K)(\mathrm{sdet}\, M)], \end{aligned} \tag{1.6.10}$$

which, subject to the boundary condition (1.6.9), integrates to

$$\mathrm{sdet}(KM) = (\mathrm{sdet}\, K)(\mathrm{sdet}\, M). \tag{1.6.11}$$

Using both the transposition and cyclic invariance of the supertrace, as well as the laws (1.5.10) and (1.5.11), one may also write

$$\begin{aligned} \delta \ln \mathrm{sdet}\, M^{\sim} &= \mathrm{str}(M^{\sim -1}\,\delta M^{\sim}) = \mathrm{str}(\delta M M^{-1})^{\sim} \\ &= \mathrm{str}(M^{-1}\delta M) = \delta \ln \mathrm{sdet}\, M, \end{aligned} \tag{1.6.12}$$

from which one may conclude that sdet M and sdet $M^{\sim}$ are equal up to a constant factor. Here one must distinguish two cases: (1) The indices on the elements of M are in opposite position (one up and one down). (2) The indices on the elements of M are in the same position (both up or both down). In the first case one may use the transposition invariance of the unit

[†] The superdeterminant will turn out to be a c-number. As long as its body is nonvanishing its logarithm may be defined by a series of the form (1.1.6).

matrix $1_{(m,n)}$ together with the boundary condition (1.6.9) to conclude that

$$\text{sdet}\, M^{\sim} = \text{sdet}\, M \quad \text{(indices in opposite positions).} \qquad (1.6.13)$$

In case (2) one must note that if the unit matrix is regarded as having both its indices in the *same* position then, from eq. (1.5.18) or eq. (1.5.20),

$$1_{(m,n)} = \text{diag}(1_m, 1_n), \qquad 1_{(m,n)}{}^{\sim} = \text{diag}(1_m, -1_n),$$

whence (see eq. (1.6.17) below) $\text{sdet}\, 1_{(m,n)}{}^{\sim} = (-1)^n$. The boundary condition (1.6.9) now yields

$$\text{sdet}\, M^{\sim} = (-1)^n \text{sdet}\, M \quad \text{(indices in the same position).} \qquad (1.6.14)$$

It turns out that one does not very often encounter matrices whose elements have both their indices in the same position unless the integer n is even. Therefore the superdeterminant is usually transposition invariant regardless of the positions of the indices.

The superdeterminant in special cases

Suppose the submatrices C and D in eq. (1.4.40) vanish. Then

$$K = \text{diag}(A, B), \quad K^{-1} = \text{diag}(A^{-1}, B^{-1}), \qquad (1.6.15)$$

and

$$\begin{aligned}\delta \ln \text{sdet}\, K &= \text{str}(K^{-1}\delta K) = \text{str}\,\text{diag}(A^{-1}\delta A, B^{-1}\delta B)\\ &= \text{tr}(A^{-1}\delta A) - \text{tr}(B^{-1}\delta B)\\ &= \delta \ln[(\det A)(\det B)^{-1}]. \end{aligned} \qquad (1.6.16)$$

The boundary condition (1.6.9) yields in this case

$$\text{sdet}\begin{pmatrix} A & 0 \\ 0 & B \end{pmatrix} = (\det A)(\det B)^{-1}. \qquad (1.6.17)$$

Note that the superdeterminant is defined only if the submatrix B is nonsingular. The inverse of the superdeterminant is defined only if the submatrix A is nonsingular. Since the space of all nonsingular square matrices of given dimension, having complex c-number elements, is a connected space, formula (1.6.17) and its inverse are obtained in all allowable cases by integrating (1.6.16) starting from the unit matrix.

Next, suppose that the matrix K has the form

$$K = \begin{pmatrix} 1_m & X \\ 0 & 1_n \end{pmatrix}, \qquad (1.6.18)$$

where X is an $m \times n$ matrix having a-number elements. Then

$$K^{-1} = \begin{pmatrix} 1_m & -X \\ 0 & 1_n \end{pmatrix} \qquad (1.6.19)$$

and, if only X is allowed to vary,

$$K^{-1}\delta K = \begin{pmatrix} 1_m & -X \\ 0 & 1_n \end{pmatrix}\begin{pmatrix} 0 & \delta X \\ 0 & 0 \end{pmatrix} = \begin{pmatrix} 0 & \delta X \\ 0 & 0 \end{pmatrix}. \qquad (1.6.20)$$

From this one obtains

$$\delta \ln \operatorname{sdet} K = \operatorname{str}(K^{-1}\delta K) = 0, \qquad (1.6.21)$$

which, for all X, yields

$$\operatorname{sdet}\begin{pmatrix} 1_m & X \\ 0 & 1_n \end{pmatrix} = 1. \qquad (1.6.22)$$

In a similar manner one can show that

$$\operatorname{sdet}\begin{pmatrix} 1_m & 0 \\ Y & 1_n \end{pmatrix} = 1, \qquad (1.6.23)$$

where Y is any $n \times m$ matrix having a-number elements.

The superdeterminant in the general case

If the submatrix B is a nonsingular then the general matrix (1.4.40) can be expressed as the following product:

$$\begin{pmatrix} A & C \\ D & B \end{pmatrix} = \begin{pmatrix} 1_m & CB^{-1} \\ 0 & 1_n \end{pmatrix}\begin{pmatrix} A - CB^{-1}D & 0 \\ 0 & B \end{pmatrix}\begin{pmatrix} 1_m & 0 \\ B^{-1}D & 1_n \end{pmatrix}. \qquad (1.6.24)$$

Equations (1.6.17), (1.6.22) and (1.6.23), together with the product law (1.6.11), allow one to infer

$$\operatorname{sdet}\begin{pmatrix} A & C \\ D & B \end{pmatrix} = \det(A - CB^{-1}D)(\det B)^{-1}. \qquad (1.6.25)$$

As in the special case (1.6.15), the superdeterminant is defined only if B is nonsingular. If the submatrix A is singular then the superdeterminant has vanishing body.

If the submatrix A is nonsingular then (1.4.40) can be expressed in an alternative product form:

$$\begin{pmatrix} A & C \\ D & B \end{pmatrix} = \begin{pmatrix} 1_m & 0 \\ DA^{-1} & 1_n \end{pmatrix}\begin{pmatrix} A & 0 \\ 0 & B - DA^{-1}C \end{pmatrix}\begin{pmatrix} 1_m & A^{-1}C \\ 0 & 1_n \end{pmatrix}, \qquad (1.6.26)$$

yielding

$$\left[\operatorname{sdet}\begin{pmatrix} A & C \\ D & B \end{pmatrix}\right]^{-1} = (\det A)^{-1}\det(B - DA^{-1}C) \qquad (1.6.27)$$

The inverse of the superdeterminant is defined only if A is nonsingular. If B is singular then this inverse has vanishing body.

If both A and B are nonsingular the matrix (1.4.40) has the following inverse:

$$\begin{pmatrix} \bar{A} & \bar{C} \\ \bar{D} & \bar{B} \end{pmatrix} \stackrel{\text{def}}{=} \begin{pmatrix} A & C \\ D & B \end{pmatrix}^{-1}$$

$$= \begin{pmatrix} (1_m - A^{-1}CB^{-1}D)^{-1}A^{-1} & -(1_m - A^{-1}CB^{-1}D)^{-1}A^{-1}CB^{-1} \\ -(1_n - B^{-1}DA^{-1}C)^{-1}B^{-1}DA^{-1} & (1_n - B^{-1}DA^{-1}C)^{-1}B^{-1} \end{pmatrix}, \tag{1.6.28a}$$

$$= \begin{pmatrix} A^{-1}(1_m - CB^{-1}DA^{-1})^{-1} & -A^{-1}CB^{-1}(1_n - DA^{-1}CB^{-1})^{-1} \\ -B^{-1}DA^{-1}(1_m - CB^{-1}DA^{-1})^{-1} & B^{-1}(1_n - DA^{-1}CB^{-1})^{-1} \end{pmatrix}, \tag{1.6.28b}$$

which allows one to express the superdeterminant in a number of alternative forms:

$$\operatorname{sdet}\begin{pmatrix} A & C \\ D & B \end{pmatrix} = (\det A)(\det B)^{-1}\det(1_m - A^{-1}CB^{-1}D), \tag{1.6.29a}$$

$$= (\det A)(\det B)^{-1}\det(1_m - CB^{-1}DA^{-1}), \tag{1.6.29b}$$

$$= (\det \bar{A})^{-1}(\det B)^{-1}, \tag{1.6.29c}$$

$$= (\det A)(\det B)^{-1}\det(1_n - B^{-1}DA^{-1}C)^{-1}, \tag{1.6.29d}$$

$$= (\det A)(\det B)^{-1}\det(1_n - DA^{-1}CB^{-1})^{-1}, \tag{1.6.29e}$$

$$= (\det A)(\det \bar{B}). \tag{1.6.29f}$$

1.7 Integration over $\mathbf{R}_c^m \times \mathbf{R}_a^n$

Notation

We shall use the symbol 'x' to denote a general point of $\mathbf{R}_c^m \times \mathbf{R}_a^n$ and shall denote the coordinates of x by x^i (or x^j, x^k, etc.). Latin indices will be understood to range over the values $-n, \ldots -1, 1, \ldots m$ (or, sometimes, $-n, \ldots -1, 0, 1, \ldots m-1$) with the negative values distinguishing the a-number coordinates from the c-number coordinates. We shall sometimes wish to focus on the a-number coordinates, or the c-number coordinates, by themselves. In that case the Latin indices will be replaced by Greek indices. Greek indices from the first part of the alphabet will be used to distinguish the a-number coordinates (e.g., $x^\alpha, x^\beta, x^\gamma$, etc.) and Greek indices from the middle of the alphabet to distinguish the c-number coordinates (e.g., $x^\mu, x^\nu x^\sigma$, etc.). The indices $\mu, \nu, \sigma, \ldots$ will then range over positive (or nonnegative) values, while the indices $\alpha, \beta, \gamma, \ldots$ range over negative values.

Sometimes we shall wish to be even more explicit. In applications of the mathematical formalism to physics, the convention has become established of using the symbol 'θ' to identify the a-number coordinates. The c-number coordinates are then denoted by $x^\mu, x^\nu, x^\sigma, \ldots$ and the a-number coordinates by $\theta^\alpha, \theta^\beta, \theta^\gamma, \ldots$. Since, in this notation, the symbol to which an index is attached identifies the coordinate type, the convention on the ranges of indices may be relaxed so that $\alpha, \beta, \gamma, \ldots$ can assume positive values (e.g., $1 \ldots n$).

In dealing with functions (or distributions) of many variables it is convenient to introduce an abbreviated notation for differentiation:

$$_{\ldots ij,}f_{,kl\ldots} \overset{\text{def}}{=} \cdots \frac{\overrightarrow{\partial}}{\partial x^i}\frac{\overrightarrow{\partial}}{\partial x^j} f \frac{\overleftarrow{\partial}}{\partial x^k}\frac{\overleftarrow{\partial}}{\partial x^l}\cdots \tag{1.7.1}$$

Here the function (or distribution) f on $\mathbf{R}_c^m \times \mathbf{R}_a^n$ is assumed to be everywhere differentiable (in the sense of eqs. (1.1.10) and (1.1.15)) with respect to each of the coordinates x^i. Because differentiability in the super sense implies that f is differentiable to arbitrarily high order, expression (1.7.1) is always meaningful. It is not difficult to show that

$$\left(\frac{\overrightarrow{\partial}}{\partial x^i} f\right)\frac{\overleftarrow{\partial}}{\partial x^j} = \frac{\overrightarrow{\partial}}{\partial x^i}\left(f\frac{\overleftarrow{\partial}}{\partial x^j}\right), \tag{1.7.2}$$

and hence it does not matter whether the left partial differentiations in (1.7.1) are performed before the right differentiations or vice versa. In other respects, however, the order of differentiations is important. For example, one always has

$$_{ij,}f = (-1)^{ij}\,_{ji,}f, \quad f_{,ij} = (-1)^{ij} f_{,ji} \tag{1.7.3}$$

and, if f is *pure* (i.e., takes its values in $\mathbf{C}_c$ or $\mathbf{C}_a$), also the rules

$$_{i,}f = (-1)^{i(f+1)} f_{,i}, \quad _{i,}f_{,j} = (-1)^{(i+j)(f+1)+ij}\,_{j,}f_{,i}, \text{etc.} \tag{1.7.4}$$

Integration

A differentiable function (or distribution) on $\mathbf{R}_c^m \times \mathbf{R}_a^n$ must be linear in each of the a-number coordinates separately (see eq. (1.3.1)). Therefore it may be expanded as a power series in the x^α's terminating at the term of nth degree. This term may be expressed in the form $g(x^1, \ldots x^m)x^{-1} \ldots x^{-n}$, g being a function (or distribution) of the c-number coordinates only, and if the volume element in $\mathbf{R}_c^m \times \mathbf{R}_a^n$ is defined to be

$$\mathrm{d}^{m,n}x \overset{\text{def}}{=} \mathrm{i}^{n(n-1)/2}\,\mathrm{d}x^1 \ldots \mathrm{d}x^m\,\mathrm{d}x^{-1} \ldots \mathrm{d}x^{-n} \tag{1.7.5}$$

(cf. eq. (1.3.19)), then it is an immediate consequence of the laws (1.3.5) and (1.3.6), together with the anticommutativity of the x^{α}'s, that

$$\int f(x)\,\mathrm{d}^{m,n}x = (2\pi \mathrm{i})^{n/2}(-\mathrm{i})^{n(n-1)/2} \int g(x^1, \ldots x^m)\,\mathrm{d}^m x, \qquad (1.7.6)$$

where

$$\mathrm{d}^m x \overset{\text{def}}{=} \mathrm{d}x^1 \ldots \mathrm{d}x^m. \qquad (1.7.7)$$

It will be observed that integration with respect to the *a*-number coordinates is equivalent to a multiple differentiation:

$$\int f(x)\,\mathrm{d}x^{-1} \ldots \mathrm{d}x^{-n} = (-1)^{n(n-1)/2}(2\pi \mathrm{i})^{n/2} f(x) \frac{\overleftarrow{\partial}}{\partial x^{-1}} \cdots \frac{\overleftarrow{\partial}}{\partial x^{-n}} \qquad (1.7.8)$$

If the integration over the *a*-number coordinates is performed first, as in eq. (1.7.6), then the integral of f over $\mathbf{R}_c^m \times \mathbf{R}_a^n$ will exist only if the integral of g over $\mathbf{R}_c^m$ is well defined and converges. A necessary (but not sufficient) condition for this is

$$f(x) \frac{\overleftarrow{\partial}}{\partial x^{-1}} \cdots \frac{\overleftarrow{\partial}}{\partial x^{-n}} \xrightarrow[x_B \to \infty]{} 0, \qquad (1.7.9)$$

where '$x_B \to \infty$' means 'as the body of any one of the *c*-number coordinates tends to $\pm\infty$'. If we wish the integral to be well defined independently of the order of integration then we must have, more generally,

$$f(x) \xrightarrow[x_B \to \infty]{} 0, \qquad (1.7.10)$$

and, in addition, the rate of approach to zero must be sufficiently rapid in all *c*-number directions. In what follows it will be assumed that these conditions hold.

Homogeneous linear transformations of the a-number coordinates

Suppose the *a*-number coordinates x^{α} are replaced by the variables

$$\bar{x}^{\alpha} = B^{\alpha}{}_{\beta} x^{\beta}, \qquad (1.7.11)$$

where B is an arbitrary nonsingular $n \times n$ matrix having real *c*-number elements. How does the volume element $\mathrm{d}^{m,n}x$ change under this transformation? The answer is determined by requiring that the fundamental definitions (1.3.5) and (1.3.6) remain invariant under the replacement $x \to \bar{x}$. First note that

$$\begin{aligned}\bar{x}^{-1}\ldots\bar{x}^{-n} &= B^{-1}{}_{\alpha_1}\ldots B^{-n}{}_{\alpha_n}x^{\alpha_1}\ldots x^{\alpha_n}\\ &= B^{-1}{}_{\alpha_1}\ldots B^{-n}{}_{\alpha_n}\varepsilon^{\alpha_1\ldots\alpha_n}x^{-1}\ldots x^{-n}\\ &= (\det B)x^{-1}\ldots x^{-n}.\end{aligned} \tag{1.7.12}$$

Since the *c*-number coordinates remain unaffected by the transformation (1.7.11) one may therefore write

$$\begin{aligned}\int f(x)\mathrm{d}^{m,n}x &= \int g(x^1\ldots x^m)x^{-1}\ldots x^{-n}\mathrm{d}^{m,n}x\\ &= \int g(x^1\ldots x^m)\bar{x}^{-1}\ldots\bar{x}^{-n}\mathrm{d}^{m,n}\bar{x}\\ &= (\det B)\int g(x^1\ldots x^m)x^{-1}\ldots x^{-n}d^{m,n}\,\bar{x},\end{aligned} \tag{1.7.13}$$

from which it follows that

$$\mathrm{d}^{m,n}\bar{x} = (\det B)^{-1}\mathrm{d}^{m,n}x. \tag{1.7.14}$$

Homogeneous linear transformations of all the coordinates

Let the coordinates x^i be replaced by the variables

$$\bar{x}^i = {}^iL_j\,x^j, \tag{1.7.15}$$

where L is a nonsingular matrix having a block form like (1.4.40) with A, B and D real and C pure imaginary. In terms of the submatrices A, B, C, D the change of variables (1.7.15) may be expressed by a three-step transformation (cf. eq. (1.6.24)):

$$\left.\begin{aligned} y^\mu &= x^\mu,\\ y^\alpha &= x^\alpha + B^{-1\alpha}{}_\beta D^\beta{}_\nu x^\nu,\end{aligned}\right\} \tag{1.7.16}$$

$$\left.\begin{aligned} \bar{y}^\mu &= (A^\mu{}_\nu - C^\mu{}_\alpha B^{-1\alpha}{}_\beta D^\beta{}_\nu)y^\nu,\\ \bar{y}^\alpha &= B^\alpha{}_\beta\, y^\beta,\end{aligned}\right\} \tag{1.7.17}$$

$$\left.\begin{aligned} \bar{x}^\mu &= \bar{y}^\mu + C^\mu{}_\alpha\, B^{-1\alpha}{}_\beta\,\bar{y}^\beta\\ \bar{x}^\alpha &= \bar{y}^\alpha.\end{aligned}\right\} \tag{1.7.18}$$

Because of the assumed boundary conditions on the integrand the integration with respect to each variable in $\int f(x)\mathrm{d}^{m,n}x$ may be carried out independently of the others. The transformations (1.7.16) and (1.7.18) involve mere shifts in the zero points. Therefore

$$\mathrm{d}^{m,n}y = \mathrm{d}^{m,n}x, \quad \mathrm{d}^{m,n}\bar{x} = \mathrm{d}^{m,n}\bar{y}. \tag{1.7.19}$$

In (1.7.17) the c-number and a-number variables transform separately. The transformation of the c-number variables can be handled by conventional integration theory, leading to the appearance of the usual Jacobian in the transformed volume element. For the transformation of the a-number variables we may use the result (1.7.14). Putting everything together and using eq. (1.6.52), therefore, we have

$$d^{m,n}\bar{y} = \det(A - CB^{-1}D)(\det B)^{-1} d^{m,n}y, \tag{1.7.20}$$
$$d^{m,n}\bar{x} = (\operatorname{sdet} L)\, d^{m,n}x. \tag{1.7.21}$$

Note that the superdeterminant sdet L is a real c-number with nonvanishing body.

Nonlinear transformations

We shall now show that under a general *nonlinear* differentiable transformation of the coordinates,

$$\bar{x}^i = \bar{x}^i(x), \tag{1.7.22}$$

the volume element suffers the transformation

$$d^{m,n}\bar{x} = J d^{m,n}x, \tag{1.7.23}$$

where J is the *super-Jacobian*:

$$J \overset{\text{def}}{=} \operatorname{sdet}(\bar{x}^i{}_{,j}) = \operatorname{sdet}({}_{j,}\bar{x}^i), \tag{1.7.24}$$

the equivalence of the two forms following from the relation

$${}_{j,}\bar{x}^i = (-1)^{j(i+j)}\bar{x}^i{}_{,j} \tag{1.7.25}$$

and the invariance of the superdeterminant under supertransposition (eqs. (1.5.16) and (1.6.13)). The transformation (1.7.22) will be assumed to be nonsingular (i.e., invertible) and to tend to the linear form (1.7.15) as the body of any c-number coordinate tends to $\pm\infty$.

We shall first prove (1.7.23) for infinitesimal transformations

$$\bar{x}^i = x^i + \delta\xi^i, \quad \delta\xi^i \xrightarrow[x_B \to \infty]{} 0. \tag{1.7.26}$$

Finite transformations may be built up from these together with (1.7.15). The form of the transformation law for the volume element is determined by the conditions

$$\int \bar{f}(\bar{x})\, d^{m,n}\bar{x} = \int f(x)\, J d^{m,n}x, \tag{1.7.27}$$

$$\bar{f}(\bar{x}) = f(x). \tag{1.7.28}$$

The function $\bar{f}$ is a different function of its arguments $\bar{x}^i$ than f is of the x^i. Since the $\bar{x}$'s on the left of eq. (1.7.27) are dummy variables we may equally well write

$$\int \bar{f}(x)\mathrm{d}^{m,n}x = \int f(x)J\mathrm{d}^{m,n}x. \tag{1.7.29}$$

Under (1.7.26) $\bar{f}$ is given, to first infinitesimal order, by

$$\bar{f}(x) = f(x) - f_{,i}(x)\delta\xi^i(x). \tag{1.7.30}$$

Writing

$$J = 1 + \varepsilon, \tag{1.7.31}$$

where ε is some infinitesimal function on $\mathbf{R}_c^m \times \mathbf{R}_a^n$, we must have

$$\begin{aligned}\int f\varepsilon\mathrm{d}^{m,n}x &= -\int f_{,i}\delta\xi^i\mathrm{d}^{m,n}x \\ &= -\int(-1)^i[(f\delta\xi^i)_{,i} - f\delta\xi^i_{,i}]\,\mathrm{d}^{m,n}x \\ &= \int(-1)^i f\delta\xi^i_{,i}\mathrm{d}^{m,n}x,\end{aligned} \tag{1.7.32}$$

in which the term in the integrand involving the total derivative has been dropped by virtue of the boundary condition (1.7.26) and the asymptotic behavior of f implied by the existence of the integral (1.7.27). Since f is arbitrary it follows that

$$\varepsilon = (-1)^i\delta\xi^i_{,i}, \tag{1.7.33}$$

$$J = 1 + (-1)^i\delta\xi^i_{,i}. \tag{1.7.34}$$

Now introduce a one-parameter family of nonlinear transformations

$$y^i(t) = f^i(x, t), \tag{1.7.35}$$

where the f^i are differentiable functions of the x^i and of an ordinary real number t between 0 and 1, such that†

$$f^i(x, 0) = x^i, \qquad f^i(x, 1) = \bar{x}^i(x). \tag{1.7.36}$$

Write

$$\mathrm{d}^{m,n}y(t) = J(x, t)\mathrm{d}^{m,n}x \tag{1.7.37}$$

† Because the $\bar{x}$'s are connected to the x's through the *homotopy* (1.7.35) we are limited here to *proper* transformations. Improper transformations can be obtained by combining proper nonlinear transformations with linear transformations (1.7.15) having super-Jacobians with negative bodies.

and introduce the abbreviations

$$M^i{}_j(x,t) \stackrel{\text{def}}{=} y^i(t)\frac{\overleftarrow{\partial}}{\partial x^j} = f^i(x,t)\frac{\overleftarrow{\partial}}{\partial x^j}, \tag{1.7.38}$$

$$M^{-1i}{}_j(x,t) = x^i\frac{\overleftarrow{\partial}}{\partial y^i(t)}. \tag{1.7.39}$$

Let δt be an infinitesimal increment in t. Then

$$y^i(t+\delta t) = y^i(t) + \delta\xi^i, \tag{1.7.40}$$

$$\delta\xi^i = \dot{f}^i(x,t)\,\delta t, \tag{1.7.41}$$

where the dot denotes differentiation with respect to t. By virtue of eq. (1.7.34) we have

$$\mathrm{d}^{m,n}y(t+\delta t) = \left[1 + (-1)^i \dot{f}^i(x,t)\frac{\overleftarrow{\partial}}{\partial y^i(t)}\delta t\right]\mathrm{d}^{m,n}y(t), \tag{1.7.42}$$

or

$$J(x,t+\delta t) = \left\{1 + (-1)^i\left[\dot{f}^i(x,t)\frac{\overleftarrow{\partial}}{\partial x^j}\right]\left[x^j\frac{\overleftarrow{\partial}}{\partial y^i(t)}\right]\delta t\right\}J(x,t) \tag{1.7.43}$$

whence

$$\begin{aligned}\frac{\mathrm{d}}{\mathrm{d}t}\ln J(x,t) &= (-1)^i \dot{M}^i{}_j(x,t)M^{-1j}{}_i(x,t)\\ &= \operatorname{str}[\dot{M}(x,t)M^{-1}(x,t)]\\ &= \frac{\mathrm{d}}{\mathrm{d}t}\ln \operatorname{sdet} M(x,t).\end{aligned} \tag{1.7.44}$$

The boundary conditions (1.7.36) imply

$$M^i{}_j(x,0) = \delta^i{}_j, \tag{1.7.45}$$

$$M^i{}_j(x,1) = \bar{x}^i{}_{,j}, \tag{1.7.46}$$

$$\operatorname{sdet} M(x,0) = 1 = J(x,0), \tag{1.7.47}$$

and therefore

$$J(x,t) = \operatorname{sdet} M(x,t), \tag{1.7.48}$$

$$J = J(x,1) = \operatorname{sdet}(\bar{x}^i{}_{,j}), \tag{1.7.49}$$

qed.

Gaussian integrals over $\mathbf{R}_c^m \times \mathbf{R}_a^n$

Let M be a supersymmetric square matrix of dimension (m, n), the elements ${}_iM_j$ of which have their indices in the lower position. It has the block form

$$M = \begin{pmatrix} A & C \\ -C^{\sim} & B \end{pmatrix}, \tag{1.7.50}$$

where the submatrices A and B have the symmetries

$$A^{\sim} = A, \qquad B^{\sim} = -B. \tag{1.7.51}$$

Suppose the elements of A are real c-numbers, the elements of B are imaginary c-numbers, and the elements of C are imaginary a-numbers. Then the quadratic form $x^i\,{}_iM_j\,x^j$, where the x's are the coordinates in $\mathbf{R}_c^m \times \mathbf{R}_a^n$, is a real c-number.[†]

As a final example in which the superdeterminant makes its appearance we shall evaluate the imaginary 'Gaussian' integral

$$I \overset{\text{def}}{=} \int \exp\left(\frac{\mathrm{i}}{2} x^i\,{}_iM_j\,x^j\right) \mathrm{d}^{m,n}x. \tag{1.7.52}$$

To make this integral well defined one must introduce into the exponent a quadratic damping term, having negative-definite body in the c-number sector, and then pass to the limit as this term goes to zero. In practice one need not carry out this procedure explicitly but merely leave it understood. However, one must assume that the submatrix A is non-singular, for otherwise the limit will not exist.

The first step is to carry out the variable transformation

$$\left.\begin{aligned} \bar{x}^\mu &= x^\mu + A^{-1\mu\nu} C_{\nu\alpha} x^\alpha, \\ \bar{x}^\alpha &= x^\alpha. \end{aligned}\right\} \tag{1.7.53}$$

Since this transformation has unit super-Jacobian we have $\mathrm{d}^{m,n}x = \mathrm{d}^{m,n}\bar{x}$, and it is not difficult to verify that the integral (1.7.52) now takes the form

$$I = \int \exp\left(\frac{\mathrm{i}}{2} \bar{x}^i\,{}_i\bar{M}_j\,\bar{x}^j\right) \mathrm{d}^{m,n}\bar{x}, \tag{1.7.54}$$

where $\bar{M}$ has the diagonal block form

$$\bar{M} = \begin{pmatrix} A & 0 \\ 0 & B + C^{\sim}A^{-1}C \end{pmatrix}. \tag{1.7.55}$$

The block A is a real symmetric c-number $m \times m$ matrix, and the block $B + C^{\sim}A^{-1}C$, like B, is an imaginary antisymmetric c-number $n \times n$ matrix.

[†] We also then have $A^\dagger = A$, $B^\dagger = B$, $C^\dagger = -C^{\sim}$, where '$\dagger$' denotes the Hermitian conjugate. M is therefore an Hermitian matrix: $M^\dagger = M$.

The matrix A is assumed to be nonsingular. The matrix $B + C^{\sim}A^{-1}C$ has the same body as B and hence is singular or nonsingular according as B is singular or nonsingular. It is *necessarily* singular if n is an odd integer. We shall assume for the present that n is even. Then there exist ordinary real orthogonal matrices O_1 and O_2, of determinant $+1$, which transform A_B and B_B respectively into the forms

$$O_1 A_B O_1^{\sim} = \mathrm{diag}(\lambda_1, \ldots, \lambda_m), \quad \lambda_\mu \neq 0, \tag{1.7.56}$$

$$O_2 B_B O_2^{\sim} = \mathrm{diag}\left(\begin{pmatrix} 0 & i\mu_1 \\ -i\mu_1 & 0 \end{pmatrix}, \ldots, \begin{pmatrix} 0 & i\mu_{n/2} \\ -i\mu_{n/2} & 0 \end{pmatrix}\right), \tag{1.7.57}$$

where the λ's and μ's are ordinary real numbers.

Next carry out the variable transformation

$$\bar{x}^i = {}^iL_j \bar{x}^j, \tag{1.7.58}$$

where

$$L = \mathrm{diag}(O_1(1_m + A_B^{-1}A_S)^{\frac{1}{2}}, O_2[1_n + B_B^{-1}(B_S + C^{\sim}A^{-1}C)]^{\frac{1}{2}}), \tag{1.7.59}$$

the square roots being defined by the binomial expansion. The super-Jacobian of this transformation is

$$J = \det(1_m + A_B^{-1}A_S)^{\frac{1}{2}} \det[1_n + B_B^{-1}(B_S + C^{\sim}A^{-1}C)]^{-\frac{1}{2}}, \tag{1.7.60}$$

and the integral (1.7.54) now takes the form

$$I = \int \exp\left\{\frac{i}{2}[\lambda_1(\bar{x}^1)^2 + \cdots + \lambda_m(\bar{x}^m)^2] - \mu_1 \bar{x}^{-1}\bar{x}^{-2} - \cdots - \mu_{n/2}\bar{x}^{-n+1}\bar{x}^{-n}\right\} J^{-1} \, d^{m,n}\bar{x}. \tag{1.7.61}$$

The integrations with respect to the c-number variables may be carried out immediately, yielding the usual Gaussian results. The exponential in the remaining a-number variables may be expanded as a power series, which terminates at order n in the x^α. One obtains

$$\begin{aligned} I &= (2\pi i)^{m/2}(\lambda_1 \ldots \lambda_m)^{-\frac{1}{2}}(-1)^{n/2}\mu_1 \ldots \mu_{n/2} J^{-1} \\ &\quad \times i^{n(n-1)/2} \int \bar{x}^{-1} \ldots \bar{x}^{-n} \, d\bar{x}^{-1} \ldots d\bar{x}^{-n} \\ &= (2\pi i)^{(m+n)/2}(\det A_B)^{-\frac{1}{2}}(-1)^{n/2}(-i)^{n(n-1)/2}\mu_1 \ldots \mu_{n/2} J^{-1}. \end{aligned} \tag{1.7.62}$$

It is not difficult to show that $(-i)^{n(n-1)/2} = (-i)^{n/2}$ when n is even. Hence, using the fact that

$$\det B_B = (-\mu_1^2) \ldots (-\mu_{n/2}^2), \tag{1.7.63}$$

we have, finally,

$$\begin{aligned} I &= (2\pi \mathrm{i})^{(m+n)/2} (\det A_{\mathrm{B}})^{-\frac{1}{2}} (\det B_{\mathrm{B}})^{\frac{1}{2}} J^{-1} \\ &= (2\pi \mathrm{i})^{(m+n)/2} (\det A)^{-\frac{1}{2}} [\det(B + \tilde{C} A^{-1} C)]^{\frac{1}{2}} \\ &= (2\pi \mathrm{i})^{(m+n)/2} (\operatorname{sdet} M)^{-\frac{1}{2}}. \end{aligned} \tag{1.7.64}$$

The rule for determining the phase of the body of $(\operatorname{sdet} M)^{-\frac{1}{2}}$ is the following: A factor $-\mathrm{i}$ for each λ having negative body. A factor i for each μ having positive body and a factor $-\mathrm{i}$ for each μ having negative body, the μ's being determined by (1.7.57) with O_2 restricted to be an orthogonal matrix of *positive* determinant.

It should be noted that the exact analogy with the classical Gaussian integral, displayed by eq. (1.7.64), depends on the choice (1.3.15) for the *c*-number Z and the choice (1.3.19) for the *a*-number contribution to the volume element, choices that have already been seen to maintain an exact analogy in Fourier transform theory. Note also that eq. (1.7.64) holds even when n is odd, for in that case both I and $\det(B + \tilde{C} A^{-1} C)$ vanish.

Exercises

1.1 Prove that (1.1.11) is the general solution of eq. (1.1.10). *Hint*: Expand v in the form (1.1.9) and $f(v)$ in the form (1.1.2), (1.1.3). Regard the coefficients of the latter expansion as functions of the c's in eq. (1.1.9). Vary the c's infinitesimally, obtaining the general form for $\mathrm{d}v$. Find the conditions on the coefficients in the expansion of $f(v)$ that are necessary for $\mathrm{d}f$ to factorize as in eq. (1.1.10). Show that these conditions lead to (1.1.11).

1.2 Outline a proof that if f is superanalytic on a smooth simply connected 2-dimensional surface-with-boundary in $\mathbf{C}_c$ then $\oint f(u)\mathrm{d}u = 0$ where the integral is over the boundary. *Hint*: Approximate the surface by a simplicial decomposition into small triangles. Show that the integral over the boundary of each triangle vanishes to second order in small quantities. Pass to the limit in which the dimensions of the triangles tend to zero. (Leave rigor to the specialists.)

1.3 Starting with the result of the preceding exercise develop the elementary parts of the theory of superanalytic functions on $\mathbf{C}_c$ in complete parallel with the theory of ordinary analytic functions of a complex variable. In particular, develop the theory of Taylor and Laurent series and show that (1.1.18), where $f_{a_1 \ldots a_n}(u)$ are functions of the form (1.1.17), is the general solution of eq. (1.1.15).

1.4 Show that $\oint f(v)\,dv$, where $f(v)$ has the general form (1.1.11), is not independent of the contour and does not generally vanish.

1.5 Show that the fundamental theorem of algebra does not hold in C_c. Give examples of finite-order polynomials in u, with coefficients in C_c, that do not have any zeros. Give examples of finite-order polynomials that have infinitely many different zeros even when their coefficients are in C, i.e., are ordinary complex numbers. (Contributed by F.L. Newman.)

1.6 If a complex c-number u has nonvanishing body, one may define its *absolute value* by $|u| \stackrel{\text{def}}{=} (u^*u)^{\frac{1}{2}}$, the square root with positive body being understood. Prove that this square root is unique. Prove that every complex c-number with nonvanishing body can be expressed in the form $u = |u|e^{i\phi}$ where ϕ is a real c-number.

1.7 Derive eq. (1.7.2) where f is an arbitrary differentiable function of the x^i, with values in Λ_∞.

1.8 Let V be a (k, l)-dimensional supervector space and let $\{L_i\}$ be a set of $(k, l) \times (k, l)$ matrices (i.e., having the block structure of the matrix K of eq. (1.4.40)) which act on V. Denote by Ker $\{L_i\}$ the *set* of sub-supervector spaces of V that remain invariant under the actions of $\{L_i\}$. The set $\{L_i\}$ is said to be *irreducible* if every element of Ker$\{L_i\}$ has the form αV for some pure supernumber α.

Let $\{L_i\}$ be an irreducible set of $(k, l) \times (k, l)$ matrices and let $\{M_i\}$ be an irreducible set of $(m, n) \times (m, n)$ matrices that can be put into one-to-one correspondence with the set $\{L_i\}$. Suppose there exists a $(k, l) \times (m, n)$ matrix A such that $L_i A = A M_i$ for all i. Prove the analog of Schur's lemma for supervector spaces. That is, prove that either $A = 0$ or else $k = m$, $l = n$ and A is a nonsingular matrix times a pure supernumber. Prove, as a corollary, that all nonsingular $(k, l) \times (k, l)$ matrices B satisfying $L_i B = B L_i$ for all i, necessarily have the form $B = \lambda 1_{(k,l)}$ where λ is a c-number with nonvanishing body.

1.9 The $(m, n) \times (m, n)$ matrix K of eq. (1.4.40) maps c-type supervectors into c-type supervectors and a-type supervectors into a-type supervectors. It may therefore be called a c-type matrix. One may also consider a-type matrices, which map c-type supervectors into a-type supervectors and vice versa. If, on the right-hand side of eq. (1.4.40), the submatrices A and B had a-type elements and the submatrices C and D had c-type elements then K would be an a-type matrix.

One may generalize the supertrace and supertranspose so that they apply to a-type matrices. This is done by defining

$$\operatorname{str} K \overset{\text{def}}{=} (-1)^{i(K+1)}{}_i K^i, \qquad \operatorname{str} L \overset{\text{def}}{=} (-1)^{i(L+1)}{}^i L_i,$$

$$ {}^i K^{\sim}{}_j \overset{\text{def}}{=} (-1)^{j(i+1)+K(i+j)}{}_j K^i, \qquad {}_i L^{\sim j} \overset{\text{def}}{=} (-1)^{i(j+1)+L(i+j)}{}^j L_i,$$

$$ {}_i M^{\sim}{}_j \overset{\text{def}}{=} (-1)^{i+j+ij+M(i+j)}{}_j M_i, \qquad {}^i N^{\sim j} \overset{\text{def}}{=} (-1)^{ij+N(i+j)}{}^j N^i.$$

Show that these definitions yield

$$K^{\sim\sim} = K, \quad L^{\sim\sim} = L, \quad M^{\sim\sim} = M, \quad N^{\sim\sim} = N, \quad \operatorname{str} K^{\sim} = \operatorname{str} K, \quad \operatorname{str} L^{\sim} = \operatorname{str} L,$$

and, for all admissible combinations of index positions,

$$(PQ)^{\sim} = (-1)^{PQ} Q^{\sim} P^{\sim},$$
$$\operatorname{str}(PQ) = (-1)^{PQ} \operatorname{str}(QP).$$

1.10 If $m = n$ one may generalize the superdeterminant so that it too applies to a-type matrices. One must then, however, distinguish between *left* and *right* superdeterminants, defined respectively by

$$\delta \ln \operatorname{sdet}_{\mathrm{L}} M = \operatorname{str}(\delta M M^{-1}),$$
$$\delta \ln \operatorname{sdet}_{\mathrm{R}} M = \operatorname{str}(M^{-1} \delta M) = (-1)^M \delta \ln \operatorname{sdet}_{\mathrm{L}} M.$$

Show that these superdeterminants satisfy

$$\operatorname{sdet}_{\mathrm{R}} M = (\operatorname{sdet}_{\mathrm{L}} M)^{(-1)^M},$$
$$\operatorname{sdet}_{\mathrm{L}}(LM) = (\operatorname{sdet}_{\mathrm{L}} L)(\operatorname{sdet}_{\mathrm{L}} M)^{(-1)^L},$$
$$\operatorname{sdet}_{\mathrm{R}}(LM) = (\operatorname{sdet}_{\mathrm{R}} L)^{(-1)^M}(\operatorname{sdet}_{\mathrm{R}} M).$$

If the matrix (1.4.40) is a-type show that its left and right superdeterminants are given by

$$\operatorname{sdet}_{\mathrm{L}}\begin{pmatrix} A & C \\ D & B \end{pmatrix} = \left[\operatorname{sdet}_{\mathrm{R}}\begin{pmatrix} A & C \\ D & B \end{pmatrix}\right]^{-1}$$
$$= [\det(C - AD^{-1}B)](\det D)^{-1} = (\det C)[\det(D - BC^{-1}A)]^{-1}$$
$$= (\det C)(\det D)^{-1}[\det(1_m - C^{-1}AD^{-1}B)]$$
$$= (\det C)(\det D)^{-1}[\det(1_m - D^{-1}BC^{-1}A)]^{-1}.$$

These last equations can, in fact, be regarded as defining the left and right superdeterminants in the yet more general case in which A, B, C, D have the dimensions $m \times n$, $n \times m$, $m \times m$, $n \times n$ respectively, with $m \neq n$, A and B having a-type elements and C and D having c-type elements.

Comments on chapter 1

The theory of analysis over supernumbers is largely due to F. A. Berezin. A number of applications of the theory will be found in Berezin's excellent book, *The Method of Second Quantization* (Academic Press, New York, 1966). Two important differences of convention should be pointed out between Berezin's work and the present volume: On the right-hand side of eq. (1.3.6) Berezin would have 1 instead of $Z(=(2\pi i)^{\frac{1}{2}})$ and the factor $i^{n(n-1)/2}$ would be missing on the right-hand side of eq. (1.3.19). These changes of convention have been introduced here in order to bring the theory of integration over *a*-numbers into exact analogy with the theory of integration over *c*-numbers, not only for Gaussian integrals but also in Fourier transform theory, of which, unlike Berezin, we make important use in the construction of the functional integral over dynamical histories (see chapter 5).

2

Supermanifolds

2.1 Definition and structure of supermanifolds

Topology of $\mathbf{R}_c^m \times \mathbf{R}_a^n$. *Differentiable mappings*

In the following sections we define and study objects called supermanifolds, which bear the same relation to $\mathbf{R}_c^m \times \mathbf{R}_a^n$ as ordinary manifolds bear to $\mathbf{R}^m$; that is, small regions of a supermanifold look like small regions of $\mathbf{R}_c^m \times \mathbf{R}_a^n$. They are said to have the same local topology.

If Λ_∞ is replaced by Λ_N then $\mathbf{R}_c^m \times \mathbf{R}_a^n$ possesses a natural topology by virtue of the fact that it is a $2^{N-1}(m+n)$-dimensional vector space. This topology, the usual vector-space topology, has nothing to do with the algebraic structure of $\mathbf{R}_c^m \times \mathbf{R}_a^n$ and cannot be extended, without modification, to hold in the limit $N \to \infty$. There is another topology, a coarser topology, that may be assigned in a natural way to $\mathbf{R}_c^m \times \mathbf{R}_a^n$, that does reflect the algebraic structure and that may be immediately imposed whether N is finite or infinite. It is the latter topology that is most appropriate for the theory of supermanifolds. It may be discovered by the following reasoning:

Let ϕ be a differentiable mapping from one region of $\mathbf{R}_c^m \times \mathbf{R}_a^n$ to another, i.e., a mapping for which the coordinates $\bar{x}^i$ of the image point $\phi(x)$ are differentiable functions (in the sense of eqs. (1.1.10), (1.1.15), (1.2.4), and (1.3.1)) of the coordinates x^i of the point x. (Cf. eq. (1.7.22).) There is no need to be precise about the domain and range of ϕ in order to remark that the explicit representation of such a mapping always has the form

$$\left.\begin{aligned} \bar{x}^\mu &= \sum_{r=0}^{n}\sum_{s=0}^{\infty} \frac{\mathrm{i}^{r(r+1)/2}}{r!s!} \frac{\partial^s X^\mu_{\beta_1\ldots\beta_r}(x_{\mathrm{B}})}{\partial x_{\mathrm{B}}^{\nu_1}\ldots\partial x_{\mathrm{B}}^{\nu_s}} x_{\mathrm{S}}^{\nu_1}\ldots x_{\mathrm{S}}^{\nu_s} x^{\beta_r}\ldots x^{\beta_1}, \\ \bar{x}^\alpha &= \sum_{r=0}^{n}\sum_{s=0}^{\infty} \frac{\mathrm{i}^{r(r-1)/2}}{r!s!} \frac{\partial^s X^\alpha_{\beta_1\ldots\beta_r}(x_{\mathrm{B}})}{\partial x_{\mathrm{B}}^{\nu_1}\ldots\partial x_{\mathrm{B}}^{\nu_s}} x_{\mathrm{S}}^{\nu_1}\ldots x_{\mathrm{S}}^{\nu_s} x^{\beta_r}\ldots x^{\beta_1}, \end{aligned}\right\} \quad (2.1.1)$$

where the X's are C^∞ functions of the x_{B}^ν, taking their values in $\mathbf{R}_c$ or $\mathbf{R}_a$ according as they bear an even or an odd number of a-type indices.

Consider the super-Jacobian matrix $(\bar{x}^i_{,j})$. When the souls of all the x's vanish this matrix reduces to

$$(\bar{x}^i_{,j})_{x^\nu_S=0,x^\beta=0} = \begin{pmatrix} \partial X^\mu(x_B)/\partial x^\nu_B & iX^\mu_\beta(x_B) \\ \partial X^\alpha(x_B)/\partial x^\nu_B & X^\alpha_\beta(x_B) \end{pmatrix}. \tag{2.1.2}$$

Suppose the submatrices $(\partial X^\mu(x_B)/\partial x^\nu_B)$ and $(X^\alpha_\beta(x_B))$ are both nonsingular for a certain range of values of the x_B's. Then the whole matrix is nonsingular for those x_B's, and it is not difficult to see that the formal series (2.1.1) can be inverted. From this fact and the fact that the x_B's determine the $\bar{x}_B$'s it follows that no matter what values may be chosen for the souls of the $\bar{x}^i$, one can, for the range of x_B's in question, always find unique souls for the x^i to yield those $\bar{x}^i$. This means that a differentiable mapping is one-to-one in a certain domain if and only if its super-Jacobian (1.7.24) is nonsingular on the projection of that domain onto the subspace $\mathbf{R}^m$ in $\mathbf{R}^m_c \times \mathbf{R}^n_a$.

The subspace $\mathbf{R}^m$ (defined as the set of points whose coordinates have vanishing souls) thus plays a dominant role in the theory of differentiable mappings. The projection in question is the mapping $\pi: \mathbf{R}^m_c \times \mathbf{R}^n_a \to \mathbf{R}^m$ that replaces each coordinate x^i by its body. π may be called the *natural projection* of $\mathbf{R}^m_c \times \mathbf{R}^n_a$ onto $\mathbf{R}^m$. Let $\mathcal{O}$ be an open set of $\mathbf{R}^m$ (in the usual sense). Then if the mapping ϕ above has nonsingular Jacobian over $\mathcal{O}$ it is one-to-one over $\pi^{-1}(\mathcal{O})$. Furthermore, if $\pi(x)$ is in $\mathcal{O}$ and $x' = \phi(x)$ then $\pi^{-1}(\pi(x')) = \phi(\pi^{-1}(\pi(x)))$. The set $\pi^{-1}(\pi(x))$ will be called the *soul subspace* over x. We see that one-to-one differentiable mappings map soul subspaces onto soul subspaces.

It is this last property that suggests the appropriate coarse topology to be assigned to $\mathbf{R}^m_c \times \mathbf{R}^n_a$ for the purposes of supermanifold theory: A subset of $\mathbf{R}^m_c \times \mathbf{R}^n_a$ will be said to be *open* if and only if it has the form $\pi^{-1}(\mathcal{O})$ where $\mathcal{O}$ is some open subset of $\mathbf{R}^m$. It will be noted that, with this topology, $\mathbf{R}^m_c \times \mathbf{R}^n_a$ is not Hausdorff. However, if x and x' are two points of $\mathbf{R}^m_c \times \mathbf{R}^n_a$ lying in distinct soul subspaces (i.e., if $\pi(x) \neq \pi(x')$) then they may be surrounded by nonintersecting neighborhoods. Such a space will be called *projectively Hausdorff*.

We now define the concept of differentiable mapping in general: Let ϕ be a mapping from an open subset $\mathcal{U}$ (in the coarse sense) of $\mathbf{R}^m_c \times \mathbf{R}^n_a$ to an open subset $\bar{\mathcal{U}}$ of $\mathbf{R}^{\bar{m}}_c \times \mathbf{R}^{\bar{n}}_a$. This mapping is said to be differentiable if the coordinates $\bar{x}^j$ $(j = 1 \ldots \bar{m}, -1 \ldots -\bar{n})$ of the image point $\phi(x)$ are differentiable functions, in the sense of eqs. (1.1.10), (1.1.15), (1.2.4) and (1.3.1), of the coordinates x^i $(i = 1 \ldots m, \ -1 \ldots -n)$ of the point x. Differentiable mappings from open subsets of $\mathbf{R}^m$ to $\mathbf{R}^{\bar{m}}_c \times \mathbf{R}^{\bar{n}}_a$ are defined similarly. Differentiable mappings from open subsets of $\mathbf{R}_c^m \times \mathbf{R}_a^n$ to $\mathbf{R}^m$ may also be defined, but, as is easily seen, they are necessarily trivial constant mappings.

It will be noted that when N is infinite, 'differentiable' means 'C^∞'. In discussing supermanifolds (see below) we shall therefore not make the distinction between C^r and C^∞ that is often made in discussing ordinary manifolds.

Supermanifolds, charts and atlases

A *Supermanifold of dimension* (m, n) is a space M, together with a collection of ordered pairs $(\mathscr{U}_A, \phi_A)$, where each $\mathscr{U}_A$ is a subset of M, and its associated ϕ_A is a one-to-one mapping of $\mathscr{U}_A$ onto an open set in $\mathbf{R}_c^m \times \mathbf{R}_a^n$ (in the coarse sense). The collection of ordered pairs is required to have the following properties:

(1) $\bigcup\limits_A \mathscr{U}_A = M$

(2) $\phi_A \circ \phi_B^{-1}$ is differentiable for all nonempty intersections $\mathscr{U}_A \cap \mathscr{U}_B$.

If p is a point in $\mathscr{U}_A$ and $\phi_A(p) = (x^1 \dots x^m, x^{-1} \dots x^{-n})$ then the x^i are called the *coordinates* of p defined by ϕ_A. The pair $(\mathscr{U}_A, \phi_A)$ is called a *chart*, or a *local coordinate patch*, or simply a *coordinate system*. Property (2) says that every pair of overlapping coordinate systems is related by a differentiable transformation.

A collection of charts $(\mathscr{U}_A, \phi_A)$ satisfying properties (1) and (2) is called an *atlas*. A given supermanifold may have more than one atlas. A second atlas is said to be *compatible* with the first one if the union of the two atlases is again an atlas. One may form the union of all atlases compatible with a given atlas. This is the *complete atlas* of the supermanifold. It is the set of all possible coordinate systems.

Scalar fields and supercurves

With the aid of an atlas one can define differentiable mappings between supermanifolds or between supermanifolds and $\mathbf{R}_c^{\bar m} \times \mathbf{R}_a^{\bar n}$ or $\mathbf{R}^{\bar m}$:

Let f be a mapping from a supermanifold M to the space $\mathbf{R}_c^{\bar m} \times \mathbf{R}_a^{\bar n}$ (which is itself a supermanifold). It is said to be differentiable if the mapping $f \circ \phi_A^{-1}$ from $\phi_A(\mathscr{U}_A)$ to $\mathbf{R}_c^{\bar m} \times \mathbf{R}_a^{\bar n}$ is differentiable for all charts $(\mathscr{U}_A, \phi_A)$ in any atlas of M. If $\bar m = 1$ and $\bar n = 0$ the mapping is called a *real c-type scalar field* over M. If $\bar m = 0$ and $\bar n = 1$ it is called a *real a-type scalar field*. One can extend these definitions in an obvious way to include scalar fields (i.e., differentiable functions) over M that take their values in $\mathbf{C}_c$ (c-type), in $\mathbf{C}_a$ (a-type), or in the full Grassmann algebra Λ_∞. Scalar fields that take their values in $\mathbf{C}_c$ or $\mathbf{C}_a$ only will be called *pure*.

Let $\mathscr{S}$ be an open subset of $\mathbf{R}_c^{\bar{m}} \times \mathbf{R}_a^{\bar{n}}$ or of $\mathbf{R}^{\bar{m}}$. A mapping λ from $\mathscr{S}$ to a supermanifold M is said to be differentiable if the mapping $\phi_A \circ \lambda$ from $\mathscr{S}$ to $\phi_A(\mathscr{U}_A)$ is differentiable for all charts $(\mathscr{U}_A, \phi_A)$ in M for which $\mathscr{U}_A \cap \lambda(\mathscr{S})$ is nonempty. If $\mathscr{S}$ is a connected open interval of the real line $\mathbf{R}$ the mapping is called a *differentiable curve*, or simply *curve*, in M and its range $\lambda(\mathscr{S})$ is called a *curved line segment* (without endpoints). If $\mathscr{S}$ is a connected open subset of $\mathbf{R}_c$ (in the coarse topology) the mapping will be called a *c-type supercurve*. If $\mathscr{S} = \mathbf{R}_a$ the mapping will be called an *a-type supercurve*. It is easy to see that every a-type supercurve lies wholly within a single soul subspace. A single coordinate patch suffices to embrace it, the supercurve being expressible as a set of linear functions over $\mathbf{R}_a$ and hence determinable by specifying $m + n$ real c-numbers, n real a-numbers and m imaginary a-numbers.

Diffeomorphisms and embeddings

Let ϕ be a mapping from one supermanifold M to another supermanifold $\bar{M}$. It is said to be differentiable if, for every chart $(\mathscr{U}_A, \phi_A)$ in M and for every chart $(\bar{\mathscr{U}}_B, \bar{\phi}_B)$ in $\bar{M}$ for which $\phi(\mathscr{U}_A) \cap \bar{\mathscr{U}}_B$ is nonempty, the mapping $\bar{\phi}_B \circ \phi \circ \phi_A^{-1}$ from $\phi_A(\mathscr{U}_A)$ to $\bar{\phi}_B(\bar{\mathscr{U}}_B)$ is differentiable. If ϕ is additionally one-to-one and onto, and its inverse is differentiable, then it is called a *diffeomorphism*, and the two supermanifolds are said to be *diffeomorphic*. If two supermanifolds are diffeomorphic, they must have the same dimension.

A subset M' of a supermanifold M of dimension (m,n) is called a *sub-supermanifold* of dimension (m',n'), $m' \leqslant m$, $n' \leqslant n$, if M' is contained in the union of a collection $\{(\mathscr{U}, \phi)\}$ of charts each of which has the following properties: (1) For all $p \in \mathscr{U} \cap M'$, $\phi(p) = (x^1, \ldots, x^{M'}, c^{M'+1}, \ldots, c^m, x^{-1}, \ldots, x^{-n'}, sa^{-n'-1}, \ldots, a^{-n})$ where $(c^{m'+1}, \ldots, c^m, a^{-n'-1}, \ldots, a^{-n})$ is a *fixed* element of $\mathbf{R}_c^{m-m'} \times \mathbf{R}_a^{n-n'}$ depending on the chart in question. (2) If one defines $\phi' : \mathscr{U} \cap M' \to \mathbf{R}_c^{m'} \times \mathbf{R}_a^{n'}$ by $\phi'(p) = (x^1, \ldots, x^{m'}, x^{-1}, \ldots, x^{-n'})$ then $\phi'(\mathscr{U} \cap M')$ is an open subset of $\mathbf{R}_c^{m'} \times \mathbf{R}_a^{n'}$. The set of pairs $(\mathscr{U} \cap M', \phi')$ constitutes an atlas for M'.

Every supermanifold of dimension (m,n) has a natural topology inherited locally from $\mathbf{R}_c^m \times \mathbf{R}_a^n$ via the mappings ϕ_A of its complete atlas: A subset of M is said to be *open* if it is the union of subsets belonging to the complete atlas of M. The complete atlas of the sub-supermanifold M' above is the union of all atlases compatible with $\{(\mathscr{U} \cap \phi')\}$. The topology that this complete atlas defines on M' is identical with the induced topology that M' inherits from M.

A mapping $\phi: M \to \bar{M}$ is said to be a *regular embedding* (or simply an *embedding*) of M in $\bar{M}$ if $\phi(M)$ is a sub-supermanifold of $\bar{M}$ and $\phi: M \to \phi(M)$ is a diffeomorphism. If ϕ is an embedding of a supermanifold M in another supermanifold $\bar{M}$ of equal dimension then $\phi(M)$ is an open subset of $\bar{M}$. Any open subset of a supermanifold is itself a supermanifold.

Any of the differentiable mappings considered above may be restricted to open subsets of the domain space. The restrictions remain differentiable. Let $(\mathscr{U}, \phi)$ be a chart in a supermanifold M. Each of the coordinates x^μ defined by ϕ may be regarded as a real c-type scalar field over $\mathscr{U}$, and each of the coordinates x^α may be regarded as a real a-type scalar field over $\mathscr{U}$. These scalar fields are known as *coordinate functions*. Their formal definition is

$$(x^1(p), \ldots, x^m(p), x^{-1}(p), \ldots, x^{-n}(p)) \stackrel{\text{def}}{=} \phi(p), \quad \text{for all } p \text{ in } \mathscr{U}. \quad (2.1.3)$$

Ordinary manifolds. Skeleton and body of a supermanifold

All of the foregoing definitions and statements about supermanifolds may be converted into corresponding definitions and statements about ordinary manifolds simply by replacing '$\mathbf{R}_c^m \times \mathbf{R}_a^n$' everywhere by '$\mathbf{R}^m$' and using the ordinary definition of 'openness' (for subsets of $\mathbf{R}^m$) and the ordinary definition of 'differentiable'. Associated with every supermanifold M of dimension (m, n) is an infinite family of ordinary manifolds, of dimensions $2^{N-1}(m+n)$, $N = 1, 2 \ldots$. Each of these ordinary manifolds is obtained by ignoring the local algebraic structure of M, replacing Λ_∞ by Λ_N and, for each chart $(\mathscr{U}_A, \phi_A)$, regarding $\phi_A(\mathscr{U}_A)$ as an open subset of the ordinary vector space $\mathbf{R}^{2^{N-1}(m+n)}$, endowed with its usual topology.† The resulting manifold will be called the Nth *skeleton* of M and denoted by $S_N(M)$. Any atlas of M is an atlas of $S_N(M)$. However, the complete atlas of M is not the complete atlas of $S_N(M)$. To get the complete atlas of a skeleton one must refine the topology by adding other charts $(\mathscr{U}'_B, \phi'_B)$, $(\mathscr{U}'_C, \phi'_C)$ whose mappings have composites $\phi'_B \circ \phi'^{-1}_C$ that need now be differentiable only in the conventional sense, and images $\phi'_B(\mathscr{U}'_B), \phi'_C(\mathscr{U}'_C)$ that are open in the usual (finer) sense.

With the aid of the family of skeletons one may classify mappings

† 'Replace Λ_∞ by Λ_N' means: 'Expand all chart-relating functions $\phi_A \circ \phi_B^{-1}$ as in eq. (2.1.1). Then expand all the coefficients as well as the souls of all the x's in power series in the Grassman generators ζ^a. Finally, omit all terms containing generators ζ^a with $a > N$.'

between supermanifolds and ordinary manifolds. It is obvious that in addition to differentiable mappings between two supermanifolds, or between two manifolds, one may also consider differentiable mappings from a manifold to a supermanifold. (An example is a differentiable curve.) However, the image of a one-to-one differentiable mapping from a manifold to a supermanifold cannot be characterized as a *submanifold* without reference to the skeletons. Such an image will be called an embedding of the manifold in the supermanifold if it is an embedding in every skeleton. The relation between the manifold and its image is then a diffeomorphism in the usual sense. Note that although differentiable mappings from supermanifolds to manifolds may in principle be defined they are all constant mappings and hence trivial.

In addition to the skeletons there is another ordinary manifold that is naturally associated with every supermanifold M. Its construction proceeds as follows: Let $(\mathcal{U}_A, \phi_A)$ and $(\mathcal{U}_B, \phi_B)$ be any two charts of M that have a nonempty intersection, and let p be a point lying in the intersection. Consider the soul subspaces, $\pi^{-1}(\pi\circ\phi_A(p))$ and $\pi^{-1}(\pi\circ\phi_B(p))$, over the image points of p under the mappings ϕ_A and ϕ_B respectively. These soul subspaces are mapped onto one another under the one-to-one differentiable mapping $\phi_A\circ\phi_B^{-1}$:

$$\pi^{-1}(\pi\circ\phi_A(p)) = \phi_A\circ\phi_B^{-1}(\pi^{-1}(\pi\circ\phi_B(p))). \qquad (2.1.4)$$

From this it follows that the images of the two soul subspaces under their respective inverse mappings, ϕ_A^{-1} and ϕ_B^{-1}, are identical.

$$\phi_A^{-1}(\pi^{-1}(\pi\circ\phi_A(p))) = \phi_B^{-1}(\pi^{-1}(\pi\circ\phi_B(p))). \qquad (2.1.5)$$

Evidently these images are chart-independent *invariants* of the supermanifold. The set $\phi_A^{-1}(\pi^{-1}(\pi\circ\phi_A(p)))$ will be called the *soul subspace over p* in the supermanifold M. Now it is easy to see that a point p of M lies in an open subset $\mathcal{U}$ of M if and only if the entire soul subspace over p lies in $\mathcal{U}$. Therefore, the soul subspaces themselves may be regarded as *points* in a new manifold, an ordinary manifold of dimension m, whose charts are the ordered pairs $(\bar{\mathcal{U}}_A, \pi\circ\phi_A)$, where, for each $A, \bar{\mathcal{U}}_A$ is the set whose elements are the soul subspaces contained in $\mathcal{U}_A$. This new manifold will be called the *body* of M and denoted by M_B.

There is a local relationship between the body and the skeletons. If $\mathcal{U}$ is a sufficiently small open set of M then

$$S_N(\mathcal{U}) \overset{\text{diff}}{=} \mathcal{U}_B \times \mathbf{R}^{2^{N-1}(m+n)-m}, \qquad (2.1.6)$$

where '$\overset{\text{diff}}{=}$' means 'is diffeomorphic to' and '×' denotes the topological product.† However, $S_N(M)$ itself need not be diffeomorphic to the product of M_B with $\mathbf{R}^{2^{N-1}(m+n)-m}$. It is a *bundle* of copies of $\mathbf{R}^{2^{N-1}(m+n)-m}$ having M_B as its base space. The bundle need not be a trivial product bundle but may be 'twisted'. Similarly, M itself may be regarded as a bundle of isomorphic soul subspaces over the base M_B. If $S_N(M)$ is a twisted bundle then so is M, but the converse is not necessarily true.

Projectively Hausdorff, compact, paracompact and orientable supermanifolds. Realizations of the body

We collect here a number of definitions and propositions involving bodies and skeletons of supermanifolds.

A supermanifold will be called *projectively Hausdorff* if its body is Hausdorff. A skeleton of a supermanifold is Hausdorff if and only if its body is Hausdorff. In this book attention is confined to projectively Hausdorff supermanifolds.

Let M and $\bar{M}$ be two supermanifolds of dimension (m, n) and $(\bar{m}, \bar{n})$ respectively. If M and $\bar{M}$ are diffeomorphic then $S_N(M)$ and $S_N(\bar{M})$ are diffeomorphic (for every N) and, furthermore, M_B and $\bar{M}_B$ are diffeomorphic and $n = \bar{n}$. However, the converses of these statements do not hold.

A topological space $\mathscr{S}$ is said to be *compact* if from every collection of open subsets $\mathscr{U}_A$ such that $\bigcup_A \mathscr{U}_A = \mathscr{S}$ it is possible to select a finite subcollection whose union also equals $\mathscr{S}$. A supermanifold is compact if and only if its body is compact. A skeleton is never compact.

A topological space $\mathscr{S}$ is said to be *paracompact* if for every collection of open subsets $\mathscr{U}_\alpha$ such that $\bigcup_\alpha \mathscr{U}_\alpha = \mathscr{S}$ there exists a collection of open subsets $\mathscr{V}_\beta$ satisfying the following conditions: (1) For every β there exists an α such that $\mathscr{V}_\beta \subset \mathscr{U}_\alpha$. (2) $\bigcup_\beta \mathscr{V}_\beta = \mathscr{S}$. (3) For every point p in $\mathscr{S}$ there exists an open subset $\mathscr{V}$ containing p and intersecting only a finite number of the $\mathscr{V}_\beta$. If M is a supermanifold then, for each N, the following three statements are equivalent, i.e., each implies the other two: (1) M is paracompact; (2) $S_N(M)$ is paracompact; (3). M_B is paracompact.

A supermanifold is said to be *orientable* if it admits an atlas of charts $(\mathscr{U}_\alpha, \phi_\alpha)$, called an *oriented atlas*, such that at every point in every nonempty

† The topological product $M \times N$ of two manifolds (or supermanifolds) M and N is obtained by adjoining to their Cartesian product as sets an atlas composed of all charts of the form $(\mathscr{U}_\alpha \times \mathscr{V}_\beta, \phi_\alpha \times \psi_\beta)$, where $(\mathscr{U}_\alpha, \phi_\alpha)$ and $(\mathscr{V}_\beta, \psi_\beta)$ are members of atlases of M and N respectively, $\mathscr{U}_\alpha \times \mathscr{V}_\beta$ denotes the set-theoretic product, and $\phi_\alpha \times \psi_\beta$ is the mapping from $\mathscr{U}_\alpha \times \mathscr{V}_\beta$ to $\phi_\alpha(\mathscr{U}_\alpha) \times \psi_\beta(\mathscr{V}_\beta)$ defined by $(\phi_\alpha \times \psi_\beta)(p, p') = (\phi_\alpha(p), \psi_\beta(p'))$ for all p in $\mathscr{U}_\alpha$ and all p' in $\mathscr{V}_\beta$.

intersection $\mathscr{U}_\alpha \cap \mathscr{U}_\beta$, the super-Jacobian matrix of the mapping $\phi_\alpha \circ \phi_\beta^{-1}$ is continuously deformable to the identity matrix without becoming singular. The same definition applies to ordinary manifolds if the prefix 'super' is omitted. If M is an orientable supermanifold then $S_N(M)$ is orientable (for every N) and M_B is orientable, but the converse statements do not hold.

Let K be a manifold embedded in a supermanifold M. Let π_M be the *natural projection from* M (or $S_N(M)$) *to* M_B, i.e., the mapping that associates to each point p of M the soul subspace over p. The restricted mapping $\pi_M|K$ from K to M_B can easily be shown to be differentiable (in the ordinary sense). If, in addition, it is one-to-one and onto and its inverse is differentiable then K is diffeomorphic to M_B and will be called a *realization* of M_B in M. Such realizations play a role in integration theory over supermanifolds (see section 2.9). They are m-dimensional analogs of integration contours in $\mathbf{R}_c$.

2.2 Supervector structures on supermanifolds

Scalar fields as supervectors

Let $\mathfrak{F}(M)$ be the set of all scalar fields over a supermanifold M, i.e., the set of all differentiable mappings $f: M \to \Lambda_\infty$. It is easy to see that $\mathfrak{F}(M)$ satisfies all the axioms of a supervector space (i.e., properties (1)–(6) of section 1.4) if one defines

$$(f+g)(p) \stackrel{\text{def}}{=} f(p) + g(p), \tag{2.2.1}$$

$$(\alpha f)(p) \stackrel{\text{def}}{=} \alpha[f(p)], \qquad (f\alpha)(p) = [f(p)]\alpha, \tag{2.2.2}$$

$$f^*(p) \stackrel{\text{def}}{=} [f(p)]^*, \tag{2.2.3}$$

for all f, g in $\mathfrak{F}(M)$, all α in Λ_∞, and all p in M. The zero supervector in $\mathfrak{F}(M)$ is the mapping that sends every point in M to 0, the negative of f is defined by $(-f)(p) = -[f(p)]$ for all p, the even and odd parts of f are defined by $f_e(p) = [f(p)]_e$, $f_o(p) = [f(p)]_o$ for all p, and so on.

Let the dimensionality of M be (m, n). If $m \neq 0$ then $\mathfrak{F}(M)$, as a supervector space, is infinite dimensional. That is, it has no basis of finite total dimension. In fact, depending on M, it may have no basis at all, unless one invokes Zorn's lemma, which, in physics at any rate, is of no value. On the other hand, to construct a finite-dimensional sub-supervector space of $\mathfrak{F}(M)$ is easy. Simply pick out a finite set $\{e^A\}$ of linearly independent scalar fields and use it as a basis. We shall call such a set a *subbasis of dimension* (p, q) if the sub-supervector space that it generates has dimension (p, q). We shall confine our attention to pure subbases.

$\mathfrak{F}(M)$ is, of course, more than a supervector space. One can multiply scalar fields together by defining

$$(fg)(p) \stackrel{\text{def}}{=} [f(p)][g(p)], \tag{2.2.4}$$

or one can build more-complicated structures. Let $\{e^A\}$ be a (p, q)-dimensional subbasis of $\mathfrak{F}(M)$. Let F be a differentiable mapping from the subspace $\times_A e^A(M)$ of $\mathbf{C}_c^p \times \mathbf{C}_a^q$ to Λ_∞. Every such mapping defines a scalar field $\bar{F}$ given by

$$\bar{F}(p) \stackrel{\text{def}}{=} F(e(p)), \quad \text{for all } p \text{ in } M. \tag{2.2.5}$$

If M can be covered by a single coordinate patch then the e^A can be chosen to be the coordinate functions x^i, and in that case the set of all scalar fields $\bar{F}$ is identical with $\mathfrak{F}(M)$.

Contravariant vector fields

A mapping $\mathbf{X}$ from $\mathfrak{F}(M)$ to itself, abbreviated by

$$\mathbf{X}(f) \stackrel{\text{def}}{=} \mathbf{X}f, \quad \text{for all } f \text{ in } \mathfrak{F}(M), \tag{2.2.6}$$

is called a *contravariant vector field* over M if it satisfies the chain rule[†]

$$(\mathbf{X}\bar{F})(p) = [(\mathbf{X}e^A)(p)]\left[\frac{\vec{\partial}}{\partial y^A} F(y)\right]_{y=e(p)}, \tag{2.2.7}$$

for all differentiable mappings $F: \times_A e^A(M) \to \Lambda_\infty$, for all pure finite dimensional subbases $\{e^A\}$ of $\mathfrak{F}(M)$, and for all p in M. By a slight abuse of notation it is customary to abbreviate (2.2.7) to

$$\mathbf{X}F = (\mathbf{X}e^A)\frac{\vec{\partial}}{\partial e^A} F, \tag{2.2.8}$$

in which reference to p has been omitted and the bar over the F on the left has been dropped. As immediate corollaries of (2.2.8) one obtains

$$\mathbf{X}(f\alpha) = (\mathbf{X}f)\alpha, \tag{2.2.9}$$

$$\mathbf{X}(f + g) = \mathbf{X}f + \mathbf{X}g, \tag{2.2.10}$$

[†] Use of the chain rule at the outset does not provide the most economical way of defining vector fields (see, for example, Choquet-Bruhat and DeWitt-Morette, with Dillard-Bleick, *Analysis, Manifolds and Physics*, Revised Edition, North Holland (1982)), but it avoids the necessity of providing a tedious proof of the mean-value theorem for differentiable functions over $\mathbf{R}_c^m \times \mathbf{R}_a^n$.

and, if f and g are pure,

$$\mathbf{X}(fg) = (\mathbf{X}f)g + (-1)^{fg}(\mathbf{X}g)f, \tag{2.2.11}$$

for all α in Λ_∞ and all f, g in $\mathfrak{F}(M)$.

Let $\mathfrak{X}(M)$ be the set of all contravariant vector fields over M. $\mathfrak{X}(M)$ can be given the structure of a supervector space by defining

$$(\mathbf{X}+\mathbf{Y})f \stackrel{\text{def}}{=} \mathbf{X}f + \mathbf{Y}f, \tag{2.2.12}$$

$$(\alpha\mathbf{X})f \stackrel{\text{def}}{=} \alpha(\mathbf{X}f), \tag{2.2.13}$$

$$(\mathbf{X}\alpha)f \stackrel{\text{def}}{=} \mathbf{X}(\alpha f), \tag{2.2.14}$$

for all $\mathbf{X}$, $\mathbf{Y}$ in $\mathfrak{X}(M)$, all α in Λ_∞ and all f in $\mathfrak{F}(M)$. The zero supervector in $\mathfrak{X}(M)$ is the contravariant vector field that maps every scalar field in $\mathfrak{F}(M)$ to the zero scalar field. The negative of a contravariant vector field $\mathbf{X}$ is defined by $(-\mathbf{X})f = -(\mathbf{X}f)$, and its even and odd parts are defined by

$$\mathbf{X} = \mathbf{U} + \mathbf{V}, \tag{2.2.15}$$

$$\mathbf{U}f = (\mathbf{X}f_e)_e + (\mathbf{X}f_o)_o, \tag{2.2.16}$$

$$\mathbf{V}f = (\mathbf{X}f_e)_o + (\mathbf{X}f_o)_e, \tag{2.2.17}$$

for all f in $\mathfrak{F}(M)$.

Let α be an a-number and $\mathbf{X}$ an a-type contravariant vector field. Then, for all f in $\mathfrak{F}(M)$ we have

$$\mathbf{X}f = (\mathbf{X}f_e)_o + (\mathbf{X}f_o)_e$$

and

$$\begin{aligned}(\alpha\mathbf{X})f &= \alpha(\mathbf{X}f) = \alpha(\mathbf{X}f_e)_o + \alpha(\mathbf{X}f_o)_e = -(\mathbf{X}f_e)_o\alpha + (\mathbf{X}f_o)_e\alpha \\ &= -[(\mathbf{X}f_e)\alpha]_e + [(\mathbf{X}f_o)\alpha]_o = -[\mathbf{X}(f_e\alpha)]_e + [\mathbf{X}(f_o\alpha)]_o \\ &= -[\mathbf{X}(\alpha f_e)]_e - [\mathbf{X}(\alpha f_o)]_o = -[\mathbf{X}(\alpha f)_o]_e - [\mathbf{X}(\alpha f)_e]_o \\ &= -\mathbf{X}(\alpha f) = -(\mathbf{X}\alpha)f = (-\mathbf{X}\alpha)f,\end{aligned}$$

whence $\alpha\mathbf{X} = -\mathbf{X}\alpha$. This is a special case of the more general result

$$\alpha\mathbf{X} = (-1)^{\alpha\mathbf{X}}\mathbf{X}\alpha, \tag{2.2.18}$$

which is proved similarly.

The complex conjugate of a contravariant vector field $\mathbf{X}$ is defined by linear extension from

$$\mathbf{X}^*f \stackrel{\text{def}}{=} (-1)^{\mathbf{X}f}(\mathbf{X}f^*)^*, \tag{2.2.19}$$

(f, $\mathbf{X}$ pure). The complex conjugate satisfies

$$(\alpha\mathbf{X})^* = \mathbf{X}^*\alpha^*, \qquad (\mathbf{X}\alpha)^* = \alpha^*\mathbf{X}^*, \tag{2.2.20}$$

for all α in Λ_∞.

Alternative presentation of contravariant vector fields

Just as the symbol 'ω' for a dual supervector can be regarded as representing two equally valid linear mappings from a supervector space to Λ_∞ (see section 1.5), so can the symbol '$\mathbf{X}$' for a contravariant vector field be regarded as representing two equally valid linear mappings from $\mathfrak{F}(M)$ to itself. One of these mappings is characterized by eqs. (2.2.6) and (2.2.8). The other, signaled by the symbol-ordering $f\mathbf{X}$ instead of $\mathbf{X}f$, is related to the first by linear extension from

$$f\mathbf{X} \overset{\text{def}}{=} (-1)^{f\mathbf{X}}\mathbf{X}f \tag{2.2.21}$$

(f, $\mathbf{X}$ pure). When the latter mapping is used eq. (2.2.8) is replaced by

$$F\mathbf{X} = F\frac{\overleftarrow{\partial}}{\partial e^A}(e^A\mathbf{X}). \tag{2.2.22}$$

The proof proceeds by linear extension from the pure case. Using obvious notation, we have

$$\begin{aligned} F\mathbf{X} &= (-1)^{\mathbf{X}F}\mathbf{X}F = (-1)^{\mathbf{X}F}(\mathbf{X}e^A)\frac{\vec{\partial}}{\partial e^A}F \\ &= (-1)^{\mathbf{X}F+\mathbf{X}A+A(F+1)+(\mathbf{X}+A)(A+F)}F\frac{\overleftarrow{\partial}}{\partial e^A}(e^A\mathbf{X}). \end{aligned}$$

Equation (2.2.22) follows because the exponent of (-1) in the last line is always an even integer.

It is straightforward to verify the following analogs of eqs. (2.2.9)–(2.2.14):

$$(\alpha f)\mathbf{X} = \alpha(f\mathbf{X}), \tag{2.2.23}$$

$$(f+g)\mathbf{X} = f\mathbf{X} + g\mathbf{X}, \tag{2.2.24}$$

$$(fg)\mathbf{X} = f(g\mathbf{X}) + (-1)^{fg}g(f\mathbf{X}), \tag{2.2.25}$$

$$f(\mathbf{X}+\mathbf{Y}) = f\mathbf{X} + f\mathbf{Y}, \tag{2.2.26}$$

$$f(\mathbf{X}\alpha) = (f\mathbf{X})\alpha, \tag{2.2.27}$$

$$f(\alpha\mathbf{X}) = (f\alpha)\mathbf{X}. \tag{2.2.28}$$

It is also easy to check that eq. (2.2.18) remains consistent with the definition

(2.2.21) and that complex conjugation may now be defined by

$$(\mathbf{X}f)^* = f^*\mathbf{X}^* \quad \text{or} \quad (f\mathbf{X})^* = \mathbf{X}^* f^*. \qquad (2.2.29)$$

We note finally the following alternative versions of eqs. (2.2.11) and (2.2.25):

$$\mathbf{X}(fg) = (\mathbf{X}f)g + (-1)^{\mathbf{X}f} f(\mathbf{X}g), \qquad (2.2.30)$$

$$(fg)\mathbf{X} = f(g\mathbf{X}) + (-1)^{g\mathbf{X}}(f\mathbf{X})g, \qquad (2.2.31)$$

$$(f\mathbf{X})g + f(\mathbf{X}g) = (-1)^{f\mathbf{X}}\mathbf{X}(fg) = (-1)^{g\mathbf{X}}(fg)\mathbf{X}. \qquad (2.2.32)$$

Components

Let $(\mathscr{U}, \phi)$ be a chart in M and let $\mathbf{X}$ be a contravariant vector field over M. The *restriction* of $\mathbf{X}$ to the open subset $\mathscr{U}$, denoted by $\mathbf{X}_{\mathscr{U}}$, is defined by

$$(\mathbf{X}_{\mathscr{U}} f_{\mathscr{U}})(p) = (\mathbf{X}f)(p), \quad \text{for all } p \text{ in } \mathscr{U} \text{ and all } f \text{ in } \mathfrak{F}(M), \qquad (2.2.33)$$

where '$f_{\mathscr{U}}$' denotes the restriction of f to $\mathscr{U}$. The set of all distinct restrictions $f_{\mathscr{U}}$ is the set $\mathfrak{F}(\mathscr{U})$ of scalar fields over $\mathscr{U}$, and the set of all distinct restrictions $\mathbf{X}_{\mathscr{U}}$ is the set $\mathfrak{X}(\mathscr{U})$ of contravariant vector fields over $\mathscr{U}$. For simplicity we shall drop the subscripts $\mathscr{U}$ and understand that, for the present, we are working only over $\mathscr{U}$.

Because of the chain rule (2.2.8) contravariant vector fields over $\mathscr{U}$ are completely determined by their actions on the coordinate functions x^i defined by the mapping ϕ. Let

$$X^i \overset{\text{def}}{=} \mathbf{X}x^i. \qquad (2.2.34)$$

Then

$$\mathbf{X}f = X^i \frac{\vec{\partial}}{\partial x^i} f = X^i{}_{,i} f, \qquad (2.2.35)$$

for all f in $\mathfrak{F}(\mathscr{U})$ and all $\mathbf{X}$ in $\mathfrak{X}(\mathscr{U})$. The X^i are scalar fields over $\mathscr{U}$, known as the *components* of $\mathbf{X}$ in the coordinate system defined by ϕ.

An alternative set of components is obtained if one uses the alternative presentation of $\mathbf{X}$ described by the chain rule (2.2.22). Thus

$$f\mathbf{X} = f \frac{\overleftarrow{\partial}}{\partial x^i}\, {}^iX = f_{,i}\, {}^iX, \qquad (2.2.36)$$

where

$${}^iX \overset{\text{def}}{=} x^i \mathbf{X}. \qquad (2.2.37)$$

When $\mathbf{X}$ is pure it follows from eq. (2.2.21) that

$$ {}^iX = (-1)^{\mathbf{X}i} X^i. \tag{2.2.38} $$

Equations (2.2.35) and (2.2.36) imply

$$ \mathbf{X} = X^i \frac{\bar{\partial}}{\partial x^i} \quad \text{or} \quad \mathbf{X} = \frac{\bar{\partial}}{\partial x^i}\, {}^iX. \tag{2.2.39} $$

The similarity of eqs. (2.2.38) and (2.2.39) to eqs. (1.5.14) and (1.5.56) suggests that the contravariant vectors $\frac{\bar{\partial}}{\partial x^i}$ or $\frac{\bar{\partial}}{\partial x^i}$ be regarded as basis supervectors for $\mathfrak{X}(\mathscr{U})$. Actually, $\mathfrak{X}(\mathscr{U})$, like $\mathfrak{F}(\mathscr{U})$, may have no basis at all (without Zorn's lemma). The X^i and ${}^iX$ in eqs. (2.2.39) are functions over $\mathscr{U}$, not constant supernumbers. In order to construct a supervector space that has a finite basis we may confine our attention to a single point.

Tangent spaces

Let $\mathbf{X}$ be an element of $\mathfrak{X}(M)$ and let p be a point of M. Denote by $\mathbf{X}_p$ the mapping from $\mathfrak{F}(M)$ to Λ_∞ defined by

$$ \mathbf{X}_p f \overset{\text{def}}{=} (\mathbf{X}f)(p) \quad \text{or, alternatively,} \quad f\mathbf{X}_p \overset{\text{def}}{=} (f\mathbf{X})(p), \tag{2.2.40} $$

for all f in $\mathfrak{F}(M)$. This mapping is called the *value of* $\mathbf{X}$ *at* p and is known as a *contravariant vector at* p. Two distinct elements of $\mathfrak{X}(M)$ may have the same value at p. Therefore it is not necessary to introduce an entire contravariant vector field over M in order to define a contravariant vector at p. An arbitrarily small neighbourhood of p will do. In fact, one sometimes defines local contravariant vectors first and then extends the contravariant vector concept to contravariant vector fields by attaching a local contravariant vector at each point and imposing differentiability requirements on the whole collection.

The set of all contravariant vectors at p is denoted by $T_p(M)$ and is called the *tangent space to M at p*. It has an obvious structure as a supervector space, stemming from the fact that each of the equations (2.2.9)–(2.2.32) and (2.2.34)–(2.2.39) has its local counterpart obtained simply by evaluating it at p. Furthermore, $T_p(M)$ has a finite basis. For each chart $(\mathscr{U}, \phi)$ containing p one may choose as basis supervectors

$$ {}_i\mathbf{e} \overset{\text{def}}{=} \left(\frac{\bar{\partial}}{\partial x^i}\right)_p \quad \text{or} \quad \mathbf{e}_i \overset{\text{def}}{=} \left(\frac{\bar{\partial}}{\partial x^i}\right)_p, \tag{2.2.41} $$

where the x^i are the coordinates defined by ϕ. The set $\{{}_i\mathbf{e}\}$ (or $\{\mathbf{e}_i\}$) is called a *coordinate basis*. It is easy to see that ${}_i\mathbf{e}$ is c-type or a-type according as x^i is

c-type or a-type. Therefore a coordinate basis is a pure basis and $T_p(M)$ is seen to have the same dimension (m, n) as M.

A coordinate basis is, in fact, a standard basis. This follows from the readily verified relation

$$({}_{i,}f)^* = f^*_{,i} = (-1)^{i(f+1)}{}_{i,}f^*, \tag{2.2.42}$$

for all pure f in $\mathfrak{F}(M)$. Thus

$$\begin{aligned} {}_i\mathbf{e}^* f &= (-1)^{if} f\,{}_i\mathbf{e}^* = (-1)^{if}({}_i\mathbf{e} f^*)^* = (-1)^{if}({}_{i,}f^*)^*(p) \\ &= (-1)^i({}_{i,}f)(p) = (-1)^i\,{}_i\mathbf{e} f, \end{aligned} \tag{2.2.43}$$

which implies the condition (1.4.61) for a standard basis.

We note also that the two presentations (2.2.41) of a coordinate basis are related by the first of eqs. (1.5.52). Thus

$${}_i\mathbf{e} f = ({}_{i,}f)(p) = (-1)^{i(f+1)}(f_{,i})(p) = (-1)^{i(f+1)} f \mathbf{e}_i = (-1)^i\, \mathbf{e}_i f. \tag{2.2.44}$$

The basis supervectors ${}_i\mathbf{e}$, $\mathbf{e}_i$ are defined at all points in the chart $(\mathscr{U}, \phi)$. The symbols ${}_i\mathbf{e}$, $\mathbf{e}_i$ are therefore often regarded as standing for the corresponding contravariant vector fields over $\mathscr{U}$, and one writes

$$\mathbf{X} = X^i\,{}_i\mathbf{e} = (-1)^{i(\mathbf{X}+i)}\,{}_i\mathbf{e} X^i = \mathbf{e}_i\,{}^iX. \tag{2.2.45}$$

This abuse of notation does no harm as long as one remembers that the e's do *not* appear here in their role as mappings, i.e., as contravariant vector fields that *act* on the scalar fields X^i, ${}^iX$. Rather, one is using the simple product mapping from $\mathfrak{X}(M) \times \mathfrak{F}(M)$ to $\mathfrak{X}(M)$ defined by

$$(\mathbf{X}f)_p \overset{\text{def}}{=} \mathbf{X}_p f(p) = (-1)^{\mathbf{X}f} f(p)\mathbf{X}_p \overset{\text{def}}{=} (-1)^{\mathbf{X}f}(f\mathbf{X})_p, \tag{2.2.46}$$

for all $\mathbf{X}$ in $\mathfrak{X}(M)$, all f in $\mathfrak{F}(M)$ and all p in M. Whether one understands the symbol $\mathbf{X}f$ in this sense or in the sense of eq. (2.2.6), in a given instance, will usually be clear from the context.

Tangents to supercurves

One sometimes wishes to define contravariant vectors over subsets other than open subsets or individual points, for example over the range $\lambda(\mathscr{S})$ of a supercurve λ (see section 2.1 for definition). Of particular interest are the *tangents* to λ, denoted by $(\vec{\partial}/\partial s)_\lambda$ and $(\overleftarrow{\partial}/\partial s)_\lambda$. These are mappings from $\mathfrak{F}(M)$ to $\mathfrak{F}(\mathscr{S})$ defined by

$$[(\vec{\partial}/\partial s)_\lambda f](\lambda(s)) \overset{\text{def}}{=} \frac{\vec{\mathrm{d}}}{\mathrm{d}s} f(\lambda(s)), \quad [f(\overleftarrow{\partial}/\partial s)_\lambda](\lambda(s)) \overset{\text{def}}{=} f(\lambda(s))\frac{\overleftarrow{\mathrm{d}}}{\mathrm{d}s}, \tag{2.2.47}$$

for all f in $\mathfrak{F}(M)$ and all s in $\mathscr{S}$. The mappings $(\vec{\partial}/\partial s)_\lambda$ and $(\overleftarrow{\partial}/\partial s)_\lambda$ stand in the same relation to each other as the contravariant vectors ${}_i\mathbf{e}$ and $\mathbf{e}_i$ of a coordinate basis. The latter are, in fact, the tangents of the *coordinate supercurves.* If λ is c-type then $(\vec{\partial}/\partial s)_\lambda$ and $(\overleftarrow{\partial}/\partial s)_\lambda$ are c-type, real and identical; if λ is a-type they are a-type, imaginary and of opposite sign.

It is sometimes useful to introduce the tangents to a supercurve at a single point p. These are mappings from $\mathfrak{F}(M)$ to Λ_∞ defined by

$$\left.\begin{aligned} (\vec{\partial}/\partial s)_p f &\stackrel{\text{def}}{=} \left[\frac{\vec{\mathrm{d}}}{\mathrm{d}s} f(\lambda(s))\right]_{s=\lambda^{-1}(p)}, \\ f(\overleftarrow{\partial}/\partial s)_p &\stackrel{\text{def}}{=} \left[f(\lambda(s))\frac{\overleftarrow{\mathrm{d}}}{\mathrm{d}s}\right]_{s=\lambda^{-1}(p)} \end{aligned}\right\} \tag{2.2.48}$$

Here it is assumed that λ is non-self-intersecting so that $\lambda:\mathscr{S}\to M$ is one-to-one. Self-intersections do not, in fact, occur with a-type supercurves. An a-type supercurve is completely determined by either of its tangents at a single point.

2.3 Super Lie brackets, local frames and covariant vector fields

Supercommutators and antisupercommutators

Contravariant vector fields are linear mappings of the supervector space $\mathfrak{F}(M)$ into itself. Each can be expressed as a sum of pure parts, i.e., parts that map pure scalar fields into pure scalar fields. Linear mappings of supervector spaces into themselves, which can be separated into pure parts, arise repeatedly in the theory of supermanifolds. Other examples include matrices (i.e., mappings of finite-dimensional supervector spaces (see exercises **1.9** and **1.10**)), Lie derivatives (section 2.5), derivations of forms (section 2.6), and operators in super Hilbert spaces and super Fock spaces (section 5.2). Such mappings can be 'multiplied' together by composition, i.e., by successive application.† The product of two such linear mappings is again a linear mapping. It is pure whenever its factors are pure, but there may be additional properties possessed by its factors that it does not share. Even when the product does not share the extra properties, other binary combinations of the two factors often do. One of these combinations is the *supercommutator*.

Let **A** and **B** be two pure linear mappings of a supervector space into

† Contravariant vector fields can also be multiplied together *via* the *tensor product* (see section 2.4).

itself. The supercommutator of $\mathbf{A}$ and $\mathbf{B}$ is defined by

$$[\mathbf{A},\mathbf{B}] \stackrel{\text{def}}{=} \mathbf{AB} - (-1)^{\mathbf{AB}}\mathbf{BA}. \tag{2.3.1}$$

It obeys the laws

$$[\mathbf{A},\mathbf{B}] = -(-1)^{\mathbf{AB}}[\mathbf{B},\mathbf{A}], \tag{2.3.2}$$

$$[\mathbf{A},\mathbf{BC}] = [\mathbf{A},\mathbf{B}]\mathbf{C} + (-1)^{\mathbf{AB}}\mathbf{B}[\mathbf{A},\mathbf{C}] \tag{2.3.3}$$

$$= \{\mathbf{A},\mathbf{B}\}\mathbf{C} - (-1)^{\mathbf{AB}}\mathbf{B}\{\mathbf{A},\mathbf{C}\}, \tag{2.3.4}$$

where, in the last line, we have introduced the *antisupercommutator*:

$$\{\mathbf{A},\mathbf{B}\} \stackrel{\text{def}}{=} \mathbf{AB} + (-1)^{\mathbf{AB}}\mathbf{BA}. \tag{2.3.5}$$

The antisupercommutator obeys the laws

$$\{\mathbf{A},\mathbf{B}\} = (-1)^{\mathbf{AB}}\{\mathbf{B},\mathbf{A}\}, \tag{2.3.6}$$

$$\{\mathbf{A},\mathbf{BC}\} = \{\mathbf{A},\mathbf{B}\}\mathbf{C} - (-1)^{\mathbf{AB}}\mathbf{B}[\mathbf{A},\mathbf{C}] \tag{2.3.7}$$

$$= [\mathbf{A},\mathbf{B}]\mathbf{C} + (-1)^{\mathbf{AB}}\mathbf{B}\{\mathbf{A},\mathbf{C}\}. \tag{2.3.8}$$

Other basic laws are the following:

$$[\mathbf{A},\{\mathbf{B},\mathbf{C}\}] = [\{\mathbf{A},\mathbf{B}\},\mathbf{C}] - (-1)^{\mathbf{AB}}[\mathbf{B},\{\mathbf{A},\mathbf{C}\}] \tag{2.3.9}$$

$$= \{[\mathbf{A},\mathbf{B}],\mathbf{C}\} + (-1)^{\mathbf{AB}}\{\mathbf{B},[\mathbf{A},\mathbf{C}]\}, \tag{2.3.10}$$

$$[\mathbf{A},[\mathbf{B},\mathbf{C}]] = \{\{\mathbf{A},\mathbf{B}\},\mathbf{C}\} - (-1)^{\mathbf{AB}}\{\mathbf{B},\{\mathbf{A},\mathbf{C}\}\} \tag{2.3.11}$$

$$= [[\mathbf{A},\mathbf{B}],\mathbf{C}] + (-1)^{\mathbf{AB}}[\mathbf{B},[\mathbf{A},\mathbf{C}]]. \tag{2.3.12}$$

All of these laws are simple identities that follow from the defining equations (2.3.1) and (2.3.5), and all may be extended to include impure mappings through use of the decompositions

$$[\mathbf{A},\mathbf{B}+\mathbf{C}] \stackrel{\text{def}}{=} [\mathbf{A},\mathbf{B}] + [\mathbf{A},\mathbf{C}], \tag{2.3.13}$$

$$\{\mathbf{A},\mathbf{B}+\mathbf{C}\} \stackrel{\text{def}}{=} \{\mathbf{A},\mathbf{B}\} + \{\mathbf{A},\mathbf{C}\}. \tag{2.3.14}$$

Equation (2.3.12), when rewritten in the form

$$[\mathbf{A},[\mathbf{B},\mathbf{C}]] + (-1)^{\mathbf{A}(\mathbf{B}+\mathbf{C})}[\mathbf{B},[\mathbf{C},\mathbf{A}]] + (-1)^{\mathbf{C}(\mathbf{A}+\mathbf{B})}[\mathbf{C},[\mathbf{A},\mathbf{B}]] = 0, \tag{2.3.15}$$

is known as the *super Jacobi identity*.

A matter of notation

It has become customary, in applying supermanifold theory to physics, for some authors to use the bracket symbol '[,}' for the supercommutator.

This is an unfortunate notation in two respects: (1) It leaves the anti-supercommutator in limbo. (2) It fails to emphasize that the supercommutator is the analog of the commutator, *not* of the anticommutator. The symbol '[,}' will not be used in this book. In those places where it is necessary or helpful to distinguish ordinary commutators and anticommutators from supercommutators and antisupercommutators, we shall identify the latter by attaching subscripts s thus: $[,]_s$ and $\{,\}_s$. In the purely mathematical parts of the book the subscripts will usually be dropped.

The super Lie bracket

The product (as mappings) of two contravariant vector fields is not itself a contravariant vector field. Although it is a linear mapping, the product does not generally satisfy the chain rule (2.2.8) or (2.2.22). The supercommutator of two contravariant vector fields, on the other hand, does. This supercommutator, known as the *super Lie bracket*, therefore *is* a contravariant vector field.

In defining the super Lie bracket it is necessary to make a decision whether to regard contravariant vector fields as acting from the left, as in eq. (2.2.8), or from the right, as in eq. (2.2.22). We shall choose the latter. The reader should be warned that the former is usually chosen in ordinary manifold theory. It leads to a super Lie bracket that is opposite in sign to that obtained here. The present choice will be convenient when we introduce super Lie algebras in chapter 3.

The proof that the super Lie bracket satisfies the chain rule is straightforward. Let $\mathbf{X}$ and $\mathbf{Y}$ be two pure contravariant vector fields. Then

$$\begin{aligned}
F[\mathbf{X},\mathbf{Y}] &= F\mathbf{X}\mathbf{Y} - (-1)^{\mathbf{XY}} F\mathbf{Y}\mathbf{X} \\
&= \left[F\frac{\overleftarrow{\partial}}{\partial e^A}(e^A\mathbf{X})\right]\mathbf{Y} - (-1)^{\mathbf{XY}}\left[F\frac{\overleftarrow{\partial}}{\partial e^B}(e^B\mathbf{Y})\right]\mathbf{X} \\
&= (-1)^{\mathbf{Y}(A+\mathbf{X})} F\frac{\overleftarrow{\partial}}{\partial e^A}\frac{\overleftarrow{\partial}}{\partial e^B}(e^B\mathbf{Y})(e^A\mathbf{X}) \\
&\quad - (-1)^{\mathbf{XY}+\mathbf{X}(B+\mathbf{Y})} F\frac{\overleftarrow{\partial}}{\partial e^B}\frac{\overleftarrow{\partial}}{\partial e^A}(e^A\mathbf{X})(e^B\mathbf{Y}) \\
&\quad + F\frac{\overleftarrow{\partial}}{\partial e^A}[e^A\mathbf{X}\mathbf{Y} - (-1)^{\mathbf{XY}} e^A\mathbf{Y}\mathbf{X}]
\end{aligned}$$

$$= [(-1)^{\mathbf{Y}(A+\mathbf{X})} - (-1)^{\mathbf{X}B+AB+(A+\mathbf{X})(B+\mathbf{Y})}] F \frac{\overleftarrow{\partial}}{\partial e^A} \frac{\overleftarrow{\partial}}{\partial e^B} (e^B \mathbf{Y})(e^A \mathbf{X})$$

$$+ F \frac{\overleftarrow{\partial}}{\partial e^A} (e^A [\mathbf{X}, \mathbf{Y}])$$

$$= F \frac{\overleftarrow{\partial}}{\partial e^A} (e^A [\mathbf{X}, \mathbf{Y}]),$$

qed. Here we have omitted some parentheses, with the understanding, for example, that $F\mathbf{XY}$ means $(F\mathbf{X})\mathbf{Y}$. The right action of the product of two contravariant vector fields is related to the left action of the same pair by

$$F\mathbf{XY} = (-1)^{\mathbf{Y}(F+\mathbf{X})} \mathbf{Y}(F\mathbf{X}) = (-1)^{\mathbf{XY}+F(\mathbf{X}+\mathbf{Y})} \mathbf{YX}F. \qquad (2.3.16)$$

To see the relation between the super Lie bracket defined here and the bracket that would have been obtained had we worked with contravariant vector fields acting from the left, let us distinguish the two by affixing to them subscripts R and L respectively. Then, using eqs. (2.2.21) and (2.3.16) we obtain

$$\begin{aligned} F[\mathbf{X}, \mathbf{Y}]_{\mathrm{L}} &= (-1)^{F(\mathbf{X}+\mathbf{Y})} [\mathbf{X}, \mathbf{Y}]_{\mathrm{L}} F \\ &= (-1)^{F(\mathbf{X}+\mathbf{Y})} [\mathbf{XY} - (-1)^{\mathbf{XY}} \mathbf{YX}] F \\ &= -F[\mathbf{XY} - (-1)^{\mathbf{XY}} \mathbf{YX}] = -F[\mathbf{X}, \mathbf{Y}]_{\mathrm{R}}, \end{aligned}$$

and hence $[\mathbf{X}, \mathbf{Y}]_{\mathrm{L}} = -[\mathbf{X}, \mathbf{Y}]_{\mathrm{R}}$. From now on it will be understood that '$[\mathbf{X}, \mathbf{Y}]$' means '$[\mathbf{X}, \mathbf{Y}]_{\mathrm{R}}$', and the subscript will be dropped unless needed for emphasis.

Expressions for the components of super Lie brackets in a coordinate basis are sometimes useful. These are readily obtained:

$$\begin{aligned} {}^i[\mathbf{X}, \mathbf{Y}] &= x^i [\mathbf{X}, \mathbf{Y}] = x^i [\mathbf{XY} - (-1)^{\mathbf{XY}} \mathbf{YX}] \\ &= {}^i X\mathbf{Y} - (-1)^{\mathbf{XY}} {}^i Y\mathbf{X} = {}^i X_{,j} {}^j Y - (-1)^{\mathbf{XY}} {}^i Y_{,j} {}^j X, \end{aligned} \qquad (2.3.17)$$

$$\begin{aligned} [\mathbf{X}, \mathbf{Y}]^i &= (-1)^{i(\mathbf{X}+\mathbf{Y})} {}^i[\mathbf{X}, \mathbf{Y}] = (-1)^{i(\mathbf{X}+\mathbf{Y})} x^i [\mathbf{X}, \mathbf{Y}]_{\mathrm{R}} \\ &= [\mathbf{X}, \mathbf{Y}]_{\mathrm{R}} x^i = -[\mathbf{X}, \mathbf{Y}]_{\mathrm{L}} x^i \\ &= -[\mathbf{XY} - (-1)^{\mathbf{XY}} \mathbf{YX}] x^i = -X^j{}_{,j} Y^i + (-1)^{\mathbf{XY}} Y^j{}_{,j} X^i. \end{aligned} \qquad (2.3.18)$$

Local frames

It is not necessary to use coordinate bases to obtain representations of contravariant vector fields in component form. At a given point p of a supermanifold M, one may use any complete set of linearly independent elements ${}_a\mathbf{e}$ (or $\mathbf{e}_a$) of $T_p(M)$ instead. Such a complete set may be expressed in terms of the coordinate basis (2.2.41) of any chart containing p:

$$ {}_a\mathbf{e} = {}_a e^i \, {}_i\mathbf{e}, \qquad \mathbf{e}_a = \mathbf{e}_i \, {}^i e_a. \tag{2.3.19} $$

The ${}_a e^i$, ${}^i e_a$ are the components, in the coordinate basis, of ${}_a\mathbf{e}$, $\mathbf{e}_a$. The linear independence and completeness of the latter imply that the matrices $({}_a e^i)$, $({}^i e_a)$ are nonsingular. Denote the inverses of these matrices by $({}_i e^a)$, $({}^a e_i)$ respectively. Then any contravariant vector $\mathbf{X}$ at p may be expressed in the form

$$ \mathbf{X} = X^a \, {}_a\mathbf{e} = \mathbf{e}_a \, {}^a X, \tag{2.3.20} $$

where

$$ X^a = X^i \, {}_i e^a, \quad {}^a X = {}^a e_i \, {}^i X. \tag{2.3.21} $$

The ${}_a\mathbf{e}$ (or $\mathbf{e}_a$) constitute a basis for $T_p(M)$. We shall assume that they constitute a pure basis and shall adopt the standard convention (cf. (1.5.52))

$$ {}_a\mathbf{e} = (-1)^a \mathbf{e}_a. \tag{2.3.22} $$

The matrices $({}_a e^i)$ and $({}^i e_a)$ are then the supertransposes of one another.

By attaching a basis $\{{}_a\mathbf{e}\}$ (or $\{\mathbf{e}_a\}$) to every point of a chart and imposing differentiability conditions on the components ${}_a e^i$, ${}^i e_a$, one can extend the e's so that they become contravariant vector fields over the chart. The e's are then said to constitute a *field of local frames.* At each point of the chart they define a *local frame* that is independent of the *coordinate frame* defined by the x^i. The e's may be expressed in the form

$$ {}_a\mathbf{e} = {}_a e^i \frac{\vec{\partial}}{\partial x^i}, \quad \mathbf{e}_a = \frac{\overleftarrow{\partial}}{\partial x^i} \, {}^i e_a \tag{2.3.23} $$

throughout the chart, and eq. (2.3.20) may be used even when $\mathbf{X}$ is a contravariant vector field and not just an element of $T_p(M)$. The X^a, ${}^a X$ are called the *components of* $\mathbf{X}$ *in the local frame.*

Except when otherwise noted we shall use Latin indices from the middle of the alphabet $(i, j, k, \ldots)$ to denote components in coordinate bases and Latin indices from the first part of the alphabet $(a, b, c, \ldots)$ to denote components in local frames.

In some supermanifolds (but by no means all) it is possible to extend the e's from a single chart to the whole manifold in such a way that the matrices $({}_a e^i)$, $({}^i e_a)$ remain nonsingular and differentiable in every chart. The e's are then said to constitute a *global* field of local frames. Even when global fields of local frames do not exist it is often possible to introduce nonsingular local frame fields over regions of the supermanifold that cannot be covered by single charts. Local frame fields are therefore useful for studies of supermanifolds in the large.

Super Lie brackets of local frame fields

Since the super Lie bracket of two contravariant vector fields is itself a contravariant vector field, the super Lie bracket of any two local frame vector fields ${}_a\mathbf{e}$ and ${}_b\mathbf{e}$ is expressible as a linear combination of the e's:

$$[{}_a\mathbf{e}, {}_b\mathbf{e}] = {}_{ab}c^c\,{}_c\mathbf{e}, \quad [\mathbf{e}_a, \mathbf{e}_b] = \mathbf{e}_c\,{}^c c_{ab}. \tag{2.3.24}$$

The coefficients ${}_{ab}c^c$ and ${}^c c_{ab}$ are related by

$$ {}_{ab}c^c = (-1)^{(a+b)(c+1)}\,{}^c c_{ab} \tag{2.3.25}$$

and are antisupersymmetric in the indices a and b:

$$ {}_{ab}c^c = -(-1)^{ab}\,{}_{ba}c^c, \quad {}^c c_{ab} = -(-1)^{ab}\,{}^c c_{ba}. \tag{2.3.26}$$

They may be expressed (and thus computed) in any chart in terms of the matrices $({}_a e^i)$, $({}^i e_a)$ and their derivatives. One simply writes

$$ {}_{ab}c^c = [\mathbf{e}_a, \mathbf{e}_b]^i\,{}_i e^c, \quad {}^c c_{ab} = {}^c e_i\,{}^i[\mathbf{e}_a, \mathbf{e}_b], \tag{2.3.27}$$

and then uses eqs. (2.3.17) and (2.3.18).

In any chart the coordinate basis itself may be viewed as a local frame field. It is a local frame field, however, with a special property. All the c-coefficients vanish:

$$[{}_i\mathbf{e}, {}_j\mathbf{e}] = \left[\frac{\vec{\partial}}{\partial x^i}, \frac{\vec{\partial}}{\partial x^j}\right] = 0, \quad [\mathbf{e}_i, \mathbf{e}_j] = \left[\frac{\overleftarrow{\partial}}{\partial x^i}, \frac{\overleftarrow{\partial}}{\partial x^j}\right] = 0. \tag{2.3.28}$$

Conversely, if all the e's of a local frame field have vanishing super Lie brackets with one another, then there exists a coordinate system, at least locally, for which the e's constitute the corresponding coordinate basis.

Covariant vector fields

Let $T_p(M)$ be the tangent space to a supermanifold M at the point p. Denote by $T_p^*(M)$ the dual to $T_p(M)$. An element of $T_p^*(M)$ is called a *covariant vector* at p, and $T_p^*(M)$ itself is called the *cotangent space* at p.

Equations (1.5.21)–(1.5.60) of section 1.5 can be immediately applied, as written, to the description of contravariant and covariant vectors and their inner products at p. If the coordinates at p are changed from x^i to $\bar{x}^i$ then the coordinate bases and the components relative to them suffer the transformations (cf. eqs. (1.4.26), (1.5.2), (1.5.36) and (1.5.39))

$$ {}_i\mathbf{e} = {}_{i,}\bar{x}^j\,{}_j\bar{\mathbf{e}}, \qquad \mathbf{e}_i = \bar{\mathbf{e}}_j\,\bar{x}^j{}_{,i}, \tag{2.3.29}$$

$$ \bar{\mathbf{e}}^i = \mathbf{e}^j{}_{j,}\bar{x}^i, \qquad {}^i\bar{\mathbf{e}} = \bar{x}^i{}_{,j}\,{}^j\mathbf{e}, \tag{2.3.30}$$

$$\bar{X}^i = X^j{}_{j,}\bar{x}^i, \quad {}^i\bar{X} = \bar{x}^i{}_{,j}{}^jX, \tag{2.3.31}$$

$$_i\omega = {}_{i,}\bar{x}^j{}_j\bar{\omega}. \quad \omega_i = \bar{\omega}_j\,\bar{x}^j{}_{,i}. \tag{2.3.32}$$

One may introduce a covariant vector $\boldsymbol{\omega}$ at every point of M in such a way that $\mathbf{X}\cdot\boldsymbol{\omega}$ is a differentiable function over M for every $\mathbf{X}$ in $\mathfrak{X}(M)$. One then has a *covariant vector field* over M, and eqs. (1.5.21)–(1.5.60) may be regarded as holding throughout the supermanifold. At every point in every chart the components of a covariant vector field are differentiable functions of the coordinates, related to one another, in overlapping charts, by eqs. (2.3.32).

Let the set of all covariant vector fields over M be denoted by $\Omega(M)$. The definitions (1.5.24)–(1.5.26) give $\Omega(M)$ the structure of a supervector space. Note that $\Omega(M)$ is *not* the set of *all* mappings from $\mathfrak{X}(M)$ to $\mathfrak{F}(M)$ that satisfy the linear laws (1.5.22) and (1.5.23). The latter is a much larger space, of which $\Omega(M)$ is a subspace consisting of only those linear mappings that can be decomposed, point-by-point, into mappings from $T_p(M)$ to Λ_∞.

In dealing with covariant vector fields, as with contravariant vector fields, it is not necessary to confine one's attention to coordinate bases. Arbitrary local frame fields $\{_a\mathbf{e}\}$ (or $\{\mathbf{e}_a\}$) can be employed instead. Dual to every field of local frames is a set of covariant vector fields $\{\mathbf{e}^a\}$ (or $\{^a\mathbf{e}\}$), and eqs. (1.5.34)–(1.5.46) and (1.5.52)–(1.5.57) may be rewritten with all the indices taken from the first part of the alphabet. Equations (1.5.36) and (1.5.39), in combination with eqs. (1.4.26) and (1.5.2) similarly rewritten, express the laws for changing from one field of local frames (and its dual) to another.† The matrix K is now a differentiable function over the domain of overlap of the two fields.

Differentials

A coordinate basis can be defined as a field of local frames $\{_i\mathbf{e}\}$ possessing the special property that the super Lie brackets of the e's with each other all vanish. The dual $\{\mathbf{e}^i\}$ to a coordinate basis also has a special property that identifies it as such. To describe this property one must introduce the concept of a *differential*.

Let f be an arbitrary scalar field over M. The *differential* or *gradient* of f, denoted by $\mathbf{d}f$, is the covariant vector field defined by

$$\mathbf{X}\cdot\mathbf{d}f \overset{\text{def}}{=} \mathbf{X}f \quad \text{or} \quad \mathbf{d}f\cdot\mathbf{X} \overset{\text{def}}{=} f\mathbf{X}, \tag{2.3.33}$$

for all $\mathbf{X}$ in $\mathfrak{X}(M)$. That $\mathbf{d}f$ thus defined *is* a covariant vector field follows

†The e's in eqs. (1.4.26) must not be regarded as acting on the elements of K.

from eqs. (2.2.12) and (2.2.13) which insure that eqs. (1.5.22) and (1.5.23), with ω replaced by $\mathbf{d}f$, are satisfied. The consistency of the two definitions (2.3.33) with each other follows from eqs. (1.5.46) and (2.2.21). Note that when f is pure $\mathbf{d}f$ is pure and is of the same type.

From eqs. (2.2.8) and (2.2.22) one easily derives the chain rule

$$\mathbf{d}F = \mathbf{d}e^A\left(\frac{\vec{\partial}}{\partial e^A}F\right) = \left(F\frac{\overleftarrow{\partial}}{\partial e^A}\right)\mathbf{d}e^A, \tag{2.3.34}$$

which justifies the terminology 'differential' for the mapping $\mathbf{d}:\mathfrak{F}(M)\rightarrow \Omega(M)$. Immediate corollaries of this rule are

$$\mathbf{d}(f+g) = \mathbf{d}f + \mathbf{d}g, \tag{2.3.35}$$

$$\mathbf{d}(\alpha f) = \alpha(\mathbf{d}f), \tag{2.3.36}$$

$$\mathbf{d}(f\alpha) = (\mathbf{d}f)\alpha, \tag{2.3.37}$$

$$\mathbf{d}(fg) = (\mathbf{d}f)g + f(\mathbf{d}g), \tag{2.3.38}$$

for all f, g in $\mathfrak{F}(M)$ and all α in Λ_∞. It is also easy to show that

$$(\mathbf{d}f)^* = \mathbf{d}f^*. \tag{2.3.39}$$

If one chooses for the functions e^A in (2.3.34) the coordinate functions in some chart one obtains

$$\mathbf{d}f = \mathbf{dx}^i{}_{i,}f = f_{,i}\mathbf{dx}^i. \tag{2.3.40}$$

The differentials $\mathbf{dx}^i$ have the important property

$$\left.\begin{aligned} {}_i\mathbf{e}\cdot\mathbf{dx}^j &= {}_i\mathbf{e}\,x^j = \frac{\vec{\partial}}{\partial x^i}x^j = {}_i\delta^j, \\ \mathbf{dx}^i\cdot\mathbf{e}_j &= x^i\,\mathbf{e}_j = x^i\frac{\overleftarrow{\partial}}{\partial x^j} = {}^i\delta_j, \end{aligned}\right\} \tag{2.3.41}$$

which reveals them as comprising the basis dual to the coordinate basis $\{{}_i\mathbf{e}\}$:

$$\mathbf{dx}^i = \mathbf{e}^i = {}^i\mathbf{e}. \tag{2.3.42}$$

Conversely, if the dual to a field of local frames consists entirely of the differentials of a set of functions, then these functions are the coordinate functions of some chart, and the field of local frames is the corresponding coordinate basis.

Using the dual to a coordinate basis one may write, for every covariant vector field ω over the corresponding chart,

$$\omega = \mathbf{dx}^i{}_i\omega = \omega_i\mathbf{dx}^i. \tag{2.3.43}$$

Because of this relation covariant vector fields are often called *differential forms of degree* 1, or, simply, 1-*forms* (see section 2.7).

2.4 Tensor fields

Tensors at a point

For brevity let us drop the reference to the supermanifold M and denote the tangent space at a point p, and its dual, simply by T_p and T_p^* respectively. Introduce the Cartesian product space

$$\Pi_r{}^s(p) \overset{\text{def}}{=} \overbrace{T_p^* \times \cdots \times T_p^*}^{r\text{ factors}} \times \overbrace{T_p \times \cdots \times T_p}^{s\text{ factors}}. \tag{2.4.1}$$

Let $\mathbf{T}$ be a mapping $\mathbf{T} : \Pi_r{}^s(p) \to \Lambda_\infty$ that sends every element $(\omega^{A_1}, \ldots, \omega^{A_r}, \mathbf{X}_{A_{r+s}}, \ldots, \mathbf{X}_{A_{r+s}})$ of $\Pi_r{}^s(p)$ into a supernumber $\mathbf{T}(\omega^{A_1}, \ldots, \mathbf{X}_{A_{r+s}})$. This mapping is said to be a *tensor of rank* (r, s) *at* p if, for all ω, σ in T_p^*, all $\mathbf{X}, \mathbf{Y}$ in T_p, and all α in Λ_∞, it satisfies the multilinear laws

$$\mathbf{T}(\ldots \omega + \sigma \ldots) = \mathbf{T}(\ldots \omega \ldots) + \mathbf{T}(\ldots \sigma \ldots), \tag{2.4.2}$$

$$\mathbf{T}(\ldots \mathbf{X} + \mathbf{Y} \ldots) = \mathbf{T}(\ldots \mathbf{X} \ldots) + \mathbf{T}(\ldots \mathbf{Y} \ldots), \tag{2.4.3}$$

$$\mathbf{T}(\ldots \omega\alpha, \sigma \ldots) = \mathbf{T}(\ldots \omega, \alpha\sigma \ldots), \tag{2.4.4}$$

$$\mathbf{T}(\ldots \omega\alpha, \mathbf{X} \ldots) = \mathbf{T}(\ldots \omega, \alpha\mathbf{X} \ldots), \tag{2.4.5}$$

$$\mathbf{T}(\ldots \mathbf{X}\alpha, \mathbf{Y} \ldots) = \mathbf{T}(\ldots \mathbf{X}, \alpha\mathbf{Y} \ldots), \tag{2.4.6}$$

$$\mathbf{T}(\ldots \mathbf{X}\alpha) = \mathbf{T}(\ldots \mathbf{X})\alpha, \tag{2.4.7}$$

where, in each equation, the dots ... represents unwritten entries that are arbitrary but the same in every term.

Let $\{\mathbf{e}_a\}$ be a pure basis of T_p and $\{\mathbf{e}^a\}$ the corresponding dual basis of T_p^*. Define

$$T^{a_1 \ldots a_r}{}_{a_{r+1} \ldots a_{r+s}} \overset{\text{def}}{=} \mathbf{T}(\mathbf{e}^{a_1}, \ldots, \mathbf{e}^{a_r}, \mathbf{e}_{a_{r+1}}, \ldots, \mathbf{e}_{a_{r+s}}). \tag{2.4.8}$$

The $T^{a_1 \cdots a_r}{}_{a_{r+n} \cdots a_{r+s}}$ are called the *components of* $\mathbf{T}$ *relative to the bases* $\{\mathbf{e}_a\}$, $\{\mathbf{e}^a\}$. Let $(\omega^{A_1}, \ldots, \omega^{A_r}, \mathbf{X}_{A_{r+1}}, \ldots, \mathbf{X}_{A_r})$ be any *pure* element of $\Pi_r{}^s(p)$, i.e., any element for which all the ω's and $\mathbf{X}$'s are pure. Then from eqs. (2.4.2)–(2.4.7) one may easily derive the relation

$$\begin{aligned} T(\omega^{A_1}, \ldots, \mathbf{X}_{A_{r+s}}) &\equiv \mathbf{T}(\mathbf{e}^{a_1}{}_{a_1}\omega^{A_1}, \ldots, \mathbf{e}_{a_{r+s}}{}^{a_{r+s}}X_{A_{r+s}}) \\ &= (-1)^{\Delta_{r+s}(a+A,a)} T^{a_1 \cdots a_r}{}_{a_{r+s} \cdots a_{r+s}}\, {}_{a_1}\omega^{A_1} \ldots {}^{a_{r+s}}X_{A_{r+s}}, \end{aligned} \tag{2.4.9}$$

where

$$\Delta_q(a, b) \overset{\text{def}}{=} \sum_{\substack{t,u=1 \\ t<u}}^{q} a_t b_u. \tag{2.4.10}$$

From (2.4.9) and its multilinear extension to the impure case it follows that a tensor is completely determined by its components relative to any basis.

The transformation law for tensor components is a special case of (2.4.9). Under the change of basis

$$\bar{\mathbf{e}}_a = \mathbf{e}_b\, {}^bL_a,\ \bar{\mathbf{e}}^a = \mathbf{e}^b\, {}_bL^{-1\sim a}, \tag{2.4.11}$$

the components of $\mathbf{T}$ suffer the change

$$\begin{aligned} T^{a_1 \dots a_r}{}_{a_{r+1} \dots a_{r+s}} &= (-)^{\Delta_{r+s}(a+b,b)}\, T^{b_1 \dots b_r}{}_{b_{r+1} \dots b_{r+s}} \\ &\quad \times {}_{b_1}L^{-1\sim a_1} \dots {}_{b_r}L^{-1\sim a_r}\, {}^{b_{r+1}}L_{a_{r+1}} \dots {}^{b_{r+s}}L_{a_{r+s}}. \end{aligned} \tag{2.4.12}$$

The supervector space $T^r{}_s(p)$

The set of all rank-(r,s) tensors at p forms a supervector space, denoted by $T^r{}_s(p)$, under the following laws of addition and multiplication:

$$(\mathbf{S}+\mathbf{T})(\omega,\dots,\mathbf{X}) \stackrel{\text{def}}{=} \mathbf{S}(\omega,\dots,\mathbf{X}) + \mathbf{T}(\omega,\dots,\mathbf{X}), \tag{2.4.13}$$

$$(\alpha\mathbf{T})(\omega,\dots,\mathbf{X}) \stackrel{\text{def}}{=} \alpha\mathbf{T}(\omega,\dots,\mathbf{X}), \tag{2.4.14}$$

$$(\mathbf{T}\alpha)(\omega,\dots,\mathbf{X}) \stackrel{\text{def}}{=} \mathbf{T}(\alpha\,\omega,\dots,\mathbf{X}), \tag{2.4.15}$$

for all $\mathbf{S}, \mathbf{T}$ in $T^r{}_s(p)$, all α in Λ_∞, and all $(\omega,\dots,\mathbf{X})$ in $\Pi_r{}^s(p)$. The zero tensor is the tensor that maps all elements of $\Pi_r{}^s(p)$ into the number zero, and the negative of a tensor is defined by $(-\mathbf{T})(\omega,\dots,\mathbf{X}) \stackrel{\text{def}}{=} -\mathbf{T}(\omega,\dots,\mathbf{X})$.

A tensor $\mathbf{T}$ is *pure* if it maps pure elements of $\Pi_r{}^s(p)$ into pure supernumbers. It is *c*-type (*a*-type) if, in the pure case, $\mathbf{T}(\omega,\dots,\mathbf{X})$ is a *c*-number (an *a*-number) when the number of *a*-type ω's and $\mathbf{X}$'s is even and an *a*-number (a *c*-number) when the number of *a*-type ω's and $\mathbf{X}$'s is odd. By multilinear extension from the case in which all the ω's and $\mathbf{X}$'s are pure one may verify, using eqs. (2.4.2)–(2.4.7), (2.4.14) and (2.4.15), the law

$$\alpha\mathbf{T} = (-1)^{\alpha\mathbf{T}}\,\mathbf{T}\alpha. \tag{2.4.16}$$

Equation (2.4.13) then extends this law to impure tensors.

The complex conjugate of a pure tensor is defined by multilinear extension from

$$\mathbf{T}^*(\omega^{A_1},\dots,\mathbf{X}_{A_{r+s}}) = (-1)^{\Delta_{r+s}(A) + \mathbf{T}(A_1 + \dots + A_{r+s})}[\mathbf{T}(\omega^{A_1 *},\dots,\mathbf{X}^*_{A_{r+s}})]^*, \tag{2.4.17}$$

where

$$\Delta_q(A) \stackrel{\text{def}}{=} \Delta_q(A,A) = \sum_{\substack{t,u=1 \\ t<u}}^{q} A_t A_u. \tag{2.4.18}$$

The complex conjugate of an impure tensor is defined by linear extension from this. If $\mathbf{T}$ is both pure and real then its components in a standard basis satisfy

$$T^{a_1 \ldots a_r}{}_{a_{r+1} \ldots a_{r+s}}{}^{*} \\ = (-1)^{\Delta_{r+s}(a) + \mathbf{T}(a_1 + \ldots a_r) + (\mathbf{T}+1)(a_{r+1} + \ldots a_{r+s})} T^{a_1 \ldots a_r}{}_{a_{r+s} \ldots a_{r+s}}. \tag{2.4.19}$$

If $\mathbf{T}$ is pure and imaginary the sign on the right of eq. (2.4.19) is reversed.

Tensor products

The supervector space $T^r{}_s(p)$ is often expressed in the form

$$T^r{}_s(p) = \overbrace{T_p \otimes \cdots \otimes T_p}^{r\text{ factors}} \otimes \overbrace{T_p^* \otimes \cdots \otimes T_p^*}^{s\text{ factors}}. \tag{2.4.20}$$

It is said to be the *tensor product* of its factors. The same notation and terminology are used to express tensors that are built out of vectors. The tensor product $\mathbf{Y}_{B_1} \otimes \cdots \otimes \mathbf{Y}_{B_r} \otimes \boldsymbol{\sigma}^{B_{r+1}} \otimes \cdots \otimes \boldsymbol{\sigma}^{B_{r+s}}$ is defined to be the tensor of rank (r, s) that maps $\Pi_r{}^s(p)$ into Λ_∞ according to the multilinear extension of the law

$$(\mathbf{Y}_{B_1} \otimes \cdots \otimes \boldsymbol{\sigma}^{B_{r+s}})(\boldsymbol{\omega}^{A_1}, \ldots, \mathbf{X}_{A_{r+s}}) = (-1)^{\Delta_{r+s}(A,B)} (\mathbf{Y}_{B_1} \cdot \boldsymbol{\omega}^{A_1}) \ldots (\boldsymbol{\sigma}^{B_{r+s}} \cdot \mathbf{X}_{A_{r+s}}). \tag{2.4.21}$$

If all the $\mathbf{Y}$'s and $\boldsymbol{\sigma}$'s are pure then the tensor product is pure. It is *c*-type or *a*-type according as the number of *a*-type $\mathbf{Y}$'s and $\boldsymbol{\sigma}$'s is even or odd.

Every tensor can be expressed in the form

$$\mathbf{T} = (-1)^{\Delta_{r+s}(a)} T^{a_1 \ldots a_r}{}_{a_{r+1} \ldots a_{r+s}} {}_{a_1}\mathbf{e} \otimes \cdots \otimes {}^{a_{r+s}}\mathbf{e}. \tag{2.4.22}$$

The set of tensors ${}_{a_1}\mathbf{e} \otimes \cdots \otimes {}^{a_{r+s}}\mathbf{e}$ evidently forms a basis for $T^r{}_s(p)$, and it is not difficult to see that $T^r{}_s(p)$ has dimension (P, Q) where

$$\left. \begin{aligned} P &= \sum_{l=0}^{\infty} \frac{(r+s)!}{(2l)!\,(r+s-2l)!} n^{2l} m^{r+s-2l}, \\ Q &= \sum_{l=0}^{\infty} \frac{(r+s)!}{(2l+1)!(r+s-2l-1)!} n^{2l+1} m^{r+s-2l-1}, \end{aligned} \right\} \tag{2.4.23}$$

$$P + Q = (m+n)^{r+s}. \tag{2.4.24}$$

If $r = 0$, $\mathbf{T}$ is called a *rank-s covariant tensor*. If $s = 0$, it is called a *rank-r contravariant tensor*. A scalar field evaluated at p may be regarded as a tensor at p for which both r and s vanish. If neither r nor s vanishes $\mathbf{T}$ is called a *mixed tensor*. Unless a tensor has special symmetry properties the order of the indices on its components cannot be changed. It is not necessary, however, to have all the upper indices standing to the left of the lower indices as above. The preceding definitions can be generalized in an obvious way to tensors defined as multilinear mappings from $\Pi_r{}^s(p) \times \Pi_t{}^u(p) \times \cdots$ to Λ_∞. An example of this more general type is the *tensor product* of a tensor **R** of rank (r, s) with a tensor **T** of rank (t, u), which, in the pure case, is defined by

$$(\mathbf{R} \otimes \mathbf{T})^{a_1 \ldots a_r}{}_{a_{r+1} \ldots a_{r+s}}{}^{b_1 \ldots b_t}{}_{b_{t+1} \ldots b_{t+u}}$$
$$\stackrel{\text{def}}{=} (-1)^{\mathbf{T}(a_1 + \ldots + a_{r+s})} R^{a_1 \ldots a_r}{}_{a_{r+1} \ldots a_{r+s}} T^{b_1 \ldots b_t}{}_{b_{t+1} \ldots b_{t+u}}, \tag{2.4.25}$$

and, in the impure case, by 2.4.25 together with

$$\mathbf{R} \otimes (\mathbf{T} + \mathbf{U}) \stackrel{\text{def}}{=} \mathbf{R} \otimes \mathbf{T} + \mathbf{R} \otimes \mathbf{U}, \quad (\mathbf{R} + \mathbf{S}) \otimes \mathbf{T} \stackrel{\text{def}}{=} \mathbf{R} \otimes \mathbf{T} + \mathbf{S} \otimes \mathbf{T}, \tag{2.4.26}$$

S and **U** having the same ranks as **R** and **T** respectively.

It is easy to verify that tensor-product multiplication, defined in this way, satisfies the laws

$$(\mathbf{R} \otimes \mathbf{T}) \otimes \mathbf{V} = \mathbf{R} \otimes (\mathbf{T} \otimes \mathbf{V}) \stackrel{\text{def}}{=} \mathbf{R} \otimes \mathbf{T} \otimes \mathbf{V}, \tag{2.4.27}$$

$$\alpha(\mathbf{R} \otimes \mathbf{T}) = (\alpha \mathbf{R}) \otimes \mathbf{T} \stackrel{\text{def}}{=} \alpha \mathbf{R} \otimes \mathbf{T}, \tag{2.4.28}$$

$$(\mathbf{R}\alpha) \otimes \mathbf{T} = \mathbf{R} \otimes (\alpha \mathbf{T}) \stackrel{\text{def}}{=} \mathbf{R}\alpha \otimes \mathbf{T} \stackrel{\text{def}}{=} \mathbf{R} \otimes \alpha \mathbf{T}, \tag{2.4.29}$$

$$(\mathbf{R} \otimes \mathbf{T})\alpha = \mathbf{R} \otimes (\mathbf{T}\alpha) \stackrel{\text{def}}{=} \mathbf{R} \otimes \mathbf{T}\alpha. \tag{2.4.30}$$

The tensor product of two pure tensors is pure, being of type c or type a according as the two are of the same type or of opposite type. The tensor product of two pure real tensors is real unless both are of type a, in which case the tensor product is imaginary.

Tensor and multitensor fields

A *tensor field* is an assignment of a tensor (of fixed rank) at each point of a supermanifold in such a way that its components with respect to every field of local frames are differentiable throughout the domain in which the field of local frames is defined. The set of all rank-(r, s) tensor fields over a supermanifold M will be denoted by $\mathcal{T}^r{}_s(M)$. This set has an obvious structure as a supervector space because all the equations of this section that involve tensors can be extended immediately, without change, to tensor fields.

An additional convention should be noted, analogous to that expressed by eq. (2.2.46). In writing the tensor product of two tensor fields one customarily omits the symbol '⊗' when one of the tensor fields has rank (0, 0), i.e., is a scalar field. Equations (2.4.28)–(2.4.30), for example, therefore continue to hold if the supernumber α is replaced by a scalar field f.

There is a generalization of the tensor field concept that is of some importance in practical applications. Instead of considering multilinear mappings from $\Pi_r{}^s(p) \times \Pi_t{}^u(p) \times \cdots$ to Λ_∞ one may take them from $\Pi_r{}^s(p) \times \Pi_t{}^u(p') \times \cdots$ to Λ_∞, where the points associated with successive factors may be distinct: $p, p', \ldots$. The rules for introducing components, changing bases, etc. are completely analogous to those for tensors. However, the components of these more general mappings bear indices that separate into groups, one group for each point. It is convenient to distinguish the various groups by putting primes on some of the indices. Thus the components of a mapping **T** now take the form

$$T^{a_1 \ldots a_r}{}_{a_{r+1} \ldots a_{r+s}}{}^{b'_1 \ldots b'_t}{}_{b'_{t+1} \ldots b'_{t+u}} \cdots \tag{2.4.31}$$

Suppose N points of the supermanifold are involved. An *N-point tensor field* is an assignment of one of these multilinear mappings to every N-uple of points in the supermanifold in such a way that its components in every N-uple of local frame fields are differentiable. Note that different points may lie in different frame domains. Thus indices bearing different numbers of primes do not have to refer to the same local frame field or, if coordinate bases are being used, to the same coordinate system.

Two-point, three-point, four-point, etc. tensor fields will be referred to collectively as *multitensor fields*. Two- and three-point tensor fields are often called *bitensor* and *tritensor* fields respectively. Important examples of bitensor fields are the auxiliary functions of super Lie groups (see section 3.1) and the covariant Green's functions of quantum field theory. Examples of multitensor fields with $N \geq 3$ are the vertex functions (actually distributions) of quantum field theory.

Index-shifting conventions. Contractions

It is not necessary to confine one's attention to tensor components for which all indices appear on the right. Any right index may be converted into a left index by rules analogous to (1.5.14) and (1.5.44). Here, however, one runs into notational problems. When several indices are present, their order may be violated when some, but not all, are shifted from right to left, unless they are shifted sequentially. One could add extra symbols to indicate the character (left or right) of each index, but in most instances it is not worth the bother. With only an occasional exception, we shall keep all the indices on the right when the *total rank*, $r + s$, of the tensor field is greater than one. The exceptions are those in which the leftmost index gets shifted according to the rules

$$\left.\begin{aligned} {}^{a}T^{b\ldots}{}_{c\ldots} &= (-1)^{a\mathbf{T}}\, T^{ab\ldots}{}_{c\ldots}, \\ {}_{a}T^{b\ldots}{}_{c\ldots} &= (-1)^{a(\mathbf{T}+1)}\, T_{a}{}^{b\ldots}{}_{c\ldots}, \text{ etc} \end{aligned}\right\} \tag{2.4.32}$$

From any mixed tensor field another tensor field may be built by the process of *contraction*, which, in terms of components, consists of setting an upper index equal to a lower index and summing, with appropriate factors of -1 thrown in. For example, starting with a rank-(r, s) tensor field $\mathbf{T}$ one may contract the pth upper index with the $(q-r)$th lower index to get a tensor field of rank $(r-1, s-1)$ having the components

$$(-1)^{a_q(1 + a_{p+1} + \ldots + a_{q-1})}\, T^{a_1 \ldots a_{p-1} a_q a_{p+1} \ldots a_r}{}_{a_{r+1} \ldots a_q \ldots a_{r+s}}. \tag{2.4.33}$$

It is a straightforward but somewhat tedious computation to show that under the change of basis (2.4.11) these components transform correctly. It will be noted that if $r = s = 1$ expression (2.4.33) becomes the supertrace $(-1)^a T^a{}_a$.

Expression (2.4.33) provides an example of a case in which it is easier to define a tensor field by its components than to define it in a basis-independent way. In physics, working in specific bases (usually coordinate bases) is an almost universal custom, and, by an abuse of strict logic, the components of tensor fields are often regarded as representing the tensor fields themselves. As long as one is careful to work with expressions that are manifestly covariant (i.e., form invariant under changes of basis) there is no harm in doing this. Use of specific bases in these circumstances does not yield basis-dependent results. The indices may, in fact, be regarded simply as devices for indicating rank and the combinatorics of contractions.

The unit tensor field

There are two rank-(1, 1) tensor fields that exist over any supermanifold. They are defined by

$$\boldsymbol{\delta} \overset{\text{def}}{=} \mathbf{e}_a \otimes {}^a\mathbf{e}, \quad \boldsymbol{\delta}^{\sim} \overset{\text{def}}{=} \mathbf{e}^a \otimes {}_a\mathbf{e}, \tag{2.4.34}$$

where $\{\mathbf{e}_a\}$ is any local frame field, with dual $\{\mathbf{e}^a\}$. It is easy to see that $\boldsymbol{\delta}$ and $\boldsymbol{\delta}^{\sim}$ are independent of the choice of **e**'s. Their components are

$$\delta^{i}{}_{j} = (-1)^{ia} e_{a}{}^{i}\,{}^{a}e_{j} = {}^{i}e_{a}\,{}^{a}e_{j} = {}^{i}\delta_{j}, \tag{2.4.35}$$

$$\delta^{\sim}{}_{i}{}^{j} = (-1)^{ia} e^{a}{}_{i}\,{}_{a}e^{j} = (-1)^{i}\,{}_{i}e^{a}\,{}_{a}e^{j} = (-1)^{i}\,{}_{i}\delta^{j}. \tag{2.4.36}$$

In component form $\boldsymbol{\delta}$ and $\boldsymbol{\delta}^{\sim}$ are the supertransposes of one another. As matrices they are identical to the unit matrix. It is customary to drop the '$^{\sim}$' on $\boldsymbol{\delta}^{\sim}$ and to call both simply the *unit tensor field*.

2.5 The Lie derivative

Definition

Denote by $\mathscr{T}(M)$ the set of all tensor fields (of all ranks) over a supermanifold M. For every $\mathbf{X}$ in $\mathfrak{X}(M)$ let $\mathfrak{L}_{\mathbf{X}}$ be the rank-preserving mapping from $\mathscr{T}(M)$ to itself defined by

$$\mathfrak{L}_{\mathbf{X}} f \overset{\text{def}}{=} \mathbf{X} f, \tag{2.5.1}$$

$$\mathfrak{L}_{\mathbf{X}}(\mathbf{Y} f) \overset{\text{def}}{=} (\mathfrak{L}_{\mathbf{X}}\mathbf{Y}) f + (-1)^{\mathbf{XY}}\mathbf{Y}(\mathfrak{L}_{\mathbf{X}} f), \tag{2.5.2}$$

$$\mathfrak{L}_{\mathbf{X}}(\mathbf{Y}\cdot\omega) \overset{\text{def}}{=} (\mathfrak{L}_{\mathbf{X}}\mathbf{Y})\cdot\omega + (-1)^{\mathbf{XY}}\mathbf{Y}\cdot(\mathfrak{L}_{\mathbf{X}}\omega), \tag{2.5.3}$$

$$\begin{aligned} \mathfrak{L}_{\mathbf{X}}[\mathbf{T}(\omega,\ldots,\mathbf{Y})] &\overset{\text{def}}{=} (\mathfrak{L}_{\mathbf{X}}\mathbf{T})(\omega,\ldots,\mathbf{Y}) + (-1)^{\mathbf{XT}}\mathbf{T}(\mathfrak{L}_{\mathbf{X}}\omega,\ldots,\mathbf{Y}) \\ &\quad + \cdots + (-1)^{\mathbf{X}(\mathbf{T}+\omega+\cdots)}\mathbf{T}(\omega,\ldots,\mathfrak{L}_{\mathbf{X}}\mathbf{Y}), \end{aligned} \tag{2.5.4}$$

$$\mathfrak{L}_{\mathbf{X}}(\mathbf{R}\otimes\mathbf{T}) \overset{\text{def}}{=} (\mathfrak{L}_{\mathbf{X}}\mathbf{R})\otimes\mathbf{T} + (-1)^{\mathbf{XR}}\mathbf{R}\otimes(\mathfrak{L}_{\mathbf{X}}\mathbf{T}), \tag{2.5.5}$$

for all f in $\mathfrak{F}(M)$, all $\mathbf{Y}$ in $\mathfrak{X}(M)$, all ω in $\boldsymbol{\Omega}(M)$, and all $\mathbf{R}$, $\mathbf{T}$ in $\mathscr{T}(M)$. The mapping $\mathfrak{L}_{\mathbf{X}}$ is called a *Lie derivation* and $\mathfrak{L}_{\mathbf{X}}\mathbf{T}$ is called the *Lie derivative of* $\mathbf{T}$ *with respect to* $\mathbf{X}$.

Explicit forms

Equation (2.5.1) defines the Lie derivative of a scalar field. Equation (2.5.2) may be rewritten in the form

$$(\mathfrak{L}_{\mathbf{X}}\mathbf{Y}) f = \mathbf{XY} f - (-1)^{\mathbf{XY}}\mathbf{YX} f = [\mathbf{X},\mathbf{Y}]_{\mathrm{L}} f, \tag{2.5.6}$$

which yields the Lie derivative of a contravariant vector field:

$$\mathfrak{L}_{\mathbf{X}}\mathbf{Y} = [\mathbf{X},\mathbf{Y}]_{\mathrm{L}} = -[\mathbf{X},\mathbf{Y}]_{\mathrm{R}} = -[\mathbf{X},\mathbf{Y}]. \tag{2.5.7}$$

The explicit form for the Lie derivative of a covariant vector field is best presented in a coordinate basis. It is obtainable from eq. (2.5.3) which, with the aid of eqs. (2.2.35), (2.3.18) and (2.5.7), may be recast in the form

$$\begin{aligned} Y^i{}_{;i}(\mathfrak{L}_{\mathbf{X}}\omega) &= (-1)^{\mathbf{XY}}\, X^j{}_{,j,}(Y^i{}_{;i}\omega) + (-1)^{\mathbf{XY}}[-X^j{}_{,j,}Y^i + (-1)^{\mathbf{XY}}\, \mathbf{Y}^j{}_{;j}\mathbf{X}^i]_{;i}\omega \\ &= Y^i[{}_{i,}X^j{}_{;j}\omega + (-1)^{\mathbf{XY}+j(\mathbf{Y}+i)+(\mathbf{X}+j)(\mathbf{Y}+i)}\, X^j{}_{,j,i}\omega]. \end{aligned} \tag{2.5.8}$$

Since the Y^i are arbitrary it follows that

$${}_{;i}(\mathfrak{L}_{\mathbf{X}}\omega) = {}_{i,}X^j{}_{;j}\omega + (-1)^{i\mathbf{X}}X^j{}_{,j,i}\omega. \tag{2.5.9}$$

With these results in hand one can use eqs. (2.5.4) and (2.5.5) to compute the Lie derivative of an arbitrary tensor field.

Lie derivations as supervectors

The set of all Lie derivations on $\mathscr{T}(M)$ may be given the structure of a supervector space by defining

$$(\mathfrak{L}_{\mathbf{X}} + \mathfrak{L}_{\mathbf{Y}})\mathbf{T} \stackrel{\text{def}}{=} \mathfrak{L}_{\mathbf{X}}\mathbf{T} + \mathfrak{L}_{\mathbf{Y}}\mathbf{T}, \tag{2.5.10}$$

$$(\alpha\mathfrak{L}_{\mathbf{X}})\mathbf{T} \stackrel{\text{def}}{=} \alpha(\mathfrak{L}_{\mathbf{X}}\mathbf{T}), \tag{2.5.11}$$

$$(\mathfrak{L}_{\mathbf{X}}\alpha)\mathbf{T} \stackrel{\text{def}}{=} \mathfrak{L}_{\mathbf{X}}(\alpha\mathbf{T}), \tag{2.5.12}$$

for all $\mathbf{X}, \mathbf{Y}$ in $\mathfrak{X}(M)$, all α in Λ_∞ and all $\mathbf{T}$ in $\mathscr{T}(M)$. The zero Lie derivation is that which maps every tensor field into the zero tensor field of the same rank. The negative of a Lie derivation $\mathfrak{L}_{\mathbf{X}}$ is defined by $(-\mathfrak{L}_{\mathbf{X}})\mathbf{T} = -(\mathfrak{L}_{\mathbf{X}}\mathbf{T})$. The complex conjugate of a Lie derivation is defined by

$$\mathfrak{L}^*_{\mathbf{X}}\mathbf{T} = (-1)^{\mathbf{XT}}(\mathfrak{L}_{\mathbf{X}}\mathbf{T}^*)^*, \tag{2.5.13}$$

($\mathbf{X}, \mathbf{T}$ pure), and so on.

Definitions (2.5.10)–(2.5.12) are what allow eqs. (2.5.1)–(2.5.5) to be extended to the impure cases. To see this note that

$$(\mathfrak{L}_{\mathbf{X}} + \mathfrak{L}_{\mathbf{Y}})f = \mathbf{X}f + \mathbf{Y}f = (\mathbf{X}+\mathbf{Y})f = \mathfrak{L}_{\mathbf{X}+\mathbf{Y}}f, \tag{2.5.14}$$

$$\begin{aligned} (\mathfrak{L}_{\mathbf{X}} + \mathfrak{L}_{\mathbf{Y}})\mathbf{Z} &= \mathfrak{L}_{\mathbf{X}}\mathbf{Z} + \mathfrak{L}_{\mathbf{Y}}\mathbf{Z} = -[\mathbf{X},\mathbf{Z}] - [\mathbf{Y},\mathbf{Z}] \\ &= -[\mathbf{X}+\mathbf{Y},\mathbf{Z}] = \mathfrak{L}_{\mathbf{X}+\mathbf{Y}}\mathbf{Z}, \end{aligned} \tag{2.5.15}$$

whence

$$\begin{aligned} [(\mathfrak{L}_{\mathbf{X}} + \mathfrak{L}_{\mathbf{Y}})\omega]\cdot\mathbf{Z} &= (\mathfrak{L}_{\mathbf{X}} + \mathfrak{L}_{\mathbf{Y}})(\omega\cdot\mathbf{Z}) - (-1)^{\omega\mathbf{Z}}[(\mathfrak{L}_{\mathbf{X}} + \mathfrak{L}_{\mathbf{Y}})\mathbf{Z}]\cdot\omega \\ &= \mathfrak{L}_{\mathbf{X}+\mathbf{Y}}(\omega\cdot\mathbf{Z}) - (-1)^{\omega\mathbf{Z}}(\mathfrak{L}_{\mathbf{X}+\mathbf{Y}}\mathbf{Z})\cdot\omega \\ &= (\mathfrak{L}_{\mathbf{X}+\mathbf{Y}}\omega)\cdot\mathbf{Z}, \end{aligned} \tag{2.5.16}$$

for all $\mathbf{X}, \mathbf{Y}, \mathbf{Z}$ in $\mathfrak{X}(M)$ and all ω in $\Omega(M)$. Equation (2.5.16) implies

$$(\mathfrak{L}_{\mathbf{X}} + \mathfrak{L}_{\mathbf{Y}})\omega = \mathfrak{L}_{\mathbf{X}+\mathbf{Y}}\omega, \tag{2.5.17}$$

which, together with eqs. (2.5.4), (2.5.5), (2.5.14) and (2.5.15), allows one to infer quite generally that $(\mathfrak{L}_{\mathbf{X}} + \mathfrak{L}_{\mathbf{Y}})\mathbf{T} = \mathfrak{L}_{\mathbf{X}+\mathbf{Y}}\mathbf{T}$, and hence

$$\mathfrak{L}_{\mathbf{X}} + \mathfrak{L}_{\mathbf{Y}} = \mathfrak{L}_{\mathbf{X}+\mathbf{Y}}, \tag{2.5.18}$$

for all $\mathbf{X}, \mathbf{Y}$ in $\mathfrak{X}(M)$. In a similar manner one can verify that

$$\alpha\mathfrak{L}_{\mathbf{X}} = \mathfrak{L}_{\alpha\mathbf{X}}, \tag{2.5.19}$$

$$\mathfrak{L}_{\mathbf{X}}\alpha = \mathfrak{L}_{\mathbf{X}\alpha}, \tag{2.5.20}$$

$$\mathfrak{L}^*_{\mathbf{X}} = \mathfrak{L}_{\mathbf{X}^*}, \tag{2.5.21}$$

for all $\mathbf{X}$ in $\mathfrak{X}(M)$ and all α in Λ_∞. Finally, it is straightforward to show that the even and odd parts of $\mathfrak{L}_{\mathbf{X}}$ are $\mathfrak{L}_{\mathbf{U}}$ and $\mathfrak{L}_{\mathbf{V}}$ respectively, where $\mathbf{U}$ and $\mathbf{V}$ are the even and odd parts of $\mathbf{X}$.

Evidently the set of all Lie derivations on $\mathcal{T}(M)$ is isomorphic, as a supervector space, to $\mathfrak{X}(M)$. The isomorphism extends even to the super Lie bracket. The product (i.e., successive application) of two Lie derivations is not generally a Lie derivation, but the supercommutator is. In defining the supercommutator of two Lie derivations one has the same choice as in defining the supercommutator of two contravariant vector fields, whether to regard the derivations as acting from the left, as above, or from the right according to the rule

$$\mathbf{T}\mathfrak{L}_{\mathbf{X}} \overset{\text{def}}{=} (-1)^{\mathbf{XT}}\,\mathfrak{L}_{\mathbf{X}}\mathbf{T}. \tag{2.5.22}$$

Using the right-hand convention we have

$$\begin{aligned}[\mathfrak{L}_{\mathbf{X}}, \mathfrak{L}_{\mathbf{Y}}]\mathbf{T} &= [\mathfrak{L}_{\mathbf{X}}, \mathfrak{L}_{\mathbf{Y}}]_{\mathrm{R}}\,\mathbf{T} = -[\mathfrak{L}_{\mathbf{X}}, \mathfrak{L}_{\mathbf{Y}}]_{\mathrm{L}}\,\mathbf{T}\\ &= -\mathfrak{L}_{\mathbf{X}}\mathfrak{L}_{\mathbf{Y}}\mathbf{T} + (-1)^{\mathbf{XY}}\mathfrak{L}_{\mathbf{Y}}\mathfrak{L}_{\mathbf{X}}\mathbf{T},\end{aligned} \tag{2.5.23}$$

and hence

$$[\mathfrak{L}_{\mathbf{X}}, \mathfrak{L}_{\mathbf{Y}}]f = [\mathbf{X}, \mathbf{Y}]f = \mathfrak{L}_{[\mathbf{X},\mathbf{Y}]}f, \tag{2.5.24}$$

$$\begin{aligned}[\mathfrak{L}_{\mathbf{X}}, \mathfrak{L}_{\mathbf{Y}}]\mathbf{Z} &= -[\mathbf{X},[\mathbf{Y},\mathbf{Z}]] + (-1)^{\mathbf{XY}}[\mathbf{Y},[\mathbf{X},\mathbf{Z}]].\\ &= -[[\mathbf{X},\mathbf{Y}],\mathbf{Z}] = \mathfrak{L}_{[\mathbf{X},\mathbf{Y}]}\mathbf{Z},\end{aligned} \tag{2.5.25}$$

$$\begin{aligned}&(\mathfrak{L}_{[\mathbf{X},\mathbf{Y}]}\mathbf{Z})\cdot\omega + (-1)^{\mathbf{Z}\omega}(\mathfrak{L}_{[\mathbf{X},\mathbf{Y}]}\omega)\cdot\mathbf{Z}\\ &\quad = \mathfrak{L}_{[\mathbf{X},\mathbf{Y}]}(\mathbf{Z}\cdot\omega) = [\mathfrak{L}_{\mathbf{X}}, \mathfrak{L}_{\mathbf{Y}}](\mathbf{Z}\cdot\omega)\\ &\quad = -\mathfrak{L}_{\mathbf{X}}[(\mathfrak{L}_{\mathbf{Y}}\mathbf{Z})\cdot\omega + (-1)^{\mathbf{Z}\omega}(\mathfrak{L}_{\mathbf{Y}}\omega)\cdot\mathbf{Z}]\\ &\qquad + (-1)^{\mathbf{XY}}\mathfrak{L}_{\mathbf{Y}}[(\mathfrak{L}_{\mathbf{X}}\mathbf{Z})\cdot\omega + (-1)^{\mathbf{Z}\omega}(\mathfrak{L}_{\mathbf{X}}\omega)\cdot\mathbf{Z}]\\ &\quad = ([\mathfrak{L}_{\mathbf{X}}, \mathfrak{L}_{\mathbf{Y}}]\mathbf{Z})\cdot\omega + (-1)^{\mathbf{Z}\omega}([\mathfrak{L}_{\mathbf{X}}, \mathfrak{L}_{\mathbf{Y}}]\omega)\cdot\mathbf{Z}\\ &\qquad - (-1)^{\omega(\mathbf{Y}+\mathbf{Z})}(\mathfrak{L}_{\mathbf{X}}\omega)\cdot(\mathfrak{L}_{\mathbf{Y}}\mathbf{Z}) - (-1)^{\mathbf{YZ}}(\mathfrak{L}_{\mathbf{X}}\mathbf{Z})\cdot(\mathfrak{L}_{\mathbf{Y}}\omega)\\ &\qquad + (-1)^{\mathbf{YZ}}(\mathfrak{L}_{\mathbf{X}}\mathbf{Z})\cdot(\mathfrak{L}_{\mathbf{Y}}\omega) + (-1)^{\omega(\mathbf{Y}+\mathbf{Z})}(\mathfrak{L}_{\mathbf{X}}\omega)\cdot(\mathfrak{L}_{\mathbf{Y}}\mathbf{Z}).\end{aligned} \tag{2.5.26}$$

The cancellation of the last two lines and the arbitrariness of **Z** allows one to infer

$$[\mathfrak{L}_{\mathbf{X}}, \mathfrak{L}_{\mathbf{Y}}]\omega = \mathfrak{L}_{[\mathbf{X},\mathbf{Y}]}\omega. \tag{2.5.27}$$

Continuing in this way with eqs. (2.5.4) and (2.5.5) one obtains, quite generally,

$$[\mathfrak{L}_{\mathbf{X}}, \mathfrak{L}_{\mathbf{Y}}] = \mathfrak{L}_{[\mathbf{X},\mathbf{Y}]}. \tag{2.5.28}$$

The derivative mapping

Let $\phi: M \to \bar{M}$ be a diffeomorphism between two supermanifolds M and $\bar{M}$. Denote by ϕ' the rank-preserving mapping from $\mathcal{T}(M)$ to $\mathcal{T}(\bar{M})$ defined by

$$\phi'(f) \overset{\text{def}}{=} f \circ \phi^{-1}, \tag{2.5.29}$$

$$\phi'(\mathbf{X}f) \overset{\text{def}}{=} \phi'(\mathbf{X})\phi'(f), \tag{2.5.30}$$

$$\phi'(\mathbf{X}\cdot\omega) \overset{\text{def}}{=} \phi'(\mathbf{X})\cdot\phi'(\omega), \tag{2.5.31}$$

$$\phi'(\mathbf{T}(\omega, \ldots \mathbf{X})) \overset{\text{def}}{=} \phi'(\mathbf{T})(\phi'(\omega), \ldots, \phi'(\mathbf{X})), \tag{2.5.32}$$

$$\phi'(\mathbf{R} \otimes \mathbf{T}) \overset{\text{def}}{=} \phi'(\mathbf{R}) \otimes \phi'(\mathbf{T}), \tag{2.5.33}$$

for all f in $\mathfrak{F}(M)$, all $\mathbf{X}$ in $\mathfrak{X}(M)$, all ω in $\Omega(M)$ and all $\mathbf{R}, \mathbf{T}$ in $\mathcal{T}(M)$. The mapping ϕ' is called the *derivative mapping* corresponding to ϕ. For every set of tensor fields on M the derivative mapping induces an isomorphic set on $\bar{M}$ which preserves all tensor operations, such as contractions, inner products, super Lie brackets, etc.†

One is often interested in the case $\bar{M} = M$ so that ϕ is a diffeomorphism from M to itself. The derivative mapping then transforms every set of tensor fields on M to an isomorphic set on M. Note that ϕ need not be the identity mapping in this case.

Integral supercurves. Congruences

Let $\mathbf{X}$ be a real c-type contravariant vector field on M and let p be an arbitrary point of M. Denote by $\lambda_{\mathbf{X},p}$ the c-type supercurve in M defined by

† It is obvious that a derivative mapping can also be defined for multitensor fields, but we shall have no occasion to exploit this possibility. For the derivative mapping of connections (section 2.7) and measure functions (section 2.9) see exercises **2.3** and **2.4**.

the equation[†]

$$(\partial/\partial s)_{\lambda_{\mathbf{X},p}(s)} = \mathbf{X}_{\lambda_{\mathbf{X},p}(s)} \tag{2.5.34}$$

and the boundary condition

$$\lambda_{\mathbf{X},p}(0) = p. \tag{2.5.35}$$

The supercurve $\lambda_{\mathbf{X},p}$ is called an *integral supercurve of* **X**, and the set of all such λ's is called the *congruence of supercurves in M generated by* **X**.

Without serious loss of generality for applications we may confine our attention to the case in which **X** has compact support in M. Equations (2.5.34) and (2.5.35) are then soluble for all s in $\mathbf{R}_c$ and we may introduce a set of mappings $X_s: M \to M$ defined by

$$X_s(p) \stackrel{\text{def}}{=} \lambda_{\mathbf{X},p}(s), \tag{2.5.36}$$

for all p in M and all s in $\mathbf{R}_c$. Note that X_0 is the identity mapping:

$$X_0(p) = p. \tag{2.5.37}$$

Note also that

$$X_s \circ X_t(p) = X_s(X_t(p)) = \lambda_{\mathbf{X},X_t(p)}(s), \tag{2.5.38}$$

$$X_{s+t}(p) = \lambda_{\mathbf{X},p}(s+t). \tag{2.5.39}$$

The right-hand sides of eqs. (2.5.38) and (2.5.39) coincide at $s = 0$:

$$\lambda_{\mathbf{X},X_t(p)}(0) = X_t(p) = \lambda_{\mathbf{X},p}(t). \tag{2.5.40}$$

Since $\lambda_{\mathbf{X},X_t(p)}$ and $\lambda_{\mathbf{X},p}$ are integral supercurves of the same contravariant vector field the right-hand sides of eqs. (2.5.38) and (2.5.39) must, in fact, coincide for all s. Hence

$$X_s \circ X_t = X_{s+t}, \tag{2.5.41}$$

which, together with (2.5.37), implies that the mapping X_s is invertible, and hence one-to-one, for all s in $\mathbf{R}_c$:

$$X_s^{-1} = X_{-s}. \tag{2.5.42}$$

Dragging of tensor fields

Because **X** is differentiable the mappings X_s and X_{-s} can be shown to be differentiable. The mapping X_s is therefore a diffeomorphism and one can

[†] Letting both sides of this equation act on the coordinate functions x^i of a chart containing $\lambda_{\mathbf{X},p}(s)$, and using eqs. (2.2.34) and (2.2.47), one can convert it to an explicit differential equation:

$$\frac{\mathrm{d}}{\mathrm{d}s} x^i(\lambda_{\mathbf{X},p}(s)) = X^i(\lambda_{\mathbf{X},p}(s)).$$

introduce the derivative mapping $X_s': \mathscr{T}(M) \to \mathscr{T}(M)$. The derivative mapping will be called a *dragging*, and under the operation $\mathbf{T} \mapsto X_s'(\mathbf{T})$ the tensor field **T** will be said to be *dragged* by the contravariant vector field **X**.

By differentiating X_s' with respect to s and setting $s = 0$, one obtains the *infinitesimal dragging operator*. Equations (2.5.29) and (2.5.42) yield

$$\left[\frac{\mathrm{d}}{\mathrm{d}s} X_s'(f)\right]_{s=0} = \left[\frac{\mathrm{d}}{\mathrm{d}s}(f \circ X_{-s})\right]_{s=0}. \tag{2.5.43}$$

Evaluating this at p and using eqs. (2.2.48) and (2.5.34) one obtains

$$\begin{aligned}\left[\frac{\mathrm{d}}{\mathrm{d}s} X_s'(f)\right]_{s=0}(p) &= \left[\frac{\mathrm{d}}{\mathrm{d}s} f(X_{-s}(p))\right]_{s=0} = -\left[\frac{\mathrm{d}}{\mathrm{d}s} f(X_s(p))\right]_{s=0} \\ &= -\left[\frac{\mathrm{d}}{\mathrm{d}s} f(\lambda_{\mathbf{X},p}(s))\right]_{s=0} = -\left[(\partial/\partial s)_{\lambda_{\mathbf{X},p}(s)} f\right]_{s=0} \\ &= -\mathbf{X}_{\lambda_{\mathbf{X},p}(0)} f = -\mathbf{X}_p f = -(\mathbf{X}f)(p),\end{aligned} \tag{2.5.44}$$

and hence

$$\left[\frac{\mathrm{d}}{\mathrm{d}s} X_s'(f)\right]_{s=0} = -\mathbf{X}f. \tag{2.5.45}$$

If, in eqs. (2.5.29)–(2.5.33), one sets $\phi' = X_s'$, differentiates with respect to s, and sets $s = 0$, one obtains a sequence of equations identical in structure to eqs. (2.5.1)–(2.5.5) but with **X** replaced by $-\mathbf{X}$ on the right of the first equation. From this it follows that

$$\left(\frac{\mathrm{d}}{\mathrm{d}s} X_s'\right)_{s=0} = \mathfrak{L}_{-\mathbf{X}} = -\mathfrak{L}_{\mathbf{X}}. \tag{2.5.46}$$

That is, the infinitesimal dragging operator is just the negative of the Lie derivative with respect to **X**. Lie derivation, as defined at the beginning of this section, generalizes the infinitesimal dragging mapping to the case in which **X** is impure and complex, not merely c-type and real.

2.6 Forms

Definition

Let **A** be a rank-r covariant tensor field on a supermanifold M, with components $A_{a_1 \ldots a_r}$ in some pure basis. Suppose the components obey the symmetry law

$$A_{\dots bc \dots} = -(-1)^{bc} A_{\dots cb \dots}. \tag{2.6.1}$$

for pairs of adjacent indices. Since any permutation of indices can be reached through interchange of adjacent indices, the behavior of the components under arbitrary index permutations is completely determined by (2.6.1). It is easy to verify that this law is invariant under changes of basis and hence describes a symmetry of the tensor field itself. Any rank-r covariant tensor field having this symmetry is called a *differential form of degree* r, or, more briefly, an r-*form*. Any covariant vector field, as we have already seen (section 2.3), is a 1-form. It will be convenient in this section to call scalar fields 0-*forms*.

If M has no a-number coordinates, i.e., if $n = 0$, then r-forms become trivial (i.e., they vanish) when $r > n$. If $n \neq 0$, then nontrivial forms of arbitrarily high degree exist.

The exterior product

Let $\nu(a_{k_1}, \dots, a_{k_r})$ be the number (mod 2) of sign changes that occur when the component $A_{a_{k_1} \dots a_{k_r}}$ is reached by permutation of adjacent indices from the component $A_{a_1 \dots a_r}$ of an r-form $\mathbf{A}$. The *exterior product* $\mathbf{A} \wedge \mathbf{B}$ of an r-form $\mathbf{A}$ and an s-form $\mathbf{B}$ is defined to be the differential form of degree $r + s$ having the components

$$(\mathbf{A} \wedge \mathbf{B})_{a_1 \dots a_{r+s}} \stackrel{\text{def}}{=} \frac{1}{r!s!} \sum_{perm} (-1)^{\nu(a_{k_1}, \dots, a_{k_{r+s}}) + \mathbf{B}(a_{k1} + \cdots + a_{kr})} \times A_{a_{k_1} \dots a_{k_r}} B_{a_{k_{r+1}} \dots a_{k_{r+s}}}, \tag{2.6.2}$$

where the sum is over the $(r+s)!$ permutations $a_{k_1}, \dots, a_{k_{r+s}}$ of the indices $a_1, \dots, a_{r+s}$. It is straightforward to verify that the definition (2.6.2) is invariant under changes of basis and obeys the laws

$$\mathbf{A} \wedge \mathbf{B} = (-1)^{\mathbf{AB} + rs} \mathbf{B} \wedge \mathbf{A}, \tag{2.6.3}$$

$$\mathbf{A} \wedge (\mathbf{B} \wedge \mathbf{C}) = (\mathbf{A} \wedge \mathbf{B}) \wedge \mathbf{C} \stackrel{\text{def}}{=} \mathbf{A} \wedge \mathbf{B} \wedge \mathbf{C}, \tag{2.6.4}$$

$$\mathbf{A} \wedge (\mathbf{B} + \mathbf{C}) = \mathbf{A} \wedge \mathbf{B} + \mathbf{A} \wedge \mathbf{C}. \tag{2.6.5}$$

If either of the two factors in an exterior product is a 0-form it is customary to omit the symbol '$\wedge$':

$$f \wedge \mathbf{A} = f\mathbf{A}, \qquad \mathbf{A} \wedge f = \mathbf{A}f. \tag{2.6.6}$$

Bases for forms

The set of all r-forms at a point is a supervector space of dimension (P, Q) where

$$P = \sum_{s=0}^{\infty} \frac{(2s+n-1)!}{(2s)!(n-1)!} \frac{m!}{(r-2s)!(m-r+2s)!},$$

$$Q = \sum_{s=0}^{\infty} \frac{(2s+n)!}{(2s+1)!(n-1)!} \frac{m!}{(r-2s-1)!(m-r+2s+1)!}. \tag{2.6.7}$$

If $\{{}^a\mathbf{e}\}$ is a pure covariant basis then the $P+Q$ linearly independent r-forms ${}^{a_1}\mathbf{e} \wedge \cdots \wedge {}^{a_r}\mathbf{e}$ at a point constitute a pure r-form basis at that point. Explicitly (cf. eq. (2.4.24))

$$\mathbf{A} = \frac{1}{r!}(-1)^{\Delta_r(a)} A_{a_1 \ldots a_r} {}^{a_1}\mathbf{e} \wedge \cdots \wedge {}^{a_r}\mathbf{e}. \tag{2.6.8}$$

In terms of a coordinate basis one may write

$$\mathbf{A} = \frac{1}{r!}(-1)^{\Delta_r(i)} A_{i_1 \ldots i_r} \mathbf{d}x^{i_1} \wedge \cdots \wedge \mathbf{d}x^{i_r}, \tag{2.6.9}$$

which explains the terminology 'differential form'. The coefficients $A_{a_1 \ldots a_r}$, $A_{i_1 \ldots i_r}$ in (2.6.8) and (2.6.9) are scalar fields over the domains of the bases $\{{}^a\mathbf{e}\}$, $\{\mathbf{d}x^i\}$ respectively.

Derivations of forms

Let $\Omega_r(M)$ be the set of all r-forms on M. Let t be an integer and D a mapping $D: \Omega_r(M) \rightarrow \Omega_{r+t}(M)$, defined for all $r \geq 0$, which satisfies the laws

$$D(\Omega_r(M)) = 0 \text{ if } r+t<0 \quad \text{(applicable when } t \text{ is negative)}, \tag{2.6.10}$$

$$D(\mathbf{A}+\mathbf{B}) = D\mathbf{A} + D\mathbf{B}, \tag{2.6.11}$$

$$D(\mathbf{A}\alpha) = (D\mathbf{A})\alpha, \tag{2.6.12}$$

$$D(\mathbf{A} \wedge \mathbf{C}) = (D\mathbf{A}) \wedge \mathbf{C} + (-1)^{\mathbf{AC}+rs}(D\mathbf{C}) \wedge \mathbf{A}, \tag{2.6.13}$$

for all $\mathbf{A}$, $\mathbf{B}$ in $\Omega_r(M)$ for all $r \geq 0$, for all $\mathbf{C}$ in $\Omega_s(M)$ for all $s \geq 0$, and for all α in Λ_∞. The mapping D is called a *derivation of degree t*. It is said to be *pure* if it maps pure forms into pure forms, being of type a if it changes the type of the form, and of type c otherwise. Moreover, it may act either from the left or the right according to the law

$$D\mathbf{A} = (-1)^{D\mathbf{A}} \mathbf{A}D. \tag{2.6.14}$$

When D is pure one may also write

$$D(\alpha\mathbf{A}) = (-1)^{D\alpha} \alpha(D\mathbf{A}), \tag{2.6.15}$$

$$D(\mathbf{A} \wedge \mathbf{C}) = (D\mathbf{A}) \wedge \mathbf{C} + (-1)^{D\mathbf{A} + rt} \mathbf{A} \wedge (D\mathbf{C}). \tag{2.6.16}$$

The Lie derivative is an example of a derivation of degree 0. It satisfies eqs. (2.6.11)–(2.6.13) already for ordinary tensors and tensor products and not merely for forms and exterior products.

The exterior derivative

Another important example of a derivation is the *exterior derivative* **d**, which is of degree 1 and type c, and is defined by

$$\mathbf{d}(\alpha\mathbf{A}) = \alpha\mathbf{d}\mathbf{A}, \qquad \mathbf{d}(\mathbf{A}+\mathbf{B}) = \mathbf{d}\mathbf{A} + \mathbf{d}\mathbf{B}, \tag{2.6.17}$$

$$\mathbf{d}(\mathbf{A} \wedge \mathbf{C}) = (\mathbf{d}\mathbf{A}) \wedge \mathbf{C} + (-1)^r \mathbf{A} \wedge (\mathbf{d}\mathbf{C}), \tag{2.6.18}$$

$$\mathbf{d}^2 = 0, \tag{2.6.19}$$

together with the statement that $\mathbf{d}f$ is just the ordinary differential (section 2.3) when f is a 0-form. That **d** is uniquely defined by these laws may be verified by applying them to eq. (2.6.9):

$$\begin{aligned}\mathbf{d}\mathbf{A} &= \frac{1}{r!}(-1)^{\Delta_r(i)}[(\mathbf{d}A_{i_1\ldots i_r}) \wedge \mathbf{d}x^{i_1} \wedge \cdots \wedge \mathbf{d}x^{i_r} \\ &\qquad + A_{i_1\ldots i_r} \wedge \mathbf{d}(\mathbf{d}x^{i_1} \wedge \cdots \wedge \mathbf{d}x^{i_r})] \\ &= \frac{1}{r!}(-1)^{\Delta_r(i)} A_{i_1\ldots i_r,j}\mathbf{d}x^j \wedge \mathbf{d}x^{i_1} \wedge \cdots \wedge \mathbf{d}x^{i_r} \\ &= \frac{1}{r!}(-1)^{r+\Delta_{r+1}(i)} A_{i_1\ldots i_r,i_{r+1}} \mathbf{d}x^{i_1} \wedge \cdots \wedge \mathbf{d}x^{i_{r+1}}.\end{aligned} \tag{2.6.20}$$

Evidently

$$(dA)_{i_1\ldots i_{r+1}} = \frac{(-1)^r}{r!} \sum_{perm} (-1)^{v(i_{k_1},\ldots,i_{k_{r+1}})} A_{i_{k_1}\ldots i_{k_r},i_{k_{r+1}}}. \tag{2.6.21}$$

Exterior derivation and Lie derivation (of forms) commute with one another. Note first that they commute when applied to 0-forms:

$$[\mathfrak{L}_{\mathbf{X}}, \mathbf{d}]f = 0. \tag{2.6.22}$$

This follows immediately from

$$\begin{aligned}\mathbf{Y}\cdot(\mathfrak{L}_{\mathbf{X}}\mathbf{d}f) &= (-1)^{\mathbf{XY}}[\mathfrak{L}_{\mathbf{X}}(\mathbf{Y}\cdot\mathbf{d}f) - (\mathfrak{L}_{\mathbf{X}}\mathbf{Y})\cdot\mathbf{d}f] \\ &= (-1)^{\mathbf{XY}}(\mathbf{XY}f - [\mathbf{X},\mathbf{Y}]_L f) = \mathbf{YX}f \\ &= \mathbf{Y}\cdot(\mathbf{d}\mathfrak{L}_{\mathbf{X}} f),\end{aligned} \tag{2.6.23}$$

where f, **X** and **Y** are arbitrary. With the aid of this result and the fact that $\mathbf{d}^2 = 0$ it is elementary to verify that if $\mathfrak{L}_{\mathbf{X}}$ and **d** are applied successively to the right-hand side of eq. (2.6.9) the order of application is immaterial.

Hence for all forms

$$[\mathfrak{L}_{\mathbf{X}}, \mathbf{d}] = 0. \tag{2.6.24}$$

If one introduces a *supercommutator for derivations* by the definition $[D, E] = DE - (-1)^{DE+tu} ED$ where D and E are derivations of degrees t and u respectively, then one may show (1) that the supercommutator is itself a derivation (of degree $t + u$) and (2) that any derivation that supercommutes with $\mathbf{d}$ when acting on 0-forms supercommutes with $\mathbf{d}$ when acting on any form, its action on forms of arbitrary degree being determined already by its action on 0-forms.

The inner product

Yet another example of a derivation, this time of degree -1, is the *interior* or *inner product* of a form with an arbitrary contravariant vector field $\mathbf{X}$, denoted by $I_{\mathbf{X}}$. It is defined by

$$I_{\mathbf{X}} f = 0, \qquad I_{\mathbf{X}} \mathbf{d} f = \mathbf{X} f \tag{2.6.25}$$

(for all f in $\mathfrak{F}(M)$) and by the requirement that its type be the same as that of $\mathbf{X}$. With the aid of the decomposition (2.6.9) and the laws (2.6.10)–(2.6.16) one readily shows that the action of $I_{\mathbf{X}}$ on an arbitrary r-form $\mathbf{A}$ is given, in terms of components, by

$$(I_{\mathbf{X}} \mathbf{A})_{i_1 \ldots i_{r-1}} = (-1)^{j(\mathbf{A}+1)} X^j A_{j i_1 \ldots i_{r-1}} = X^j{}_j A_{i_1 \ldots i_{r-1}}. \tag{2.6.26}$$

From this it is easy to verify that

$$I_{\mathbf{X}}^2 = (-1)^{\mathbf{X}+1} I_{\mathbf{X}}^2, \tag{2.6.27}$$

and hence that $I_{\mathbf{X}}^2 = 0$ when $\mathbf{X}$ is c-type.

The inner product may also be considered in combination with other derivations. For example, it is straightforward to show that the supercommutator, $[\mathbf{d}, I_{\mathbf{X}}]$, of $I_{\mathbf{X}}$ with exterior derivation is itself a derivation. Moreover, it is a derivation that (super)commutes with $\mathbf{d}$:

$$[\mathbf{d}, [\mathbf{d}, I_{\mathbf{X}}]] = \mathbf{d}^2 I_{\mathbf{X}} + \mathbf{d} I_{\mathbf{X}} \mathbf{d} - \mathbf{d} I_{\mathbf{X}} \mathbf{d} - I_{\mathbf{X}} \mathbf{d}^2 = 0. \tag{2.6.28}$$

Therefore its action on any form is determined by its action on 0-forms. It is elementary to verify that

$$[\mathbf{d}, I_{\mathbf{X}}] = \mathfrak{L}_{\mathbf{X}}. \tag{2.6.29}$$

Another simple relation is the following:

$$[\mathfrak{L}_{\mathbf{X}}, I_{\mathbf{Y}}] = I_{[\mathbf{X}, \mathbf{Y}]}, \tag{2.6.30}$$

which may be derived by comparing the action of both sides of the equation on 0-forms and 1-forms.

2.7 Connections

Definition

A *connection* on a supermanifold M is a mapping $\boldsymbol{\nabla}$ from the set $\mathscr{T}(M)$ of all tensor fields on M to itself, which satisfies the following conditions:

(1) It maps each rank-(r, s) tensor field into a rank-$(r, s+1)$ tensor field of the same type. In terms of components relative to some basis one may regard $\boldsymbol{\nabla}$ as adding one more lower index. It will be convenient to think of this index as appearing on the left (as in the second of eqs. (2.4.32)) when $\boldsymbol{\nabla}$ is regarded as acting on the left, and on the right when $\boldsymbol{\nabla}$ is regarded as acting on the right.

(2) For every $\mathbf{X}$ in $\mathfrak{X}(M)$, $\boldsymbol{\nabla}$ may be converted into a mapping $\boldsymbol{\nabla}_{\mathbf{X}}$ by contraction with $\mathbf{X}$. If the added index is a then one multiplies the tensor components in which a appears by X^a on the left when $\boldsymbol{\nabla}$ acts on the left, and by ${}^aX$ on the right when $\boldsymbol{\nabla}$ acts on the right. Alternative notations for $\boldsymbol{\nabla}_{\mathbf{X}}$ are $\mathbf{X}\cdot\boldsymbol{\nabla}$ and $\boldsymbol{\nabla}\cdot\mathbf{X}$ in the two cases. Addition of the index a itself is effected by the operators $\boldsymbol{\nabla}_{{}_a\mathbf{e}}$ and $\boldsymbol{\nabla}_{\mathbf{e}_a}$ respectively.

(3) For every $\mathbf{X}$ in $\mathfrak{X}(M)$, $\boldsymbol{\nabla}_{\mathbf{X}}$ obeys all the laws of the Lie derivative *except* eq. (2.5.2). That is, eqs. (2.5.1), (2.5.3)–(2.5.5), (2.5.10)–(2.5.13), and (2.5.18)–(2.5.22) all remain valid if the symbol $\mathfrak{L}_{\mathbf{X}}$ is replaced by $\boldsymbol{\nabla}_{\mathbf{X}}$. A connection is therefore a derivation. For every tensor field $\mathbf{T}$, $\boldsymbol{\nabla}\mathbf{T}$ is called the *covariant derivative of* $\mathbf{T}$, and $\boldsymbol{\nabla}_{\mathbf{X}}\mathbf{T}$ is called the *covariant derivative of* $\mathbf{T}$ *with respect to* (or *in the direction of*) $\mathbf{X}$.

Equation (2.5.2) *cannot* be satisfied by the covariant derivative because $\boldsymbol{\nabla}_{\mathbf{X}}\mathbf{Y}$ must be obtainable simply by contracting $\mathbf{X}$ with $\boldsymbol{\nabla}\mathbf{Y}$. This means that when $\boldsymbol{\nabla}_{\mathbf{X}}\mathbf{Y}$ is expressed in terms of components, derivatives of the components of $\mathbf{X}$ cannot appear, a restriction that does not hold for the Lie derivative (see eqs. (2.5.7) and (2.3.18)). One must specify the covariant derivative of a contravariant vector field independently instead of using (2.5.2) to define it. Thus there is not just a single unique covariant derivative on a supermanifold, but an infinite family of them, one for each connection.

Note, however, that the covariant derivative of the unit tensor field vanishes no matter which connection is chosen. This follows from eqs. (2.5.3) and (2.5.4) (with $\mathfrak{L}_{\mathbf{X}}$ replaced by $\boldsymbol{\nabla}_{\mathbf{X}}$) and the fact that

$$\boldsymbol{\delta}(\omega, \mathbf{Y}) = (\mathbf{e}_a \otimes {}^a\mathbf{e})(\omega, \mathbf{Y}) = \omega\cdot\mathbf{Y}. \tag{2.7.1}$$

Thus

$$(\nabla_{\mathbf{X}}\delta)(\omega,\mathbf{Y}) = \nabla_{\mathbf{X}}[\delta(\omega,\mathbf{Y})] - \delta(\nabla_{\mathbf{X}}\omega,\mathbf{Y}) - (-1)^{X\omega}\delta(\omega,\nabla_{\mathbf{X}}\mathbf{Y})$$
$$= \nabla_{\mathbf{X}}(\omega\cdot\mathbf{Y}) - (\nabla_{\mathbf{X}}\omega)\cdot\mathbf{Y} - (-1)^{X\omega}\omega\cdot(\nabla_{\mathbf{X}}\mathbf{Y}) = 0. \quad (2.7.2)$$

For completely identical reasons we also have

$$\mathfrak{L}_{\mathbf{X}}\delta = \mathbf{0}. \quad (2.7.3)$$

The connection components

The mapping ∇, like the mapping $\mathfrak{L}_{\mathbf{X}}$, is linear. Therefore, to fix the action of ∇ on contravariant vector fields it suffices to specify the actions of the mappings $\nabla_{\mathbf{e}_a}$ on an arbitrary local frame field $\{\mathbf{e}_a\}$. We define

$$\Gamma^a{}_{bc} \overset{\text{def}}{=} {}^a(\mathbf{e}_b\nabla_{\mathbf{e}_c}) = {}^a\mathbf{e}\cdot(\mathbf{e}_b\nabla_{\mathbf{e}_c}). \quad (2.7.4)$$

The $\Gamma^a{}_{bc}$ are called the *components of the connection* ∇.

Strictly speaking the $\Gamma^a{}_{bc}$ are components of the contravariant vector fields $\mathbf{e}_b\nabla_{\mathbf{e}_c}$. Since, however, these fields themselves get transformed under the change of basis (2.4.11), the transformation law for the $\Gamma^a{}_{bc}$ is more complicated than that for the components of a single contravariant vector field. We have

$$\bar{\Gamma}^a{}_{bc} = {}^aL^{-1}{}_d{}^d[(\mathbf{e}_f{}^fL_b)\nabla_{\mathbf{e}_g}]{}^gL_c$$
$$= {}^aL^{-1}{}_d{}^d[(-1)^{g(f+b)}(\mathbf{e}_f\nabla_{\mathbf{e}_g}){}^fL_b + \mathbf{e}_f(\mathbf{d}{}^fL_b)\cdot\mathbf{e}_g]{}^gL_c$$
$$= (-1)^{g(f+b)}{}^aL^{-1}{}_d\Gamma^d{}_{fg}{}^fL_b{}^gL_c + {}^aL^{-1}{}_d(\mathbf{d}{}^dL_b)_g{}^gL_c, \quad (2.7.5)$$

where we have used the fact that

$$\mathbf{X}\cdot(\mathbf{d}f) = \mathbf{X}f = \nabla_{\mathbf{X}}f = \mathbf{X}\cdot(\nabla f),\ (\mathbf{d}f)\cdot\mathbf{X} = f\mathbf{X} = f\nabla_{\mathbf{X}} = (f\nabla)\cdot\mathbf{X}, \quad (2.7.6)$$

and

$$\nabla f = f\nabla = \mathbf{d}f, \quad (2.7.7)$$

for all f in $\mathfrak{F}(M)$. When the transformation is from one coordinate basis to another eq. (2.7.5) takes the form

$$\bar{\Gamma}^i{}_{jk} = (-1)^{n(m+j)}\bar{x}^i{}_{,l}\Gamma^l{}_{mn}\left(x^m\frac{\bar{\partial}}{\partial\bar{x}^j}\right)\left(x^n\frac{\bar{\partial}}{\partial\bar{x}^k}\right) + \bar{x}^i{}_{,l}\left(x^l\frac{\bar{\partial}}{\partial\bar{x}^j}\frac{\bar{\partial}}{\partial\bar{x}^k}\right). \quad (2.7.8)$$

Note that the $\Gamma^a{}_{bc}$ do *not* transform as the components of a mixed tensor field.

An alternative expression for the $\Gamma^a{}_{bc}$ may be obtained by taking the covariant derivative of the equation ${}^a\mathbf{e}\cdot\mathbf{e}_b = {}^a\delta_b$ and using the analog of eq. (2.5.3):

$$0 = (\mathbf{d}{}^a\delta_b)\cdot\mathbf{e}_c = ({}^a\mathbf{e}\cdot\mathbf{e}_b)\nabla_{\mathbf{e}_c} = (-1)^{bc}({}^a\mathbf{e}\,\nabla_{\mathbf{e}_c})\cdot\mathbf{e}_b + {}^a\mathbf{e}\cdot(\mathbf{e}_b\nabla_{\mathbf{e}_c}),$$

whence

$$\Gamma^a{}_{bc} = (-1)^{bc}({}^a\mathbf{e}\nabla_{\mathbf{e}_c})_b. \quad (2.7.9)$$

Explicit forms

In calculating covariant derivatives in explicit component form it is convenient to introduce the following adaptation of a notation frequently used in ordinary manifold theory:

$$\begin{aligned} {}_{b\,;}T^{a_1 \ldots a_r}{}_{a_{r+1} \ldots a_{r+s}} &\stackrel{\text{def}}{=} {}_b(\nabla \mathbf{T})^{a_1 \ldots a_r}{}_{a_{r+1} \ldots a_{r+s}} \\ &= ({}_{b}\nabla_{\mathbf{e}} \mathbf{T})^{a_1 \ldots a_r}{}_{a_{r+1} \ldots a_{r+s}} = (-1)^{b(\mathbf{T}+1)}(\mathbf{T}\nabla_{\mathbf{e}_b})^{a_1 \ldots a_r}{}_{a_{r+1} \ldots a_{r+s}} \\ &\stackrel{\text{def}}{=} (-1)^{b(\mathbf{T}+1+a_1+\cdots+a_{r+s})}\, T^{a_1 \ldots a_r}{}_{a_{r+1} \ldots a_{r+s};b}, \end{aligned} \quad (2.7.10)$$

where $\mathbf{T}$ is any rank-(r, s) tensor field. When a coordinate basis is used $\nabla_{\mathbf{e}_i}$ and ${}_i\nabla_{\mathbf{e}}$ are often written simply ∇_i and ${}_i\nabla$ respectively.

If f is a scalar field, $\mathbf{X}$ a contravariant vector field, and $\boldsymbol{\omega}$ a covariant vector field, one finds

$$f_{;a} = f\nabla_{\mathbf{e}_a} = (\mathbf{d}f)\cdot\mathbf{e}_a = f_{,i}{}^i e_a, \quad (2.7.11)$$

$$\begin{aligned} X^a{}_{;b} &= (-1)^{ab}(\mathbf{X}\nabla_{\mathbf{e}_b})^a = (-1)^{ab}[(X^c{}_c\mathbf{e})\nabla_{\mathbf{e}_b}]^a \\ &= (-1)^{b(a+c)}[(\mathbf{d}X^c)_b\,{}_c\mathbf{e}]^a + (-1)^{c(a+1)}X^c\,{}^a(\mathbf{e}_c\nabla_{\mathbf{e}_b}) \\ &= (\mathbf{d}X^a)_b + (-1)^{c(a+1)}X^c\Gamma^a{}_{cb}, \end{aligned} \quad (2.7.12)$$

$$\begin{aligned} \omega_{a;b} &= (-1)^{ab}(\boldsymbol{\omega}\nabla_{\mathbf{e}_b})_a = (-1)^{ab}[(\omega_c{}^c\mathbf{e})\nabla_{\mathbf{e}_b}]_a \\ &= (-1)^{b(a+c)}[(\mathbf{d}\omega_c)_b\,{}^c\mathbf{e}]_a + (-1)^{ab}\omega_c({}^c\mathbf{e}\nabla_{\mathbf{e}_b})_a \\ &= (\mathbf{d}\omega_a)_b - \omega_c\Gamma^c{}_{ab}, \end{aligned} \quad (2.7.13)$$

or, in a coordinate basis,

$$f_{;i} = f\nabla_i = f_{,i}, \quad (2.7.14)$$

$$X^i{}_{;j} = (-1)^{ij}(\mathbf{X}\nabla_j)^i = X^i{}_{,j} + (-1)^{k(i+1)}X^k\Gamma^i{}_{kj}, \quad (2.7.15)$$

$$\omega_{i;j} = (-1)^{ij}(\boldsymbol{\omega}\nabla_j)_i = \omega_{i,j} - \omega_k\Gamma^k{}_{ij}. \quad (2.7.16)$$

From these equations one can compute the covariant derivative of any tensor field in terms of the tensor components, their ordinary derivatives, and the connection components. By an abuse of notation one sometimes writes also $X^i{}_{;j} = X^i\nabla_j = (-1)^{j(\mathbf{X}+1+i)}{}_j\nabla X^i$, $\omega_{i;j} = \omega_i\nabla_j = (-1)^{j(\omega+1+i)}{}_j\nabla\omega_i$.

Multiple covariant derivatives. The torsion

A supermanifold possessing a connection possesses also a number of fundamental tensor structures induced by the connection. These can be constructed by applying the mapping ∇ repeatedly to arbitrary tensor fields. When working with components we shall use the following notation:

$$(\mathbf{S}\nabla\nabla\ldots)^{a_1\ldots a_r}{}_{a_{r+1}\ldots a_{r+s}bc\ldots} \stackrel{\text{def}}{=} S^{a_1\ldots a_r}{}_{a_{r+1}\ldots a_{r+s};bc\ldots}, \tag{2.7.17}$$

where $\mathbf{S}$ is an arbitrary rank-(r,s) tensor.

Let $\mathbf{X}$ and $\mathbf{Y}$ be arbitrary contravariant vector fields and f an arbitrary scalar field. Then

$$\begin{aligned}(f\nabla\nabla)(\mathbf{X},\mathbf{Y}) &= (-1)^{XY}[(f\nabla)\nabla_{\mathbf{Y}}]\cdot\mathbf{X}\\ &= [(\mathbf{d}f)\cdot\mathbf{X}]\nabla_{\mathbf{Y}} - (\mathbf{d}f)\cdot(\mathbf{X}\nabla_{\mathbf{Y}}) = f\mathbf{X}\mathbf{Y} - (\mathbf{d}f)\cdot(\mathbf{X}\nabla_{\mathbf{Y}})\end{aligned} \tag{2.7.18}$$

and hence

$$\begin{aligned}&(f\nabla\nabla)(\mathbf{X},\mathbf{Y}) - (-1)^{XY}(f\nabla\nabla)(\mathbf{Y},\mathbf{X})\\ &\qquad = (\mathbf{d}f)\cdot\{[\mathbf{X},\mathbf{Y}] - \mathbf{X}\nabla_{\mathbf{Y}} + (-1)^{XY}\mathbf{Y}\nabla_{\mathbf{X}}\}.\end{aligned} \tag{2.7.19}$$

The left-hand side of this equation is linear in $\mathbf{X}$ and in $\mathbf{Y}$. The right-hand side is linear in $\mathbf{d}f$. Therefore there must exist a rank-(1, 2) tensor field $\mathbf{T}$ such that both sides are equal to $-\mathbf{T}(\mathbf{d}f,\mathbf{X},\mathbf{Y})$. This tensor field is known as the *torsion*. Explicitly

$$\mathbf{T}(\omega,\mathbf{X},\mathbf{Y}) \stackrel{\text{def}}{=} \omega\cdot\{\mathbf{X}\nabla_{\mathbf{Y}} - (-1)^{XY}\mathbf{Y}\nabla_{\mathbf{X}} - [\mathbf{X},\mathbf{Y}]\}. \tag{2.7.20}$$

In terms of components eq. (2.7.19) takes the form

$$f_{;ab} - (-1)^{ab} f_{;ba} = -f_{;c}\, T^c{}_{ab}, \tag{2.7.21}$$

where

$$\begin{aligned}T^a{}_{bc} = \mathbf{T}(\mathbf{e}^a,\mathbf{e}_b,\mathbf{e}_c) &= {}^a\mathbf{e}\cdot\{\mathbf{e}_b\nabla_{\mathbf{e}_c} - (-1)^{bc}\mathbf{e}_c\nabla_{\mathbf{e}_b} - [\mathbf{e}_b,\mathbf{e}_c]\}\\ &= \Gamma^a{}_{bc} - (-1)^{bc}\,\Gamma^a{}_{cb} - {}^a c_{bc},\end{aligned} \tag{2.7.22}$$

the ${}^a c_{bc}$ being the coefficients in eqs. (2.3.23). In a coordinate basis the c's vanish and we have

$$T^i{}_{jk} = \Gamma^i{}_{jk} - (-1)^{jk}\Gamma^i{}_{kj}. \tag{2.7.23}$$

From eq. (2.7.8) it is easy to see that the $T^i{}_{jk}$ indeed transform as the components of a tensor field.

The Riemann tensor field

Let $\mathbf{S}$ be an arbitrary rank-(r,s) tensor field. Then

$$\begin{aligned}(\mathbf{S}\nabla\nabla)(\omega,\ldots,\mathbf{Z},\mathbf{X},\mathbf{Y}) &= (-1)^{Y(\omega+\cdots+Z+X)}[(\mathbf{S}\nabla)\nabla_{\mathbf{Y}}](\omega\ldots\mathbf{Z},\mathbf{X})\\ &= (-1)^{X(\omega+\cdots+Z)}\{[(\mathbf{S}\nabla_{\mathbf{X}})(\omega,\ldots,\mathbf{Z})]\nabla_{\mathbf{Y}}\\ &\quad - (-1)^{Y(\cdots+Z)}(\mathbf{S}\nabla_{\mathbf{X}})(\omega\nabla_{\mathbf{Y}},\ldots,\mathbf{Z})\\ &\quad - \cdots - (\mathbf{S}\nabla_{\mathbf{X}})(\omega,\ldots,\mathbf{Z}\nabla_{\mathbf{Y}})\} - (\mathbf{S}\nabla)(\omega,\ldots,\mathbf{Z},\mathbf{X}\nabla_{\mathbf{Y}})\}\end{aligned}$$

$$= (-1)^{(\mathbf{X}+\mathbf{Y})(\omega+\cdots+\mathbf{Z})}(\mathbf{S}\nabla_\mathbf{X}\nabla_\mathbf{Y})(\omega,\ldots,\mathbf{Z}) - (\mathbf{S}\nabla)(\omega,\ldots,\mathbf{Z},\mathbf{X}\nabla_\mathbf{Y}). \quad (2.7.24)$$

From this result one easily obtains

$$(\mathbf{S}\nabla\nabla)(\omega,\ldots,\mathbf{Z},\mathbf{X},\mathbf{Y}) - (-1)^{\mathbf{XY}}(\mathbf{S}\nabla\nabla)(\omega,\ldots,\mathbf{Z},\mathbf{Y},\mathbf{X})$$
$$= (-1)^{(\mathbf{X}+\mathbf{Y})(\omega+\cdots+\mathbf{Z})}\{\mathbf{S}([\nabla_\mathbf{X},\nabla_\mathbf{Y}] - \nabla_{[\mathbf{X},\mathbf{Y}]}) - (\mathbf{S}\nabla)\cdot\mathbf{T}(\cdot,\mathbf{X},\mathbf{Y})\}(\omega,\ldots,\mathbf{Z}), \quad (2.7.25)$$

where $\mathbf{T}(\cdot,\mathbf{X},\mathbf{Y})$ is the contravariant vector field obtained by contracting the torsion with $\mathbf{X}$ and $\mathbf{Y}$, and $[\nabla_\mathbf{X},\nabla_\mathbf{Y}]$ is the supercommutator of $\nabla_\mathbf{X}$ and $\nabla_\mathbf{Y}$.

The left-hand side of eq. (2.7.25) is linear in $\mathbf{X}$ and in $\mathbf{Y}$. The term in the torsion on the right has this property. Hence the tensor $\mathbf{S}([\nabla_\mathbf{X},\nabla_\mathbf{Y}] - \nabla_{[\mathbf{X},\mathbf{Y}]})$ must share this property also. Using the fact that the operator $[\nabla_\mathbf{X},\nabla_\mathbf{Y}] - \nabla_{[\mathbf{X},\mathbf{Y}]}$ gives zero when applied to a scalar field, and showing that terms in which $\nabla_\mathbf{X}$ and $\nabla_\mathbf{Y}$ act on different arguments all cancel, one readily verifies that

$$\{\mathbf{S}([\nabla_\mathbf{X},\nabla_\mathbf{Y}] - \nabla_{[\mathbf{X},\mathbf{Y}]})\}(\omega,\ldots,\mathbf{Z})$$
$$= -(-1)^{\omega(\mathbf{X}+\mathbf{Y})}\mathbf{S}(\omega([\nabla_\mathbf{X},\nabla_\mathbf{Y}] - \nabla_{[\mathbf{X},\mathbf{Y}]}),\ldots,\mathbf{Z})$$
$$-\cdots-(-1)^{(\mathbf{X}+\mathbf{Y})(\omega+\cdots+\mathbf{Z})}\mathbf{S}(\omega,\ldots,\mathbf{Z}([\nabla_\mathbf{X},\nabla_\mathbf{Y}] - \nabla_{[\mathbf{X},\mathbf{Y}]})). \quad (2.7.26)$$

The left-hand side of this equation is linear separately in $\omega,\ldots,\mathbf{Z}$. The right-hand side is linear in the components of $\mathbf{S}$. From this one may conclude that there exists a rank-(1, 3) tensor field $\mathbf{R}$ such that

$$\mathbf{R}(\omega,\mathbf{Z},\mathbf{X},\mathbf{Y}) = -\omega\cdot\{\mathbf{Z}([\nabla_\mathbf{X},\nabla_\mathbf{Y}] - \nabla_{[\mathbf{X},\mathbf{Y}]})\} \quad (2.7.27a)$$
$$= (-1)^{\omega\mathbf{Z}}\mathbf{Z}\cdot\{\omega([\nabla_\mathbf{X},\nabla_\mathbf{Y}] - \nabla_{[\mathbf{X},\mathbf{Y}]})\}, \quad (2.7.27b)$$

for all ω in $\Omega(M)$ and all $\mathbf{X},\mathbf{Y},\mathbf{Z}$ in $\mathfrak{X}(M)$. This tensor is known as the *Riemann tensor field*.

The components of the Riemann tensor field are readily calculated:

$$R^a{}_{bcd} = -\mathbf{e}^a\cdot\{\mathbf{e}_b([\nabla_{\mathbf{e}_c},\nabla_{\mathbf{e}_d}] - \nabla_{[\mathbf{e}_c,\mathbf{e}_d]})\}$$
$$= -(\mathbf{d}\Gamma^a{}_{bc})_d + (-1)^{cd}(\mathbf{d}\Gamma^a{}_{bd})_c + (-1)^{c(f+b)}\Gamma^a{}_{fc}\Gamma^f{}_{bd}$$
$$-(-1)^{d(f+b+c)}\Gamma^a{}_{fd}\Gamma^f{}_{bc} + \Gamma^a{}_{bf}{}^f c_{cd}. \quad (2.7.28)$$

In a coordinate basis this reduces to

$$R^i{}_{jkl} = -\Gamma^i{}_{jk,l} + (-1)^{kl}\Gamma^i{}_{jl,k} + (-1)^{k(m+j)}\Gamma^i{}_{mk}\Gamma^m{}_{jl}$$
$$-(-1)^{l(m+j+k)}\Gamma^i{}_{ml}\Gamma^m{}_{jk}. \quad (2.7.29)$$

In terms of components eq. (2.7.25) yields the following special cases:

$$X^a{}_{;bc} - (-1)^{bc}X^a{}_{;cb} = -(-1)^{d(a+1)}X^d R^a{}_{dbc} - X^a{}_{;d}T^d{}_{bc}, \quad (2.7.30)$$
$$\omega_{a;bc} - (-1)^{bc}\omega_{a;cb} = \omega_d R^d{}_{abc} - \omega_{a;d}T^d{}_{bc}. \quad (2.7.31)$$

With the aid of eq. (2.7.26) these results may be used to determine the rules for interchanging indices induced by covariant differentiation of any tensor field.

The super Bianchi identity

Let **W**, **X**, **Y**, **Z** be arbitrary contravariant vector fields and ω an arbitrary covariant vector field. Then

$$
\begin{aligned}
&(\mathbf{R}\nabla)(\omega, \mathbf{W}, \mathbf{X}, \mathbf{Y}, \mathbf{Z}) \\
&= (-1)^{\mathbf{Z}(\omega+\mathbf{W}+\mathbf{X}+\mathbf{Y})}(\mathbf{R}\nabla_{\mathbf{Z}})(\omega, \mathbf{W}, \mathbf{X}, \mathbf{Y}) \\
&= [\mathbf{R}(\omega, \mathbf{W}, \mathbf{X}, \mathbf{Y})]\nabla_{\mathbf{Z}} - (-1)^{\mathbf{Z}(\mathbf{W}+\mathbf{X}+\mathbf{Y})}\mathbf{R}(\omega\nabla_{\mathbf{Z}}, \mathbf{W}, \mathbf{X}, \mathbf{Y}) \\
&\quad - (-1)^{\mathbf{Z}(\mathbf{X}+\mathbf{Y})}\mathbf{R}(\omega, \mathbf{W}\nabla_{\mathbf{Z}}, \mathbf{X}, \mathbf{Y}) - (-1)^{\mathbf{Z}\mathbf{Y}}\mathbf{R}(\omega, \mathbf{W}, \mathbf{X}\nabla_{\mathbf{Z}}, \mathbf{Y}) \\
&\quad - \mathbf{R}(\omega, \mathbf{W}, \mathbf{X}, \mathbf{Y}\nabla_{\mathbf{Z}}) \\
&= -\omega\cdot\{\mathbf{W}([[\nabla_{\mathbf{X}}, \nabla_{\mathbf{Y}}], \nabla_{\mathbf{Z}}] - [\nabla_{[\mathbf{X},\mathbf{Y}]}, \nabla_{\mathbf{Z}}])\} \\
&\quad + (-1)^{\mathbf{Z}\mathbf{Y}}\omega\cdot\{\mathbf{W}([\nabla_{\mathbf{X}\nabla_{\mathbf{Z}}}, \nabla_{\mathbf{Y}}] - \nabla_{[\mathbf{X}\nabla_{\mathbf{Z}},\mathbf{Y}]})\} \\
&\quad + \omega\cdot\{\mathbf{W}([\nabla_{\mathbf{X}}, \nabla_{\mathbf{Y}\nabla_{\mathbf{Z}}}] - \nabla_{[\mathbf{X},\mathbf{Y}\nabla_{\mathbf{Z}}]})\}.
\end{aligned}
\tag{2.7.32}
$$

in which eq. (2.7.27a) has been used in obtaining the final form. From this, together with the super Jacobi identity (2.3.15), it is straightforward to obtain

$$
\begin{aligned}
&(\mathbf{R}\nabla)(\omega, \mathbf{W}, \mathbf{X}, \mathbf{Y}, \mathbf{Z}) + (-1)^{\mathbf{X}(\mathbf{Y}+\mathbf{Z})}(\mathbf{R}\nabla)(\omega, \mathbf{W}, \mathbf{Y}, \mathbf{Z}, \mathbf{X}) \\
&\quad + (-1)^{\mathbf{Z}(\mathbf{X}+\mathbf{Y})}(\mathbf{R}\nabla)(\omega, \mathbf{W}, \mathbf{Z}, \mathbf{X}, \mathbf{Y}) \\
&= \omega\cdot\{\mathbf{W}([\nabla_{\mathbf{X}}, \nabla_{\mathbf{T}(\cdot,\mathbf{Y},\mathbf{Z})}] - \nabla_{[\mathbf{X},\mathbf{T}(\cdot,\mathbf{Y},\mathbf{Z})]}) \\
&\quad + (-1)^{\mathbf{X}(\mathbf{Y}+\mathbf{Z})}\mathbf{W}([\nabla_{\mathbf{Y}}, \nabla_{\mathbf{T}(\cdot,\mathbf{Z},\mathbf{X})}] - \nabla_{[\mathbf{Y},\mathbf{T}(\cdot,\mathbf{Z},\mathbf{X})]}) \\
&\quad + (-1)^{\mathbf{Z}(\mathbf{X}+\mathbf{Y})}\mathbf{W}([\nabla_{\mathbf{Z}}, \nabla_{\mathbf{T}(\cdot,\mathbf{X},\mathbf{Y})}] - \nabla_{[\mathbf{Z},\mathbf{T}(\cdot,\mathbf{X},\mathbf{Y})]})\} \\
&= -\mathbf{R}(\omega, \mathbf{W}, \mathbf{X}, \mathbf{T}(\cdot, \mathbf{Y}, \mathbf{Z})) - (-1)^{\mathbf{X}(\mathbf{Y}+\mathbf{Z})}\mathbf{R}(\omega, \mathbf{W}, \mathbf{Y}, \mathbf{T}(\cdot, \mathbf{Z}, \mathbf{X})) \\
&\quad - (-1)^{\mathbf{Z}(\mathbf{X}+\mathbf{Y})}\mathbf{R}(\omega, \mathbf{W}, \mathbf{Z}, \mathbf{T}(\cdot, \mathbf{X}, \mathbf{Y})).
\end{aligned}
\tag{2.7.33}
$$

This relation is known as the *super Bianchi identity*. In terms of components it takes the form

$$
\begin{aligned}
&R^a{}_{bcd;e} + (-1)^{c(d+e)} R^a{}_{bde;c} + (-1)^{e(c+d)} R^a{}_{bec;d} \\
&= -R^a{}_{bcf} T^f{}_{de} - (-1)^{c(d+e)} R^a{}_{bdf} T^f{}_{ec} - (-1)^{e(c+d)} R^a{}_{bef} T^f{}_{cd}.
\end{aligned}
\tag{2.7.34}
$$

It is satisfied by every Riemann tensor field.

Parallel transport. Supergeodesics

Let λ be a supercurve (of either c- or a-type) and **S** a tensor field in a supermanifold with a connection ∇. One defines the *covariant derivative* *of* **S** *along* λ by

$$\left(\frac{\vec{\mathrm{D}}}{\mathrm{ds}}\right)_\lambda \mathbf{S} \stackrel{\text{def}}{=} \nabla_{(\vec{\partial}/\partial s)\lambda}\mathbf{S} \quad \text{or} \quad \mathbf{S}\left(\frac{\overleftarrow{\mathrm{D}}}{\mathrm{ds}}\right)_\lambda \stackrel{\text{def}}{=} \mathbf{S}\nabla_{(\overleftarrow{\partial}/\partial s)\lambda}, \tag{2.7.35}$$

s being the supercurve parameter and $(\vec{\partial}/\partial s)_\lambda$, $(\overleftarrow{\partial}/\partial s)_\lambda$ the tangents (see section 2.2). Suppose $\mathbf{S}(\overleftarrow{\mathrm{D}}/\mathrm{ds})_\lambda = 0$ for all s in the domain $\mathscr{S}$ of λ. Then **S** at any point in $\lambda(\mathscr{S})$ is said to be obtained from **S** at any other point in $\lambda(\mathscr{S})$ by *parallel transport* along the supercurve.

If **S** is a contravariant vector field **X** then, using eq. (2.7.15) and the fact that ${}^i(\overleftarrow{\partial}/\partial s)_{\lambda(s)} = x^i(\lambda(s))(\overleftarrow{\mathrm{d}}/\mathrm{ds})$, we have

$$\begin{aligned} {}^i[\mathbf{X}(\overleftarrow{\mathrm{D}}/\mathrm{ds})_{\lambda(s)}] &= (-1)^{i\mathbf{X}}[X^i{}_{;j}{}^j(\overleftarrow{\partial}/\partial s)]_{\lambda(s)} \\ &= (-1)^{i\mathbf{X}}\left\{X^i(\lambda(s))\frac{\overleftarrow{\mathrm{d}}}{\mathrm{ds}} + (-1)^{k(i+1)}(X^k\Gamma^i{}_{kj})_{\lambda(s)}\left[x^j(\lambda(s))\frac{\overleftarrow{\mathrm{d}}}{\mathrm{ds}}\right]\right\}. \end{aligned} \tag{2.7.36}$$

Suppose **X** is the tangent $(\overleftarrow{\partial}/\partial s)_\lambda$ itself.[†] Then eq. (2.7.36) takes the form

$$\begin{aligned} {}^i\left[\left(\frac{\overleftarrow{\partial}}{\partial s}\right)_\lambda\left(\frac{\overleftarrow{\mathrm{D}}}{\mathrm{ds}}\right)_{\lambda(s)}\right] &= x^i(\lambda(s))\frac{\overleftarrow{\mathrm{d}}}{\mathrm{ds}}\frac{\overleftarrow{\mathrm{d}}}{\mathrm{ds}} + (-1)^{j(k+s)}\Gamma^i{}_{kj}(\lambda(s)) \\ &\quad \times \left[x^k(\lambda(s))\frac{\overleftarrow{\mathrm{d}}}{\mathrm{ds}}\right]\left[x^j(\lambda(s))\frac{\overleftarrow{\mathrm{d}}}{\mathrm{ds}}\right] \\ &= x^i(\lambda(s))\frac{\overleftarrow{\mathrm{d}}}{\mathrm{ds}}\frac{\overleftarrow{\mathrm{d}}}{\mathrm{ds}} + \tfrac{1}{2}(-1)^{s(j+1)}[\Gamma^i{}_{jk} + (-1)^{jk+s}\Gamma^i{}_{kj}]_{\lambda(s)} \\ &\quad \times \left[x^k(\lambda(s))\frac{\overleftarrow{\mathrm{d}}}{\mathrm{ds}}\right]\left[x^j(\lambda(s))\frac{\overleftarrow{\mathrm{d}}}{\mathrm{ds}}\right], \end{aligned} \tag{2.7.37}$$

in which use has been made of the symmetry properties of the product $[x^k(\lambda(s))\overleftarrow{\mathrm{d}}/\mathrm{ds}][x^j(\lambda(s))\overleftarrow{\mathrm{d}}/\mathrm{ds}]$ in passing to the last line.

The supercurve λ is said to be a *supergeodesic* or a *self-parallel supercurve* if the tangent to λ is parallel to its own covariant derivative along λ, i.e., if

$$\left(\frac{\overleftarrow{\partial}}{\partial s}\right)_\lambda\left(\frac{\overleftarrow{\mathrm{D}}}{\mathrm{ds}}\right)_{\lambda(s)} = \left(\frac{\overleftarrow{\partial}}{\partial s}\right)_{\lambda(s)} f(s), \tag{2.7.38}$$

for some differentiable function f of s.

Consider first the case in which λ is of type *a*. The function f is then

[†] $(\overleftarrow{\partial}/\partial s)_\lambda$ is, of course, not a contravariant vector field. However, it can be extended to a contravariant vector field by letting λ be one of a congruence of supercurves. Expression 2.7.37 is independent of the congruence chosen.

a-number valued and, being differentiable, necessarily has the linear form

$$f(s) = \alpha + \beta s, \tag{2.7.39}$$

where α is some a-number and β is some c-number. The functions $x^i(\lambda(s))$ likewise have the form

$$x^i(\lambda(s)) = a^i + b^i s, \tag{2.7.40}$$

where a^i and b^i are constants. In this case ${}^i(\bar{\partial}/\partial s)_\lambda = b^i$, and eqs. (2.7.37) and (2.7.38) together imply

$$b^i f(s) = -\tfrac{1}{2}(-1)^j T^i{}_{jk}(\lambda(s)) b^k b^j. \tag{2.7.41}$$

Expanding both sides of this equation in powers of s, one finds

$$\left.\begin{aligned} b^i\alpha &= -\tfrac{1}{2}(-1)^j T^i{}_{jk}(\lambda(0)) b^k b^j, \\ b^i\beta &= -\tfrac{1}{2}(-1)^k T^i{}_{jk,l}(\lambda(0)) b^l b^k b^j. \end{aligned}\right\} \tag{2.7.42}$$

Suppose the connection ∇ is such that for every point of the supermanifold it is possible to have an a-type supergeodesic containing that point and having an arbitrary imaginary a-type tangent there. Then eqs. (2.7.42) must hold for arbitrary b^i at every point. This can happen only if the torsion vanishes (which implies also $f(s) = 0$). Since a-type supercurves exist even in supermanifolds having no a-number coordinates, this condition requires vanishing torsion even in such supermanifolds. When the torsion vanishes every a-type supercurve is automatically a supergeodesic. A nonvanishing torsion imposes constraints on the possible existence of a-type supergeodesics.

Turn now to c-type supergeodesics. They are given by solutions of the equations

$$\frac{d^2 s^i(\lambda(s))}{ds^2} + \Gamma^i{}_{jk}(\lambda(s)) \frac{dx^k(\lambda(s))}{ds}\frac{dx^j(\lambda(s))}{ds} = \frac{dx^i(\lambda(s))}{ds} f(s), \tag{2.7.43}$$

where f is any differentiable real c-number valued function. The torsion is seen to play no role here; only the supersymmetric part of the connection enters. It is readily verified that if, instead of s, one employs the variable

$$\bar{s} \overset{\text{def}}{=} \int e^{\int f(s)ds} ds, \tag{2.7.44}$$

then eqs. (2.7.43) take the simpler form

$$\frac{d^2 x^i}{d\bar{s}^2} + \Gamma^i{}_{jk}\frac{dx^k}{d\bar{s}}\frac{dx^j}{d\bar{s}} = 0. \tag{2.7.45}$$

Equations (2.7.45) are known as the *supergeodesic equations*, and the variable $\bar{s}$ is called an *affine parameter*. When the supergeodesic λ is parametrized by means of an affine parameter one may obtain the tangent at any point on λ by simple parallel transport from any other point on λ. Note that the bodies of s and $\bar{s}$ are monotonically related by eq. (2.7.44) regardless of the sign of the body of $f(s)$. Note also that an affine parameter is unique up to a c-number scale factor (with nonvanishing body) and a shift of the zero point.

Distant parallelism

Suppose the Riemann tensor field vanishes. Then it is possible to set up a unique *parallelism at a distance* in any simply connected open subset $\mathscr{U}$. Given an arbitrary tensor $\mathbf{S}_p$ at a point p of $\mathscr{U}$, one can construct a unique tensor field $\mathbf{S}$ that takes the value $\mathbf{S}_p$ at p and has vanishing covariant derivative throughout $\mathscr{U}$. The local integrability condition for the equation $\mathbf{S}\boldsymbol{\nabla} = 0$ is the statement that the right-hand side of eq. (2.7.25) must vanish for all $\mathbf{X}$ and $\mathbf{Y}$. But the first term inside the curly brackets vanishes because the Riemann tensor field vanishes, and the second term vanishes if $(\mathbf{S}\boldsymbol{\nabla})\cdot\mathbf{T} = 0$, which is not a new condition independent of $\mathbf{S}\boldsymbol{\nabla} = 0$. Global integrability throughout $\mathscr{U}$ follows from the simple connectedness of $\mathscr{U}$. If λ is *any* supercurve in $\mathscr{U}$ through p then the result of transporting $\mathbf{S}_p$ along λ in a parallel fashion to any other point on λ coincides with the value at $\mathbf{S}$ at that point.

Let $\{\mathbf{e}_a\}_p$ be a standard basis for T_p. If the Riemann tensor field vanishes then $\{\mathbf{e}_a\}_p$ can be extended to a field of *parallel* local frames satisfying $\mathbf{e}_a\boldsymbol{\nabla} = 0$ through $\mathscr{U}$. The dual basis field $\{\mathbf{e}^a\}$ then also satisfies $\mathbf{e}^a\boldsymbol{\nabla} = 0$, and the necessary and sufficient condition that a tensor field $\mathbf{S}$ satisfy $\mathbf{S}\boldsymbol{\nabla} = 0$ in $\mathscr{U}$ is that the components of $\mathbf{S}$ relative to these bases be everywhere constant in $\mathscr{U}$. It is sometimes possible to extend the field of parallel local frames throughout the whole supermanifold even when it is not itself simply connected.

The torsion tensor has a simple expression in terms of a field of parallel local frames. Referring to eq. (2.7.20) one sees that

$$\mathbf{T}(\cdot, \mathbf{e}_a, \mathbf{e}_b) = -[\mathbf{e}_a, \mathbf{e}_b]. \tag{2.7.46}$$

Evidently, a field of parallel local frames cannot be a coordinate basis if the torsion is nonvanishing.

There is an alternative definition of distant parallelism that avoids this difficulty. It is applicable not when the Riemann tensor field vanishes but when an *associated* Riemann tensor field $\bar{\mathbf{R}}$ vanishes. This tensor field is constructed from an associated connection $\bar{\boldsymbol{\nabla}}$, which is defined by

$$\mathbf{X}\bar{\boldsymbol{\nabla}} \stackrel{\text{def}}{=} \mathbf{X}\boldsymbol{\nabla} - \tfrac{1}{2}\mathbf{T}(\cdot, \mathbf{X}, \cdot), \tag{2.7.47}$$

or, equivalently,

$$\omega\bar{\nabla} \stackrel{\text{def}}{=} \omega\nabla + \tfrac{1}{2}\mathbf{T}(\omega,\cdot,\cdot), \tag{2.7.48}$$

for all contravariant vector fields **X** and all covariant vector fields ω. It is easy to verify that the torsion $\bar{\mathbf{T}}$ associated with $\bar{\nabla}$ vanishes, and that if a c-type supercurve is a supergeodesic with respect to ∇ it is also a supergeodesic with respect to $\bar{\nabla}$.

When $\bar{\mathbf{R}}$ vanishes tensor fields **S** may be introduced which take arbitrary values at a given point and satisfy $\mathbf{S}\bar{\nabla} = 0$ throughout simply connected open sets $\mathscr{U}$. A field of local frames satisfying $\mathbf{e}_a\bar{\nabla} = 0$ *is* a coordinate basis.

2.8 Riemannian supermanifolds

The metric tensor field

A supermanifold M is called a *Riemannian* supermanifold if it is endowed with a real c-type range-(0, 2) tensor field g having the following properties:

(1) $g(\mathbf{X},\mathbf{Y}) = (-1)^{\mathbf{XY}} g(\mathbf{Y},\mathbf{X})$, for all $\mathbf{X}$, $\mathbf{Y}$ in $\mathfrak{X}(M)$

(2) $g(\cdot,\mathbf{X}) = \mathbf{0}$ if and only if $\mathbf{X} = \mathbf{0}$.

This field is called the *metric tensor field*.

In terms of components one has

$$g(\mathbf{X},\mathbf{Y}) = (-1)^{b(a+\mathbf{X})} g_{ab}\, {}^aX\, {}^bY = X^a\, {}_ag_b\, {}^bY. \tag{2.8.1}$$

Property (1) says that the matrix $({}_ag_b)$ is supersymmetric. Property (2) says that it is nonsingular. It is therefore like the matrix M of eq. (1.7.51). Note that the number n of a-number coordinates of a Riemannian supermanifold is always even. If n were odd g would result necessarily be singular.

The elements of the matrix inverse to $({}_ag_b)$ are the components of a real c-type rank-(2, 0) tensor field denoted by g^{-1}. When writing the components of g^{-1} it is customary to omit the exponent -1. These components satisfy

$${}^ag^b = g^{ab} = (-1)^{ab} g^{ba}, \tag{2.8.2}$$

$${}_ag_c\, g^{cb} = {}_a\delta^b = \delta^b{}_a, \quad g^{ac}\, {}_cg_b = {}^a\delta_b = \delta^a{}_b. \tag{2.8.3}$$

By contraction with $g(g^{-1})$ every contravarient (covariant) vector field may be converted into an associated covariant (contravariant) vector field of the same type. In component language this contraction process is called the *raising* or *lowering of indices*. It follows the rules

$$\left.\begin{aligned} {}_aX &= {}_ag_b\, {}^bX, & X_a &= X^b\, {}_bg_a, \\ {}^aX &= g^{ab}\, {}_bX, & X^a &= X_b\, g^{ba}. \end{aligned}\right\} \tag{2.8.4}$$

When a covariant vector field is obtained from a contravariant one through contraction with the metric tensor field the two vector fields may be regarded as alternative embodiments of the same object. Equations (2.8.4) define an isomorphic mapping between the supervector spaces T_p and T_p^* at each point, which enables expression (2.8.1) to be regarded as an inner product.

$$g(\mathbf{X}, \mathbf{Y}) = X_a{}^a Y = X^a{}_a Y \stackrel{\text{def}}{=} \mathbf{X}\cdot\mathbf{Y} = (-1)^{\mathbf{XY}}\mathbf{Y}\cdot\mathbf{X}. \tag{2.8.5}$$

The tensor fields g and g^{-1} can also be used to lower and raise indices denoting tensor field components. The rules are obvious:

$$T_{a_1\ldots a_r}{}^c{}_{b_1\ldots b_s} = (-1)^{(e+c)(b_1+\cdots+b_s)}\, T_{a_1\ldots a_r e b_1\ldots b_s}\, g^{ec}, \tag{2.8.6}$$

etc.

Canonical form of the metric tensor at a point

Let x^i be the coordinates of some chart containing a point p of the supermanifold. Let $({}_ig_j)_p$ be the matrix of components of the metric tensor relative to the coordinates x^i at p. We have already remarked that $({}_ig_j)_p$ is like the matrix M of eq. (1.7.51). In fact, let us denote it by M and use the block decomposition shown in eq. (1.7.51). Let

$$X = \begin{pmatrix} 1_m & -A^{-1}C \\ 0 & 1_n \end{pmatrix} \tag{2.8.7}$$

and let L be the matrix defined by eqs. (1.7.56)–(1.7.59). We have already seen that

$$L^{\sim}X^{\sim}MXL = \operatorname{diag}\left(\lambda_1, \ldots, \lambda_m, \begin{pmatrix} 0 & \mathrm{i}\mu_1 \\ -\mathrm{i}\mu_1 & 0 \end{pmatrix}, \ldots, \begin{pmatrix} 0 & \mathrm{i}\mu_{n/2} \\ -\mathrm{i}\mu_{n/2} & 0 \end{pmatrix}\right), \tag{2.8.8}$$

where the λ's and μ's are real and '$\sim$' denotes, as usual, the supertranspose. Let us relax the condition that the matrix O_2 of eq. (1.7.57) have unit determinant and instead choose it in such a way that all the μ's are positive. Furthermore let

$$Y = \operatorname{diag}(|\lambda_1|^{-\frac{1}{2}}, \ldots |\lambda_m|^{-\frac{1}{2}}, {\mu_1}^{-\frac{1}{2}}, {\mu_1}^{-\frac{1}{2}}, \ldots, {\mu_{n/2}}^{-\frac{1}{2}}, {\mu_{n/2}}^{-\frac{1}{2}}) \tag{2.8.9}$$

Then

$$N^{\sim}MN = \eta \stackrel{\text{def}}{=} \operatorname{diag}\left(-1, \ldots, -1, 1, \ldots, 1, \begin{pmatrix} 0 & \mathrm{i} \\ -\mathrm{i} & 0 \end{pmatrix}, \ldots, \begin{pmatrix} 0 & \mathrm{i} \\ -\mathrm{i} & 0 \end{pmatrix}\right), \tag{2.8.10}$$

where

$$N = XLYZ, \tag{2.8.11}$$

Z being any permutation matrix that may be needed to bring the sequence $\lambda_1, \ldots, \lambda_m$ into an order such that the negative λ's (if any) stand at the head of the line. Now introduce new coordinates $\bar{x}^i$ at p, defined by

$$\bar{x}^i = {}^i N^{-1}{}_j x^j. \tag{2.8.12}$$

In these coordinates the components of the metric tensor at p become

$$({}_i\bar{g}_j)_p = \eta. \tag{2.8.13}$$

The matrix η is known as the *canonical form* of the metric tensor.

The canonical form – in particular, the number of -1's in the canonical form – is an invariant of the supermanifold if the supermanifold is connected. That is, the canonical form at a given point is the same as that at any other point. This may be proved as follows. First note that

$$[\operatorname{sdet}({}_ig_j)_p]_B = (-1)^{n/2} \frac{\lambda_1 \ldots \lambda_m}{(\mu_1 \ldots \mu_{n/2})^2}. \tag{2.8.14}$$

The λ's and μ's are continuous functions of the bodies of the coordinates x^i. As the x^i are varied none of the λ's can change sign without passing through zero. But a zero λ indicates that $({}_ig_j)$ has become singular, which violates property (2) of a Riemannian supermanifold. Therefore the number of negative λ's is a constant throughout a given coordinate chart.

That this number is the same in any overlapping chart, and hence, by extension, throughout the supermanifold, may be inferred by noting that at any overlap point the ${}_ig_j$'s in the two charts are related by transformations of the form

$$({}_i\bar{g}_j) = K^{\sim}({}_ig_j)K, \tag{2.8.15}$$

where K is nonsingular. The matrix K can always be expressed as the product of a matrix that inverts a certain number (possibly zero) of coordinates and a matrix that can be deformed continuously and nonsingularly to the identity matrix. Both coordinate inversion and continuous nonsingular deformation leave the number of negative λ's unchanged.

Canonical or orthosymplectic bases

In coordinate bases it is generally possible to bring the metric components into canonical form only at isolated points. By using fields of local frames $\{\mathbf{e}_a\}$, on the other hand, the canonical form can be maintained over open sets. Relative to a coordinate basis the components ${}^i\mathbf{e}_a$ of the contravariant vector fields $\mathbf{e}_a$ at any point are just the elements of the matrix N of eq. (2.8.11) at that point:

$$ {}_a\eta_b = {}_a e^i{}_i g_j{}^j e_b = {}_a\mathbf{e}\cdot\mathbf{e}_b. \tag{2.8.16}$$

A local frame field satisfying eq. (2.8.16) is said to constitute a *canonical basis* or an *orthosymplectic basis*.

An orthosymplectic basis is not unique. Given one orthosymplectic basis an infinity of others can be obtained by carrying out transformations of the form

$$\bar{\mathbf{e}}_a = \mathbf{e}_b{}^b L_a, \tag{2.8.17}$$

where the ${}^bL_a$ are scalar fields satisfying

$$ {}_a L^{\sim c}{}_c\eta_d{}^d L_b = {}_a\eta_b. \tag{2.8.18}$$

At each point the set of all transformations (2.8.18) constitutes a representation of the so-called *generalized orthosymplectic group* (see section 4.2). The $\mathbf{e}_a$ of an orthosymplectic basis are often referred to as *(m, n)-ads*, *(m, n)-beine* or *vielbeine*, and the group of all differentiable transformations (2.8.17) satisfying (2.8.18) over the domain of definition of the e's, is called the *(m, n)-bein group*.

The dual $\{{}^a\mathbf{e}\}$ to an orthosymplectic basis $\{\mathbf{e}_a\}$ is given by the simple rule

$$ {}^a\mathbf{e} = {}^a\eta^b{}_b\mathbf{e}, \tag{2.8.19}$$

where $({}^a\eta^b)$ is the matrix η^{-1} inverse to η.† It is not difficult to see that the dual basis satisfies

$$ {}_i e^a{}_a\eta_b{}^b e_j = {}_i g_j, \tag{2.8.20}$$

and hence it is sometimes regarded as the 'square root' of the metric tensor field.

Riemannian connections

Every Riemannian supermanifold possesses a natural connection called a *Riemannian connection*, which is uniquely determined by the conditions

$$ g\nabla = 0, \tag{2.8.21}$$

$$ \mathbf{T} = 0. \tag{2.8.22}$$

In terms of components these conditions take the forms

$$ 0 = g_{ab;c} = (\mathrm{d}g_{ab})_c - (-1)^{b(a+d)} g_{db}\,\Gamma^d{}_{ac} - g_{ac}\,\Gamma^d{}_{bc}, \tag{2.8.23}$$

$$ \Gamma^a{}_{bc} - (-1)^{bc}\,\Gamma^a{}_{cb} = {}^a c_{bc}, \tag{2.8.24}$$

† η and η^{-1} are, in fact, identical.

(see eqs. (2.7.13) and (2.7.22)). These equations, together with the supersymmetry of g, imply

$$\Gamma^a{}_{bc} = (-1)^d g^{ad} \Gamma_{dbc}, \tag{2.8.25}$$

$$\Gamma_{abc} \overset{\text{def}}{=} \tfrac{1}{2}[(\mathbf{d}g_{ab})_c + (-1)^{bc}(\mathbf{d}g_{ac})_b - (-1)^{a(b+c)}(\mathbf{d}g_{bc})_a + c_{abc} - (-1)^{ab} c_{bac} - (-1)^{c(a+b)} c_{cab}]. \tag{2.8.26}$$

$$c_{abc} \overset{\text{def}}{=} g_{ad}{}^d c_{bc}. \tag{2.8.27}$$

In a coordinate basis the c's vanish and eq. (2.8.26) reduces to

$$\Gamma_{ijk} = \tfrac{1}{2}[g_{ij,k} + (-1)^{jk} g_{ik,j} - (-1)^{i(j+k)} g_{jk,i}] = (-1)^{jk}\Gamma_{ikj}. \tag{2.8.28}$$

In an orthosymplectic basis we have ${}_a g_b = {}_a\eta_b$, $\mathbf{d}g_{ab} = 0$, and eq. (2.8.26) reduces to

$$\Gamma_{abc} = \tfrac{1}{2}[c_{abc} - (-1)^{ab} c_{bac} - (-1)^{c(a+b)} c_{cab}] = -(-1)^{ab}\Gamma_{bac}. \tag{2.8.29}$$

The curvature tensor field

The Riemann tensor field constructed from a Riemannian connection is known as the *curvature tensor field.* The curvature tensor field possesses a number of special symmetries that can be discovered as follows. Working in a coordinate basis and referring to eq. (2.7.29) write

$$\begin{aligned} R_{ijkl} &= g_{im} R^m{}_{jkl} \\ &= g_{im}[-\Gamma^m{}_{jk,l} + (-1)^{kl}\Gamma^m{}_{jl,k} + (-1)^{k(n+j)}\Gamma^m{}_{nk}\Gamma^n{}_{jl} \\ &\qquad - (-1)^{l(n+j+k)}\Gamma^m{}_{nl}\Gamma^n{}_{jk}]. \end{aligned} \tag{2.8.30}$$

Convert the first two terms on the right into total derivatives plus remainders, and make use of the supersymmetry of the connection components (eq. (2.8.28)) together with

$$g_{ij,k} = \Gamma_{ijk} + (-1)^{ij}\Gamma_{jik}, \tag{2.8.31}$$

to express (2.8.30) in the form

$$\begin{aligned} R_{ijkl} = \tfrac{1}{2}[&-(-1)^{jk} g_{ik,jl} - (-1)^{i(j+l)+kl} g_{jl,ik} \\ &+ (-1)^{l(j+k)} g_{il,jk} + (-1)^{i(j+k)} g_{jk,il}] \\ &- (-1)^{m(i+k)+jk}\Gamma_{mik}\Gamma^m{}_{jl} + (-1)^{m(i+l)+l(j+k)}\Gamma_{mil}\Gamma^m{}_{jk}. \end{aligned} \tag{2.8.32}$$

It is then straightforward to verify that

$$R_{ijkl} = -(-1)^{ij} R_{jikl} = -(-1)^{kl} R_{ijkl} = (-1)^{(i+j)(k+l)} R_{klij}, \tag{2.8.33}$$

$$R_{ijkl} + (-1)^{j(k+l)} R_{iklj} + (-1)^{l(j+k)} R_{iljk} = 0. \tag{2.8.34}$$

Since $\mathbf{R}$ is a tensor field these symmetries hold in any basis.

By carefully counting the number of independent constraints that eqs. (2.8.33) and (2.8.34) impose on **R** one may show that in a Riemannian supermanifold of dimension (m,n) the curvature tensor at each point has N_{curv} algebraically independent components where

$$N_{\text{curv}} = \tfrac{1}{12}(m+n)^2[(m+n)^2 - 1] - mn. \tag{2.8.35}$$

The Ricci tensor field

Because of the symmetries (2.8.38) and (2.8.34) and the supersymmetry of the metric tensor field there is essentially only one way to obtain a new tensor field by contraction of a pair of indices of the curvature tensor field. Contraction of the first pair of indices, or the last pair, gives zero. Contraction of the first index with the third yields the so-called *Ricci tensor field*:

$$R_{ij} \overset{\text{def}}{=} (-1)^{k(i+1)} R^k{}_{ikj} = (-1)^{k(i+1)+l} g^{kl} R_{likj}. \tag{2.8.36}$$

The symmetries (2.8.33) imply that R_{ij} is supersymmetric:

$$R_{ij} = (-1)^{ij} R_{ji}. \tag{2.8.37}$$

The Ricci tensor field in turn may be contracted to yield the *curvature scalar field*:

$$R \overset{\text{def}}{=} R_i{}^i = R_{ij} g^{ji}. \tag{2.8.38}$$

The Ricci tensor and curvature scalar fields appear together in a differential identity that may be obtained from the super Bianchi identity (2.7.34) by raising suitable indices and performing contractions, namely

$$0 \equiv R_{ij;}{}^j - \tfrac{1}{2} R_{;i} \equiv (R_{ij} - \tfrac{1}{2} R g_{ij})_;{}^j. \tag{2.8.39}$$

or equivalently,

$$(-1)^j (R^{ij} - \tfrac{1}{2} R g^{ij})_{;j} \equiv 0. \tag{2.8.40}$$

Because the torsion vanishes the super Bianchi identity itself now takes the form

$$R^i{}_{jkl;m} + (-1)^{k(l+m)} R^i{}_{jlm;k} + (-1)^{m(k+l)} R^i{}_{jmk;l} \equiv 0. \tag{2.8.41}$$

Because of the supersymmetry (2.8.37) the Ricci tensor at a point has only N_{Ricc} algebraically independent components where

$$N_{\text{Ricc}} = \tfrac{1}{2}(m+n)(m+n+1) - n, \qquad (m,n) \neq (1,0) \quad \text{or} \quad (2,0) \tag{2.8.42}$$

The cases $(m,n) = (1,0)$ or $(2,0)$ are excluded because the Ricci tensor at a point cannot have more independent components than the curvature tensor from which it is derived. When $(m,n)=(1,0)$ both the curvature and Ricci tensor fields vanish. When $(m,n)=(2,0)$ they each have only one independent component at each point and are expressible in terms of the curvature scalar:

$$\left.\begin{aligned} R_{ijkl} &= \tfrac{1}{2}R[(-1)^{jk}g_{ik}g_{jl} - (-1)^{l(j+k)}g_{il}g_{jk}], \\ R_{ij} &= \tfrac{1}{2}Rg_{ij}. \end{aligned}\right\} \quad (m,n)=(2,0) \quad (2.8.43)$$

When $(m,n)=(0,2)$ the two fields again have only one independent component at each point. In this case they are expressible in the forms

$$\left.\begin{aligned} R_{ijkl} &= \tfrac{1}{6}R[(-1)^{jk}g_{ik}g_{jl} - (-1)^{l(j+k)}g_{il}g_{ik}], \\ R_{ij} &= -\tfrac{1}{2}Rg_{ij}. \end{aligned}\right\} \quad (m,n)=(0,2) \quad (2.8.44)$$

The only other cases in which the two fields have the same number of independent components are $(m,n)=(1,2)$ and $(m,n)=(3,0)$, the numbers being 4 and 6 respectively. In these cases we have

$$\begin{aligned} R_{ijkl} = &\frac{1}{m-n-2}[(-1)^{jk}(g_{ik}R_{jl} + R_{ik}g_{jl}) - (-1)^{l(j+k)}(g_{il}R_{jk} + R_{il}g_{jk})] \\ &- \frac{1}{(m-n-1)(m-n-2)}R[(-1)^{jk}g_{ik}g_{jl} - (-1)^{l(j+k)}g_{il}g_{jk}], \\ &\qquad (m,n)=(1,2) \quad \text{or} \quad (3,0). \qquad (2.8.45) \end{aligned}$$

Flat Riemannian supermanifolds

If the curvature tensor field vanishes everywhere the Riemannian supermanifold is said to be *flat*. Because the torsion vanishes the curvature tensor field $\mathbf{R}$ and the associated tensor field $\bar{\mathbf{R}}$ (see end of section 2.7) coincide. Therefore if one introduces an orthosymplectic basis $\{\mathbf{e}_a\}_p$ at a point p in a flat Riemannian supermanifold and extends it to a field of local frames $\{\mathbf{e}_a\}$ satisfying $\mathbf{e}_a\nabla = 0$ throughout any simply connected open set $\mathscr{U}$ containing p, then the result is a coordinate basis throughout $\mathscr{U}$. The coordinates x^a, defined as solutions of the equations $\mathbf{d}x^a = \mathbf{e}^a$, are called *super-Cartesian coordinates*. In a super-Cartesian coordinate basis the components of the metric tensor field assume the canonical form everywhere.

Conformally related Riemannian supermanifolds.
The Weyl tensor field

Two Riemannian supermanifolds M and $\bar{M}$ are said to be *conformally equivalent* if there exists a diffeomorphism $\phi: M \to \bar{M}$ between them such that their respective metric tensor fields g and $\bar{g}$ satisfy

$$\bar{g} = e^{\mu}\phi'(g), \tag{2.8.46}$$

where ϕ' is the derivative mapping associated with ϕ (see section 2.5) and μ is a real c-type scalar field over $\bar{M}$.

Instead of regarding M and $\bar{M}$ as two distinct supermanifolds one may view the tensor fields g and $\bar{g}$ as alternative metric tensor fields imposed on the same supermanifold. These tensor fields are then said to be conformally equivalent. No generality is lost in this case if ϕ is assumed to be the identity mapping so that eq. (2.8.46) becomes

$$\bar{g} = e^{\mu}g. \tag{2.8.47}$$

The transformation (multiplication by e^{μ}) that converts g into $\bar{g}$ is called a *local conformal transformation.* The set of all local conformal transformations forms an infinite-dimensional Abelian group known as the *local conformal group.*

It is a straightforward but tedious computation to determine the transformation laws for the components of the curvature and Ricci tensor fields, as well as the curvature scalar field, under local conformal transformations. When $m - n$ is not equal to 1 or 2 it is found that the tensor field **C**, with components

$$\begin{aligned} C^i{}_{jkl} \stackrel{\text{def}}{=} R^i{}_{jkl} &- \frac{1}{m-n-2}[(-1)^{jk}(\delta^i{}_k R_{jl} + R^i{}_k g_{jl}) \\ &\qquad - (-1)^{l(j+k)}(\delta^i{}_l R_{jk} + R^i{}_l g_{jk})] \\ &+ \frac{1}{(m-n-1)(m-n-2)} R[(-1)^{jk}\delta^i{}_k g_{jl} - (-1)^{l(j+k)}\delta^i{}_l g_{jk}], \end{aligned} \tag{2.8.48}$$

remains unchanged. It is said to be *conformally invariant.* **C** is known as the *Weyl tensor field.* In its covariant form (C_{ijkl}) it has all the algebraic symmetries (2.8.33) and (2.8.34) of the curvature tensor field. In addition, all its contractions vanish:

$$(-1)^i C^i{}_{ikl} = 0, \qquad (-1)^{i(j+1)} C^i{}_{jik} = 0. \tag{2.8.49}$$

Conformally flat Riemannian supermanifolds

A Riemannian supermanifold with metric tensor field g is said to be *conformally flat* if it is locally conformally equivalent to a flat Riemannian supermanifold, i.e., if there exists an atlas in every (local) chart of which a scalar field μ exists such that the metric tensor field $\bar{g}$ of eq. (2.8.47) yields a vanishing curvature tensor field. The determination whether the field μ exists locally can be reduced to a finite series of algebraic and differential tests. We state without proof (see exercise **2.12**) the following theorem which gives the necessary and sufficient conditions for conformal flatness in all cases:

Theorem

(1) *If $m-n$ is not equal to 1 or 2 and if $(m,n) \neq (3,0)$ the necessary and sufficient condition for an (m,n)-dimensional Riemannian supermanifold to be conformally flat is that the Weyl tensor field vanish. (In the cases $(m,n) = (0,2)$ and $(m,n) = (1,2)$ the Weyl tensor field vanishes identically. Therefore all Riemannian supermanifolds with these dimensions are automatically conformally flat.)*

(2) *If $m-n$ is equal to 1 or 2 and if $n \neq 0$ the Riemann tensor field must have the form*

$$R^i{}_{jkl} = (-1)^{jk}(\delta^i{}_k A_{jl} + A^i{}_k g_{jl}) - (-1)^{l(j+k)}(\delta^i{}_l A_{jk} + A^i{}_l g_{jk}). \tag{2.8.50}$$

The easiest way to check whether this form is valid is to compute

$$\left.\begin{aligned} A^\mu{}_\nu &= \frac{1}{m-2}\left[R^{\mu\sigma}{}_{\nu\sigma} - \frac{1}{2(m-1)}\delta^\mu{}_\nu R^{\sigma\tau}{}_{\sigma\tau}\right], \\ A^\mu{}_\alpha &= \frac{1}{m-1}R^{\mu\nu}{}_{\alpha\nu}, \quad A^\alpha{}_\mu = \frac{1}{m-1}R^{\alpha\nu}{}_{\mu\nu}, \\ A^\alpha{}_\beta &= \frac{1}{n+2}\left[R^{\alpha\nu}{}_{\beta\nu} - \frac{1}{2(n+1)}\delta^\alpha{}_\beta R^{\nu\delta}{}_{\nu\delta}\right], \end{aligned}\right\} \tag{2.8.51}$$

and then to substitute back in (2.8.50).

(3) *If $(m,n) = (3,0)$ the tensor field $\mathbf{D}$ with components*

$$D_{ijk} \stackrel{\text{def}}{=} (R_{ij} - \tfrac{1}{4}Rg_{ij})_{;k} - (-1)^{jk}(R_{ik} - \tfrac{1}{4}Rg_{ik})_{;j} \tag{2.8.52}$$

must vanish. (In this case the Weyl tensor vanishes identically.)

(4) *Every (2, 0)-dimensional Riemannian supermanifold is conformally flat.*

(5) *Every (1, 0)-dimensional Riemannian supermanifold is flat.*

It may happen (but need not) that the scalar fields μ, which convert a conformally flat Riemannian supermanifold locally into a flat Riemannian supermanifold, can be chosen in every chart in such a way that they agree in all the overlap regions. A single scalar field μ then exists over the whole supermanifold, and the supermanifold is *globally* conformally equivalent to some flat Riemannian supermanifold.

Killing vector fields

A *Killing vector field* in a Riemannian supermanifold is a contravariant vector field **K** for which

$$\mathfrak{L}_{\mathbf{K}}g = 0. \tag{2.8.53}$$

By virtue of eq. (2.5.18) and the fact that g is c-type and real, the condition (2.5.53) must hold separately for the real, imaginary, even and odd parts of **K**. Moreover, by virtue of eq. (12.5.19), if **K** is an a-type Killing vector field then, for every a-number α, $\alpha\mathbf{K}$ is a c-type Killing vector field.

We have seen in section 2.5 that the Lie derivative with respect to a real c-type vector field is an infinitesimal dragging operation. Therefore the existence of a Killing vector field in a Riemannian supermanifold indicates that there is at least one direction, at every point, in which the metric tensor can be dragged without changing the local geometry at that point.

Most Riemannian supermanifolds possess no Killing vector fields at all. The existence of a Killing vector field signals a geometrical symmetry of the supermanifold. One can think of the Killing vector field as pointing in a direction in which the supermanifold can be 'moved into itself'. Such motions can be compounded to form a *group of motions*. This group is a super Lie group (see Chapter 3) with an associated super Lie algebra defined as follows. Consider the set of all Killing vector fields that the supermanifold admits. By virtue of eqs. (2.5.18)–(2.5.21) this set forms a supervector space. The super Lie algebra in question is just this supervector space with an added bracket operation, namely the super Lie bracket for vector fields. Because of eq. (2.5.28) the supervector space is closed under this bracket operation.

Conformal Killing vector fields

A *conformal Killing vector field* in a Riemannian supermanifold is a contravariant vector field **C** for which

$$\mathfrak{L}_{\mathbf{C}}g = \varphi g, \tag{2.8.54}$$

where φ is some scalar field having the same type and reality properties as C. (If $\varphi = 0$ then C is a Killing vector field.) The set of all conformal Killing vector fields admitted by the supermanifold, like the set of all Killing vector fields, forms a supervector space. It is easy to verify that this supervector space too is closed under the super Lie bracket operation. Therefore the set of all conformal Killing vector fields defines a super Lie algebra, of which the super Lie algebra defined by the set of all Killing vector fields is a subalgebra.

Let g and $\bar{g}$ be two metric tensor fields related by eq. (2.8.47), and let C be a vector field satisfying eq. (2.8.54). Then

$$\mathfrak{L}_{\mathbf{C}}\bar{g} = \mathfrak{L}_{\mathbf{C}}(e^{\mu}g) = (\varphi + \mathbf{C}\mu)\bar{g}. \tag{2.8.55}$$

Evidently a conformal Killing vector field remains a conformal Killing vector field under the actions of the local conformal group. Moreover, any two conformally equivalent Riemannian supermanifolds have isomorphic sets of conformal Killing vector fields, forming identical super Lie algebras.

The global conformal group

A real c-type conformal Killing vector field C defines a congruence of integral supercurves and a set of associated diffeomorphisms C_s of the supermanifold into itself (see section 2.5) which preserve the metric tensor field up to a conformal factor e^{μ}. It is of interest to study these diffeomorphisms in the case in which the geometry of the supermanifold has one of the maximum possible symmetries, namely flatness with global topology $\mathbf{R}_c^m \times \mathbf{R}_a^n$.

In terms of components relative to a coordinate basis eq. (2.8.54) takes the form

$$C^k{}_{k,i}g_j + {}_{i,}C^k{}_k g_j + {}_i g_k{}^k C_{,j} = \varphi\, {}_i g_j, \tag{2.8.56}$$

which, in a global super-Cartesian coordinate system, reduces to

$$C_{i,j} + (-1)^{ij} C_{j,i} = \varphi \eta_{ij}, \tag{2.8.57}$$

$$C_i \overset{\text{def}}{=} C^j{}_j \eta_i. \tag{2.8.58}$$

We state without proof (see exercise **2.13**) the following theorem which gives the general solution of eq. (2.8.57).

Theorem

(1) *If* $(m, n) = (1, 0)$ *then every contravariant vector field satisfies eq.* (2.8.57) *for some* φ.

(2) *If* $(m, n) = (2, 0)$ *the general solution of eq.* (2.8.57) *is*

$$C_1 = \tfrac{1}{2}\eta_{11} \int_0 \varphi \, dx^1 - \int b \, dx^2, \quad C_2 = \tfrac{1}{2}\eta_{22} \int_0 \varphi \, dx^2 - \int a \, dx^1, \tag{2.8.59}$$

where $\varphi(x^1, x^2)$ *is any harmonic function* (i.e., *solution of* $\varphi_{,i}{}^{i} = 0$) *and the functions* $a(x^1)$ *and* $b(x^2)$ *are defined by*

$$\tfrac{1}{2}\eta_{11} \int_0 \varphi_{,2} \, dx^1 + \tfrac{1}{2}\eta_{22} \int_0 \varphi_{,1} \, dx^2 = a(x^1) + b(x^2), \tag{2.8.60}$$

the necessity of the quantity on the left having the form given on the right following from the harmonicity of φ.

(3) *In all other cases the general solution of eq.* (2.8.57) *is*

$$C_i = \varepsilon_i + \varepsilon_{ij} x^j + \zeta x_i + \zeta_i \mathbf{x}^2 - 2(\boldsymbol{\zeta} \cdot \mathbf{x}) x_i, \tag{2.8.61}$$

where

$$x_i \overset{\text{def}}{=} \eta_{ij} x^j = x^j{}_j\eta_i, \qquad \mathbf{x}^2 \overset{\text{def}}{=} x_i x^i, \qquad \boldsymbol{\zeta} \cdot \mathbf{x} \overset{\text{def}}{=} \zeta_i x^i, \tag{2.8.62}$$

$$\varepsilon_{ij} = -(-1)^{ij} \varepsilon_{ji}, \tag{2.8.63}$$

ε_i, ε_{ij}, ζ *and* ζ_i *being arbitrary pure supernumbers having the type indicated by their indices.*

In part (3) of the theorem the ε's and the ζ's constitute $\frac{1}{2}(m+1)(m+2) + \frac{1}{2}n(n+1)$ *c*-type and $(m+2)n$ *a*-type independent parameters. For Killing vector fields the ζ's vanish and only $\frac{1}{2}m(m+1) + \frac{1}{2}n(n+1)$ *c*-type and $(m+1)n$ *a*-type independent parameters remain. The ε's and ζ's are the parameters of a super Lie group known as the *global conformal group*. The ε's alone are the parameters of a subgroup that we shall call the *super Killing group*. We now outline the explicit construction of these groups.

Let $\bar{x}(x, s)$ denote the point of $\mathbf{R}_c^m \times \mathbf{R}_a^n$ to which the point x is dragged under the mapping $\mathbf{C}_s$ generated by the conformal Killing vector field $\mathbf{C}$. In terms of coordinates, $\bar{x}(x, s)$ satisfies

$$\frac{\partial}{\partial s} \bar{x}^i(x, s) = C^i(\bar{x}(x, s)), \qquad \bar{x}^i(x, 0) = x^i. \tag{2.8.64}$$

We confine our attention to the following four special cases:

Case (1). $\varepsilon^i \neq 0$; all other parameters vanish. The solution of (2.8.64) is then

$$\bar{x}^i(x, s) = x^i + \varepsilon^i s. \tag{2.8.65}$$

This transformation is just a parallel displacement of the coordinates. It leaves the components of the metric tensor field η_{ij} unchanged.

Case (2). $\varepsilon_{ij} \neq 0$; all other parameters vanish. The solution of (2.8.64) in this case is

$$\bar{x}^i(x, s) = (e^{\boldsymbol{\varepsilon} s})^i{}_j x^j, \tag{2.8.66}$$

$$\boldsymbol{\varepsilon} \stackrel{\text{def}}{=} (\varepsilon^i{}_j), \qquad \varepsilon^i{}_j \stackrel{\text{def}}{=} \eta^{ik} \varepsilon_{kj}. \tag{2.8.67}$$

It is easily verified that the matrix $\boldsymbol{\varepsilon}$ satisfies

$$\boldsymbol{\varepsilon}^{\sim} = -\boldsymbol{\eta}\boldsymbol{\varepsilon}\boldsymbol{\eta}^{-1}, \tag{2.8.68}$$

$$(e^{\boldsymbol{\varepsilon} s})^{\sim} \boldsymbol{\eta} e^{\boldsymbol{\varepsilon} s} = e^{\boldsymbol{\varepsilon}^{\sim} s} e^{\boldsymbol{\eta}\boldsymbol{\varepsilon}\boldsymbol{\eta}^{-1} s} \boldsymbol{\eta} = \boldsymbol{\eta}, \tag{2.8.69}$$

and hence that the transformation (2.8.66) again leaves the metric components unchanged.

The set of all case-(2) transformations forms a super Lie group, which is a generalization, to the case in which the diagonal elements of $\boldsymbol{\eta}$ need not all be nonnegative, of the *orthosymplectic group* $\mathrm{Osp}(m, n)$ defined in section 4.2 (or rather, it is the proper or connected subgroup of this generalization). The group of all case-(1) and case-(2) transformations compounded together forms a larger group, the *proper super Killing group*, which is the group of motions of the flat supermanifold, i.e., the group generated by the super Lie algebra of all Killing vector fields.

Case (3). $\zeta \neq 0$; all other parameters vanish. The solution of (2.8.64) now takes the form

$$\bar{x}^i(x, s) = e^{\zeta s} x^i, \tag{2.8.70}$$

which represents a *dilation* transformation. Under this transformation the metric components suffer the change

$$\bar{g}_{ij} = e^{-2\zeta s} \eta_{ij}. \tag{2.8.71}$$

Case (4). $\zeta^i \neq 0$; all other parameters vanish. In this case the solution of (2.8.64) is

$$\bar{x}^i(x, s) = \frac{x^i + \zeta^i \mathbf{x}^2 s}{1 + 2\boldsymbol{\zeta}\cdot\mathbf{x}s + \boldsymbol{\zeta}^2\mathbf{x}^2 s^2}, \tag{2.8.72}$$

which represents a transformation that can be obtained by compounding the following three transformations:

$$\left.\begin{aligned} y^i &= x^i/\mathbf{x}^2, \\ \bar{y}^i &= y^i + \zeta^i s, \\ \bar{x}^i &= y^i/\mathbf{y}^2. \end{aligned}\right\} \tag{2.8.73}$$

Under this transformation the metric components suffer the change

$$\bar{g}_{ij} = (1 + 2\boldsymbol{\zeta}\cdot\mathbf{x}s + \boldsymbol{\zeta}^2\mathbf{x}^2 s^2)^2 \eta_{ij}. \tag{2.8.74}$$

The set of all case-(1)–(4) transformations compounded together forms the *proper global conformal group*. Combining these transformations with coordinate inversions one obtains the full global conformal group.

The global conformal group must not be confused with the local conformal group. The latter is Abelian and infinite dimensional. The former is non-Abelian and finite dimensional. The global conformal group reflects a geometrical symmetry possessed by the supermanifold. The local conformal group makes no statement about the geometry at all. Every Riemannian supermanifold possesses its local conformal group, but unless it has at least one nonvanishing conformal Killing vector field its global conformal group is trivial.

2.9 Integration over supermanifolds

Integration over $\mathbf{R}_c^m \times \mathbf{R}_a^n$. Measure functions

We have seen in section 1.7 that under nonlinear transformations of coordinates the volume element for integrals over $\mathbf{R}_c^m \times \mathbf{R}_a^n$ suffers the transformation

$$d^{m,n}\bar{x} = J\, d^{m,n}x, \qquad J = \operatorname{sdet}(\bar{x}^i{}_{,j}). \tag{2.9.1}$$

Let μ be a differentiable function on $\mathbf{R}_c^m \times \mathbf{R}_a^n$ that obeys the coordinate transformation law

$$\bar{\mu} = J^{-1}\mu. \tag{2.9.2}$$

Then it is evident that the integral $\int f\mu\, d^{m,n}x$ over $\mathbf{R}_c^m \times \mathbf{R}_a^n$ is coordinate invariant for all (integrable) scalar functions f on $\mathbf{R}_c^m \times \mathbf{R}_a^n$. If μ takes its values in $\mathbf{R}_c$ and has a nonvanishing body it is called a *density* or *measure function* over $\mathbf{R}_c^m \times \mathbf{R}_a^n$.

Locally finite atlases and partitions of unity

In section 1.7 it was assumed that the coordinates x^i in $\mathbf{R}_c^m \times \mathbf{R}_a^n$ were global. This assumption is unnecessary; $\mathbf{R}_c^m \times \mathbf{R}_a^n$, like any other supermanifold, can be covered with an atlas containing many charts, not just one. One may, in fact, consider the complete atlas of $\mathbf{R}_c^m \times \mathbf{R}_a^n$. A function like μ is then seen to be merely one of a collection of measure functions μ_α, one for each chart $(\mathscr{U}_\alpha, \phi_\alpha)$ in the complete atlas, and eq. (2.9.2) is replaced by

$$\mu_\beta = J_{\beta\alpha}^{-1}\,\mu_\alpha, \quad \text{for all } p \in \mathscr{U}_\alpha \cap \mathscr{U}_\beta \text{ for all } \alpha, \beta \text{ with } \mathscr{U}_\alpha \cap \mathscr{U}_\beta \text{ nonempty,} \tag{2.9.3}$$

where $J_{\beta\alpha}$ is the super-Jacobian of the mapping $\phi_\beta \circ \phi_\alpha^{-1}$ from $\phi_\alpha(\mathscr{U}_\alpha \cap \mathscr{U}_\beta)$ to $\phi_\beta(\mathscr{U}_\alpha \cap \mathscr{U}_\beta)$.

Now $\mathbf{R}_c^m \times \mathbf{R}_a^n$ is paracompact (see section 1.8), and hence it can be covered by a *locally finite atlas*, i.e., an atlas such that each point of $\mathbf{R}_c^m \times \mathbf{R}_a^n$ is contained in only a finite number of charts. Associated with every locally finite atlas $\{(\mathscr{U}_\alpha, \phi_\alpha)\}$ one can introduce a *partition of unity*, i.e., a set of real c-number valued C^∞ functions θ_α on $\mathbf{R}_c^m \times \mathbf{R}_a^n$, having the following properties:

(1) $[\theta_\alpha(p)]_\mathrm{B} > 0$ for all $p \in \mathscr{U}_\alpha$, for all α.

(2) $\theta_\alpha(p) = 0$ for all $p \notin \mathscr{U}_\alpha$, for all α.

(3) $\sum_\alpha \theta_\alpha(p) = 1$ for all $p \in \mathbf{R}_c^m \times \mathbf{R}_a^n$.

Conditions (1) and (2) are in practice not difficult to reproduce. For example, $\mathbf{R}_c^m \times \mathbf{R}_a^n$ can be covered by a collection of *open balls* of radius a 'centered' at appropriate points p_α:

$$\mathscr{U}_\alpha \overset{\text{def}}{=} \left\{ p \in \mathbf{R}_c^m \times \mathbf{R}_a^n : \left(\sum_\mu [x^\mu(p) - x^\mu(p_\alpha)]^2 \right)_\mathrm{B} < a^2 \right\}.$$

The functions

$$\bar{\theta}_\alpha(p) \overset{\text{def}}{=} \begin{cases} \exp\left(-a^2/\{a^2 - \sum_\mu [x^\mu(p) - x^\mu(p_\alpha)]^2\} \right), & \left(\sum_\mu [x^\mu(p) - x^\mu(p_\alpha)]^2 \right)_\mathrm{B} < a^2, \\ 0, & \left(\sum_\mu [x^\mu(p) - x^\mu(p_\alpha)^2] \right)_\mathrm{B} \geq a^2, \end{cases} \tag{2.9.4}$$

are C^∞ and satisfy conditions (1) and (2) relative to these balls. Given any set of C^∞ functions $\bar{\theta}_\alpha$ satisfying conditions (1) and (2) one obtains a set of C^∞ functions θ_α satisfying all three conditions by defining

$$\theta_\alpha(p) \overset{\text{def}}{=} \bar{\theta}_\alpha(p) / \sum_\beta \bar{\theta}_\beta(p). \tag{2.9.5}$$

Given a partition of unity associated with a locally finite atlas $\{(\mathscr{U}_\alpha, \phi_\alpha)\}$ one can express the integral $\int f \mu \, \mathrm{d}^{m,n}x$ over $\mathbf{R}_c^m \times \mathbf{R}_a^n$ in the alternative form $\sum_\alpha \int_{\phi_\alpha(\mathscr{U}_\alpha)} \theta_\alpha f \mu_\alpha \, \mathrm{d}^{m,n}x$, i.e., as a sum of integrals over the individual open sets $\mathscr{U}_\alpha$, the coordinates x^i in each set being those assigned by the mapping ϕ_α. The choice of atlas is immaterial because of the relations (2.9.2), and the choice of functions θ_α is immaterial as long as conditions (1)–(3) above are respected.

The individual integrals in the sum deserve some comment. Since every soul subspace that intersects an open set $\mathscr{U}_\alpha$ lies wholly in that set, the values of the a-number coordinates in each integral are unrestricted. The integration over these coordinates is carried out according to the rules given in section 1.3. As for the integration over the c-number coordinates, it can be carried out over any submanifold of $\mathbf{R}_c^m \times \mathbf{R}_a^n$ that has the same natural projection onto $\mathbf{R}^m$ as $\phi_\alpha(\mathscr{U}_\alpha)$ has. Because the function θ_α vanishes on the boundary of $\mathscr{U}_\alpha$ the behaviour of the souls of the c-number coordinates over $\mathscr{U}_\alpha$ is immaterial (see section 1.2).

Integration over paracompact orientable supermanifolds

The above construction can obviously be extended to any paracompact orientable supermanifold M endowed with a collection of measure functions μ_α. Again one introduces a locally finite atlas $\{(\mathscr{U}_\alpha, \phi_\alpha)\}$ and an associated partition of unity with C^∞ functions θ_α satisfying conditions (1)–(3) above. Then one writes

$$\int_M f\mu\, d^{m,n}x = \sum_\alpha \int_{\phi_\alpha(\mathscr{U}_\alpha)} \theta_\alpha f \mu_\alpha\, d^{m,n}x, \tag{2.9.6}$$

the quantity on the right defining what is meant by the symbol on the left. In this case it is sometimes useful to regard the integrations over the c-number coordinates as carried out over submanifolds of the form $\phi_\alpha(\mathscr{U}_\alpha \cap K)$ where K is any realization of the body M_{B} (see section 2.1), it does not matter which. Integrations over the a-number coordinates are, as before, carried out according to the rules given in section 1.3.

Integration over Riemannian supermanifolds

If the supermanifold M is Riemannian it has a canonical measure function given by

$$\mu = g^{1/2}, \tag{2.9.7}$$

$$g \overset{\text{def}}{=} |\mathrm{sdet}({}_i g_j)|. \tag{2.9.8}$$

We do not attach a chart-identifying index to each side of this equation, but it must be understood of course that the superdeterminant g, like the function μ, has no meaning other than relative to a specific chart. One is not *obliged* to adopt $g^{1/2}$ as the measure function for a Riemannian supermanifold, but this choice differs from any other measure function by a scalar factor that is *independent* of coordinate systems. Since such a factor can always be absorbed into the rest of an integrand there is no particular advantage in *not* making the choice (2.9.7).

Integrals of total divergences

Of frequent occurrence in integration theory are integrals of the form $\int_M (-1)^i (\mu X^i)_{,i} \mathrm{d}^{m,n}x$ where μ is a measure function and $\mathbf{X}$ is a contravariant vector field. Such integrals are atlas independent. This follows from the coordinate transformation law of the integrand. Using eqs. (1.6.8) and (1.7.25) and the transformation law for vector components, we find

$$\begin{aligned}
(-1)^i(\bar{\mu}\bar{X}^i)_{,\bar{i}} &= (-1)^i (J^{-1}\mu X^j{}_{,j}\bar{x}^i)\frac{\bar{\partial}}{\partial \bar{x}^i} \\
&= (-1)^{i+j+ij}(J^{-1}\mu X^j \bar{x}^i{}_{,j})\frac{\bar{\partial}}{\partial \bar{x}^i} \\
&= (-1)^j (J^{-1}\mu X^j)_{,j} + (-1)^{i+j+ij} J^{-1}\mu X^j \bar{x}^i{}_{,jk}\left(x^k \frac{\bar{\partial}}{\partial \bar{x}^i}\right) \\
&= J^{-1}(-1)^i(\mu X^i)_{,i} + J^{-1}\mu X^j\left[-(-1)^{j+k}\left(x^k\frac{\bar{\partial}}{\partial \bar{x}^i}\right)\bar{x}^i{}_{,kj}\right. \\
&\qquad\qquad \left. + (-1)^{i+j+ij}\bar{x}^i{}_{,jk}\left(x^k\frac{\bar{\partial}}{\partial \bar{x}^i}\right)\right] \\
&= J^{-1}(-1)^i(\mu X^i)_{,i}.
\end{aligned} \tag{2.9.9}$$

For Riemannian supermanifolds, with the measure function chosen according to (2.9.7), the invariance of the integral follows from the identity

$$(-1)^i (g^{1/2} X^i)_{,i} = (-1)^i g^{1/2} X^i{}_{;i} = g^{1/2} X_{i;}{}^i, \tag{2.9.10}$$

which can be derived with the aid of eq. (2.7.15) and the readily verified relation

$$(-1)^i \Gamma^i{}_{ij} = (\ln g^{1/2})_{,j}. \tag{2.9.11}$$

The compact case

The value of the integral $\int_M (-1)^i (\mu X^i)_{,i} \mathrm{d}^{m,n}x$ depends on boundary conditions. If M is compact, or if $\mathbf{X}$ vanishes outside some open set whose closure is compact (i.e., if the *support* of $\mathbf{X}$ is compact) then the integral vanishes:

$$\int_M (-1)^i (\mu X^i)_{,i} \mathrm{d}^{m,n}x = 0 \quad \text{if supp}\,\mathbf{X} \text{ is compact.} \tag{2.9.12}$$

The proof is as follows. Introduce, as usual, a locally finite atlas and a corresponding partition of unity, and write

$$\int_M (-1)^i (\mu X^i)_{,i} \mathrm{d}^{m,n}x \stackrel{\text{def}}{=} \sum_\alpha \int_{\phi_\alpha(\mathscr{U}_\alpha)} \theta_\alpha (-1)^i (\mu_\alpha X^i)_{,i} \mathrm{d}^{m,n}x. \tag{2.9.13}$$

Only those charts for which $\mathscr{U}_\alpha \cap \operatorname{supp} \mathbf{X}$ is nonempty need be included in the sum. Because of the compactness of supp **X** these may be chosen finite in number. The sum is therefore effectively a finite sum and one may rewrite the right-hand side of (2.9.13) in the form

$$\sum_\alpha \int_{\phi_\alpha(\mathscr{U}_\alpha)} (-1)^i (\theta_\alpha \mu_\alpha X^i)_{,i} \, \mathrm{d}^{m,n}x - \sum_\alpha \int_{\phi_\alpha(\mathscr{U}_\alpha)} \theta_{\alpha,i} \mu_\alpha X^i \, \mathrm{d}^{m,n}x. \quad (2.9.14)$$

Each term in the first sum vanishes either because of the rules given in section 1.3 (when the index i is a-type) or because of the fact that $\theta_\alpha \mu_\alpha X^i$ has compact support and vanishes on the boundary of $\mathscr{U}_\alpha$ (when i is c-type).

To show that the second sum also vanishes define the following sets:

$$\left.\begin{aligned}
V_\alpha &\overset{\text{def}}{=} \mathscr{U}_\alpha \cap \Bigl(M - \bigcup_{\substack{\delta \\ \delta \neq \alpha}} \mathscr{U}_\delta \Bigr), \\
V_{\alpha\beta} &\overset{\text{def}}{=} \mathscr{U}_\alpha \cap \mathscr{U}_\beta \cap \Bigl(M - \bigcup_{\substack{\delta \\ \delta \neq \alpha,\beta}} \mathscr{U}_\delta \Bigr), \qquad \alpha < \beta, \\
V_{\alpha\beta\gamma} &\overset{\text{def}}{=} \mathscr{U}_\alpha \cap \mathscr{U}_\beta \cap \mathscr{U}_\gamma \cap \Bigl(M - \bigcup_{\substack{\delta \\ \delta \neq \alpha,\beta,\gamma}} \mathscr{U}_\delta \Bigr), \qquad \alpha < \beta < \gamma, \\
&\cdots\cdots\cdots\cdots\cdots\cdots
\end{aligned}\right\} \quad (2.9.15)$$

Here the symbol '<' indicates an arbitrary ordering of the index set of the finitely many $\mathscr{U}_\alpha$ contributing to the sum, and

$$M - \mathscr{U} \overset{\text{def}}{=} \{ p \in M \mid p \notin \mathscr{U} \}. \quad (2.9.16)$$

The V's are mutually disjoint, cover supp **X**, and have the following properties:

$$\left.\begin{aligned}
&p \in V_\alpha \rightarrow \theta_\alpha(p) = 1, \quad \theta_{\alpha,i}(p) = 0, \\
&p \in V_{\alpha\beta} \rightarrow \theta_\alpha(p) + \theta_\beta(p) = 1, \quad \theta_{\alpha,i}(p) + \theta_{\beta,i}(p) = 0, \\
&p \in V_{\alpha\beta\gamma} \rightarrow \theta_\alpha(p) + \theta_\beta(p) + \theta_\gamma(p) = 1, \quad \theta_{\alpha,i}(p) + \theta_{\beta,i}(p) + \theta_{\gamma,i}(p) = 0 \\
&\vdots
\end{aligned}\right\} \quad (2.9.17)$$

The second sum in (2.9.14) may be rewritten in the form

$$\sum_\alpha \int_{\phi_\alpha(V_\alpha)} \theta_{\alpha,i} \mu_\alpha X^i \, \mathrm{d}^{m,n}x + \sum_{\alpha<\beta} \int_{\phi_\alpha(V_{\alpha\beta})} (\theta_\alpha + \theta_\beta)_{,i} \mu_\alpha X^i \, \mathrm{d}^{m,n}x$$
$$+ \sum_{\alpha<\beta<\gamma} \int_{\phi_\alpha(V_{\alpha\beta\gamma})} (\theta_\alpha + \theta_\beta + \theta_\gamma)_{,i} \mu_\alpha X^i \, \mathrm{d}^{m,n}x + \cdots, \quad (2.9.18)$$

which vanishes by eqs. (2.9.17).

An example

As an example of the use of eq. (2.9.12) consider the integral $\int_M Rg^{\frac{1}{2}}\mathrm{d}^{m,n}x$ where M is a compact Riemannian supermanifold and R is the curvature scalar field. Suppose the metric tensor field is subjected to an infinitesimal change δg. The corresponding change in the integral may be obtained by first observing that the changes in the curvature and Ricci tensor fields are given respectively by

$$\delta R^l{}_{ijk} = -\delta\Gamma^l{}_{ij;k} + (-1)^{jk}\delta\Gamma^l{}_{ik;j}, \tag{2.9.19}$$

$$\delta R_{ij} = -(-1)^k\delta\Gamma^k{}_{ki;j} + (-1)^{k(i+j+1)}\delta\Gamma^k{}_{ij;k}, \tag{2.9.20}$$

as may be readily verified by varying eqs. (2.7.29) and (2.8.36). Here $\delta\Gamma^l{}_{ij}$ denotes the change in the Riemannian connection components, and we remember that since the difference between any two connections is a tensor field, the $\delta\Gamma^l{}_{ij}$ are in fact tensor components so that the notation $\delta\Gamma^l{}_{ij;k}$ makes good sense. Equation (2.9.20), together with eqs (1.6.8) and (2.8.38), yields

$$\begin{aligned}\delta(g^{\frac{1}{2}}R) = g^{\frac{1}{2}}(-1)^i\{&[-(-1)^k\delta\Gamma^k{}_{kj}g^{ji} + \delta\Gamma^i{}_{kj}g^{jk}]_{;i}\\ &-(R^{ij}-\tfrac{1}{2}g^{ij}R)\delta{}_jg_i\}.\end{aligned} \tag{2.9.21}$$

Because of eqs. (2.9.10) and (2.9.12) the integral over M of the first term in the braces in (2.9.21) vanishes, and we are left with

$$\delta\int_M g^{\frac{1}{2}}R\,\mathrm{d}^{m,n}x = -\int_M g^{\frac{1}{2}}(-1)^i(R^{ij}-\tfrac{1}{2}g^{ij}R)\delta_jg_i\,\mathrm{d}^{m,n}x. \tag{2.9.22}$$

There is a useful consistency check on this result. Suppose we choose

$$\delta g = \mathfrak{L}_{\delta\xi}g \quad\text{or}\quad \delta g_{ij} = \delta\zeta_{i;j} + (-1)^{ij}\delta\zeta_{j;i}, \tag{2.9.23}$$

where $\delta\xi$ is an arbitrary infinitesimal real *c*-type contravariant vector field. Since the Lie derivative is an infinitesimal dragging operation that leaves all invariants intact the corresponding change in $\int_M g^{\frac{1}{2}}R\,\mathrm{d}^{m,n}x$ should vanish. Because of the supersymmetry of R^{ij} and g^{ij}, we may write in this case

$$\begin{aligned}\delta\int_M g^{\frac{1}{2}}R\,\mathrm{d}^{m,n}x &= -2\int_M g^{\frac{1}{2}}(-1)^{i+j+ij}(R^{ij}-\tfrac{1}{2}g^{ij}R)\delta\xi_{i;j}\mathrm{d}^{m,n}x\\ &= 2\int_M g^{\frac{1}{2}}(-1)^{i+j}(R^{ij}-\tfrac{1}{2}g^{ij}R)_{;j}\delta\xi_i\,\mathrm{d}^{m,n}x,\end{aligned} \tag{2.9.24}$$

where an integration by parts, based on eqs. (2.9.10) and (2.9.12), has been carried out in passing to the last line. The final expression vanishes identically for all $\delta\xi$ in view of eq. (2.8.40).

Exercises

2.1 Construct a supermanifold that is nonorientable. Construct a supermanifold which, when regarded as a bundle of isomorphic soul subspaces over its body, is a twisted bundle.

2.2 Prove that the supercommutator of two derivations (of forms), of degrees t and u respectively, is a derivation of degree $t+u$.

2.3 Let $\phi: M \to \bar{M}$ be a diffeomorphism between two supermanifolds M and $\bar{M}$ and let $\phi': \mathscr{T}(M) \to \mathscr{T}(\bar{M})$ be the associated derivative mapping. The derivative mapping can be extended so that it applies to connections as well as to tensor fields. The relevant definition is

$$\phi'(\mathbf{\nabla S}) \stackrel{\text{def}}{=} \phi'(\mathbf{\nabla})\phi'(\mathbf{S}) \quad \text{or} \quad \phi'(\mathbf{S\nabla}) = \phi'(\mathbf{S})\phi'(\mathbf{\nabla}),$$

for all tensor fields $\mathbf{S}$ and all connections $\mathbf{\nabla}$. The Lie derivative can be similarly generalized:

$$\mathfrak{L}_{\mathbf{X}}(\mathbf{\nabla S}) \stackrel{\text{def}}{=} (\mathfrak{L}_{\mathbf{X}}\mathbf{\nabla})\mathbf{S} + \mathbf{\nabla}(\mathfrak{L}_{\mathbf{X}}\mathbf{S}),$$

$$(\mathbf{S\nabla})\mathfrak{L}_{\mathbf{X}} \stackrel{\text{def}}{=} \mathbf{S}(\mathbf{\nabla}\mathfrak{L}_{\mathbf{X}}) + (\mathbf{S}\mathfrak{L}_{\mathbf{X}})\mathbf{\nabla}.$$

Using these definitions show that the effect of Lie differentiation on the connection components is given by

$$\begin{aligned} {}^a[\mathbf{e}_b(\mathbf{\nabla}\mathfrak{L}_{\mathbf{X}})\cdot\mathbf{e}_c] = (-1)^{\mathbf{X}c}\{\Gamma^a{}_{bc}\mathbf{X} - (-1)^{\mathbf{X}(b+c)}\,{}^a[\mathbf{X},\mathbf{e}_d]\Gamma^d{}_{bc} \\ -(-1)^{c(d+b)}\Gamma^a{}_{dc}\,{}^d[\mathbf{e}_b,\mathbf{X}] - \Gamma^a{}_{bd}\,{}^d[\mathbf{e}_c,\mathbf{X}] \\ -(\mathbf{d}\,{}^a[\mathbf{e}_b,\mathbf{X}])_c\} \end{aligned}$$

or, in a coordinate basis,

$$\begin{aligned} {}^i[\mathbf{e}_j(\mathbf{\nabla}\mathfrak{L}_{\mathbf{X}})\cdot\mathbf{e}_k] = (-1)^{\mathbf{X}k}\{\Gamma^i{}_{jk,l}\,{}^lX - (-1)^{\mathbf{X}(j+k)}\,{}^iX_{,l}\Gamma^l{}_{jk} + (-1)^{k(l+j)+\mathbf{X}j} \\ \times\Gamma^i{}_{lk}\,{}^lX_{,j} + (-1)^{\mathbf{X}k}\Gamma^i{}_{jl}\,{}^lX_{,k} + (-1)^{\mathbf{X}(j+k)}\,{}^iX_{,jk}\}. \end{aligned}$$

By an abuse of notation the quantity inside the final braces is sometimes written as $\Gamma^i{}_{jk}\,\mathfrak{L}_{\mathbf{X}}$.

2.4 The derivative mapping can also be extended so that it applies to measure functions. Let $(\mathscr{U}_\alpha, \psi_\alpha)$ and $(\bar{\mathscr{U}}_\beta, \bar{\psi}_\beta)$ be charts in M and $\bar{M}$ respectively such that $\phi(\mathscr{U}_\alpha) \cap (\bar{\mathscr{U}}_\beta)$ is nonempty. Let μ_α be a measure function for the chart $(\mathscr{U}_\alpha, \psi_\alpha)$. One may define a corresponding measure function in the restriction of the chart $(\bar{\mathscr{U}}_\beta, \bar{\psi}_\beta)$ to $\phi(\mathscr{U}_\alpha) \cap \bar{\mathscr{U}}_\beta$ by

$$\phi'(\mu)_\beta \stackrel{\text{def}}{=} J^{-1}(\bar{\psi}_\beta \circ \phi \circ \psi_\alpha^{-1})(\mu_\alpha \circ \phi^{-1}).$$

By mapping to $\bar{M}$ in this way measure functions satisfying eq. (2.9.3), from other charts in M, one can extend $\phi'(\mu_\beta)$ to the whole chart $(\bar{\mathscr{U}}_\beta; \bar{\psi}_\beta)$.

Now let $M = \bar{M}$ and $(\bar{\mathscr{U}}_\beta, \bar{\psi}_\beta) = (\mathscr{U}_\alpha, \psi_\alpha)$, and let ϕ be the mapping X_s introduced in section 2.5. By differentiating the equation

$$X'_s(\mu)_\alpha = J^{-1}(\psi_\alpha \circ X_s \circ \psi_\alpha^{-1})(\mu_\alpha \circ X_{-s})$$

with respect to s and setting $s = 0$ obtain the Lie derivative of μ with respect to the vector field $\mathbf{X}$. Show that it is given by

$$\mathfrak{L}_{\mathbf{X}}\mu = (-1)^i (X^i \mu)_{,i},$$

a formula that admits of immediate extension to the case in which $\mathbf{X}$ is complex and impure. *Hint*: Introduce the coordinate functions $(x^i) = \psi_\alpha(p)$ and show that $(\psi_\alpha \circ X_{\delta s})(p) = (x^i + X^i \delta s)$ when δs is infinitesimal.

2.5 Let $\phi: M \to \bar{M}$, $\psi: \bar{M} \to \bar{\bar{M}}$ be any two diffeomorphisms. Prove that $(\psi \circ \phi)' = \psi' \circ \phi'$ and $\phi^{-1\prime} = \phi'^{-1}$. If $\{\mathbf{e}_{ap}\}$ is a basis for $T_p(M)$ prove that $\{\phi'(\mathbf{e}_a)_{\phi(p)}\}$ is a basis for $T_{\phi(p)}(\bar{M})$: Let $\mathbf{X}$ and $\mathbf{Y}$ be any two contravariant vector fields on M. Prove that $\phi'([\mathbf{X}, \mathbf{Y}]) = [\phi'(\mathbf{X}), \phi'(\mathbf{Y})]$.

Let α be an arbitrary supernumber, and let $\mathbf{S}$ and $\mathbf{V}$ be any two tensor fields of equal rank on M. Prove that $\phi'(\alpha\mathbf{S}) = \alpha\phi'(\mathbf{S})$ and $\phi'(\mathbf{S} + \mathbf{V}) = \phi'(\mathbf{S}) + \phi'(\mathbf{V})$. Prove that $\phi'(\mathbf{dA}) = \mathbf{d}\phi'(\mathbf{A})$ where $\mathbf{A}$ is any r-form on M.

2.6 Derive the following identity, which generalizes eq. (2.8.34) to the case of non-Riemannian connections:

$$\begin{aligned}
&\mathbf{R}(\cdot, \mathbf{X}, \mathbf{Y}, \mathbf{Z}) + (-1)^{\mathbf{X}(\mathbf{Y}+\mathbf{Z})}\mathbf{R}(\cdot, \mathbf{Y}, \mathbf{Z}, \mathbf{X}) + (-1)^{\mathbf{Z}(\mathbf{X}+\mathbf{Y})}\mathbf{R}(\cdot, \mathbf{Z}, \mathbf{X}, \mathbf{Y}) \\
&\quad = -(\mathbf{T}\nabla)(\cdot, \mathbf{X}, \mathbf{Y}, \mathbf{Z}) - (-1)^{\mathbf{X}(\mathbf{Y}+\mathbf{Z})}(\mathbf{T}\nabla)(\cdot, \mathbf{Y}, \mathbf{Z}, \mathbf{X}) - (-1)^{\mathbf{Z}(\mathbf{X}+\mathbf{Y})} \\
&\qquad \times (\mathbf{T}\nabla)(\cdot, \mathbf{Z}, \mathbf{X}, \mathbf{Y}) - \mathbf{T}(\cdot, \mathbf{X}, \mathbf{T}(\cdot, \mathbf{Y}, \mathbf{Z})) - (-1)^{\mathbf{X}(\mathbf{Y}+\mathbf{Z})}\mathbf{T}(\cdot, \mathbf{Y}, \mathbf{T}(\cdot, \mathbf{Z}, \mathbf{X})) \\
&\qquad - (-1)^{\mathbf{Z}(\mathbf{X}+\mathbf{Y})}\mathbf{T}(\cdot, \mathbf{Z}, \mathbf{T}(\cdot, \mathbf{X}, \mathbf{Y})).
\end{aligned}$$

In a coordinate basis this identity takes the form

$$\begin{aligned}
&R^i{}_{jkl} + (-1)^{j(k+l)} R^i{}_{klj} + (-1)^{l(j+k)} R^i{}_{ljk} \\
&\quad = -T^i{}_{jk;l} - (-1)^{j(k+l)} T^i{}_{kl;j} - (-1)^{l(j+k)} T^i{}_{lj;k} \\
&\qquad - T^i{}_{jm} T^m{}_{kl} - (-1)^{j(k+l)} T^i{}_{km} T^m{}_{lj} - (-1)^{l(j+k)} T^i{}_{lm} T^m{}_{jk}.
\end{aligned}$$

2.7 Let $\{\mathbf{e}_a\}$ be a field of local frames on a supermanifold M, and let $\{\mathbf{e}^a\}$ be the set of dual basis fields. The $\mathbf{e}^a$ are 1-forms. Show that their exterior derivatives are given by

$$\mathbf{d}\mathbf{e}^a = -\tfrac{1}{2}{}^a[\mathbf{e}_b, \mathbf{e}_c]\mathbf{e}^c \wedge \mathbf{e}^b = -\tfrac{1}{2}{}^a c_{bc}\mathbf{e}^c \wedge \mathbf{e}^b,$$

where the ${}^a c_{bc}$ are the coefficients in eq. (2.3.23). Let M have a connection

∇ and let the connection components in the local frames be $\Gamma^a{}_{bc}$. Define

$$\omega^a{}_b \stackrel{\text{def}}{=} \Gamma^a{}_{bc}\mathbf{e}^c.$$

The $\omega^a{}_b$ are known as *connection* 1-*forms*. Define also

$$\mathbf{T}^a \stackrel{\text{def}}{=} \mathbf{d}\mathbf{e}^a + \omega^a{}_b \wedge \mathbf{e}^b,$$

$$\mathbf{\Omega}^a{}_b \stackrel{\text{def}}{=} \mathbf{d}\omega^a{}_b + \omega^a{}_c \wedge \omega^c{}_b.$$

Show that

$$\mathbf{T}^a = \tfrac{1}{2}T^a{}_{bc}\mathbf{e}^c \wedge \mathbf{e}^b,$$
$$\mathbf{\Omega}^a{}_b = -\tfrac{1}{2}R^a{}_{bcd}\mathbf{e}^d \wedge \mathbf{e}^c,$$

where $T^a{}_{bc}$ and $R^a{}_{bcd}$ are the components of the torsion and Riemannian tensor fields respectively in the basis $\{\mathbf{e}_a\}$. The $\mathbf{T}^a$ and $\mathbf{\Omega}^a{}_b$ are known as *torsion* and *curvature* 2-*forms* respectively. Finally, prove the identity

$$\mathbf{d}\mathbf{\Omega}^a{}_b + \omega^a{}_c \wedge \mathbf{\Omega}^c{}_b - \mathbf{\Omega}^a{}_c \wedge \omega^c{}_b \equiv 0$$

and show that it is equivalent to the Bianchi identity.

2.8 Let λ be an *a*-type supercurve in a supermanifold M. Prove that for any connection on M the range of λ contains an infinity of *c*-type supergeodesics, each of which would also be a supergeodesic with any other connection. *Hint*: Consider supergeodesics with tangents equal to $\alpha(\vec{\partial}/\partial s)_\lambda$ where α is any real *a*-number.

Show that the range of a *c*-type geodesic is not a sub-supermanifold if its tangent has vanishing body.

2.9 Let λ be a *c*-type supercurve in a Riemannian supermanifold M. Suppose $(\partial/\partial s)_\lambda \cdot (\partial/\partial s)_\lambda$ has nonvanishing body everywhere on λ, the inner product being defined by the metric tensor field. Let s_1 and s_2 be any two values of the supercurve parameter s (in the domain of λ) such that $s_{1\mathrm{B}} < s_{2\mathrm{B}}$. Then the *arc length of* λ *between* s_1 *and* s_2 is defined to be

$$L(\lambda, s_1, s_2) \stackrel{\text{def}}{=} \int_{s_1}^{s_2} |(\partial/\partial s)_\lambda \cdot (\partial/\partial s)_\lambda|^{1/2}\, \mathrm{d}s.$$

If s_1 and s_2 are close enough together so that the range of λ in the domain $\{s : s_{1\mathrm{B}} \le s_\mathrm{B} \le s_{2\mathrm{B}}\}$ can be covered by a single coordinate patch, one may write

$$L(\lambda, s_1, s_2) = \int_{s_1}^{s_2} |\dot{x}^i{}_i g_j \dot{x}^j|^{1/2}\, \mathrm{d}s.$$

Show that if $L(\lambda, s_1, s_2)$ remains stationary under infinitesimal variations of λ that leave the points $\lambda(s_1)$ and $\lambda(s_2)$ fixed, then λ is a supergeodesic with respect to the Riemannian connection on M. Show furthermore that if s is chosen so that $\dot{x}^i{}_i g_j \dot{x}^j$ is everywhere constant on λ, then s is an affine parameter.

2.10 Let λ_1 and λ_2 be any two supercurves passing through a point p of a supermanifold. Show that if the tangents of λ_1 and λ_2 are linearly independent at p then there exists an open set containing p but containing no other common point of λ_1 and λ_2.

2.11 Derive eqs. (2.8.32) and (2.8.34).

2.12 Compute the transformation laws for the components of the curvature and Ricci tensor fields, as well as the curvature scalar field, under the local conformal transformation (2.8.47). Show that if the transformed fields are to vanish the Riemann tensor field must have the form (2.8.50) and the scalar field μ must satisfy the differential equation

$$\mu_{;ij} = 2A_{ij} + \tfrac{1}{2}\mu_{;i}\,\mu_{;j} - \tfrac{1}{4}g_{ij}\,\mu_{;k}\,\mu_{;}{}^{k}.$$

This equation can be solved in a local open region if and only if it respects the identity

$$\mu_{;ijk} - (-1)^{jk}\,\mu_{;ikj} - \mu_{;l}\,R^{l}{}_{ijk} = 0.$$

Show that this identity will be respected if and only if the following integrability condition is satisfied:

$$A_{ij;k} - (-1)^{jk}\,A_{ik;j} = 0.$$

If $m - n$ is not equal to 1 or 2 then the Riemann tensor field automatically has the form (2.8.50) whenever the Weyl tensor field vanishes. If $m - n$ is equal to 1 or 2 and if $n \neq 0$ then the determination whether the form (2.8.50) is valid can be carried out by performing the computations indicated in eqs. (2.8.51). Suppose the form (2.8.50) *is* valid. Substitute it into the Bianchi identity and perform a sequence of partial contractions (analogous to those in eqs. (2.8.51)) on the result, distinguishing the various possible cases in which the indices are c-type or a-type and starting with the case in which all indices are a-type. Show that except when $(m, n) = (3, 0)$ the Bianchi identity *implies* the above integrability condition. In the case $(m, n) = (3, 0)$ the integrability condition must be imposed separately. In this case the form (2.8.50) automatically holds, with $A_{ij} = R_{ij} - \tfrac{1}{4}Rg_{ij}$, and the integrability condition requires the vanishing of the tensor field $\mathbf{D}$ of eq. (2.8.52).

Complete the proof of the theorem on conformal flatness by showing that every (1, 0)-dimensional manifold is trivially flat and that for every (2, 0)-dimensional manifold the scalar field μ need merely satisfy $\mu_{;i}{}^{i} = R$, which is always soluble locally.

2.13 Prove the theorem giving the general solution of eq. (2.8.57) by proceeding as follows. Consider eq. (2.8.57) together with its first and second partial derivatives. By distinguishing, in these three equations, the various possible cases in which the indices are c-type or a-type, and by performing partial contractions, show that except when $(m, n) = (1, 0)$ or $(2, 0)$, the function φ must satisfy $\varphi_{,ij} = 0$ and hence must have the form

$$\varphi = 2\xi - 4\zeta_i x^i,$$

where ζ and ζ^i are arbitrary constants. Now consider the first derivative of eq. (2.8.57):

$$C_{i,jk} + (-1)^{ij} C_{j,ik} = (-1)^{k(i+j)} \varphi_{,k} \eta_{ij}.$$

Interchange the indices j and k in this equation and multiply by $(-1)^{jk}$. Subtract from this the result of making a further interchange of the indices i and j with a factor $(-1)^{ij}$ inserted. Obtain

$$C_{i,jk} - (-1)^{ij} C_{j,ik} = -\varphi_{,i} \eta_{jk} + (-1)^{ij} \varphi_{,j} \eta_{ik}.$$

Added to the original equation this yields

$$C_{i,jk} = \tfrac{1}{2}[-\varphi_{,i} \eta_{jk} + (-1)^{ij} \varphi_{,j} \eta_{ik} + (-1)^{k(i+j)} \varphi_{,k} \eta_{ij}].$$

Together with $\varphi_{,ij} = 0$ this implies immediately

$$C_{i,jkl} = 0,$$

and hence

$$C_i = \varepsilon_i + \alpha_{ij} x^j + \beta_{ijk} x^k x^j$$

where the ε's, α's and β's are constants with $\beta_{ijk} = (-1)^{jk} \beta_{ikj}$. By substituting this and the expression for φ into eq. (2.8.57) and the equation or $C_{i,jk}$ show that

$$\beta_{ijk} = \zeta_i \eta_{jk} - (-1)^{ij} \zeta_j \eta_{ik} - (-1)^{k(i+j)} \zeta_k \eta_{ij},$$
$$\alpha_{ij} = \varepsilon_{ij} + \zeta \eta_{ij},$$

where the ε_{ij} are constants satisfying (2.8.63). This yields the general solution (2.8.61).

The proofs of parts 1 and 2 of the theorem, for the cases $(m, n) = (1, 0)$ and $(2, 0)$ respectively, are elementary and should present no particular difficulty.

Comments on chapter 2

It is evident from this chapter that nearly everything in ordinary manifold theory has its analog in the theory of supermanifolds. The chief exception is the de Rham theory of integration. Although the concepts of sub-supermanifolds and boundaries of sub-supermanifolds can be defined readily enough, integration over these subspaces cannot be defined in a natural way for two reasons. Firstly, the Berezin integral over *a*-number coordinates has no fixed endpoints and is not a measure theoretical construct. Therefore the only de Rham cohomology that a supermanifold possesses is (at best) that possessed by its body. Secondly, although integration over the *c*-number coordinates is always to be understood as an integration over a realization of the body, there is no natural (i.e., chart-independent) way of restricting the integration to an open subset $\mathscr{U}_{\mathrm{B}}$ of the realization unless the value of the integrand on the bundle of soul subspaces over the boundary $\partial\mathscr{U}_{\mathrm{B}}$ is independent of both the *a*-number coordinates and the souls of the *c*-number coordinates, a restriction that requires the integrand to be trivial if it is to be differentiable. Berezin's attempt to define an analog of de Rham theory for supermanifolds (F. A. Berezin, *Differential Forms on Supermanifolds*, *Sov. J. Nucl. Phys.*, **30**, 605 (1979)) does not avoid these difficulties.

The absence of a natural de Rham theory for supermanifolds means that the theory of differential forms is much less interesting for supermanifolds than it is for manifolds. Unless there are no *a*-number coordinates, forms do not play a role even in integration over the whole supermanifold. This is because when *a*-number coordinates are present, nontrivial forms of arbitrarily high rank exist, and there is no analog of the Levi–Civita ε symbol.

An interesting attempt to refine the topology of supermanifolds by giving the soul subspaces more topological structure has been made by Rogers (A. Rogers, *A Global Theory of Supermanifolds*, *J. Math. Phys.*, **21**, 1352 (1980)). It remains to be seen whether this additional structure can interact in an interesting way with differentiability criteria and the algebraic properties of the *a*-number coordinates.

Mention should be made of the work of Kostant† on *graded* manifolds, which is directed largely to mathematicians (B. Kostant, in *Differential Geometrical Methods in Mathematical Physics*, *Lecture Notes in Mathe-*

† A useful shorter work is F. A. Berezin & D. A. Leites, *Sov. Math. Dokl.*, **16**, 1218 (1975).

matics No. 570, Dold and Eckman, eds., Springer, 1977). Loosely speaking, graded manifolds are supermanifolds for which the *a*-number coordinates are regarded as fixed generators of a Grassmann algebra rather than as variable anticommuting elements of an infinite dimensional Grassmann algebra. In the theory of graded manifolds there is no room for skeletons (section 2.1), for *a*-type supercurves (section 2.1), for the infinite family of ordinary Lie groups associated with every super Lie group (section 3.1), or for unconventional super Lie groups (section 4.1).

3
Super Lie groups. General theory

3.1 Definition and structure of super Lie groups

Definition

We first recall the definition of a group. A *group* is a set G endowed with a binary-operation mapping $F: G \times G \to G$, abbreviated by $F(x, y) = xy$ for all x, y in G and called *multiplication*, which has the following properties:

(1) $(xy)z = x(yz) \stackrel{\text{def}}{=} xyz$, for all x, y, z in G.

(2) There exists an element e of G, called the *identity*, such that $ex = xe = x$ for all x in G.

(3) For every x in G there exists an element of G called the *inverse* of x, written x^{-1}, such that $x^{-1}x = xx^{-1} = e$.

It is easy to verify that e is a unique element of G and that every element has a unique inverse. In particular, $e^{-1} = e$ and $(x^{-1})^{-1} = x$ for all x in G.

A super Lie group is a group G that has also the following additional properties:

(4) It is a supermanifold, the points of which are the group elements.

(5) The multiplication mapping F is differentiable.

If, in the last two sentences, one replaces 'supermanifold' by 'manifold' and 'differentiable' by 'C^∞' one has the definition of an ordinary Lie group.

Associated with every super Lie group G is an infinite family of ordinary Lie groups, each of which is obtained by replacing G with one of its skeletons (see section 2.1). If G is replaced by its Nth skeleton the corresponding ordinary Lie group will be denoted by $S_N(G)$. Differentiability of the multiplication mapping in G assures that it is C^∞ in $S_N(G)$ for all N.

Another ordinary Lie group associated with every super Lie group G is its body G_B. Because the multiplication mapping in G maps soul subspaces onto soul subspaces, multiplication in G_B may be consistently (and C^∞ differentiably) defined by $\pi_G(x)\pi_G(y) = \pi_G(xy)$, where π_G is the mapping that sends each point of G to the soul subspace over that point (see section 2.1). This multiplication in G_B is easily verified to satisfy the group axioms, with $\pi_G(e)$ being the identity element of G_B and $[\pi_G(x)]^{-1} = \pi_G(x^{-1})$.

Canonical diffeomorphisms

Associated with every element x of G are two natural (canonical) one-to-one mappings, x_L and x_R, of G into itself, defined by

$$x_L(y) = xy, \qquad x_R(y) = yx, \tag{3.1.1}$$

for all y in G. The mapping x_L is called a *left translation*; the mapping x_R is called a *right translation.* Because the multiplication mapping is differentiable x_L and x_R are differentiable. They are in fact diffeomorphisms, for each is one-to-one and has a differentiable inverse given by

$$x_L^{-1} = (x^{-1})_L, \quad x_R^{-1} = (x^{-1})_R. \tag{3.1.2}$$

Note also the following relations:

$$(xy)_L = x_L \circ y_L, \quad (xy)_R = y_R \circ x_R, \tag{3.1.3}$$

$$x_L \circ y_R = y_R \circ x_L. \tag{3.1.4}$$

The last relation says that left translations commute with right translations.

Since x_L and x_R are diffeomorphisms one may introduce the corresponding derivative mappings x_L' and x_R' (see section 2.5). These mappings will be called *left* and *right draggings* respectively. Let f be an arbitrary scalar field on G. Then

$$\left.\begin{aligned} [x_L'(f)](y) &= (f \circ x_L^{-1})(y) = f(x_L^{-1}(y)) = f(x^{-1}y), \\ [x_R'(f)](y) &= (f \circ x_R^{-1})(y) = f(x_R^{-1}(y)) = f(yx^{-1}), \end{aligned}\right\} \tag{3.1.5}$$

for all y in G. Let $\mathbf{X}$ be an arbitrary contravariant vector field on G. Then

$$\left.\begin{aligned} [x_L'(\mathbf{X})]_y x_L'(f) = [x_L'(\mathbf{X}) x_L'(f)](y) &= [x_L'(\mathbf{X}f)](y) \\ &= (\mathbf{X}f)(x^{-1}y) = \mathbf{X}_{x^{-1}y} f, \\ [x_R'(\mathbf{X})]_y x_R'(f) = [x_R'(\mathbf{X}) x_R'(f)](y) &= [x_R'(\mathbf{X}f)](y) \\ &= (\mathbf{X}f)(yx^{-1}) = \mathbf{X}_{yx^{-1}} f, \end{aligned}\right\} \tag{3.1.6}$$

for all y in G. Continuing in this way one may compute the actions of x_L' and x_R' on any tensor field on G:

$$\left.\begin{aligned} [x_L'(\mathbf{X}) \cdot x_L'(\omega)](y) &= (\mathbf{X} \cdot \omega)(x^{-1}y), \\ [x_R'(\mathbf{X}) \cdot x_R'(\omega)](y) &= (\mathbf{X} \cdot \omega)(yx^{-1}), \end{aligned}\right\} \tag{3.1.7}$$

$$\left.\begin{aligned} [(x_L'(\mathbf{T}))(x_L'(\omega), \ldots, x_L'(\mathbf{X}))](y) &= [\mathbf{T}(\omega, \ldots, \mathbf{X})](x^{-1}y), \\ [x_R'(\mathbf{T}))(x_R'(\omega), \ldots, x_R'(\mathbf{X}))](y) &= [\mathbf{T}(\omega, \ldots, \mathbf{X})](yx^{-1}). \end{aligned}\right\} \tag{3.1.8}$$

Left- and right-invariant vector fields

Let $\mathbf{X}_L$ and $\mathbf{X}_R$ be contravariant vector fields on G that satisfy

$$\mathbf{X}_{Le} = \mathbf{X}_{Re} \stackrel{\text{def}}{=} \mathbf{X}, \tag{3.1.9}$$

$$\mathbf{X}_{Lx} = [x'_L(\mathbf{X}_L)]_x, \quad \mathbf{X}_{Rx} = [x'_R(\mathbf{X}_R)]_x, \tag{3.1.10}$$

for all x in G. By setting $y = x$ in eqs. (3.1.6) one sees that $\mathbf{X}_L$ and $\mathbf{X}_R$ are completely determined by their common value $\mathbf{X}$ at the identity. (Note that $\mathbf{X}$ is here a vector at e, not a vector field.)

Let y be an arbitrary element of G and f an arbitrary scalar field on G. Then, for all x in G,

$$\begin{aligned}[y'_L(\mathbf{X}_L)]_x y'_L(f) &= [y'_L(\mathbf{X}_L f)](x) = (\mathbf{X}_L f)(y^{-1}x) = \mathbf{X}_{Ly^{-1}x} f \\ &= [(y^{-1}x)'_L(\mathbf{X}_L)]_{y^{-1}x} f = [y'_L((y^{-1}x)'_L(\mathbf{X}_L))]_x y'_L(f) \\ &= [x'_L(\mathbf{X}_L)]_x y'_L(f) = \mathbf{X}_{Lx} y'_L(f),\end{aligned} \tag{3.1.11}$$

whence

$$y'_L(\mathbf{X}_L) = \mathbf{X}_L. \tag{3.1.12}$$

The vector field $\mathbf{X}_L$ is evidently invariant under left draggings. Similarly $\mathbf{X}_R$ is invariant under right draggings:

$$y'_R(\mathbf{X}_R) = \mathbf{X}_R. \tag{3.1.13}$$

$\mathbf{X}_L$ and $\mathbf{X}_R$ are called *left-* and *right-invariant vector fields* respectively.

Denote by $\mathfrak{X}_L(G)$ the set of all left-invariant vector fields on G and by $\mathfrak{X}_R(G)$ the set of all right-invariant vector fields on G. Using the elementary properties of derivative mappings proved in exercise **2.5** one may easily show that $\mathfrak{X}_L(G)$ and $\mathfrak{X}_R(G)$ are closed under the operations of addition, multiplication by supernumbers, and forming super Lie brackets. Each is therefore a supervector space endowed with a bracket operation that satisfies the super Jacobi identity. Since every member of $\mathfrak{X}_L(G)$ or $\mathfrak{X}_R(G)$ is completely determined by its value at e it follows that $\mathfrak{X}_L(G)$ and $\mathfrak{X}_R(G)$ are isomorphic to $T_e(G)$, the tangent space to G at the identity. They are thus finite-dimensional supervector spaces, their dimensions being the same as that of G.

Left- and right-invariant local frame fields

Let $\{\mathbf{e}_a\}$ be a basis for $T_e(G)$. Let $\mathbf{e}_{aL}$ and $\mathbf{e}_{aR}$ be, respectively, the left- and right-invariant vector fields having the value $\mathbf{e}_a$ at the identity. The sets $\{\mathbf{e}_{aL}\}$ and $\{\mathbf{e}_{aR}\}$ are bases for $\mathfrak{X}_L(G)$ and $\mathfrak{X}_R(G)$ respectively. They are more than that. Since the property of linear independence at a point is maintained under a derivative mapping (see exercise 2.5), $\{\mathbf{e}_{aLx}\}$ and $\{\mathbf{e}_{aRx}\}$ are bases for $T_x(G)$. This means that $\{\mathbf{e}_{aL}\}$ and $\{\mathbf{e}_{aR}\}$ are in fact local frame fields on U. They are defined everywhere on G. Therefore each is a global field of local frames.

Theorem

Every super Lie group admits a left-invariant global field of local frames and a right-invariant global field of local frames.

Left- and right-invariant congruences

Let $\mathbf{X}_L$ be a real c-type left-invariant contravariant vector field on G. Introduce the congruence of c-type supercurves $\lambda_{\mathbf{X}_L,x}$ in G generated by $\mathbf{X}_L$ (see section 2.5). Explicitly

$$(\partial/\partial s)_{\lambda_{\mathbf{X}_L,x}(s)} = \mathbf{X}_{L\lambda_{\mathbf{X}_L,x}(s)}, \qquad \lambda_{\mathbf{X}_L,x}(0) = x. \tag{3.1.14}$$

Define

$$x_{\mathbf{X}_L} \stackrel{\text{def}}{=} \lambda_{\mathbf{X}_L,e}, \tag{3.1.15}$$

which is the c-type supercurve generated by $\mathbf{X}_L$, passing through the identity:

$$x_{\mathbf{X}_L}(0) = \lambda_{\mathbf{X}_L,e}(0) = e. \tag{3.1.16}$$

Let y be an arbitrary element of G. Let $yx_{\mathbf{X}_L}$ be the c-type supercurve in G defined by

$$(yx_{\mathbf{X}_L})(s) = y(x_{\mathbf{X}_L}(s)) \stackrel{\text{def}}{=} yx_{\mathbf{X}_L}(s). \tag{3.1.17}$$

Let f be an arbitrary scalar field on G. Then

$$\begin{aligned}
(\partial/\partial s)_{yx_{\mathbf{X}_L}(s)} y'_L(f) &= (\partial/\partial s)_{yx_{\mathbf{X}_L}(s)}(f \circ y_L^{-1}) \\
&= \frac{\mathrm{d}}{\mathrm{d}s}(f \circ y_L^{-1})(yx_{\mathbf{X}_L}(s)) \\
&= \frac{\mathrm{d}}{\mathrm{d}s} f(x_{\mathbf{X}_L}(s)) = (\partial/\partial s)_{x_{\mathbf{X}_L}(s)} f \\
&= \mathbf{X}_{Lx_{\mathbf{X}_L}(s)} f = \{[x_{\mathbf{X}_L}(s)]'_L(\mathbf{X}_L)\}_{x_{\mathbf{X}_L}(s)} f \\
&= \{[x_{\mathbf{X}_L}(s)]'_L(\mathbf{X}_L)\}_{y^{-1}yx_{\mathbf{X}_L}(s)} f \\
&= (y'_L\{[x_{\mathbf{X}_L}(s)]'_L(\mathbf{X}_L)\})_{yx_{\mathbf{X}_L}(s)} y'_L(f) \\
&= \{[yx_{\mathbf{X}_L}(s)]'_L(\mathbf{X}_L)\}_{yx_{\mathbf{X}_L}(s)} y'_L(f) \\
&= \mathbf{X}_{Lyx_{\mathbf{X}_L}(s)} y'_L(f),
\end{aligned} \tag{3.1.18}$$

which implies

$$(\partial/\partial s)_{yx_{\mathbf{X}_L}(s)} = \mathbf{X}_{Lyx_{\mathbf{X}_L}(s)}. \tag{3.1.19}$$

That is, $yx_{\mathbf{X}_L}$ is an integral supercurve of $\mathbf{X}_L$. It has the same boundary value as $\lambda_{\mathbf{X}_L,y}$:

$$yx_{\mathbf{X}_L}(0) = y = \lambda_{\mathbf{X}_L,y}(0). \tag{3.1.20}$$

Therefore it must be identical with $\lambda_{\mathbf{X}_L,y}$:

$$yx_{\mathbf{X}_L} = \lambda_{\mathbf{X}_L,y}. \tag{3.1.21}$$

Evidently the congruence of c-type supercurves generated by $\mathbf{X}_L$ is obtained by multiplying the integral supercurve that passes through the identity on the left by the elements of G. This congruence is said to be *left invariant*.

In a similar manner one can show that the congruence of c-type supercurves generated by a real c-type right-invariant contravariant vector field $\mathbf{X}_R$ is obtained by multiplying the integral supercurve passing through the identity on the right by the elements of G:

$$\lambda_{\mathbf{X}_R,y} = x_{\mathbf{X}_R} y, \tag{3.1.22}$$

$$x_{\mathbf{X}_R} = \lambda_{\mathbf{X}_R,e}, \qquad x_{\mathbf{X}_R}(0) = e. \tag{3.1.23}$$

This congruence is said to be *right invariant.*

One-parameter Abelian subgroups

Let $X_{Ls}: G \to G$ be the diffeomorphism defined by

$$X_{Ls}(y) \stackrel{\text{def}}{=} \lambda_{\mathbf{X}_L,y}(s) = yx_{\mathbf{X}_L}(s) \tag{3.1.24}$$

(see section 2.5). Note that this diffeomorphism, which is generated by the *left*-invariant vector field $\mathbf{X}_L$, is a *right* translation:

$$X_{Ls} = [x_{\mathbf{X}_L}(s)]_R. \tag{3.1.25}$$

The corresponding derivative mapping X'_{Ls} is therefore a *right* dragging:

$$X'_{Ls} = [x_{\mathbf{X}_L}(s)]'_R. \tag{3.1.26}$$

Similarly, the diffeomorphism $X_{Rs}: G \to G$,

$$y \mapsto X_{Rs}(y) \stackrel{\text{def}}{=} \lambda_{\mathbf{X}_R,y}(s) = x_{\mathbf{X}_R}(s)y, \tag{3.1.27}$$

generated by the *right*-invariant vector field $\mathbf{X}_R$, is a *left* translation:

$$X_{Rs} = [x_{\mathbf{X}_R}(s)]_L, \tag{3.1.28}$$

and the corresponding derivative mapping is a *left* dragging:

$$X'_{Rs} = [x_{\mathbf{X}_R}(s)]'_L. \tag{3.1.29}$$

We are now in a position to prove the following

Theorem

If $\mathbf{X}_L$ and $\mathbf{X}_R$ are related by eq. (3.1.9) then the supercurves $x_{\mathbf{X}_L}$ and $x_{\mathbf{X}_R}$ are identical and their range is a one-parameter Abelian subgroup of G.

We begin by proving that $x_{\mathbf{X}_L}(\mathbf{R}_c)$ is a one-parameter Abelian subgroup of G. Note that eqs. (3.1.15) and (3.1.24) imply

$$x_{\mathbf{X}_L}(s) = \lambda_{\mathbf{X}_{L,e}}(s) = X_{Ls}(e). \tag{3.1.30}$$

Therefore, if s and t are any two points of $\mathbf{R}_c$, we obtain, with the aid of eqs. (2.5.41) and (3.1.24),

$$\begin{aligned} x_{\mathbf{X}_L}(s)x_{\mathbf{X}_L}(t) &= X_{Lt}(x_{\mathbf{X}_L}(s)) = (X_{Lt}\circ X_{Ls})(e) = X_{Lt+s}(e) \\ &= x_{\mathbf{X}_L}(t+s) = x_{\mathbf{X}_L}(s+t) = x_{\mathbf{X}_L}(t)x_{\mathbf{X}_L}(s), \end{aligned} \tag{3.1.31}$$

qed. In a similar manner we find

$$\begin{aligned} x_{\mathbf{X}_R}(s) &= X_{Rs}(e), \\ x_{\mathbf{X}_R}(s)x_{\mathbf{X}_R}(t) &= X_{Rs}(x_{\mathbf{X}_R}(t)) = (X_{Rs}\circ X_{Rt})(e) = X_{Rs+t}(e) \\ &= x_{\mathbf{X}_R}(s+t) = x_{\mathbf{X}_R}(t+s) = x_{\mathbf{X}_R}(t)x_{\mathbf{X}_R}(s). \end{aligned} \tag{3.1.32}$$

In view of eqs. (3.1.16) and (3.1.23) we have the corollaries

$$[x_{\mathbf{X}_L}(s)]^{-1} = x_{\mathbf{X}_L}(-s), \qquad [x_{\mathbf{X}_R}(s)]^{-1} = x_{\mathbf{X}_R}(-s). \tag{3.1.33}$$

The proof that $x_{\mathbf{X}_L}$ and $x_{\mathbf{X}_R}$ are identical proceeds as follows. Let f be an arbitrary scalar field on G. Then

$$\begin{aligned} \frac{\mathrm{d}}{\mathrm{d}s}X'_{Ls}(f) &= \left[\frac{\mathrm{d}}{\mathrm{d}t}X'_{Ls+t}(f)\right]_{t=0} = \left[\frac{\mathrm{d}}{\mathrm{d}t}(X_{Lt}\circ X_{Ls})'(f)\right]_{t=0} \\ &= \left[\frac{\mathrm{d}}{\mathrm{d}t}X'_{Lt}(X'_{Ls}(f))\right]_{t=0} = -\mathbf{X}_L X'_{Ls}(f), \end{aligned} \tag{3.1.34}$$

in which eq. (2.5.45) has been used in passing to the final form. Similarly,

$$\frac{\mathrm{d}}{\mathrm{d}s}X'_{Rs}(f) = -\mathbf{X}_R X'_{Rs}(f). \tag{3.1.35}$$

Alternative versions of (3.1.34) and (3.1.35) are

$$\frac{\mathrm{d}}{\mathrm{d}s}f\circ[x_{\mathbf{X}_L}(s)]_R = \frac{\mathrm{d}}{\mathrm{d}s}[x_{\mathbf{X}_L}(-s)]'_R(f) = \mathbf{X}_L[x_{\mathbf{X}_L}(-s)]'_R(f), \tag{3.1.36}$$

$$\frac{\mathrm{d}}{\mathrm{d}s}f\circ[x_{\mathbf{X}_R}(s)]_L = \frac{\mathrm{d}}{\mathrm{d}s}[x_{\mathbf{X}_R}(-s)]'_L(f) = \mathbf{X}_R[x_{\mathbf{X}_R}(-s)]'_L(f). \tag{3.1.37}$$

Evaluating eq. (3.1.36) at the identity and making use of eq. (3.1.4) and the invariance of $\mathbf{X}_\mathrm{R}$ under right draggings, one obtains

$$\begin{aligned}\frac{\mathrm{d}}{\mathrm{d}s}f(x_{\mathbf{X}_\mathrm{L}}(s)) &= \mathbf{X}_{\mathrm{L}e}[x_{\mathbf{X}_\mathrm{L}}(-s)]'_\mathrm{R}(f) = \mathbf{X}_{\mathrm{R}e}[x_{\mathbf{X}_\mathrm{L}}(-s)]'_\mathrm{R}(f)\\ &= \{\mathbf{X}_\mathrm{R}[x_{\mathbf{X}_\mathrm{L}}(-s)]'_\mathrm{R}(f)\}(e) = \{[x_{\mathbf{X}_\mathrm{L}}(-s)]'_\mathrm{R}(\mathbf{X}_\mathrm{R}f)\}(e)\\ &= (\mathbf{X}_\mathrm{R}f)(x_{\mathbf{X}_\mathrm{L}}(s)).\end{aligned} \tag{3.1.38}$$

But, by definition,

$$\frac{\mathrm{d}}{\mathrm{d}s}f(x_{\mathbf{X}_\mathrm{L}}(s)) = (\partial/\partial s)_{x\mathbf{X}_\mathrm{L}(s)}f = \mathbf{X}_{\mathrm{L}x\mathbf{X}_\mathrm{L}(s)}f = (\mathbf{X}_\mathrm{L}f)(x_{\mathbf{X}_\mathrm{L}}(s)). \tag{3.1.39}$$

In a similar manner one also obtains

$$\frac{\mathrm{d}}{\mathrm{d}s}f(x_{\mathbf{X}_\mathrm{R}}(s)) = (\mathbf{X}_\mathrm{L}f)(x_{\mathbf{X}_\mathrm{R}}(s)) = (\mathbf{X}_\mathrm{R}f)(x_{\mathbf{X}_\mathrm{R}}(s)). \tag{3.1.40}$$

Since $f(x_{\mathbf{X}_\mathrm{L}}(s))$ and $f(x_{\mathbf{X}_\mathrm{R}}(s))$ satisfy identical differential equations with identical boundary conditions, and since f is arbitrary, it follows that $x_{\mathbf{X}_\mathrm{L}}$ and $x_{\mathbf{X}_\mathrm{R}}$ are identical:

$$x_{\mathbf{X}_\mathrm{L}} = x_{\mathbf{X}_\mathrm{R}} \stackrel{\mathrm{def}}{=} x_\mathbf{X}. \tag{3.1.41}$$

Equations (3.1.39) and (3.1.40) furthermore imply that $\mathbf{X}_\mathrm{L}$ and $\mathbf{X}_\mathrm{R}$ coincide not only at e but everywhere on $x_\mathbf{X}$:

$$\mathbf{X}_{\mathrm{L}x\mathbf{X}(s)} = \mathbf{X}_{\mathrm{R}x\mathbf{X}(s)}, \quad \text{for all } s \text{ in } \mathbf{R}_c. \tag{3.1.42}$$

We close this section by rewriting several earlier key equations in the notation of eq. (3.1.41).

$$x_\mathbf{X}(s)x_\mathbf{X}(t) = x_\mathbf{X}(t)x_\mathbf{X}(s) = x_\mathbf{X}(s+t), \tag{3.1.43}$$

$$X_{\mathrm{L}s} = [x_\mathbf{X}(s)]_\mathrm{R}, \quad X_{\mathrm{R}s} = [x_\mathbf{X}(s)]_\mathrm{L}, \tag{3.1.44}$$

$$\left.\begin{aligned}\lambda_{\mathbf{X}_\mathrm{L},y}(s) = X_{\mathrm{L}s}(y) = yx_\mathbf{X}(s),\\ \lambda_{\mathbf{X}_\mathrm{R},y}(s) = X_{\mathrm{R}s}(y) = x_\mathbf{X}(s)y,\end{aligned}\right\} \tag{3.1.45}$$

$$\frac{\mathrm{d}}{\mathrm{d}s}f(x_\mathbf{X}(s)) = (\mathbf{X}_\mathrm{L}f)(x_\mathbf{X}(s)) = (\mathbf{X}_\mathrm{R}f)(x_\mathbf{X}(s)). \tag{3.1.46}$$

The exponential mapping. Canonical coordinates

Repeated differentiation of eq. (3.1.46) yields

$$\left(\frac{\mathrm{d}}{\mathrm{d}s}\right)^n f(x_\mathbf{X}(s)) = (\mathbf{X}_\mathrm{L}^n f)(x_\mathbf{X}(s)) = (\mathbf{X}_\mathrm{R}^n f)(x_\mathbf{X}(s)). \tag{3.1.47}$$

Suppose the scalar field f is chosen in such a way that $f(x_\mathbf{X}(s))$ is a superanalytic function of s (see section 1.1). Then one may write

$$f(x_{\mathbf{X}}(s)) = [\exp(s\partial/\partial t)f(x_{\mathbf{X}}(t))]_{t=0}$$
$$= [\exp(s\mathbf{X}_L)f](e) = [\exp(s\mathbf{X}_R)f](e). \tag{3.1.48}$$

This suggests the introduction of the following notation:

$$\exp(s\mathbf{X}) \overset{\text{def}}{=} x_{\mathbf{X}}(s), \tag{3.1.49}$$

for all s in $\mathbf{R}_c$ and all $\mathbf{X}$ in $E_e(G)$, where $E_e(G)$ is the subspace of $T_e(G)$ consisting of all the real c-type contravariant vectors at e. The mapping *exp* from $E_e(G)$ to G is known as the *exponential mapping*.

If G has dimension (p, q) then $E_e(G)$ is isomorphic to $\mathbf{R}_c^p \times \mathbf{R}_a^q$ and inherits a topology therefrom. If *exp* is restricted to a sufficiently small open set $\mathscr{U}$ of $E_e(G)$, containing the zero vector, then it will generally be one-to-one, and $\exp(\mathscr{U})$ will be an open set of G containing the identity. Let $\{\mathbf{e}_a\}$ be a standard basis for $T_e(G)$ and let $\{\mathbf{e}^a\}$ be the dual basis. Let ϕ be the mapping from $\exp(\mathscr{U})$ to $\mathbf{R}_c^p \times \mathbf{R}_a^q$ defined by

$$\phi(x) = (\mathbf{e}^a \cdot \exp^{-1}(x)). \tag{3.1.50}$$

Then $(\exp(\mathscr{U}), \phi)$ is a chart in G containing the identity. The coordinates $\mathbf{e}^a \cdot \exp^{-1}(x)$ of this chart are known as *canonical coordinates*. Canonical coordinates are unique up to a linear transformation of the form (2.4.13). Note that they vanish at the identity.

The super Lie algebra

We have previously remarked that $\mathfrak{X}_L(G)$ and $\mathfrak{X}_R(G)$ are finite-dimensional supervector spaces endowed with bracket operations (super Lie brackets) that satisfy the super Jacobi identity. Such supervector spaces are known as *super Lie algebras*.

Since the elements of $\mathfrak{X}_L(G)$ and $\mathfrak{X}_R(G)$ are completely determined by their values at the identity, the super Lie bracket operations on $\mathfrak{X}_L(G)$ and $\mathfrak{X}_R(G)$ define corresponding bracket operations on $T_e(G)$. For definiteness choose $\mathfrak{X}_R(G)$ and set

$$[\mathbf{X}, \mathbf{Y}] \overset{\text{def}}{=} [\mathbf{X}_R, \mathbf{Y}_R]_e, \tag{3.1.51}$$

for all $\mathbf{X}, \mathbf{Y}$ in $T_e(G)$, the quantity on the right being the super Lie bracket evaluated at the identity. The supervector space $T_e(G)$, endowed with the bracket operation (3.1.51), is called the *super Lie algebra of* G.

If $\mathfrak{X}_L(G)$ were chosen instead of $\mathfrak{X}_R(G)$ then the brackets in $T_e(G)$ would be reversed in sign. This may be seen by first restricting attention to real

c-type supervectors $\mathbf{X}$, $\mathbf{Y}$ in $T_e(G)$. One may then write, for all f in $\mathfrak{F}(G)$ (see eqs. (2.5.45), (3.1.3) and (3.1.44)),

$$\begin{aligned}\mathbf{X}_{\mathrm{L}}\mathbf{Y}_{\mathrm{L}}f &= \left[\frac{\partial^2}{\partial s\,\partial t}(X_{\mathrm{L}\,-s}{}'\circ Y_{\mathrm{L}\,-t}{}')(f)\right]_{s=t=0} \\ &= \left\{\frac{\partial^2}{\partial s \partial t}([x_{\mathbf{X}}(-s)]_{\mathrm{R}}'\circ[x_{\mathbf{Y}}(-t)]_{\mathrm{R}}')(f)\right\}_{s=t=0} \\ &= \left\{\frac{\partial^2}{\partial s \partial t}[x_{\mathbf{Y}}(-t)x_{\mathbf{X}}(-s)]_{\mathrm{R}}'(f)\right\}_{s=t=0} \\ &= \left\{\frac{\partial^2}{\partial s \partial t} f\circ[x_{\mathbf{Y}}(-t)x_{\mathbf{X}}(-s)]_{\mathrm{R}}^{-1}\right\}_{s=t=0} \\ &= \left\{\frac{\partial^2}{\partial s \partial t} f\circ[x_{\mathbf{X}}(s)x_{\mathbf{Y}}(t)]\right\}_{s=t=0} \end{aligned} \tag{3.1.52}$$

and, similarly,

$$\mathbf{X}_{\mathrm{R}}\mathbf{Y}_{\mathrm{R}}f = \left\{\frac{\partial^2}{\partial s \partial t} f\circ[x_{\mathbf{Y}}(t)x_{\mathbf{X}}(s)]\right\}_{s=t=0}. \tag{3.1.53}$$

From this it follows that

$$(\mathbf{X}_{\mathrm{L}}\mathbf{Y}_{\mathrm{L}}f)(e) = \left[\frac{\partial^2}{\partial s \partial t} f(x_{\mathbf{X}}(s)x_{\mathbf{Y}}(t))\right]_{s=t=0} = (\mathbf{Y}_{\mathrm{R}}\mathbf{X}_{\mathrm{R}}f)(e) \tag{3.1.54}$$

and hence (see eq. (2.3.16))

$$\begin{aligned}[\mathbf{X}_{\mathrm{L}}, \mathbf{Y}_{\mathrm{L}}]_e f &= -(\mathbf{X}_{\mathrm{L}}\mathbf{Y}_{\mathrm{L}}f)(e) + (\mathbf{Y}_{\mathrm{L}}\mathbf{X}_{\mathrm{L}}f)(e) \\ &= -(\mathbf{Y}_{\mathrm{R}}\mathbf{X}_{\mathrm{R}}f)(e) + (\mathbf{X}_{\mathrm{R}}\mathbf{Y}_{\mathrm{R}}f)(e) \\ &= -[\mathbf{X}_{\mathrm{R}}, \mathbf{Y}_{\mathrm{R}}]_e f. \end{aligned} \tag{3.1.55}$$

Now let $\mathbf{X}$ and $\mathbf{Y}$ be arbitrary elements of $T_e(G)$. By decomposing them into their real, imaginary, even and odd parts, multiplying the odd parts by arbitrary generators ζ^a of Λ_∞, and then using (3.1.55) and the fact that if $\zeta^a\mathbf{Z} = 0$ for all a then $\mathbf{Z} = 0$, one may conclude quite generally that

$$[\mathbf{X}_{\mathrm{L}}, \mathbf{Y}_{\mathrm{L}}]_e = -[\mathbf{X}_{\mathrm{R}}, \mathbf{Y}_{\mathrm{R}}]_e. \tag{3.1.56}$$

The structure constants

Let $\{\mathbf{e}_{a\mathrm{R}}\}$ be a right-invariant global field of local frames over G. Since $\{\mathbf{e}_{a\mathrm{R}}\}$ is a basis for $\mathfrak{X}_{\mathrm{R}}(G)$ and since $\mathfrak{X}_{\mathrm{R}}(G)$ is closed under the super Lie bracket operation there must exist a set of constant supernumbers ${}^c c_{ab}$ such that

$$[\mathbf{e}_{a\mathrm{R}}, \mathbf{e}_{b\mathrm{R}}] = \mathbf{e}_{c\mathrm{R}}\,{}^c c_{ab}. \tag{3.1.57}$$

The ${}^c c_{ab}$ are known as the *structure constants* of G (or of the super Lie algebra of G). In view of eq. (3.1.56) the super Lie brackets of the corresponding left-invariant global field of local frames are given by

$$[\mathbf{e}_{aL}, \mathbf{e}_{bL}] = -\mathbf{e}_{cL}\,{}^c c_{ab}. \tag{3.1.58}$$

Equation (3.1.57) may be evaluated at the identity, yielding the bracket operation of the super Lie algebra in terms of the basis $\{\mathbf{e}_a\}$ of $T_e(G)$:

$$[\mathbf{e}_a, \mathbf{e}_b] = \mathbf{e}_c\,{}^c c_{ab}. \tag{3.1.59}$$

From this one may compute the bracket of any two elements $\mathbf{X}, \mathbf{Y}$ of $T_e(G)$ in terms of their components relative to the basis $\{\mathbf{e}_a\}$:

$$\begin{aligned} {}^a[\mathbf{X}, \mathbf{Y}] &= {}^a\mathbf{e}\cdot[\mathbf{e}_b\,{}^bX, \mathbf{e}_c\,{}^cY] \\ &= (-1)^{c(\mathbf{X}+b)}\,{}^a c_{bc}\,{}^bX\,{}^cY. \end{aligned} \tag{3.1.60}$$

By applying the super Jacobi identity to any three basis vectors and using the linear independence of the $\mathbf{e}_a$, one discovers that the structure constants must satisfy the identity

$${}^d c_{ae}\,{}^e c_{bc} + (-1)^{a(b+c)}\,{}^d c_{be}\,{}^e c_{ca} + (-1)^{c(a+b)}\,{}^d c_{ce}\,{}^e c_{ab} = 0. \tag{3.1.61}$$

The problem of classifying all super Lie algebras is the problem of finding, modulo nonsingular linear transformations of the form

$${}^a\bar{c}_{bc} = (-1)^{f(e+b)}\,{}^a\mathrm{L}^{-1}{}_d\,{}^d c_{ef}\,{}^e\mathrm{L}_b\,{}^f\mathrm{L}_c, \tag{3.1.62}$$

all sets of pure supernumbers ${}^a c_{bc}$ satisfying (3.1.61) and the symmetry condition

$${}^a c_{bc} = -(-1)^{bc}\,{}^a c_{cb}. \tag{3.1.63}$$

The right and left auxiliary functions

It is sometimes convenient to express the relations satisfied by right- or left-invariant vector fields in coordinate language. In what follows the coordinates of points x, y, z in G will be indicated by affixing indices from the first part of the alphabet: x^a, y^b, z^c.† The mere affixing of indices is a rather casual convention, for although it assumes the existence of a global atlas it does not specify which chart is being used at any given point. The convention does no harm as long as it is understood that although x^a and y^b may refer to different charts when x and y are far apart, the same chart is to be used for x^a and y^b whenever x and y are brought into coincidence.

† In this chapter therefore we drop the restriction of reserving such indices for the labelling of basis vectors of local frames and components therein.

In coordinate language the group properties (1)–(3) take the form

$$F^a(xy, z) = F^a(x, yz), \tag{3.1.64}$$

$$F^a = (e, x) = F^a(x, e) = x^a, \tag{3.1.65}$$

$$F^a(x^{-1}, x) = F^a(x, x^{-1}) = e^a, \tag{3.1.66}$$

the multiplication mapping symbol F being employed here through the definition

$$F^a(x, y) \stackrel{\text{def}}{=} (xy)^a. \tag{3.1.67}$$

It is convenient also to introduce the following abbreviations:

$$_{\dots bc}F^a{}_{de\dots}(x, y) \stackrel{\text{def}}{=} \dots \frac{\vec{\partial}}{\partial x^b}\frac{\vec{\partial}}{\partial x^c} F^a(x, y) \frac{\overleftarrow{\partial}}{\partial y^d}\frac{\overleftarrow{\partial}}{\partial y^e}\dots. \tag{3.1.68}$$

Consider now the second of eqs. (3.1.6), with y replaced by yx. In coordinate language it takes the form

$$\begin{aligned} [x'_{\mathbf{R}}(\mathbf{X})]_{yx}{}^a \frac{\vec{\partial}}{\partial (yx)^a} f(y) &= X_y{}^a \frac{\vec{\partial}}{\partial y^a} f(y) \\ &= X_y{}^b \left[\frac{\vec{\partial}}{\partial y^b}(yx)^a\right] \frac{\vec{\partial}}{\partial (yx)^a} f(y). \end{aligned} \tag{3.1.69}$$

Since f is arbitrary one may infer that†

$$[x'_{\mathbf{R}}(\mathbf{X})]_{yx}{}^a = X_y{}^b{}_bF^a(y, x). \tag{3.1.70}$$

In a similar manner one finds†

$$^a[x'_{\mathbf{L}}(\mathbf{X})]_{xy} = F^a{}_b(x, y)^b X_y. \tag{3.1.71}$$

With y set equal to e the coordinate versions of eqs. (3.1.10) immediately follow

$$X_{\mathbf{R}}{}^a(x) = X^b{}_bR^a(x), \qquad {}^aX_{\mathbf{L}}(x) = {}^aL_b(x)^b X, \tag{3.1.72}$$

$${}_bR^a(x) \stackrel{\text{def}}{=} {}_bF^a(e, x), \qquad {}^aL_b(x) \stackrel{\text{def}}{=} F^a{}_b(x, e). \tag{3.1.73}$$

Equations (3.1.72) provide explicit constructions of right- and left-invariant vector fields in terms of the functions ${}_bR^a(x)$ and ${}^aL_b(x)$ respectively. These functions are known as the *right* and *left auxiliary functions* of G. They are, in fact, the components of the global local frame fields $\{{}_b\mathbf{e}_{\mathbf{R}}\}$ and $\{\mathbf{e}_{b\mathbf{L}}\}$ respectively. Because of the pointwise linear independence of these local frame fields the matrices

$$R(x) \stackrel{\text{def}}{=} ({}_bR^a(x)), \qquad L(x) \stackrel{\text{def}}{=} ({}^aL_b(x)) \tag{3.1.74}$$

† ${}_bF^a(y, x)$ and $F^a{}_b(x, y)$ are evidently components of bitensor fields (see section 2.5), the lower index referring to the point y and the upper index to the points yx and xy respectively.

are necessarily nonsingular in every admissible chart. They reduce to the unit matrix at the identity:

$$R(e) = 1^{\sim}_{(p,q)}, \qquad L(e) = 1_{(p,q)}. \tag{3.1.75}$$

Identities satisfied by the auxiliary functions

The auxiliary functions satisfy a number of important identities, which can be obtained by differentiating eq. (3.1.64) with respect to the arguments appearing in it:

$$_bF^c(x, y)\,_cF^a(xy, z) = {}_bF^a(x, yz), \tag{3.1.76}$$

$$(-1)^{b(c+1)}\,F^c{}_b(x, y)\,_cF^a(xy, z) = (-1)^{c(a+1)}\,_bF^c(y, z)\,F^a{}_c(x, yz), \tag{3.1.77}$$

$$F^a{}_b(xy, z) = F^a{}_c(x, yz)\,F^c{}_b(y, z). \tag{3.1.78}$$

Setting $x = e$ in the first of these equations, $y = e$ in the second, and $z = e$ in the third, we get

$$_bR^c(y)\,_cF^a(y, z) = {}_bR^a(yz), \tag{3.1.79}$$

$$(-1)^{b(c+1)}\,{}^cL_b(x)\,_cF^a(x, z) = (-1)^{c(a+1)}\,_bR^c(x)\,F^a{}_c(x, z), \tag{3.1.80}$$

$${}^aL_b(xy) = F^a{}_c(x, y)\,{}^cL_b(y), \tag{3.1.81}$$

whence

$$_bF^a(x, y) = {}_bR^{-1c}(x)\,_cR^a(xy), \tag{3.1.82}$$

$$F^a{}_b(x, y) = {}^aL_c(x, y)\,{}^cL_b^{-1}(y), \tag{3.1.83}$$

and

$$L^{\sim}(x)R^{-1}(x)R(xy) = R(y)L^{\sim -1}(y)L^{\sim}(xy). \tag{3.1.84}$$

Equations (3.1.82) and (3.1.83), when rewritten in the forms

$$\frac{\vec{\partial}}{\partial x^b}F^a = {}_bR^{-1c}(x)\,_cR^a(F), \tag{3.1.85}$$

$$F^a\frac{\overleftarrow{\partial}}{\partial y^b} = {}^aL_c(F)\,{}^cL_b^{-1}(y), \tag{3.1.86}$$

show that a knowledge of either R or L alone suffices to determine the multiplication mapping F. The functions $F^a(x, y)$ may be determined by integrating either (3.1.85) or (3.1.86) subject to the boundary conditions (3.1.65).

Another set of identities satisfied by the auxiliary functions is obtained by expressing eqs. (3.1.57) and (3.1.58) in coordinate component form:

$${}^dR^{\sim}{}_{a,c}\,{}^cR^{\sim}{}_b - (-1)^{ab}\,{}^dR^{\sim}{}_{b,c}\,{}^cR^{\sim}{}_a = {}^dR^{\sim}_c\,{}^cc_{ab}, \tag{3.1.87}$$

$${}^dL_{a,c}\,{}^cL_b - (-1)^{ab}\,{}^dL_{b,c}\,{}^cL_a = -{}^dL_c\,{}^cc_{ab}. \tag{3.1.88}$$

These equations, in which the structure constants appear, may be shown to be the integrability conditions for eqs. (3.1.85) and (3.1.86). The structure constants themselves are obtained by evaluating expression (3.1.87) or (3.1.88) at the identity:

$$
\begin{aligned}
{}^{c}c_{ab} &= {}^{c}R^{\sim}{}_{a,b}(e) - (-1)^{ab}\,{}^{c}R^{\sim}{}_{b,a}(e) && \text{(3.1.89a)}\\
&= -{}^{c}L_{a,b}(e) + (-1)^{ab}\,{}^{c}L_{b,a}(e) && \text{(3.1.89b)}\\
&= (-1)^{a(c+1)}\,{}_{a}F^{c}{}_{b}(e,e) - (-1)^{b(c+a+1)}\,{}_{b}F^{c}{}_{a}(e,e). && \text{(3.1.89c)}
\end{aligned}
$$

Construction of a super Lie group from its super Lie algebra

We have seen that a knowledge of either of the auxiliary functions suffices to determine the multiplication mapping and hence the group G itself. If the auxiliary functions are known only in a chart containing the identity then the multiplication mapping can still be determined for all x, y in that chart such that xy is also in the chart. It turns out that in a canonical coordinate system the auxiliary functions are completely determined by the structure constants. Therefore the group itself is completely determined, in a canonical chart, by its own super Lie algebra.

To see this begin by writing eq. (3.1.46) in coordinate language:

$$
\begin{aligned}
\dot{x}_{\mathbf{X}}{}^{a}(s) &= X_{\mathrm{L}}{}^{a}(x_{\mathbf{X}}(s)) = X_{\mathrm{R}}{}^{a}(x_{\mathbf{X}}(s))\\
&= {}^{a}R^{\sim}{}_{b}(x_{\mathbf{X}}(s))\,X^{b} = {}^{a}L_{b}(x_{\mathbf{X}}(s))\,X^{b}. \qquad \text{(3.1.90)}
\end{aligned}
$$

Here the dot denotes differentiation with respect to s and, in passing to the last line, we have used eqs. (3.1.72) and the fact that $X^{b} = {}^{b}X$. Now suppose the coordinate system is canonical. Then from eqs. (3.1.49) and (3.1.50) it follows that

$$x_{\mathbf{X}}{}^{a}(s) = \mathbf{e}^{a}\cdot\exp^{-1}(x_{\mathbf{X}}(s)) = sX^{a}. \qquad \text{(3.1.91)}$$

Substitute this into the left-hand side of eq. (3.1.90) and multiply the result by s, obtaining

$$x^{a} = {}^{a}R^{\sim}{}_{b}(x)x^{b} = {}^{a}L_{b}(x)x^{b} \qquad \text{(3.1.92)}$$

for all x in the canonical chart.

The relation (3.1.92) is a condition that completely characterizes canonical coordinates. It will be convenient to have it in the alternative form

$$x^{a} = {}^{a}R^{\sim -1}{}_{b}(x)x^{b} = {}^{a}L^{-1}{}_{b}(x)x^{b}. \qquad \text{(3.1.93)}$$

It will also be convenient to rewrite eqs. (3.1.87) and (3.1.88) in the forms

$$
\begin{aligned}
{}^{a}R^{\sim -1}{}_{b,c} - (-1)^{bc}\,{}^{a}R^{\sim -1}{}_{c,b} &= -(-1)^{e(b+d)}\,{}^{a}c_{de}\,{}^{d}R^{\sim -1}{}_{b}\,{}^{e}R^{\sim -1}{}_{c}, && \text{(3.1.94)}\\
{}^{a}L^{-1}{}_{b,c} - (-1)^{bc}\,{}^{a}L^{-1}{}_{c,b} &= (-1)^{e(b+d)}\,{}^{a}c_{de}\,{}^{d}L^{-1}{}_{b}\,{}^{e}L^{-1}{}_{c}. && \text{(3.1.95)}
\end{aligned}
$$

Now multiply these last two equations by x^c and use (3.1.93), obtaining

$$ {}^aR^{\sim-1}{}_{b,c}x^c - (-1)^{bc}\,{}^aR^{\sim-1}{}_{c,b}x^c = -(-1)^{c(b+d)}\,{}^ac_{dc}\,{}^dR^{\sim-1}{}_bx^c, \quad (3.1.96)$$

$$ {}^aL^{-1}{}_{b,c}x^c - (-1)^{bc}\,{}^aL^{-1}{}_{c,b}x^c = (-1)^{c(b+d)}\,{}^ac_{dc}\,{}^dL^{-1}{}_bx^c. \quad (3.1.97)$$

Next differentiate eqs. (3.1.93) with respect to x^b:

$$ {}^a\delta_b = (-1)^{bc}\,{}^aR^{\sim-1}{}_{c,b}x^c + {}^aR^{\sim-1}{}_b, \quad (3.1.98)$$

$$ {}^a\delta_b = (-1)^{bc}\,{}^aL^{-1}{}_{c,b}x^c + {}^aL^{-1}{}_b. \quad (3.1.99)$$

Finally, add eqs. (3.1.96) and (3.1.97) to eqs. (3.1.98) and (3.1.99) respectively. The results may be expressed in matrix form:

$$ 1_{(p,q)} = R^{\sim-1} + R^{\sim-1}{}_{,a}x^a - \mathbf{c}\cdot\mathbf{x}\,R^{\sim-1}, \quad (3.1.100a)$$

$$ = L^{-1} + L^{-1}{}_{,a}x^a + \mathbf{c}\cdot\mathbf{x}\,L^{-1}, \quad (3.1.100b)$$

where

$$ {}^a(\mathbf{c}\cdot\mathbf{x})_b \overset{\text{def}}{=} (-1)^{bc}\,{}^ac_{cb}\,x^c. \quad (3.1.101)$$

Equations (3.1.100) have unique solutions satisfying the boundary conditions (3.1.75). These are

$$ R^{\sim-1} = 1_{(p,q)} + \frac{1}{2!}\mathbf{c}\cdot\mathbf{x} + \frac{1}{3!}(\mathbf{c}\cdot\mathbf{x})^2 + \cdots = \frac{\mathrm{e}^{\mathbf{c}\cdot\mathbf{x}} - 1_{(p,q)}}{\mathbf{c}\cdot\mathbf{x}}, \quad (3.1.102)$$

$$ L^{-1} = 1_{(p,q)} - \frac{1}{2!}\mathbf{c}\cdot\mathbf{x} + \frac{1}{3!}(\mathbf{c}\cdot\mathbf{x})^2 - \cdots = \frac{1_{(p,q)} - \mathrm{e}^{-\mathbf{c}\cdot\mathbf{x}}}{\mathbf{c}\cdot\mathbf{x}}. \quad (3.1.103)$$

Because $\mathbf{c}\cdot\mathbf{x}$ is a singular matrix (applied to (x^a) it gives zero) the closed expressions are meant only in a formal sense as standing for the infinite series. The series themselves are everywhere convergent, and hence R^{-1} and L^{-1} exist for all possible values of the canonical coordinates. The matrices L and R, however, do not necessarily exist for all x^a since, for certain values of the x^a, R^{-1} and L^{-1} may be singular. Canonical coordinates may cover the group. Indeed, if the group is compact and connected they may cover it more than once. However, they provide an acceptable chart only in a one-to-one singularity-free neighbourhood of the identity. In this neighbourhood expressions (3.1.102) and (3.1.103) may be used in eqs. (3.1.85) and (3.1.86) to compute the multiplication mapping functions $F^a(x, y)$. It is easy to obtain the first few terms of the series expansion of these functions:

$$ (xy)^a = F^a(x, y) = x^a + y^a - \tfrac{1}{2}\,{}^ac_{bc}\,x^c\,y^b + 0(x^2y, xy^2). \quad (3.1.104)$$

We record here also the following additional relations that hold when canonical coordinates are employed:

$$ e^a = 0, \qquad x^{-1a} = -x^a, \quad (3.1.105)$$

$$\left.\begin{aligned} L^{-1}(x) &= e^{-c\cdot x}R^{-1\sim}(x) = R^{-1\sim}(x)e^{-c\cdot x}, \\ L(x) &= e^{c\cdot x}R^{\sim}(x) = R^{\sim}(x)e^{c\cdot x}, \end{aligned}\right\} \tag{3.1.106}$$

$$L(x) = R^{\sim}(x^{-1}). \tag{3.1.107}$$

In the theory of ordinary Lie groups it is well known that two distinct Lie groups may have the same Lie algebra (e.g. SU(2) and SO(3)). Therefore the complete group is *not* determined by the structure constants alone. Since ordinary Lie groups are but special cases of super Lie groups the same must be true of the latter. For many important classes of ordinary Lie groups (e.g., compact or semisimple Lie groups) there exists a complete classification of the distinct global groups having the same Lie algebra. In the case of super Lie groups and super Lie algebras the analogous theory has yet to be worked out.

3.2 Realizations of super Lie groups

Definition

Let Φ be a supermanifold and G a super Lie group. Let the points of Φ be denoted by Greek letters φ, χ, ψ, etc. Let $R: G \times \Phi \to \Phi$ be a differentiable mapping from $G \times \Phi$ to Φ which, for all x, y in G and all φ in Φ, satisfies the following conditions:

$$R(e, \varphi) = \varphi, \tag{3.2.1}$$

$$R(xy, \varphi) = R(x, R(y, \varphi)). \tag{3.2.2}$$

The mapping R is called a *realization* of G, and G is said to *act on* Φ.

Super Lie groups are commonly encountered in the guise of realizations, and therefore it is important to study the general properties of such mappings. A realization serves to *define* the super Lie group provided it is *faithful*. The realization R is said to be faithful if and only if $R(x, \varphi) = R(y, \varphi)$ for all φ implies $x = y$. Note that '$R(x, \varphi) = R(y, \varphi)$ for all φ' implies '$R(xy^{-1}, \varphi) = R(x, R(y^{-1}, \varphi)) = R(y, R(y^{-1}, \varphi)) = \varphi$ for all φ' and conversely. Hence an equivalent statement of the faithfulness criterion is: $R(x, \varphi) = \varphi$ for all φ implies $x = e$.

Orbits

In discussing realizations of super Lie groups it is convenient to introduce the following subsets:

$$\mathrm{Orb}(\chi) \overset{\text{def}}{=} \{\psi \in \Phi \,|\, \psi = R(x, \chi) \quad \text{for some } x \text{ in } G\}, \tag{3.2.3}$$

$$K(\varphi, \chi) \stackrel{\text{def}}{=} \{x \in G | \varphi = R(x, \chi)\}. \tag{3.2.4}$$

Here φ and χ are any two points of Φ. $\mathrm{Orb}(\chi)$ is called the *orbit of G through* χ. The following statements are easily seen to be equivalent: (1) $\mathrm{Orb}(\varphi) = \mathrm{Orb}(\chi)$; (2) $\varphi \in \mathrm{Orb}(\chi)$; (3) $\chi \in \mathrm{Orb}(\varphi)$; (4) $K(\varphi, \chi)$ is nonempty.

If, for some χ in Φ, the orbit of G through χ does not contain all points of Φ (i.e., if $\mathrm{Orb}(\chi) \neq \Phi$) then some of the sets $K(\varphi, \chi)$ are empty. In this case the group decomposes Φ into more than one orbit and the realization is said to be *intransitive*. If the realization is faithful then the generic orbit will typically be a sub-supermanifold of Φ on which the group G acts both faithfully and *transitively*.†

Transitive realizations

The realization R is said to be *transitive* if $K(\varphi, \chi)$ is nonempty for all φ, χ in Φ. Said in another way, the realization is transitive if, for any two points φ and χ of Φ, there exists at least one x in G such that $\varphi = R(x, \chi)$. Although we shall not prove it, if R is transitive then x can be chosen to tend smoothly to e as φ tends to χ.

For the next several paragraphs we shall confine our attention to transitive realizations. The sets $K(\varphi, \chi)$, besides being then nonempty, have also other important properties. To obtain these it is helpful to introduce some simple notation. Let A, B and C be any subsets of G and x an element of G. Define

$$A^{-1} \stackrel{\text{def}}{=} \{x \in G | x^{-1} \in A\}, \tag{3.2.5}$$

$$xA \stackrel{\text{def}}{=} \{z \in G | z = xy \quad \text{for some } y \in A\}, \tag{3.2.6}$$

$$Ax \stackrel{\text{def}}{=} \{z \in G | z = yx \quad \text{for some } y \in A\}, \tag{3.2.7}$$

$$AB \stackrel{\text{def}}{=} \{z \in G | z = xy \quad \text{for some } x \in A \text{ and some } y \in B\}. \tag{3.2.8}$$

The following statements are easily verified.

$$(A^{-1})^{-1} = A, \tag{3.2.9}$$

$$A \subset B \quad \text{implies} \quad A^{-1} \subset B^{-1} \quad \text{and} \quad AC \subset BC, \tag{3.2.10}$$

$$(AB)C = A(BC) \stackrel{\text{def}}{=} ABC. \tag{3.2.11}$$

† There may be degenerate orbits for which this is not true. Degenerate orbits need not be sub-supermanifolds.

Moreover, if $AB \subset C$ then $ABB^{-1} \subset CB^{-1}$, which, because the identity is obviously an element of BB^{-1}, implies $A \subset CB^{-1}$. That is to say,

$$AB \subset C \quad \text{implies} \quad A \subset CB^{-1}. \tag{3.2.12}$$

Now suppose x is an element of $K(\varphi, \chi)$. Then $R(x^{-1}, \varphi) = R(x^{-1}, R(x, \chi)) = \chi$, which implies that x^{-1} is an element of $K(\chi, \varphi)$. Thus

$$K(\varphi, \chi)^{-1} \subset K(\chi, \varphi). \tag{3.2.13}$$

Taking the inverses of both sides of this relation and interchanging φ and χ (which are arbitrary), we obtain also

$$K(\chi, \varphi) \subset K(\varphi, \chi)^{-1}, \tag{3.2.14}$$

whence

$$K(\varphi, \chi)^{-1} = K(\chi, \varphi). \tag{3.2.15}$$

Next, let φ, χ, ψ be any three points of Φ and let x be an element of $K(\varphi, \chi)$ and y an element of $K(\chi, \psi)$. Then $R(xy, \psi) = R(x, R(y, \psi)) = R(x, \chi) = \varphi$, which implies that xy is an element of $K(\varphi, \psi)$. An equivalent statement is

$$K(\varphi, \chi)K(\chi, \psi) \subset K(\varphi, \psi), \tag{3.2.16}$$

which, in view of (3.2.12) and (3.2.15), implies

$$K(\varphi, \chi) \subset K(\varphi, \psi)K(\chi, \psi)^{-1} = K(\varphi, \psi)K(\psi, \chi). \tag{3.2.17}$$

Interchanging χ and ψ we get

$$K(\varphi, \psi) \subset K(\varphi, \chi)K(\chi, \psi), \tag{3.2.18}$$

which, together with (3.2.16), yields

$$K(\varphi, \chi)K(\chi, \psi) = K(\varphi, \psi). \tag{3.2.19}$$

Isotropy subgroups

If the points φ, χ, ψ are taken to coincide then eqs. (3.2.15) and (3.2.19) reduce to

$$K(\varphi, \varphi)^{-1} = K(\varphi, \varphi), \tag{3.2.20}$$

$$K(\varphi, \varphi)K(\varphi, \varphi) = K(\varphi, \varphi). \tag{3.2.21}$$

which, together with the obvious fact that $K(\varphi, \varphi)$ contains the identity, implies that $K(\varphi, \varphi)$ is a subgroup of G. This subgroup is called the *isotropy subgroup at* φ. It is the set of elements of G that leave φ invariant.

Although the isotropy subgroup at each point of Φ is distinct from that at every other point, all the isotropy subgroups are isomorphic. This may be shown as follows. Let φ and χ be any two points of Φ and let x be any element of $K(\varphi, \chi)$. Then x^{-1} is an element of $K(\chi, \varphi)$ and

$$xK(\chi, \chi)x^{-1} \subset K(\varphi, \chi)K(\chi, \chi)K(\chi, \varphi) = K(\varphi, \varphi), \tag{3.2.22}$$

whence

$$K(\chi, \chi) \subset x^{-1}K(\varphi, \varphi)x. \tag{3.2.23}$$

Since φ and χ are arbitrary (3.2.23) must hold with φ and χ interchanged and x replaced by x^{-1}. Thus

$$K(\varphi, \varphi) \subset xK(\chi, \chi)x^{-1}, \tag{3.2.24}$$

which, together with (3.2.22), implies

$$xK(\chi, \chi)x^{-1} = K(\varphi, \varphi). \tag{3.2.25}$$

From this it follows that the mapping $x_A : K(\chi, \chi) \to K(\varphi, \varphi)$, defined by $x_A(y) = xyx^{-1}$ for all y in $K(\chi, \chi)$, is one-to-one. Moreover, this mapping is readily seen to be an isomorphism: $x_A(e) = e$, $x_A(y^{-1}) = x_A(y)^{-1}$, $x_A(yz) = x_A(y)x_A(z)$ for all y, z in $K(\chi, \chi)$, qed.

Coset spaces

Let x again be any element of $K(\varphi, \chi)$. Then

$$xK(\chi, \chi) \subset K(\varphi, \chi)K(\chi, \chi) = K(\varphi, \chi), \tag{3.2.26}$$

$$x^{-1}K(\varphi, \chi) \subset K(\chi, \varphi)K(\varphi, \chi) = K(\chi, \chi). \tag{3.2.27}$$

Relation (3.2.27) may be rewritten

$$K(\varphi, \chi) \subset xK(\chi, \chi). \tag{3.2.28}$$

which, together with (3.2.26) implies

$$K(\varphi, \chi) = xK(\chi, \chi). \tag{3.2.29}$$

For fixed χ the subsets $K(\varphi, \chi)$ are seen to be the *left cosets* of the isotropy subgroup $K(\chi, \chi)$. It is easy to verify, similarly, that for fixed φ the $K(\varphi, \chi)$ are the *right cosets* of the isotropy subgroup $K(\varphi, \varphi)$.

The left cosets $K(\varphi, \chi)$ of the isotropy subgroup $K(\chi, \chi)$ of any chosen fixed point χ may themselves be regarded as 'points' in another space. This space is known as a *left coset space* and is denoted by $[G/K(\chi, \chi)]_L$. Because all the isotropy subgroups are isomorphic, all the left coset spaces (for different choices of χ) are isomorphic. Thus $[G/K(\psi, \psi)]_L$ is obtained from $[G/K(\chi, \chi)]_L$ by multiplying all the elements of the latter on the right by $K(\chi, \psi)$. In a similar manner the right cosets $K(\chi, \varphi)$ of $K(\chi, \chi)$ may be assembled into a space $[G/K(\chi, \chi)]_R$ known as a *right coset space*. Again all the right coset spaces are isomorphic to one another.

The points φ of Φ are in one-to-one correspondence with the points

$K(\varphi,\chi)$ of $[G/K(\chi,\chi)]_L$ as well as with the points $K(\chi,\varphi)$ of $[G/K(\chi,\chi)]_R$, χ fixed. Indeed, for every χ, Φ may be *identified* with either of these coset spaces, for the realization $R:G\times\Phi\to\Phi$ is itself 'realized' either by the mapping $R^\chi_L:G\times[G/K(\chi,\chi)]_L\to[G/K(\chi,\chi)]_L$ defined by $R^\chi_L(y,K(\varphi,\chi))=yK(\varphi,\chi)$ or by the mapping $R^\chi_R:G\times[G/K(\chi,\chi)]_R\to[G/K(\chi,\chi)]_R$ defined by $R^\chi_R(y,K(\chi,\varphi))=K(\chi,\varphi)y^{-1}$. To show this let x be an element of $K(\varphi,\chi)$. Then x^{-1} is an element of $K(\chi,\varphi)$. Moreover $R(yx,\chi)=R(y,R(x,\chi))=R(y,\varphi)$, which implies that yx belongs to the left coset $K(R(y,\varphi),\chi)$ and hence that $(yx)^{-1}$ belongs to the right coset $K(\chi,R(y,\varphi))$. From this it follows that

$$R^\chi_L(y,K(\varphi,\chi))=yK(\varphi,\chi)=yxK(\chi,\chi)=K(R(y,\varphi),\chi), \tag{3.2.30}$$

$$R^\chi_R(y,K(\chi,\varphi))=K(\chi,\varphi)y^{-1}=K(\chi,\chi)(yx)^{-1}=K(\chi,R(y,\varphi)), \tag{3.2.31}$$

for all φ, χ in Φ and all y in G.

The coset spaces $[G/K(\chi,\chi)]_L$ and $[G/K(\chi,\chi)]_R$ in G obviously inherit a supermanifold structure from the supermanifold Φ. It is an important result of the theory that each also inherits a supermanifold structure from G itself, which is consistent with that inherited from Φ. To show this it is necessary to introduce a special coordinate system in G, determined in part by the subgroup $K(\chi,\chi)$, and to introduce the concept of *Killing flows* on Φ. We begin with the latter.

Killing flows

Let $x_{\mathbf{X}}(\mathbf{R}_c)$ be the one-parameter Abelian subgroup of G generated by the real c-type contravariant vector $\mathbf{X}$ at the identity. Let $\lambda_{\mathbf{X},\varphi}$ be the c-type supercurve in Φ defined by

$$\lambda_{\mathbf{X},\varphi}(s)\overset{\text{def}}{=}R(x_{\mathbf{X}}(s),\varphi). \tag{3.2.32}$$

The set of supercurves $\lambda_{\mathbf{X},\varphi}$ for all φ, with fixed $\mathbf{X}$, forms a congruence in Φ. Denote by $\mathbf{Q}_{\mathbf{X}}$ the real c-type contravariant vector field on Φ that generates this congruence:

$$\mathbf{Q}_{\mathbf{X}\lambda_{\mathbf{X},\varphi}(s)}\overset{\text{def}}{=}(\partial/\partial s)_{\lambda_{\mathbf{X},\varphi}(s)}. \tag{3.2.33}$$

This vector field is called the *Killing flow on* Φ *generated by* $\mathbf{X}$.

Let f be an arbitrary scalar field on Φ. Then

$$\mathbf{Q}_{\mathbf{X}\lambda_{\mathbf{X},\varphi}(s)}f=\frac{\mathrm{d}}{\mathrm{d}s}f(\lambda_{\mathbf{X},\varphi}(s))=\frac{\mathrm{d}}{\mathrm{d}s}f(R(x_{\mathbf{X}}(s),\varphi)), \tag{3.2.34}$$

$$\mathbf{Q}_{\mathbf{X}\varphi}f=\left[\frac{\mathrm{d}}{\mathrm{d}s}f(R(x_{\mathbf{X}}(s),\varphi))\right]_{s=0}. \tag{3.2.35}$$

Since the mappings $f:\Phi\to\Lambda_\infty$, $R:G\times\Phi\to\Phi$ and $x_{\mathbf{X}}:\mathbb{R}_c\to G$ are all differentiable, the quantity on the right of eq. (3.2.35) is a differentiable function of φ. Hence $\mathbf{Q_X}$ is a differentiable vector field.

Let $\mathbf{Q_Y}$ be a second Killing flow on Φ, generated by a real c-type element $\mathbf{Y}$ of $T_e(G)$. Then

$$\begin{aligned}(\mathbf{Q_X Q_Y}f)(\varphi) = \mathbf{Q}_{\mathbf{X}\varphi}(\mathbf{Q_Y}f) &= \left[\frac{\mathrm{d}}{\mathrm{d}s}(\mathbf{Q_Y}f)(R(x_{\mathbf{X}}(s),\varphi))\right]_{s=0}\\ &= \left[\frac{\partial^2}{\partial s\,\partial t}f(R(x_{\mathbf{Y}}(t),R(x_{\mathbf{X}}(s),\varphi)))\right]_{s=t=0}\\ &= \left[\frac{\partial^2}{\partial s\,\partial t}f(R(x_{\mathbf{Y}}(t)x_{\mathbf{X}}(s),\varphi))\right]_{s=t=0}. \end{aligned} \tag{3.2.36}$$

Define

$$g_\varphi(x) \overset{\text{def}}{=} f(R(x,\varphi)). \tag{3.2.37}$$

Use of eq. (3.1.54) yields

$$\begin{aligned}(\mathbf{Q_X Q_Y}f)(\varphi) &= \left[\frac{\partial^2}{\partial s\,\partial t}g_\varphi(x_{\mathbf{Y}}(t)x_{\mathbf{X}}(s))\right]_{s=t=0}\\ &= (\mathbf{X_R Y_R}g_\varphi)(e),\end{aligned} \tag{3.2.38}$$

and hence (see eqs. (3.1.46) and (3.1.51))

$$\begin{aligned}[\mathbf{Q_X},\mathbf{Q_Y}]_\varphi f &= [\mathbf{X_R},\mathbf{Y_R}]_e g_\varphi = [\mathbf{X},\mathbf{Y}]g_\varphi\\ &= \left[\frac{\mathrm{d}}{\mathrm{d}s}g_\varphi(x_{[\mathbf{X},\mathbf{Y}]}(s))\right]_{s=0} = \left[\frac{\mathrm{d}}{\mathrm{d}s}f(R(x_{[\mathbf{X},\mathbf{Y}]}(s),\varphi))\right]_{s=0}\\ &= \mathbf{Q}_{[\mathbf{X},\mathbf{Y}]\varphi}f.\end{aligned} \tag{3.2.39}$$

Since f and φ are arbitrary it follows that

$$[\mathbf{Q_X},\mathbf{Q_Y}] = \mathbf{Q}_{[\mathbf{X},\mathbf{Y}]}. \tag{3.2.40}$$

The concept of Killing flows can be extended to the case in which $\mathbf{X}$ and $\mathbf{Y}$ are complex and/or impure by defining

$$\mathbf{Q}_{\alpha\mathbf{X}} \overset{\text{def}}{=} \alpha\mathbf{Q_X}, \qquad \mathbf{Q}_{\mathbf{X}+\mathbf{Y}} \overset{\text{def}}{=} \mathbf{Q_X}+\mathbf{Q_Y}, \tag{3.2.41}$$

for all α in Λ_∞ and all $\mathbf{X}$, $\mathbf{Y}$ in $T_e(G)$. Equation (3.2.40) continues to hold for such generalized flows, and the set $\mathscr{Q}(\Phi)$ of all Killing flows on Φ is closed under the operations of addition, multiplication by supernumbers, and forming super Lie brackets. If the realization R is faithful then $\mathscr{Q}(\Phi)$ is isomorphic to $T_e(G)$ and constitutes a faithful realization of the super Lie algebra of G. This follows from the fact that the faithfulness condition for R implies $\mathbf{Q_X}=\mathbf{0}$ if and only if $\mathbf{X}=\mathbf{0}$.

Let $\{\mathbf{e}_a\}$ be a standard basis for $T_e(G)$. Introduce the abbreviation

$$\mathbf{Q}_a \overset{\text{def}}{=} \mathbf{Q}_{\mathbf{e}_a}. \tag{3.2.42}$$

If R is faithful then $\{\mathbf{Q}_a\}$ is a standard basis for $\mathscr{Q}(\Phi)$. Whether R is faithful or not the following equations hold:

$$\mathbf{Q}_{\mathbf{X}} = \mathbf{Q}_a{}^a X, \tag{3.2.43}$$

$$[\mathbf{Q}_a, \mathbf{Q}_b] = \mathbf{Q}_c{}^c c_{ab}. \tag{3.2.44}$$

Properties of the coordinate components of the $\mathbf{Q}_a$

Let us now introduce in Φ a chart containing the fixed point χ. Let the coordinates of the variable point φ in this chart be denoted by φ^i, and let functions R^i be introduced such that the equation $\varphi = R(x, \chi)$ may be written in the coordinate form $\varphi^i = R^i(x, \chi)$. Let x^a be canonical coordinates in G and let $\mathbf{X}$ be a real c-type element of $T_e(G)$. If f is an arbitrary scalar field on Φ then

$$\begin{aligned} f_{,i}(\varphi)^i Q_a(\varphi)^a X &= f\mathbf{Q}_{\mathbf{X}\varphi} = \mathbf{Q}_{\mathbf{X}\varphi} f \\ &= \left[\frac{\mathrm{d}}{\mathrm{d}s} f(R(x_{\mathbf{X}}(s), \varphi))\right]_{s=0} = f_{,i}(\varphi) R^i{}_{,a}(e, \varphi)\dot{x}_{\mathbf{X}}{}^a(0) \\ &= f_{,i}(\varphi) R^i{}_{,a}(e, \varphi)^a X, \end{aligned} \tag{3.2.45}$$

where

$$R^i{}_{,a}(x, \varphi) \overset{\text{def}}{=} R^i(x, \varphi)\frac{\overleftarrow{\partial}}{\partial x^a}. \tag{3.2.46}$$

Since f and $\mathbf{X}$ are arbitrary it follows that the components of the $\mathbf{Q}_a$ in the coordinate system φ^i are given by

$${}^i Q_a(\varphi) = R^i{}_{,a}(e, \varphi). \tag{3.2.47}$$

Let x be infinitesimally close to the identity and let $\varphi^i = R^i(x, \chi)$. Then $\varphi^i = \chi^i + \delta\varphi^i$, where the $\delta\varphi^i$ are the components of the infinitesimal vector

$$\delta\varphi = \mathbf{Q}_{a\chi}\, x^a. \tag{3.2.48}$$

Denote by Q the matrix $({}^i Q_a(\chi))$. Let the dimensions of Φ and G be (m, n) and (p, q) respectively. Then the body of Q has the block form

$$Q_B = \begin{pmatrix} A & 0 \\ 0 & B \end{pmatrix}, \tag{3.2.49}$$

where A is an ordinary $m \times p$ matrix and B is an ordinary $n \times q$ matrix. We shall show that the rows of Q_B (and hence of A and B) are linearly independent.

Suppose they were not. Then there would exist a real covariant vector ω at χ, of pure type and with nonvanishing body, such that $(\omega \cdot \mathbf{Q}_{a\chi})_{\mathrm{B}} = [\omega_i {}^i Q_a(\chi)]_{\mathrm{B}} = 0$ for all a. If ω were c-type one could choose a real c-type infinitesimal contravariant vector $\delta\varphi$ at χ such that $(\omega \cdot \delta\varphi)_{\mathrm{B}} \neq 0$, and there would exist no x^a's that could be inserted into eq. (3.2.48) yielding that vector. But this contradicts the transitivity of the realization R. Therefore ω, if it exists at all, must be of type a.

If ω is of type a then $\omega \cdot \mathbf{Q}_{a\chi}$ is a real a-number or a real c-number according as the index a is c-type or a-type. Let these bodiless numbers be expanded in powers of the generators of Λ_∞, as in eq. (1.1.3), and let $(\omega \cdot \mathbf{Q}_{a\chi})_1$ denote the result of dropping all but the linear terms. Evidently $(\omega \cdot \mathbf{Q}_{a\chi})_1$ vanishes when the index a is of type a and it lies in the infinite-dimensional real vector space $(\Lambda_\infty)_1$, having the generators of Λ_∞ as basis vectors, when a is of type c. The a-numbers $(\omega \cdot \mathbf{Q}_{a\chi} x^a)_1$, for all x^a, span a subspace of $(\Lambda_\infty)_1$ that is at most p-dimensional, p being the number of c-type x^a's. On the other hand, since ω has nonvanishing body, one can choose an imaginary a-type contravariant vector $\mathbf{V}$ at χ such that $(\omega \cdot \mathbf{V})_{\mathrm{B}} \neq 0$. If one also chooses $\delta\varphi = \mathbf{V}\varepsilon$, where ε is an arbitrary real a-number, then $(\omega \cdot \delta\varphi)_1 = (\omega \cdot \mathbf{V})_{\mathrm{B}}(\varepsilon)_1$. But (ε_1), and hence $(\omega \cdot \delta\varphi)_1$, can range over the entire vector space $(\Lambda_\infty)_1$. It follows that one can choose a $\delta\varphi$ that cannot be produced by any x^a's, leading once again to a contradiction. Evidently ω can be of neither type c nor type a, and the rows of the matrix Q_{B} must be linearly independent. This implies, as a corollary, $m \leq p$ and $n \leq q$.

A special canonical coordinate system

The linear independence of the rows of the matrices A and B of eq. (3.2.49) implies the existence of ordinary matrices K, L, M and N, of dimensions $p \times m$, $p \times (p-m)$, $q \times n$ and $q \times (q-n)$ respectively, such that the square matrices $(K\, L)$, $(M\, N)$, AK and BM are nonsingular and such that $AL = 0$ and $BN = 0$. If we define

$$Z_{\mathrm{B}} \stackrel{\text{def}}{=} \begin{pmatrix} K & L & 0 & 0 \\ 0 & 0 & M & N \end{pmatrix} \tag{3.2.50}$$

then

$$Q_{\mathrm{B}} Z_{\mathrm{B}} = \begin{pmatrix} AK & 0 & 0 & 0 \\ 0 & 0 & BM & 0 \end{pmatrix}. \tag{3.2.51}$$

Let the soul of the matrix Q be expressed in the block form

$$Q_S = \begin{pmatrix} C & E \\ F & D \end{pmatrix}, \tag{3.2.52}$$

and let Z be the square matrix having body Z_B and soul given by

$$Z_S = \begin{pmatrix} 0 & S & 0 & U \\ 0 & T & 0 & V \end{pmatrix}, \tag{3.2.53}$$

where

$$\begin{pmatrix} S \\ T \end{pmatrix} = -\begin{pmatrix} K \\ M \end{pmatrix} \begin{pmatrix} (A+C)K & EM \\ FK & (B+D)M \end{pmatrix}^{-1} \begin{pmatrix} C & L \\ F & L \end{pmatrix}, \tag{3.2.54}$$

$$\begin{pmatrix} U \\ V \end{pmatrix} = -\begin{pmatrix} K \\ M \end{pmatrix} \begin{pmatrix} (A+C)K & EM \\ FK & (B+D)M \end{pmatrix}^{-1} \begin{pmatrix} E & N \\ D & N \end{pmatrix}. \tag{3.2.55}$$

Then it is easy to verify that

$$QZ = \begin{pmatrix} (A+C)K & 0 & EM & 0 \\ FK & 0 & (B+D)M & 0 \end{pmatrix}. \tag{3.2.56}$$

If we carry out the basic transformation

$$\bar{\mathbf{e}}_a = \mathbf{e}_b{}^b Z_a \tag{3.2.57}$$

in $T_e(G)$, which is equivalent to transforming to new canonical coordinates $\bar{x}^a$ given by

$$\bar{x}^a = x^b{}_b Z^{-1a}, \tag{3.2.58}$$

then the components of the new Killing flow vectors $\bar{\mathbf{Q}}_a$ will be just the elements of the matrix (3.2.56). Let us suppose that this transformation has already been carried out, so that we may omit the bars on the e's, x's and Q's. Introduce capital Latin indices from the first of the alphabet to identify the coordinates corresponding to the vanishing columns of the matrix (3.2.56) (e.g., x^A, x^B, etc.) and use lower case Latin indices from the middle of the alphabet to identify the remaining coordinates (e.g., x^i, x^j, etc.).
Then

$$({}^iQ_j(\chi)) = \begin{pmatrix} (A+C)K & EM \\ FK & (B+D)M \end{pmatrix}, \tag{3.2.59}$$

which is a nonsingular square matrix, and

$${}^iQ_A(\chi) = 0. \tag{3.2.60}$$

If, in eq. (3.2.48), all the coordinates x^i vanish, then $\delta\varphi = 0$, which implies that the point x lies in the subgroup $K(\chi, \chi)$. If x lies in $K(\chi, \chi)$ then any power of x also lies in $K(\chi,\chi)$, and from this it follows that even if the canonical coordinates of x are finite, x will lie in $K(\chi,\chi)$ if all the x^i vanish. The x^A are evidently canonical coordinates for $K(\chi,\chi)$, and $K(\chi,\chi)$ is seen to be a sub-supermanifold of dimension $(p-m, q-n)$. We note that the subgroup property of $K(\chi,\chi)$ implies

$$^{i}c_{AB}=0 \tag{3.2.61}$$

in this special canonical coordinate system.

Coordinates for the coset spaces

Let $(\mathscr{U},\phi)$ be a chart such that

$$\phi(x)=(x^i,x^A), \tag{3.2.62}$$

for all x in a neighborhood $\mathscr{U}$ of the identity. Let $(\mathscr{V},\psi_{\mathrm{L}})$ be another chart satisfying

$$\psi_{\mathrm{L}}^{-1}(\xi,\eta)=\phi^{-1}(\xi,0)\phi^{-1}(0,\eta) \tag{3.2.63}$$

throughout an appropriate (obviously nonempty) neighborhood $\mathscr{V}$ of the identity, with $\mathscr{V}\subset\mathscr{U}$. Within $\mathscr{V}$ the new coordinates ξ^i identify the left cosets of $K(\chi,\chi)$, while the new coordinates η^A identify the points within each coset. By group multiplication we can generate, from the chart $(\mathscr{V},\psi_{\mathrm{L}})$, an atlas of coordinates having these properties throughout G. Every chart $(\bar{\mathscr{V}},\bar{\psi}_{\mathrm{L}})$ in this atlas is related to $(\mathscr{V},\psi_{\mathrm{L}})$ by

$$\bar{\psi}_{\mathrm{L}}^{-1}(\xi,\eta)=\bar{x}\psi_{\mathrm{L}}^{-1}(\xi,\eta)\bar{y}, \tag{3.2.64}$$

$$\bar{\mathscr{V}}=\bar{x}\mathscr{V}\bar{y}, \tag{3.2.65}$$

where $\bar{x}$ is an element of G and $\bar{y}$ is an element of $K(\chi,\chi)$, both elements being fixed for each chart. Because group multiplication is differentiable, all overlapping charts are differentiably related. Finally, by ignoring the coordinates η^A in each chart we arrive at a differentiable structure – and hence a supermanifold structure – for $[G/K(\chi,\chi)]_{\mathrm{L}}$. This supermanifold structure is determined entirely by G and $K(\chi,\chi)$.

By working with the alternative mapping ψ_{R}, defined by

$$\psi_{\mathrm{R}}^{-1}(\xi,\eta)=\phi^{-1}(0,\eta)\phi^{-1}(\xi,0), \tag{3.2.66}$$

we obtain in a similar manner a supermanifold structure for $[G/K(\chi,\chi)]_{\mathrm{R}}$. Note that for both ψ_{L} and ψ_{R} the coordinates ξ^i, η^A coincide with the canonical coordinates x^i, x^A at the identity:

$$(\xi,\eta)_e=(0,0), \tag{3.2.67}$$

$$(\partial/\partial\xi^i)_e=(\partial/\partial x^i)_e, \qquad (\partial/\partial\eta^A)_e=(\partial/\partial x^A)_e. \tag{3.2.68}$$

It is now straightforward to show that the above supermanifold structures are consistent with the supermanifold structures that $[G/K(\chi,\chi)]_{\mathrm{L}}$ and $[G/K(\chi,\chi)]_{\mathrm{R}}$ already inherit from Φ. In an appropriate neighborhood of χ in Φ, and in an appropriate neighborhood of the identity in G, one may express the condition that $\psi_{\mathrm{L}}^{-1}(\xi,\eta)$ belong to the left coset $K(\varphi,\chi)$ of $K(\chi,\chi)$ in the form

$$\varphi^i = R^i(\psi_{\mathrm{L}}^{-1}(\xi,\eta),\chi) = R^i(\phi^{-1}(\xi,0)\phi^{-1}(0,\eta),\chi) = R^i(\phi^{-1}(\xi,0),\chi), \tag{3.2.69}$$

which yields

$$\varphi^i \frac{\overleftarrow{\partial}}{\partial\eta^A} = 0, \quad \varphi^i \frac{\overleftarrow{\partial}}{\partial\xi^j} = R^i(\phi^{-1}(\xi,0),\chi)\frac{\overleftarrow{\partial}}{\partial\xi^j} \underset{\xi\to 0}{\longrightarrow} Q^i{}_j(\chi). \tag{3.2.70}$$

Since $(Q^i{}_j(\chi))$ is a nonsingular matrix and all functions are differentiable, the super-Jacobian $\mathrm{sdet}(\varphi^i\overleftarrow{\partial}/\partial\xi^i)$ must exist in a finite neighborhood of $\xi = 0$. The coordinates ξ^i therefore stand in a one-to-one differentiable relationship to the coordinates φ^i in this neighborhood, and may be used *in place of* the φ^i as coordinates in Φ. The chart thereby constructed may be displaced by the actions of appropriate elements of G to generate an atlas for Φ that is consistent not only with its complete atlas but also with the atlas previously constructed for $[G/K(\chi,\chi)]_{\mathrm{L}}$.

In a similar manner, one may express the condition that $\psi_{\mathrm{R}}^{-1}(\xi,\eta)$ belong to the right coset $K(\chi,\varphi)$ of $K(\chi,\chi)$ in the form

$$\begin{aligned}\varphi^i = R^i((\psi_{\mathrm{R}}^{-1}(\xi,\eta))^{-1},\chi) &= R^i(\phi^{-1}(-\xi,0)\phi^{-1}(0,-\eta),\chi)\\ &= R^i(\phi^{-1}(-\xi,0),\chi),\end{aligned} \tag{3.2.71}$$

which yields

$$\varphi^i \frac{\overleftarrow{\partial}}{\partial\eta^A} = 0, \qquad \varphi^i \frac{\overleftarrow{\partial}}{\partial\xi^j} = R^i(\phi^{-1}(-\xi,0),\chi)\frac{\overleftarrow{\partial}}{\partial\xi^j} \underset{\xi\to 0}{\longrightarrow} -Q^i{}_j(\chi). \tag{3.2.72}$$

Again the ξ^i may be used in place of the φ^i as coordinates in ξ.

Classification of transitive realizations

Instead of starting with a transitive realization of a super Lie group, which is given *a priori*, it is clear from the above results that one could simply look for a subgroup that is simultaneously a sub-supermanifold (possibly of dimension (0, 0)) and then construct the coset space (either left or right) of that subgroup. The actions of the group on the cosets then yield a realization of the group. Indeed it is clear that the problem of classifying all possible transitive realizations of a given super Lie group is identical with the problem of finding all possible subgroups that are also sub-supermanifolds.

There is an additional constraint that must be imposed on these subgroups if the corresponding realizations are to be faithful as well as transitive. Suppose we have an unfaithful realization. Then the subset

$$H \stackrel{\text{def}}{=} \{x \in G | R(x, \varphi) = \varphi \quad \text{for all } \varphi \text{ in } \Phi\} \tag{3.2.73}$$

will contain more than just the identity element e. It is elementary to verify that H is a subgroup of G. Moreover, it is a subgroup of every isotropy group $K(\chi, \chi)$. Let z be an arbitrary element of H and let x be an arbitrary element of G. Then, for all φ in Φ,

$$R(xzx^{-1}, \varphi) = R(x, R(z, R(x^{-1}, \varphi))) = R(x, R(x^{-1}, \varphi)) = \varphi,$$

which implies that xzx^{-1} is an element of H. That is, H is an invariant subgroup of G. From this it follows that a sufficient condition for a realization based on a given subgroup of G to be faithful is that that subgroup contain no nontrivial invariant subgroups of G. This condition is also necessary. For suppose the subgroup in question *has* a nontrivial subgroup H that is invariant in G. Then for every y in H and every x in G there exists a z in H such that $xyx^{-1} = z$. Let the basic subgroup be identified with $K(\chi, \chi)$ and let x be an element of the left coset $K(\varphi, \chi)$. Then

$$R(z, \varphi) = R(zx, \chi) = R(xy, \chi) = R(x, \chi) = \varphi,$$

and this must hold for all φ, because φ ranges over Φ as x ranges over G. But if $y \neq e$ then $z \neq e$, which contradicts the faithfulness condition. The problem of classifying all realizations that are both transitive and faithful therefore reduces to finding all subgroups of a given super Lie group G that (1) are sub-supermanifolds and (2) contain no nontrivial invariant subgroups of G.

Every super Lie group has at least one realization that is both transitive and faithful, namely that which is *provided by the group itself*. In this realization Φ is simply the group supermanifold, and all the isotropy subgroups are trivial. The action of the group on itself may be taken to be either left multiplication, or right multiplication by the inverse, corresponding respectively to the choices $F(x, y)$ or $F(y, x^{-1})$ for the realization function R. With the aid of eqs. (3.1.73), (3.1.75) and (3.1.105) it is straightforward to verify that the components ${}^i Q_a$ of the Killing flow vectors $\mathbf{Q}_a$ become respectively ${}^b R^{\sim}{}_a$ or $-{}^b L_a$.

Matrix representations of super Lie groups

Suppose Φ is diffeomorphic to $\mathbf{R}_c^m \times \mathbf{R}_a^n$, and suppose there exists a global coordinate system in Φ such that the functions $R^i(x, \varphi)$ become linear homogeneous in the φ^i. That is,

$$R^i(x, \varphi) = {}^iD_j(x)\varphi^j, \tag{3.2.74}$$

for certain functions ${}^iD_j(x)$. Then R is called a *matrix representation* of G.† If the ${}^iD_j(x)$ are assembled into a square matrix,

$$D(x) \stackrel{\text{def}}{=} ({}^iD_j(x)), \tag{3.2.75}$$

then conditions (3.2.1) and (3.2.2) require the matrix function $D(x)$ to satisfy

$$D(e) = 1_{(m,n)}, \tag{3.2.76}$$

$$D(xy) = D(x)D(y), \tag{3.2.77}$$

for all x, y in G. An immediate corollary of (3.2.76) and (3.2.77) is

$$D(x^{-1}) = D(x)^{-1}, \tag{3.2.78}$$

for all x in G. The matrix $D(x)$ must therefore be nonsingular for all x.

Although the realization space Φ has been introduced here as a diffeomorph of $\mathbf{R}_c^m \times \mathbf{R}_a^n$, nothing prevents it from being extended to the status of a full (m, n)-dimensional supervector space. Moreover, the concept of a matrix representation is customarily extended to allow complex mappings. That is, the matrices $D(x)$ need not preserve the reality properties of supervectors in Φ. The only restrictions on the $D(x)$ are that they satisfy eqs. (3.2.76) and (3.2.77) and that they preserve the type properties of supervectors, i.e., that they be c-type matrices, in the terminology of exercise **1.9**.

Contragredient representations

Corresponding to every matrix representation $D(x)$ there exists another, given by the matrices $D(x)^{-1\sim}$. The two representations are said to be *contragredient* to one another, and their corresponding supervector spaces are said to transform contragrediently under the action of G. Let Φ be the supervector space on which $D(x)$ acts and Ψ the supervector space on which $D(x)^{-1\sim}$ acts. The two supervector spaces may be regarded as dual to one another. If φ and ψ are elements of Φ and Ψ respectively then the inner product $\psi \cdot \varphi$ is invariant under the action of G.

† Matrix representations are usually intransitive realizations.

It often happens that the representations $D(x)$ and $D(x)^{-1\sim}$ are equivalent. That is, there exists a fixed nonsingular matrix η such that

$$D(x)^{-1\sim} = \eta D(x)\eta^{-1}, \quad \text{for all } x \text{ in } G, \tag{3.2.79}$$

which implies

$$[\eta^{-1}\eta^{\sim}, D(x)] = 0. \tag{3.2.80}$$

If $D(x)$ is an irreducible representation $\eta^{-1}\eta^{\sim}$ must be of the form

$$\eta^{-1}\eta^{\sim} = \lambda 1_{(m,n)}, \tag{3.2.81}$$

where λ is a c-number with nonvanishing body (see exercise **1.8**). But this implies $\eta^{\sim} = \lambda\eta$, $\eta = \lambda\eta^{\sim} = \lambda^2\eta$, whence

$$\lambda = \pm 1, \qquad \eta^{\sim} = \pm\eta. \tag{3.2.82}$$

That is, η must be either supersymmetric or antisupersymmetric. The matrix η plays a role similar to that of the canonical metric tensor in a flat Riemannian supermanifold. It defines a canonical mapping of Φ into its dual. We note that if φ and ψ are any two elements of Φ then the inner product $\varphi\cdot\psi \overset{\text{def}}{=} \varphi^i{}_i\eta_j{}^j\psi$, where $\varphi^i = (-1)^{\varphi i}{}^i\varphi$, is invariant under the action of G.

Inner automorphisms. The adjoint representation

For each x in G let $x_A : G \to G$ be the mapping

$$x_A \overset{\text{def}}{=} x_L \circ x_R^{-1} = x_R^{-1} \circ x_L. \tag{3.2.83}$$

Explicitly

$$x_A(y) = xyx^{-1}, \tag{3.2.84}$$

for all y in G. It is easy to verify that

$$x_A \circ y_A = (xy)_A \tag{3.2.85}$$

and hence that the mappings x_A constitute a realization of G. Since $x_A(e) = e$ for all x in G this realization is never transitive (unless G is trivial). It may or may not be faithful.

Since $x_A(e) = e$ and $x_A(y^{-1}) = [x_A(y)]^{-1}$, for all x, y in G, the mappings x_A are *automorphisms* of G. They are known as *inner automorphisms*. It is easy to verify that

$$x_A^{-1} = (x^{-1})_A, \tag{3.2.86}$$

for all x in G. Hence inner automorphisms are diffeomorphisms, and one may introduce the associated derivative mappings x'_A. If f is an arbitrary scalar field on G and $\mathbf{X}$ is an arbitrary contravariant vector field on G

then

$$[x'_{\mathrm{A}}(f)](y) = f(x^{-1}yx), \tag{3.2.87}$$

$$[x'_{\mathrm{A}}(\mathbf{X})]_y x'_{\mathrm{A}}(f) = \mathbf{X}_{x^{-1}yx} f. \tag{3.2.88}$$

Using the fact that left translations commute with right translations (see eq. (3.1.4)) one may show without difficulty that if $\mathbf{X}_{\mathrm{L}}$ is a left-invariant vector field then so is $x'_{\mathrm{A}}(\mathbf{X}_{\mathrm{L}})$, and if $\mathbf{X}_{\mathrm{R}}$ is a right-invariant vector field so is $x'_{\mathrm{A}}(\mathbf{X}_{\mathrm{R}})$. That is, left- and right-invariance properties remain intact under inner automorphisms.

Since inner automorphisms leave the identity invariant the associated derivative mappings map $T_e(G)$ into itself. The latter mappings are, in fact, linear, and hence inner automorphisms generate a matrix representation of G, with $T_e(G)$ as the realization space. This representation is called the *adjoint representation.*

With the aid of eqs. (3.1.70)–(3.1.75) together with eq. (3.1.82), it is easy to derive the explicit form of the adjoint representation. If $\mathbf{X}$ is an arbitrary vector field on G then, for every x in G,

$$\begin{aligned} {}^a[x'_{\mathrm{A}}(\mathbf{X})]_e &= {}^a[(x_{\mathrm{R}}^{-1\prime} \circ x'_{\mathrm{L}})(\mathbf{X})]_e = (-1)^{c(a+1)}{}_cF^a(x, x^{-1})\,{}^c[x'_{\mathrm{L}}(\mathbf{X})]_x \\ &= {}^aR^{\sim -1}{}_c(x)F^c{}_b(x, e)\,{}^bX_e = {}^aD_{\mathrm{ad}\,b}(x)\,{}^bX_e, \end{aligned} \tag{3.2.89}$$

where ${}^aD_{\mathrm{ad}\,b}(x)$ are the elements of the matrix

$$D_{\mathrm{ad}}(x) \stackrel{\mathrm{def}}{=} R^{\sim -1}(x)L(x). \tag{3.2.90}$$

That the matrices $D_{\mathrm{ad}}(x)$ do indeed provide a matrix representation of G follows immediately from eq. (3.1.84). The representation contragredient to the adjoint representation, namely

$$D_{\mathrm{ad}}^{\sim -1}(x) = R(x)\mathrm{L}^{\sim -1}(x), \tag{3.2.91}$$

is called the *coadjoint representation.*

Matrix representations of the super Lie algebra

For every matrix representation of a super Lie group G there is a corresponding representation of the associated super Lie algebra. If we define

$${}^iG_{aj} \stackrel{\mathrm{def}}{=} \left[(-1)^{aj}\,{}^iD_j(x)\frac{\overleftarrow{\partial}}{\partial x^a}\right]_{x=e}, \tag{3.2.92}$$

then it follows from eqs. (3.2.44), (3.2.47) and (3.2.74) that

$${}^iG_{ak}\,{}^kG_{bj} - (-1)^{ab}\,{}^iG_{bk}\,{}^kG_{aj} = (-1)^{j(a+b+c)}\,{}^iG_{cj}\,{}^cc_{ab}. \tag{3.2.93}$$

For every pure $\mathbf{X}, \mathbf{Y}$ in $T_e(G)$ define

$$\left.\begin{aligned} {}^iG_{\mathbf{X}j} &\overset{\text{def}}{=} (-1)^{j(a+\mathbf{X})}\,{}^iG_{aj}\,{}^aX, \\ {}^iG_{\mathbf{Y}j} &\overset{\text{def}}{=} (-1)^{j(a+\mathbf{Y})}\,{}^iG_{aj}\,{}^aY, \end{aligned}\right\} \tag{3.2.94}$$

and let $G_{\mathbf{X}}$ and $G_{\mathbf{Y}}$ be the matrices formed out of the ${}^iG_{\mathbf{X}j}$ and ${}^iG_{\mathbf{Y}j}$. Then

$$[G_{\mathbf{X}}, G_{\mathbf{Y}}] = G_{\mathbf{X}}G_{\mathbf{Y}} - (-1)^{\mathbf{XY}}G_{\mathbf{Y}}G_{\mathbf{X}} = G_{[\mathbf{X},\mathbf{Y}]}. \tag{3.2.95}$$

That is, the bracket operation of the super Lie algebra is mapped into the supercommutator of the matrices $G_{\mathbf{X}}, G_{\mathbf{Y}}$.

In the case of the adjoint representation we have, using eqs. (3.1.89c) and (3.2.90),

$$\begin{aligned} &\left[(-1)^{bc}\,{}^aD_{\text{ad}c}(x)\frac{\overleftarrow{\partial}}{\partial x^b}\right]_{x=e} \\ &= \left\{(-1)^{bc}[{}^aR^{\sim-1}{}_d(x)\,{}^dL_c(x)]\frac{\overleftarrow{\partial}}{\partial x^b}\right\}_{x=e} \\ &= (-1)^{bc}[-{}^aR_{c,b}(e) + {}^aL_{c,b}(e)] \\ &= (-1)^{b(a+1)}{}_bF^a{}_c(e,e) - (-1)^{c(a+b+1)}{}_cF^a{}_b(e,e) = {}^ac_{bc}, \end{aligned} \tag{3.2.96}$$

and the matrices $({}^iG_{aj})$ are seen to be just the matrices formed from the structure constants by treating their first and last indices as matrix indices. It is easy to see that eq. (3.2.93) in this case is just an alternative version of the identity (3.1.61). It is also easy to see, from eqs. (3.1.106), that in canonical coordinates the matrices of the adjoint representation take the exponential form

$$D_{\text{ad}}(x) = e^{\mathbf{c}\cdot\mathbf{x}}. \tag{3.2.97}$$

More generally, an arbitrary representation may, in canonical coordinates, be written in the form

$$D(x) = e^{\mathbf{G}\cdot\mathbf{x}}, \tag{3.2.98}$$

$${}^i(\mathbf{G}\cdot\mathbf{x})_j \overset{\text{def}}{=} (-1)^{ja}\,{}^iG_{aj}\,x^a. \tag{3.2.99}$$

The matrices $({}^iG_{aj})$ may thus be called the *generators* of the representation.

3.3 Geometry of coset spaces

Invariant tensor fields

Let Φ be a coset space of a super Lie group G, and let R be the realization mapping that describes the action of G on Φ. For every x in G let x_Φ:

$\Phi \to \Phi$ be the mapping defined by

$$x_\Phi(\varphi) \stackrel{\text{def}}{=} R(x, \varphi). \qquad (3.3.1)$$

The mappings x_Φ are one-to-one and differentiable, and each possesses a differentiable inverse $x_\Phi^{-1} = (x^{-1})_\Phi$. They are therefore diffeomorphisms and one may introduce the associated derivative mappings x'_Φ. Let **T** be a tensor field on Φ. Suppose

$$x'_\Phi(\mathbf{T}) = \mathbf{T}, \qquad (3.3.2)$$

for all x in G. Then **T** is called an *invariant tensor field* on Φ. By extending the concept of derivative mapping to connections ∇ and measure functions μ, as done in exercises **2.3** and **2.4**, one can also speak of *invariant connections* and *invariant measure functions* on Φ, satisfying

$$x'_\Phi(\nabla) = \nabla, \qquad x'_\Phi(\mu) = \mu. \qquad (3.3.3)$$

The question whether a given coset space Φ possesses invariant tensor fields, connections or measure functions is nontrivial. Such fields, if they exist, are special – one may say *canonical.* A supermanifold endowed with canonical tensor fields, connections or measure functions can be regarded as having special or characteristic *geometrical* properties.

For purposes of computing invariant geometrical structures for given coset spaces it is useful to have a more practical characterization of invariance than that given by eqs. (3.3.2) and (3.3.3). Denote by $Q_{\mathbf{X}s}$ the diffeomorphisms defined by the congruence of c-type supercurves generated by the Killing flow $\mathbf{Q}_\mathbf{X}$ of eqs. (3.2.32) and (3.2.33). Explicitly,

$$Q_{\mathbf{X}s}(\varphi) = \lambda_{\mathbf{X},\varphi}(s) = R(x_\mathbf{X}(s), \varphi) = [x_\mathbf{X}(s)]_\Phi(\varphi). \qquad (3.3.4)$$

If **T** is an invariant tensor field on Φ then

$$0 = -\left\{\frac{\mathrm{d}}{\mathrm{d}s}[x_\mathbf{X}(s)]'_\Phi(\mathbf{T})\right\}_{s=0} = -\left[\frac{\mathrm{d}}{\mathrm{d}s}Q'_{\mathbf{X}s}(\mathbf{T})\right]_{s=0} = \mathfrak{L}_{\mathbf{Q}_\mathbf{X}}\mathbf{T}. \qquad (3.3.5)$$

That is to say, if **T** is invariant then its Lie derivative with respect to every Killing flow vanishes. If G is a connected group then, because any group action can be compounded of infinitesimal group actions, the converse statement is also true. In this case the vanishing of their Lie derivatives with respect to Killing flows completely characterizes invariant tensor fields.

Similarly (assuming G connected) a measure function μ or a connection ∇ on Φ is invariant under the actions of G if and only if

$$\mathfrak{L}_{\mathbf{Q}_\mathbf{X}}\mu = 0 \quad \text{or} \quad \mathfrak{L}_{\mathbf{Q}_\mathbf{X}}\nabla = 0, \qquad (3.3.6)$$

for all **X** in $T_e(G)$.

Differential equations for geometrical structures

We shall confine our attention to three types of group-invariant geometrical structures on Φ: measure functions μ, metric tensor fields $\boldsymbol{g}$, and connections $\boldsymbol{\nabla}$. It is not difficult to show that if $\boldsymbol{\nabla}$ is invariant then the torsion and Riemann tensor fields constructed out of $\boldsymbol{\nabla}$ are invariant. Similarly, if $\boldsymbol{g}$ is invariant then the measure function, Riemannian connection and curvature tensor field constructed out of $\boldsymbol{g}$ are invariant. Note that $\boldsymbol{g}$ is invariant if and only if it is a metric tensor field for which the Killing flows become Killing vector fields (see section 2.8).

Referring to exercises **2.3** and **2.4** and to the Lie derivation rules described in section 2.5 one sees that the necessary and sufficient conditions for $\mu, \boldsymbol{g}$ amd $\boldsymbol{\nabla}$ to be group invariant are that their components, in each chart of Φ, satisfy the differential equations

$$0 = (-1)^{i(a+1)} (\mu\, {}^i Q_a)_{,i}, \tag{3.3.7}$$

$$0 = {}_i g_{j,k}\, {}^k Q_a + (-1)^{(i+k)(j+1)+ia}\, {}_k g_j\, {}^k Q_{a,i} + (-1)^{ja}\, {}_i g_k\, {}^k Q_{a,j}, \tag{3.3.8}$$

$$\begin{aligned} 0 = {} & \Gamma^i{}_{jk,l}\, {}^l Q_a - (-1)^{a(j+k)}\, {}^i Q_{a,l}\, \Gamma^l{}_{jk} + (-1)^{k(l+j)+ja}\, \Gamma^i{}_{lk}\, {}^l Q_{a,j} \\ & + (-1)^{ka}\, \Gamma^i{}_{jl}\, {}^l Q_{a,k} + (-1)^{a(j+k)}\, {}^i Q_{a,jk}. \end{aligned} \tag{3.3.9}$$

The question immediately arises whether these differential equations can be integrated. We shall first show that if $\mu, {}_i g_j, \Gamma^i{}_{jk}$ and their first derivatives with respect to the coordinates can be chosen at a fixed point χ of a given chart, in such a way that eqs. (3.3.7)–(3.3.9) are satisfied at that point, then these equations can be satisfied throughout the chart. Since eqs. (3.3.7)–(3.3.9) are form invariant under transformations to overlapping charts it will follow (see eqs. (3.3.50)–(3.3.52) below) that if they can be satisfied at a point they can, in fact, be satisfied throughout Φ.

Integrability of the differential equations

Consider first eq. (3.3.7), which can be rewritten in the form

$$\mu_{,i}\, {}^i Q_a = -(-1)^{i(a+1)}\, \mu\, {}^i Q_{a,i}. \tag{3.3.10}$$

Differentiating this equation with respect to φ^j and multiplying by ${}^j Q_b$, one finds

$$\begin{aligned} (-1)^{j(i+a)}\, \mu_{,ij}\, {}^i Q_a\, {}^j Q_b = {} & -\mu_{,i}\, {}^i Q_{a,j}\, {}^j Q_b - (-1)^{i(a+1)+ab}\, \mu_{,j}\, {}^j Q_b\, {}^i Q_{a,i} \\ & - (-1)^{i(a+b+1)}\, \mu ({}^i Q_{a,j}\, {}^j Q_b)_{,i} \\ & + (-1)^{i(a+b+1)}\, \mu\, {}^i Q_{a,j}\, {}^j Q_{b,i} \\ = {} & -\mu_{,i}\, {}^i Q_{a,j}\, {}^j Q_b - (-1)^{i(a+b+1)}\, \mu ({}^i Q_{a,j}\, {}^j Q_b)_{,i} \\ & + \mu[(-1)^{i(a+1)+j(b+1)}\, {}^i Q_{a,i}\, {}^j Q_{b,j} \\ & \quad + (-1)^{i(a+b+1)}\, {}^i Q_{a,j}\, {}^j Q_{b,i}], \end{aligned} \tag{3.3.11}$$

in which (3.3.10) itself has been used in obtaining the final form. Now if the indices a and b are interchanged the left-hand side of (3.3.11) is changed only by a factor $(-1)^{ab}$. Therefore if one subtracts from the expression on the right-hand side of (3.3.11) the same expression with a and b interchanged, multiplied by $(-1)^{ab}$, one must obtain zero. This is the only condition that must be satisfied in order that eq. (3.3.10) be integrable (subject to the given boundary conditions at χ). Carrying out the stated operation and making use of the component version of eq. (3.2.44), one in fact obtains

$$-(-1)^{i(c+1)}(\mu\,{}^iQ_c)_{,i}\,{}^c c_{ab},$$

which vanishes by virtue of (3.3.7).

Proceeding in a similar fashion with eq. (3.3.8), one finds

$$\begin{aligned}
&(-1)^{l(k+a)}\,{}_ig_{j,kl}\,{}^kQ_a\,{}^lQ_b \\
&\quad = -{}_ig_{j,k}\,{}^kQ_{a,l}\,{}^lQ_b - (-1)^{(i+k)(j+1)+i(a+b)}\,{}_kg_j({}^kQ_{a,l}\,{}^lQ_b)_{,i} \\
&\qquad -(-1)^{j(a+b)}\,{}_ig_k({}^kQ_{a,l}\,{}^lQ_b)_{,j} \\
&\qquad +(-1)^{(i+a)(j+1)+i(a+b)}\,{}_kg_j[{}^kQ_{a,l}\,{}^lQ_{b,i} + (-1)^{ab}\,{}^kQ_{b,l}\,{}^lQ_{a,i}] \\
&\qquad +(-1)^{j(a+b)}\,{}_ig_k[{}^kQ_{a,l}\,{}^lQ_{b,j} + (-1)^{ab}\,{}^kQ_{b,l}\,{}^lQ_{a,i}] \\
&\qquad +(-1)^{(i+k)(j+1)+i(a+b)}\,{}_kg_l[(-1)^{a(j+k)}\,{}^lQ_{a,j}\,{}^kQ_{b,i} \\
&\qquad\qquad +(-1)^{b(j+k)+ab}\,{}^lQ_{b,j}\,{}^kQ_{a,i}].
\end{aligned} \tag{3.3.12}$$

Subtracting from the right-hand side of this equation the same expression with a and b interchanged, multiplied by $(-1)^{ab}$, one gets

$$-[{}_ig_{j,k}\,{}^kQ_i + (-1)^{(i+k)(j+1)+ic}\,{}_kg_j\,{}^kQ_{c,i} + (-1)^{jc}\,{}_ig_k\,{}^kQ_{c,j}]\,{}^c c_{ab},$$

which vanishes by virtue of (3.3.8).

Finally, going through the same routine with eq. (3.3.9), one finds

$$\begin{aligned}
&(-1)^{m(l+a)}\,\Gamma^i{}_{jk,lm}\,{}^lQ_a\,{}^mQ_b \\
&\quad = -\Gamma^i{}_{jk,l}\,{}^lQ_{a,m}\,{}^mQ_b + (-1)^{(a+b)(j+k)}({}^iQ_{a,m}\,{}^mQ_b)_{,l}\,\Gamma^l{}_{jk} \\
&\qquad -(-1)^{k(l+j)+j(a+b)}\,\Gamma^i{}_{lk}({}^lQ_{a,m}\,{}^mQ_b)_{,j} \\
&\qquad -(-1)^{k(a+b)}\,\Gamma^i{}_{jl}({}^lQ_{a,m}\,{}^mQ_b)_{,k} \\
&\qquad -(-1)^{(a+b)(j+k)}({}^iQ_{a,m}\,{}^mQ_b)_{,jk} \\
&\qquad +(-1)^{k(a+b)}[(-1)^{ja}\,{}^iQ_{a,jm}\,{}^mQ_{b,k} + (-1)^{jb+ab}\,{}^iQ_{b,jm}\,{}^mQ_{a,k}] \\
&\qquad +(-1)^{j(a+b)}[(-1)^{k(a+j)}\,{}^iQ_{a,km}\,{}^mQ_{b,j} \\
&\qquad\qquad +(-1)^{k(b+j)+ab}\,{}^iQ_{b,km}\,{}^mQ_{a,j}] \\
&\qquad -(-1)^{j(a+b)+k(m+j)}[(-1)^{ka}\,{}^iQ_{a,l}\,\Gamma^l{}_{mk}\,{}^mQ_{b,j} \\
&\qquad\qquad +(-1)^{kb+ab}\,{}^iQ_{b,l}\,\Gamma^l{}_{mk}\,{}^mQ_{a,j}]
\end{aligned}$$

$$
\begin{aligned}
&-(-1)^{k(a+b)}[(-1)^{ja}\,{}^iQ_{a,l}\,\Gamma^l{}_{jm}\,{}^mQ_{b,k} \\
&\qquad +(-1)^{jb+ab}\,{}^iQ_{b,l}\,\Gamma^l{}_{jm}\,{}^mQ_{a,k}] \\
&+(-1)^{k(l+j)+j(a+b)}\,\Gamma^i{}_{lk}[{}^lQ_{a,m}\,{}^mQ_{b,j}+(-1)^{ab}\,{}^lQ_{b,m}\,{}^mQ_{a,j}] \\
&+(-1)^{k(a+b)}\,\Gamma^i{}_{jl}[{}^lQ_{a,m}\,{}^mQ_{b,k}+(-1)^{ab}\,{}^lQ_{b,m}\,{}^mQ_{a,k}] \\
&+(-1)^{k(a+b)+m(j+l)}\,\Gamma^i{}_{lm}[{}^lQ_{a,j}\,{}^mQ_{b,k}+(-1)^{ab}\,{}^lQ_{b,j}\,{}^mQ_{a,k}].
\end{aligned} \tag{3.3.13}
$$

Subtracting from the right-hand side of this equation the same expression with a and b interchanged, multiplied by $(-1)^{ab}$, one gets

$$
\begin{aligned}
&-[\Gamma^i{}_{jk,l}\,{}^lQ_c-(-1)^{c(j+k)}\,{}^iQ_{c,l}\,\Gamma^l{}_{jk} \\
&\qquad +(-1)^{k(l+j)+jc}\,\Gamma^i{}_{lk}\,{}^lQ_{c,j} \\
&\qquad +(-1)^{kc}\,\Gamma^i{}_{jl}\,{}^lQ_{c,k}+(-1)^{c(j+k)}\,{}^iQ_{c,jk}]\,{}^cc_{ab},
\end{aligned}
$$

which vanishes by virtue of eq. (3.3.9).

A special coordinate system

To determine whether eqs. (3.3.7)–(3.3.9) can be satisfied at a fixed point χ of Φ it is convenient, first, to make use of the canonical group coordinates x^A, x^i, based on the isotropy subgroup $K(\chi, \chi)$, which were introduced in section 3.2, and, second, to introduce a special coordinate system in Φ in the neighborhood of χ.

When the group coordinates x^A, x^i are used the component version of eq. (3.2.44) decomposes into three separate equations:

$$ {}^iQ_{A,j}\,{}^jQ_B-(-1)^{AB}\,{}^iQ_{B,j}\,{}^jQ_A={}^iQ_C\,{}^Cc_{AB}, \tag{3.3.14} $$

$$ {}^iQ_{A,k}\,{}^kQ_j-(-1)^{Aj}\,{}^iQ_{j,k}\,{}^kQ_A={}^iQ_B\,{}^Bc_{Aj}+{}^iQ_k\,{}^kc_{Aj}, \tag{3.3.15} $$

$$ {}^iQ_{j,l}\,{}^lQ_k-(-1)^{jk}\,{}^iQ_{k,l}\,{}^lQ_j={}^iQ_A\,{}^Ac_{jk}+{}^iQ_l\,{}^lc_{jk}. \tag{3.3.16} $$

Because of the special property (3.2.61) possessed by the structure constants in these coordinates, only one term appears on the right-hand side of eq. (3.3.14). The structure constant identity (3.1.61) itself decomposes into seven separate equations, of which we need at this point only the following two:

$$ {}^Ac_{BE}\,{}^Ec_{CD}+(-1)^{B(C+D)}\,{}^Ac_{CE}\,{}^Ec_{DB}+(-1)^{D(B+C)}\,{}^Ac_{DE}\,{}^Ec_{BC}=0, \tag{3.3.17} $$

$$ {}^ic_{Ak}\,{}^kc_{Bj}-(-1)^{AB}\,{}^ic_{Bk}\,{}^kc_{Aj}=(-1)^{j(A+B+C)}\,{}^ic_{Cj}\,{}^Cc_{AB}. \tag{3.3.18} $$

The ${}^Ac_{BC}$ are the structure constants of the isotropy subgroup $K(\chi, \chi)$, and eq. (3.3.18) shows that the matrices $({}^ic_{Aj})$ generate a representation of the super Lie algebra of this subgroup (see eq. (3.2.93)).

We shall need also eqs. (3.3.15) and (3.3.16) evaluated at the point χ. In view of eq. (3.2.60) we find

$$ {}^{i}Q_{A,k}(\chi)\,{}^{k}Q_{j}(\chi) = {}^{i}Q_{k}(\chi)\,{}^{k}c_{Aj}, \tag{3.3.19} $$

$$ {}^{i}Q_{j,l}(\chi)\,{}^{l}Q_{k}(\chi) - (-1)^{jk}\,{}^{i}Q_{k,l}(\chi)\,{}^{l}Q_{j}(\chi) = {}^{i}Q_{l}(\chi)\,{}^{l}c_{jk}. \tag{3.3.20} $$

Now introduce special coordinates $\bar{\varphi}^{i}$ in the neighborhood of χ, which are related to an arbitrary set of coordinates φ^{i} by

$$ \bar{\varphi}^{i} = \chi^{i} + {}^{i}Q^{-1}{}_{j}(\chi)(\varphi^{j} - \chi^{j}) + \tfrac{1}{2}\,{}^{i}Q^{-1}{}_{j,k}(\chi)(\varphi^{k} - \chi^{k})(\varphi^{j} - \chi^{j}) + O(\varphi - \chi)^{3}. \tag{3.3.21} $$

Then

$$ \left(\bar{\varphi}^{i}\frac{\overleftarrow{\partial}}{\partial\varphi^{j}}\right)_{\chi} = {}^{i}Q^{-1}{}_{j}(\chi), \tag{3.3.22} $$

$$ \left(\bar{\varphi}^{i}\frac{\overleftarrow{\partial}}{\partial\varphi^{j}}\frac{\overleftarrow{\partial}}{\partial\varphi^{k}}\right)_{\chi} = \tfrac{1}{2}[{}^{i}Q^{-1}{}_{j,k}(\chi) + (-1)^{jk}\,{}^{i}Q^{-1}{}_{k,j}(\chi)], \tag{3.3.23} $$

and in these new coordinates we have

$$ \begin{aligned} {}^{i}\bar{Q}_{j}(\chi) &= \left(\bar{\varphi}^{i}\frac{\overleftarrow{\partial}}{\partial\varphi^{k}}\right)_{\chi}{}^{k}Q_{j}(\chi) = \delta^{i}_{j}, \\ {}^{i}\bar{Q}_{j,k}(\chi) &= \left[\left(\bar{\varphi}^{i}\frac{\overleftarrow{\partial}}{\partial\varphi^{l}}\right){}^{l}Q_{j}\frac{\overleftarrow{\partial}}{\partial\bar{\varphi}^{k}}\right]_{\chi} \\ &= \left[(-1)^{m(l+j)}\left(\bar{\varphi}^{i}\frac{\overleftarrow{\partial}}{\partial\varphi^{l}}\frac{\overleftarrow{\partial}}{\partial\varphi^{m}}\right)_{\chi}{}^{l}Q_{j}(\chi)\right. \\ &\qquad \left. + \left(\bar{\varphi}^{i}\frac{\overleftarrow{\partial}}{\partial\varphi^{l}}\right)_{\chi}{}^{l}Q_{j,m}(\chi)\right]\left(\varphi^{m}\frac{\overleftarrow{\partial}}{\partial\bar{\varphi}^{k}}\right)_{\chi} \\ &= \{\tfrac{1}{2}(-1)^{m(l+j)}[{}^{i}Q^{-1}{}_{l,m}(\chi) + (-1)^{lm}\,{}^{i}Q^{-1}{}_{m,l}(\chi)]\,{}^{l}Q_{j}(\chi) \\ &\qquad + {}^{i}Q^{-1}{}_{l}(\chi)\,{}^{l}Q_{j,m}(\chi)\}\,{}^{m}Q_{k}(\chi) \\ &= \tfrac{1}{2}\,{}^{i}Q^{-1}{}_{l}(\chi)[{}^{l}Q_{j,m}(\chi)\,{}^{m}Q_{k}(\chi) - (-1)^{jk}\,{}^{l}Q_{k,m}(\chi)\,{}^{m}Q_{j}(\chi)] \\ &= \tfrac{1}{2}\,{}^{i}c_{jk}, \end{aligned} \tag{3.3.24} $$

in which eq. (3.3.20) is used in passing to the last line.

Let us assume that this coordinate transformation has already been carried out so that we may omit the bars over the φ's and Q's. We then have, using eq. (3.3.19),

$$ {}^{i}Q_{j}(\chi) = \delta^{i}{}_{j}, \qquad {}^{i}Q_{j,k}(\chi) = \tfrac{1}{2}\,{}^{i}c_{jk}, \tag{3.3.25} $$

$$ {}^{i}Q_{A}(\chi) = 0, \qquad {}^{i}Q_{A,j}(\chi) = {}^{i}c_{Aj}. \tag{3.3.26} $$

Condition for the existence of a group-invariant measure function

Consider now eq. (3.3.10) at χ. With the above choice of coordinates it decomposes into the following pair of equations:

$$\mu_{,i}(\chi) = -\tfrac{1}{2}(-1)^{j(i+1)}\,\mu(\chi)\,{}^{j}c_{ij}, \tag{3.3.27}$$

$$0 = -(-1)^{i(A+1)}\,\mu(\chi)\,{}^{i}c_{Ai}. \tag{3.3.28}$$

Equation (3.3.27) shows that once $\mu(\chi)$ is chosen, $\mu_{,i}(\chi)$ is completely determined. Equation (3.3.28) shows, however, that eq. (3.3.10) cannot be consistently integrated unless

$$(-1)^{i(A+1)}\,{}^{i}c_{Ai} = 0. \tag{3.3.29}$$

This is the necessary and sufficient condition for the coset space Φ to admit a group-invariant measure function. Equivalent to this condition is the requirement that the representation matrices of $K(\chi, \chi)$, which the ${}^{i}c_{Aj}$ generate by exponentiation (see eqs. (3.2.98) and (3.2.99)), have unit superdeterminant.

Although this condition was obtained by choosing a fixed point χ in Φ, it is in fact independent of χ and depends only on the nature of G and of its subgroup $K(\chi, \chi)$. This follows from the fact that all isotropy subgroups are related by inner automorphisms (see eq. (3.2.5)), and that group-coordinate transformations generated by inner automorphisms leave the structure constants unchanged (see exercise **3.3**).

Condition for the existence of a group-invariant metric tensor field

Turn next to eq. (3.3.8). At the point χ it decomposes into the following pair of equations:

$${}_{i}g_{j,k}(\chi) = -\tfrac{1}{2}(-1)^{(i+l)(j+1)+ik}\,{}_{l}g_{k}(\chi)\,{}^{l}c_{ki} - \tfrac{1}{2}(-1)^{jk}\,{}_{i}g_{l}(\chi)\,{}^{l}c_{kj}, \tag{3.3.30}$$

$$0 = -(-1)^{(i+l)(j+1)+iA}\,{}_{l}g_{j}(\chi)\,{}^{l}c_{Ai} - (-1)^{jA}\,{}_{i}g_{l}(\chi)\,{}^{l}c_{Aj}. \tag{3.3.31}$$

From these equations one sees that eq. (3.3.8) can be consistently integrated if and only if there exists a nonsingular supersymmetric matrix $({}_{i}\eta_{j})$ satisfying

$$(-1)^{(i+l)(j+1)+iA}\,{}_{l}\eta_{j}\,{}^{l}c_{Ai} + (-1)^{jA}\,{}_{i}\eta_{l}\,{}^{l}c_{Aj} = 0, \tag{3.3.32}$$

and if one chooses the metric tensor at χ to be a multiple of this matrix. Without loss of generality the multiple can be chosen to be unity.

Equation (3.3.32) is equivalent to the statement

$$\eta\mathbf{c}\cdot\mathbf{x}\eta^{-1} = -(\mathbf{c}\cdot\mathbf{x})^{\sim}, \quad \text{for all } x^{A}, \tag{3.3.33}$$

where

$$ {}^{i}(\mathbf{c}\cdot\mathbf{x})_j \overset{\text{def}}{=} (-1)^{jA}\,{}^{i}c_{Aj}\,x^A. \tag{3.34}$$

It follows that the necessary and sufficient condition for the coset space to admit a group-invariant metric tensor field is that the representation of $K(\chi,\chi)$ generated by the matrices $({}^{i}c_{Aj})$ be equivalent to its contragredient representation and that the nonsingular matrix connecting the two representations be supersymmetric.

We note that, because of the invariance of the supertrace under supertransposition, condition (3.3.29) is satisfied whenever (3.3.33) holds. Therefore if Φ admits a group-invariant metric tensor field it also admits a group-invariant measure function. Of course, we already knew this because, when a group-invariant metric tensor field exists, the square root of its superdeterminant in each chart can be chosen as the measure function. It is straightforward to verify that eq. (3.3.8) implies $(-1)^{i(a+1)}(g^{\frac{1}{2}}\,{}^{i}Q_a)_{,i} = 0$.

Condition for the existence of a group-invariant connection

If Φ admits a group-invariant metric tensor field it also admits a group-invariant connection, namely the torsionless Riemannian connection (see exercise 3.7). However, Φ may admit other group-invariant connections, or it may admit a group-invariant connection even when it does not admit a group-invariant metric tensor field. It turns out that there exists no simple single necessary and sufficient condition for the existence of a group-invariant connection. We shall describe here just one of the known sufficient conditions, which guarantees the existence of an invariant connection even when there is no invariant metric tensor field.

Note that the canonical coordinates x^A, x^i decompose $T_e(G)$ into a direct sum† $V_1 \oplus V_2$ where V_1 is the super Lie algebra of $K(\chi,\chi)$ and V_2 is a complementary supervector space. One may speak of the *action* of V_1 on the full super Lie algebra $V_1 \oplus V_2$ as being the collection of binary operations (3.1.60) where $\mathbf{X}$ is an element of V_1 and $\mathbf{Y}$ is an element of $V_1 \oplus V_2$. This action is described by the matrices

$$\begin{pmatrix} {}^{B}c_{AC} & {}^{B}c_{Aj} \\ 0 & {}^{i}c_{Aj} \end{pmatrix},$$

† The direct sum of two supervector spaces V_1 and V_2 is the set of all formal sums $\mathbf{X}_1 + \mathbf{X}_2$ where $\mathbf{X}_1 \in V_1$ and $\mathbf{X}_2 \in V_2$ and where all the standard supervector laws are extended to such sums.

which generate a representation of V_1. This representation is partially reduced by virtue of eq. (3.2.61). That is, the action of V_1 leaves V_1 invariant, although it does not necessarily leave V_2 invariant.

Suppose the above matrix representation is *fully reducible*; i.e., suppose the complementary vector space V_2 (and hence the coordinates x^i) can be chosen in such a way that it too remains invariant under the action of V_1. The coset space Φ is then said to be *reductive*, and one has, in addition to eq. (3.2.61), also

$$ {}^B c_{Aj} = 0. \tag{3.3.35}$$

We shall prove that reductivity is a sufficient condition for Φ to admit a group-invariant connection.

Observe that eq. (3.3.15) now gets modified to

$$ {}^iQ_{A,k}\,{}^kQ_j - (-1)^{Aj}\,{}^iQ_{j,k}\,{}^kQ_A = {}^iQ_k\,{}^kc_{Aj}. \tag{3.3.36}$$

Note also that the seven equations into which (3.1.61) previously decomposed reduce to six. Two of these are eqs. (3.3.17) and (3.3.18) as before. A third, which plays an essential role, is

$$ (-1)^{A(j+k)}\,{}^ic_{Al}\,{}^lc_{jk} - (-1)^{k(l+j)+jA}\,{}^ic_{lk}\,{}^lc_{Aj} - (-1)^{kA}\,{}^ic_{jl}\,{}^lc_{Ak} = 0. \tag{3.3.37}$$

Now differentiate eq. (3.3.36) with respect to φ^k and make use of eqs. (3.3.25), (3.3.26) and (3.3.37), obtaining

$$ \begin{aligned} {}^iQ_{A,jk}(\chi) &= \tfrac{1}{2}[-\,{}^ic_{Al}\,{}^lc_{jk} + (-1)^{Aj}\,{}^ic_{jl}\,{}^lc_{Ak} + (-1)^{k(l+A+j)}\,{}^ic_{lk}\,{}^lc_{Aj}] \\ &= 0. \end{aligned} \tag{3.3.38}$$

Then examine eq. (3.3.9) at the point χ in the special coordinate system. It is clear that the derivatives $\Gamma^i{}_{jk,l}(\chi)$ are fixed once $\Gamma^i{}_{jk}(\chi)$ and ${}^iQ_{j,kl}(\chi)$ have been determined. Moreover, $\Gamma^i{}_{jk}(\chi)$ itself must satisfy

$$ \begin{aligned} &-(-1)^{A(j+k)}\,{}^ic_{Al}\,\Gamma^l{}_{jk}(\chi) + (-1)^{k(l+j)+jA}\,\Gamma^i{}_{lk}(\chi)\,{}^lc_{Aj} \\ &\quad + (-1)^{kA}\,\Gamma^i{}_{jl}(\chi)\,{}^lc_{Ak} = 0. \end{aligned} \tag{3.3.39}$$

By virtue of eq. (3.3.37) one sees that this equation is satisfied by

$$ \Gamma^i{}_{jk}(\chi) = \lambda\,{}^ic_{jk}, \tag{3.3.40}$$

where λ is any real c-number.

Theorem

If Φ is reductive then it admits at least a one-parameter family of group-invariant connections. For one member of this family ($\lambda = 0$) the torsion vanishes.

It is of interest to determine the Riemann tensor to which the solution (3.3.40) leads. First differentiate eq. (3.3.16) and evaluate the result at χ, obtaining

$$\begin{aligned} {}^iQ_{j,kl}(\chi) - (-1)^{jk}\,{}^iQ_{k,jl}(\chi) &= -\tfrac{1}{4}{}^ic_{jm}{}^mc_{kl} + \tfrac{1}{4}(-1)^{jk}\,{}^ic_{km}{}^mc_{jl} \\ &\quad - \tfrac{1}{2}(-1)^{l(j+k)}\,{}^ic_{lm}{}^mc_{jk} - (-1)^{l(j+k)}\,{}^ic_{lA}{}^Ac_{jk}. \end{aligned} \tag{3.3.41}$$

Next evaluate eq. (3.3.9) at χ with the index a set equal to l:

$$\begin{aligned} \Gamma^i{}_{jk,l}(\chi) &= \tfrac{1}{2}\lambda(-1)^{l(j+k)}\,{}^ic_{lm}{}^mc_{jk} + \tfrac{1}{2}\lambda(-1)^{j(k+l)}\,{}^ic_{km}{}^mc_{lj} \\ &\quad + \tfrac{1}{2}\lambda\,{}^ic_{jm}{}^mc_{kl} - (-1)^{l(j+k)}\,{}^iQ_{l,jk}(\chi). \end{aligned} \tag{3.3.42}$$

Then insert this together with (3.3.40) into eq. (2.7.29) and use (3.3.41), finally obtaining

$$\begin{aligned} R^i{}_{jkl}(\chi) &= (\tfrac{1}{2} - \lambda)\,{}^ic_{jm}{}^mc_{kl} + (\tfrac{1}{2} - \lambda)^2(-1)^{j(k+l)}\,{}^ic_{km}{}^mc_{lj} \\ &\quad + (\tfrac{1}{2} - \lambda)^2(-1)^{l(j+k)}\,{}^ic_{lm}{}^mc_{jk} + {}^ic_{jA}{}^Ac_{kl}. \end{aligned} \tag{3.3.43}$$

Solutions of the differential equations

Having chosen acceptable values for μ, ${}_ig_j$ or $\Gamma^i{}_{jk}$ in the special coordinate system at χ, one does not actually have to solve the differential equations (3.3.7)–(3.3.9) to obtain their values elsewhere. Consider the action of the dragging mappings x'_Φ on the components of a contravariant vector field $\mathbf{X}$ on Φ. For every scalar field f in Φ we have

$$\begin{aligned} &[x'_\Phi(\mathbf{X})_{x_\Phi(\chi)}]^i \frac{\vec{\partial}}{\partial[x_\Phi(\chi)]^i} f(\chi) = x'_\Phi(\mathbf{X})_{x_\Phi(\chi)}\, x'_\Phi(f) \\ &= \mathbf{X}_\chi f = X^j(\chi)\frac{\vec{\partial}}{\partial\chi^j} f(\chi) = X^j(\chi)\left\{\frac{\vec{\partial}}{\partial\chi^j}[x_\Phi(\chi)]^i\right\}\frac{\vec{\partial}}{\partial[x_\Phi(\chi)]^i} f(\chi), \end{aligned}$$

whence

$$[x'_\Phi(\mathbf{X})_{x_\Phi(\chi)}]^i = X^j(\chi)\,{}_jR^{\sim i}(x,\chi) \tag{3.3.44a}$$

or, equivalently,

$$^i[x'_\Phi(\mathbf{X})_{x_\Phi(\chi)}] = {}^iR_j(x,\chi)\,{}^jX(\chi), \tag{3.3.44b}$$

where

$$^iR_j(x,\chi) \overset{\text{def}}{=} [x_\Phi(\chi)]^i \frac{\overleftarrow{\partial}}{\partial\chi^j} = R^i(x,\chi)\frac{\overleftarrow{\partial}}{\partial\chi^j}. \tag{3.3.45}$$

By considering contractions with contravariant vector fields one readily derives from this the corresponding dragging law for the components of a metric tensor field g (not necessarily group-invariant) on Φ:

$${}_i[x'_\Phi(g)_{x_\Phi(\chi)}]_j = {}_iR^{\sim -1\,k}(x,\chi)\,{}_kg_l(\chi)\,{}^lR^{-1}{}_j(x,\chi). \tag{3.3.46}$$

Because of the invariance of linear independence under derivative mappings, the matrix $R(x,\chi)$ formed out of the $^iR_j(x,\chi)$ is nonsingular in every pair of charts containing χ and $x_\Phi(\chi)$ respectively. Therefore its inverse is always well defined.

Using the definitions given in exercises **2.3** and **2.4**, one can in a similar fashion obtain the dragging laws for measure functions μ and connection components $\Gamma^i{}_{jk}$:

$$x'_\Phi(\mu)_{x_\Phi(\chi)} = \operatorname{sdet} R^{-1}(x,\chi)\mu(\chi). \tag{3.3.47}$$

$$\begin{aligned}[x'_\Phi(\Gamma)_{x_\Phi(\chi)}]^i{}_{jk} = {} & (-1)^{n(m+j)}\,{}^iR_l(x,\chi)\Gamma^l{}_{mn}(\chi)\,{}^mR^{-1}{}_j(x,\chi)\,{}^nR^{-1}{}_k(x,\chi) \\ & -(-1)^{jm}\,{}^iR_{ml}(x,\chi)\,{}^lR^{-1}{}_j(x,\chi)\,{}^mR^{-1}{}_k(x,\chi),\end{aligned} \tag{3.3.48}$$

where

$$^iR_{ml}(x,\chi) \overset{\text{def}}{=} R^i(x,\chi)\frac{\overleftarrow{\partial}}{\partial\chi^m}\frac{\overleftarrow{\partial}}{\partial\chi^l} = (-1)^{ml}\,{}^iR_{lm}(x,\chi). \tag{3.3.49}$$

If condition (3.3.29), (3.3.32) or (3.3.35) is satisfied, and *if* ${}_ig_j$ or $\Gamma^i{}_{jk}$ is chosen at χ as previously specified (namely, equal to ${}_i\eta_j$ or $\lambda\,{}^ic_{jk}$) it is clear that one obtains a group-invariant measure function, metric tensor field or connection by choosing μ, ${}_ig_j$ or $\Gamma^i{}_{jk}$ at every other point φ of Φ to be given respectively by

$$\mu(\varphi) = \operatorname{sdet} R^{-1}(x,\chi)\mu(\chi), \tag{3.3.50}$$

$${}_ig_j(\varphi) = {}_iR^{\sim -1\,k}(x,\chi)\,{}_k\eta_l\,{}^lR^{-1}{}_k(x,\chi), \tag{3.3.51}$$

$$\begin{aligned}\Gamma^i{}_{jk}(\varphi) = {} & (-1)^{n(m+j)}\lambda\,{}^iR_l(x,\chi)\,{}^lc_{mn}\,{}^mR^{-1}{}_j(x,\chi)\,{}^nR^{-1}{}_k(x,\chi) \\ & -(-1)^{jm}\,{}^iR_{ml}(x,\chi)\,{}^lR^{-1}{}_j(x,\chi)\,{}^mR^{-1}{}_k(x,\chi),\end{aligned} \tag{3.3.52}$$

where x is any element of the coset $K(\varphi,\chi)$. These are the solutions of the differential equations (3.3.7)–(3.3.9). They are unique as well as global since the ^iR_j are unique for every pair of charts. The conditions (3.3.29), (3.3.32) and (3.3.35) respectively guarantee that the right-hand sides of eqs. (3.3.50)–(3.3.52) remain unchanged if x is replaced by any other element of $K(\varphi,\chi)$.

Geometry of the group supermanifold

A special case of great interest is that in which Φ is the group supermanifold. The super Lie algebra V_1 is then trivial (of dimension zero), so the group itself *always* admits group-invariant measure functions, metric tensor fields and connections. However, these geometrical structures may be different depending on whether the action of the group on itself is taken to be left multiplication, or right multiplication by the inverse.

Consider first the case in which the action is left multiplication. The realization mapping function $R^i(x, \varphi)$ becomes $R^a(x, y) = F^a(x, y)$, and the components ${}^iQ_a(\varphi)$ of the Killing flows $\mathbf{Q}_a$ become ${}^bR^{\sim}_a(y)$. If the fixed point χ is chosen to be the identity then the functions ${}^iR_j(x, \chi)$ of eq. (3.3.45) become ${}^aL_b(x)$, and eqs. (3.3.50)–(3.3.52) yield the following left-invariant geometrical structures on G:

$$\mu_L = \text{const} \times \text{sdet}\, L^{-1}, \tag{3.3.53}$$

$$ {}_a g^L{}_b = {}_a L^{\sim -1c}\, {}_c\eta_d\, {}^dL^{-1}{}_b, \tag{3.3.54}$$

$$\begin{aligned}\Gamma_L{}^a{}_{bc} = &(-1)^{f(e+b)}\, {}^aL_d\, \Gamma_L{}^d{}_{ef}(\mathrm{e})\, {}^eL^{-1}{}_b\, {}^fL^{-1}{}_c \\ &-(-1)^{bc}\, {}^aL_{ed}\, {}^dL^{-1}{}_b\, {}^eL^{-1}{}_c,\end{aligned} \tag{3.3.55}$$

where

$$ {}^aL_{ed}(x) \overset{\text{def}}{=} F^a{}_{ed}(x, e). \tag{3.3.56}$$

In eq. (3.3.54) $({}_c\eta_d)$ is an arbitrary nonsingular supersymmetric matrix, and in eq. (3.3.55) the components $\Gamma_L{}^d{}_{ef}(e)$ are arbitrary. It is of interest to examine the Riemann tensor field generated by the connection (3.3.55), and in order to be able to use eq. (3.3.43) for this purpose let us make the choice $\Gamma_L{}^d{}_{ef}(e) = \lambda\, {}^dc_{ef}$. We must then also evaluate (3.3.55) in a canonical coordinate system so that eqs. (3.3.25) will be satisfied in the present context.

Differentiating eq. (3.1.81) with respect to y^c we find

$$\begin{aligned}{}^aL_{b,d}(xy)\, F^d{}_c(x, y) = &(-1)^{c(d+b)}\, F^a{}_{dc}(x, y)\, {}^dL_b(y) \\ &+ F^a{}_d(x, y)\, {}^dL_{b,c}(y),\end{aligned} \tag{3.3.57}$$

whence

$$\begin{aligned}{}^aL_{bc}(x) &= {}^aL_{b,d}(x)\, {}^dL_c(x) - {}^aL_d(x)\, {}^dL_{b,c}(e) \\ &= L_{b,d}(x)\, {}^dL_c(x) + \tfrac{1}{2}\, {}^aL_d(x)\, {}^dc_{bc},\end{aligned} \tag{3.3.58}$$

and therefore

$$\begin{aligned}\Gamma_L{}^a{}_{bc} = &(\lambda + \tfrac{1}{2})(-1)^{f(e+b)}\, {}^aL_d\, {}^dc_{ef}\, {}^eL^{-1}{}_b\, {}^fL^{-1}{}_c \\ &-(-1)^{bd}\, {}^aL_{d,b}\, {}^dL^{-1}{}_c,\end{aligned} \tag{3.3.59}$$

yielding

$$R_{\mathrm{L}}{}^a{}_{bcd}(e) = (\tfrac{1}{2} - \lambda)\,{}^a c_{be}\,{}^e c_{cd} + (\tfrac{1}{2} - \lambda)^2(-1)^{b(c+d)}\,{}^a c_{ce}\,{}^e c_{db} + (\tfrac{1}{2} - \lambda)^2(-1)^{d(b+c)}\,{}^a c_{de}\,{}^e c_{bc} = (\tfrac{1}{4} - \lambda^2)\,{}^a c_{be}\,{}^e c_{cd}. \tag{3.3.60}$$

We observe that if $\lambda = \pm\frac{1}{2}$ the Riemann tensor field vanishes at e. Since it is left invariant it must, in fact, vanish everywhere. Hence the supermanifold of every non-Abelian (${}^a c_{bc} \not\equiv 0$) super Lie group admits two distinct left-invariant parallelisms at a distance. Before examining these let us derive the corresponding right-invariant structures.

The realization mapping function in this case is $R^a(x, y) = F^a(y, x^{-1})$ and the components of the Killing flows $\mathbf{Q}_a$ are $-{}^b L_a(y)$. With the fixed point again chosen to be the identity the functions ${}^i R_j(x, \chi)$ become ${}^a R^{\sim}{}_b(x^{-1})$. The coset $K(y, e)$ has only the single element y^{-1}. To reach the variable point y, therefore, we must set $x = y^{-1}$ and ${}^a R^{\sim}{}_b(x^{-1}) = {}^a R^{\sim}{}_b(y)$. The right-invariant geometrical structures on G are evidently

$$\mu_{\mathrm{R}} = \text{const} \times \operatorname{sdet} R^{-1}, \tag{3.3.61}$$

$${}_a g^{\mathrm{R}}{}_b = {}_a R^{-1c}{}_c\eta_d\,{}^d R^{\sim -1}{}_b, \tag{3.3.62}$$

$$\Gamma_{\mathrm{R}}{}^a{}_{bc} = (-1)^{f(e+b)}\,{}^a R^{\sim}{}_d\,\Gamma_{\mathrm{R}}{}^d{}_{ef}(e)\,{}^e R^{\sim -1}{}_b\,{}^f R^{\sim -1}{}_c - (-1)^{be}\,{}^a R^{\sim}{}_{ed}\,{}^d R^{\sim -1}{}_b\,{}^e R^{\sim -1}{}_c, \tag{3.3.63}$$

where

$${}^a R^{\sim}{}_{ed}(x) = (-1)^{e(a+1)+d(a+e+1)}\,{}_{de}F^a(e, x) \tag{3.3.64}$$

Differentiating eq. (3.1.79) with respect to y^c we find

$$(-1)^{c(d+a)}\,{}_b R^d{}_{,e}(y)\,{}_d F^a(y, z) + (-1)^{c(d+a+1)}\,{}_b R^d(y)\,{}_{cd}F^a(y, z) = (-1)^{c(d+1)}\,{}_b R^a{}_{,d}(yz)\,{}_c F^d(y, z), \tag{3.3.65}$$

which, after rearrangement of the order of some of the indices, yields

$${}^a R^{\sim}{}_{bc}(z) = {}^a R^{\sim}{}_{b,d}(z)\,{}^d R^{\sim}{}_c(z) - {}^a R^{\sim}{}_d(z)\,{}^d R^{\sim}{}_{b,c}(e) = {}^a R^{\sim}{}_{b,d}(z)\,{}^d R^{\sim}{}_c(z) - \tfrac{1}{2}\,{}^a R^{\sim}{}_d(z)\,{}^d c_{bc} \tag{3.3.66}$$

(cf. eq. (3.3.58)). Setting $\Gamma_{\mathrm{R}}{}^d{}_{ef}(e) = \lambda\,{}^d c_{ef}$ in eq. (3.3.63), we therefore get

$$\Gamma_{\mathrm{R}}{}^a{}_{bc} = (\lambda - \tfrac{1}{2})(-1)^{f(e+b)}\,{}^a R^{\sim}{}_d\,{}^d c_{ef}\,{}^e R^{\sim -1}{}_b\,{}^f R^{\sim -1}{}_c - (-1)^{bd}\,{}^a R^{\sim}{}_{d,b}\,{}^d R^{\sim -1}{}_c \tag{3.3.67}$$

and, of course, once again

$$R_{\mathrm{R}}{}^a{}_{bcd}(e) = (\tfrac{1}{4} - \lambda^2)\,{}^a c_{be}\,{}^e c_{cd}. \tag{3.3.68}$$

Evidently every group supermanifold also has two right-invariant parallelisms at a distance.

Identity of the left- and right-invariant connections

The existence of two right-invariant as well as two left-invariant parallelisms at a distance, together with the identity of the two Riemann tensors (3.3.60) and (3.3.68), makes one suspect a deeper relation than mere analogy between the left- and right-invariant connections (3.3.59) and (3.3.67). It turns out that these two connections are, in fact, identical, and therefore, that each one is simultaneously left *and* right invariant for all values of λ.

Note first that the associated Riemann tensor fields are identical. Although eqs. (3.3.60) and (3.3.68) give the values of these fields only at e, their values elsewhere in G are obtained simply through multiplication by factors ${}^aL_b$, ${}^aL^{-1}{}_b$ (for $\mathbf{R}_L$) or ${}^aR^{\sim}{}_b$, ${}^aR^{\sim-1}{}_b$ (for $\mathbf{R}_R$). But

$$(-1)^{f(e+b)}\,{}^aL_d\,{}^dc_{ef}\,{}^eL^{-1}{}_b\,{}^fL^{-1}{}_c = (-1)^{f(e+b)}\,{}^aR^{\sim}{}_d\,{}^dc_{ef}\,{}^eR^{\sim-1}{}_b\,{}^fR^{\sim-1}{}_c, \tag{3.3.69}$$

as follows from the invariance of the structure constants under coordinate transformations generated by inner automorphisms (see exercise **3.3**). Therefore both left-dragging and right-dragging of the tensors (3.3.60) and (3.3.68) yield identical tensor fields.

To show that the connections themselves are identical one needs the following relation between the auxiliary functions:

$${}^aL_d\,{}^dL^{-1}{}_{b,c} = (-1)^{bc}\,{}^aR^{\sim}{}_d\,{}^dR^{\sim-1}{}_{c,b}. \tag{3.3.70}$$

This relation, which holds in any coordinate system, may be obtained by differentiating eq. (3.1.82) with respect to y^f, setting $y=e$, and then rearranging factors and performing some integrations by parts. Using this relation together with (3.1.87) and (3.1.88), one obtains

$$\begin{aligned}\Gamma_L{}^a{}_{bc} &= (\tfrac{1}{2}+\lambda)\,{}^aL_d\,{}^dL^{-1}{}_{b,c} + (-1)^{bc}(\tfrac{1}{2}-\lambda)\,{}^aL_d\,{}^dL^{-1}{}_{c,b}\\ &= (\tfrac{1}{2}-\lambda)\,{}^aR^{\sim}{}_d\,{}^dR^{\sim-1}{}_{b,c} + (-1)^{bc}(\tfrac{1}{2}+\lambda)\,{}^aR^{\sim}{}_d\,{}^dR^{\sim-1}{}_{c,b}\\ &= \Gamma_R{}^a{}_{bc}.\end{aligned} \tag{3.3.71}$$

Parallelism at a distance in the group supermanifold

Ignoring the above results for a moment let us ask the question: How would one go about setting up a natural parallelism at a distance in the group supermanifold? The answer is obvious. One would choose the connection in such a way that every left-invariant vector field has everywhere vanishing covariant derivative, or else one would choose it in such a way that every right-invariant vector field has everywhere vanishing covariant derivative. Suppose we make the latter choice. Then every right dragging is a parallel displacement and, since the product of two right draggings is itself a right dragging, parallel displacement becomes path independent. In view of the definition (2.7.4) the connection components must vanish everywhere if evaluated in the field of local frames $\{\mathbf{e}_{\mathrm{R}a}\}$. To get them in an arbitrary coordinate frame one has only to remember that

$$\mathbf{e}_{\mathrm{R}a} = \frac{\bar{\partial}}{\partial x^b}\, {}^bR^{\sim}{}_a, \qquad \frac{\bar{\partial}}{\partial x^a} = \mathbf{e}_{\mathrm{R}_b}\, {}^bR^{\sim -1}{}_a, \tag{3.3.72}$$

and then use the transformation law (2.7.5), with L replaced by $R^{\sim -1}$:

$$\Gamma^a{}_{bc} = {}^aR^{\sim}{}_d\, {}^dR^{\sim -1}{}_{b,c}. \tag{3.3.73}$$

If parallel displacements are defined to be left draggings instead of right draggings, equation (3.3.73) gets replaced by

$$\Gamma^a{}_{bc} = {}^aL_d\, {}^dL^{-1}{}_{b,c}. \tag{3.3.74}$$

Reference to eq. (3.3.71) shows that expressions (3.3.73) and (3.3.74) correspond to the choices $\lambda = -\frac{1}{2}$ and $\lambda = \frac{1}{2}$ respectively.

The connection (3.3.73), which has the property that every right-invariant vector field has vanishing covariant derivative, is obviously right invariant. It is also left invariant. This follows from the fact that if any right-invariant vector field is subjected to a left dragging the result is again a right-invariant vector field, as is easily seen by making use of the commutativity of left draggings with right draggings. Similarly, the connection (3.3.74) is both left and right invariant. Finally, expression (3.3.69), with an arbitrary coefficient, may be added to either (3.3.73) or (3.3.74) without affecting the invariance properties of these connections. From this follows the simultaneous left and right invariance of the whole family of connections (3.3.71).

Integration over the group

Although every super Lie group admits a family of connections that are simultaneously left and right invariant, it does not, except in special cases, admit even one metric tensor field or measure function that is both left and right invariant, let alone a whole family. In the case of the measure function there is nevertheless a relation between the left and right versions (3.3.53) and (3.3.61). This may be obtained as follows. First differentiate eq. (3.1.66) with respect to x^b and then make use of the following corollaries of eqs. (3.1.82) and (3.1.83),

$${}_bF^a(x^{-1}, x) = {}_bR^{-1a}(x^{-1}), \tag{3.3.75}$$

$$F^a{}_b(x^{-1}, x) = {}^aL^{-1}{}_b(x), \tag{3.3.76}$$

obtaining

$$x^{-1a}\frac{\overleftarrow{\partial}}{\partial x^b} = -{}^aR^{\sim}{}_c(x^{-1})^cL^{-1}{}_b(x). \tag{3.3.77}$$

If the 'const' is assumed to be the same in eqs. (3.3.53) and (3.3.61) then one finds, upon taking the superdeterminant of both sides of eq. (3.3.77),

$$\frac{\mu_L(x)}{\mu_R(x^{-1})} = \left|\operatorname{sdet}\left(x^{-1a}\frac{\overleftarrow{\partial}}{\partial x^b}\right)\right|. \tag{3.3.78}$$

Relation (3.3.78) finds a use in the theory of integration over group supermanifolds. Define

$$d_L x \stackrel{\text{def}}{=} \mu_L(x)\, d^{p,q}x, \tag{3.3.79}$$

$$d_R x \stackrel{\text{def}}{=} \mu_R(x)\, d^{p,q}x, \tag{3.3.80}$$

where $d^{p,q}x$ is defined as in eq. (1.7.5). Then, if y is a fixed element of G, the respective left- and right-invariance properties of the measure functions μ_L and μ_R insure that

$$d_L(yx) = d_L x, \qquad d_R(xy) = d_R x, \tag{3.3.81}$$

as may also be verified directly by taking superdeterminants of eqs. (3.1.82) and (3.1.83). Equation (3.3.78) yields the additional relations

$$d_R x^{-1} = d_L x, \qquad d_L x^{-1} = d_R x, \tag{3.3.82}$$

which are easily seen to be consistent with (3.3.81).

Now let f be a scalar function on G. For every fixed y in G, equations (3.3.81) and (3.3.82) permit one to write

$$\int f(yx)d_L x = \int f(x)d_L x, \tag{3.3.83}$$

$$\int f(xy)d_R x = \int f(x)d_R x, \tag{3.3.84}$$

$$\int f(x^{-1})d_L x = \int f(x)d_R x, \tag{3.3.85}$$

$$\int f(x^{-1})d_R x = \int f(x)d_L x. \tag{3.3.86}$$

The integration here is over the whole group. If the group is not compact f is assumed to have such properties (e.g., compact support) as will ensure convergence of the integrals.

A special class of super Lie groups

There is a special class of super Lie groups, of great importance in practice, for which metric tensor fields and measure functions *do* exist that are simultaneously left and right invariant. This is the class of super Lie groups whose adjoint representations are equivalent to their co-adjoint representations, with the connecting matrix η being supersymmetric. Explicitly (see eqs. (3.1.101) and (3.2.97)),

$$\eta \mathbf{c}\cdot\mathbf{x}\eta^{-1} = -(\mathbf{c}\cdot\mathbf{x})^{\sim}, \quad \text{for all } x^a, \tag{3.3.87}$$

or, equivalently,

$$_a\eta_d\,{}^d c_{bc} = -(-1)^{a(b+1)+d(a+b)}\,{}^d c_{ba}\,{}_d\eta_c. \tag{3.3.88}$$

It is easy to verify that eq. (3.3.88), together with the supersymmetry of η, implies that the tensor

$$c_{abc} \overset{\text{def}}{=} (-1)^a\,{}_a\eta_d\,{}^d c_{bc} \tag{3.3.89}$$

is a 3-form at e, i.e., is antisupersymmetric in all its indices.

When condition (3.3.88) holds every coset space, and not merely the group supermanifold, admits an invariant metric tensor field. In contravariant form this field is given by†

$$g^{ij} \overset{\text{def}}{=} {}^iQ_a\,\eta^{ab}\,{}_bQ^{\sim j}, \qquad (\eta^{ab}) = \eta^{-1}, \tag{3.3.90}$$

or, equivalently,

$$g^{-1} = \mathbf{Q}_a \otimes \mathbf{Q}_b \eta^{ba} \tag{3.3.91}$$

Its group invariance follows from eq. (3.2.44) and the antisupersymmetry of c_{abc}:

$$\begin{aligned}
\mathfrak{L}_{\mathbf{Q}_a} g^{-1} &= \mathfrak{L}_{\mathbf{Q}_a}(\mathbf{Q}_b \otimes \mathbf{Q}_c \eta^{cb}) \\
&= -\{[\mathbf{Q}_a, \mathbf{Q}_b] \otimes \mathbf{Q}_c + (-1)^{ab}\,\mathbf{Q}_b \otimes [\mathbf{Q}_a, \mathbf{Q}_c]\}\eta^{cb} \\
&= -[(\mathbf{Q}_d\,{}^d c_{ab}) \otimes \mathbf{Q}_c + (-1)^{ab}\,\mathbf{Q}_b \otimes \mathbf{Q}_d\,{}^d c_{ac}]\eta^{cb} \\
&= c_{abc}[{}^c\mathbf{Q} \otimes {}^b\mathbf{Q} + (-1)^{bc}\,{}^b\mathbf{Q} \otimes {}^c\mathbf{Q}] = 0,
\end{aligned} \tag{3.3.92}$$

where

$${}^a\mathbf{Q} = \mathbf{Q}^a \overset{\text{def}}{=} \mathbf{Q}_b \eta^{ba}. \tag{3.3.93}$$

† (g^{ij}) is nonsingular because η is nonsingular and, as we have seen in section 3.2, the rows of the matrix $({}^iQ_a)_B$ are linearly independent at every point of the realization space when the realization is transitive.

Remembering that the components of the left-invariant Killing flows on the group manifold are ${}^bR^{\sim}{}_a$, one obtains the following left-invariant metric tensor field on the group:

$$g^{ab} = {}^aR^{\sim}{}_c\eta^{cd}{}_dR^b \quad \text{or} \quad {}_ag_b = {}_aR^{-1c}{}_c\eta_d{}^dR^{\sim -1}{}_b. \tag{3.3.94}$$

But eq. (3.3.62) shows that this metric tensor field is also *right* invariant. Similarly, using right-invariant Killing flows, one can show that the metric tensor field (3.3.54) is both left and right invariant, provided the matrix η is again chosen to be that which connects the adjoint and co-adjoint representations. With this choice the metric tensor fields (3.3.54) and (3.3.62) coincide at the identity. Since each is both left and right invariant they satisfy identical differential equations, and hence they must be identical everywhere:

$${}_ag^{\mathrm{L}}{}_b = {}_ag^{\mathrm{R}}{}_b \overset{\text{def}}{=} {}_ag_b. \tag{3.3.95}$$

Note that the identity of these tensor fields implies the identity of $\operatorname{sdet}(L^{-1})$ and $\operatorname{sdet}(R^{-1})$ and hence of the left- and right-invariant measure functions:

$$\mu_{\mathrm{L}} = \mu_{\mathrm{R}} = \text{const} \times g^{1/2}. \tag{3.3.96}$$

For the special class of super Lie groups in question, therefore,

$$d_{\mathrm{L}}x = d_{\mathrm{R}}x \overset{\text{def}}{=} dx = dx^{-1}. \tag{3.3.97}$$

By making use of eqs. (2.8.28), (3.1.94) and (3.1.95), together with the antisupersymmetry of c_{abc}, it is straightforward to show that the Riemannian connection to which the special choice of η leads is given by

$$\Gamma^a{}_{bc} = \tfrac{1}{2}{}^aL_d[{}^dL^{-1}{}_{b,c} + (-1)^{bc}\,{}^dL^{-1}{}_{c,b}] \tag{3.3.98a}$$

$$= \tfrac{1}{2}{}^aR^{\sim}{}_d[{}^dR^{\sim -1}{}_{b,c} + (-1)^{bc}\,{}^dR^{\sim -1}{}_{c,b}]. \tag{3.3.98b}$$

This corresponds to the choice $\lambda = 0$ in eq. (3.3.71) and yields, for the curvature tensor,

$$R^a{}_{bcd}(e) = \tfrac{1}{4}{}^ac_{be}{}^ec_{cd}. \tag{3.3.99}$$

We finally note that, by making use of the structure-constant identity (3.1.61), one can show that the matrix

$${}_a\eta_b \overset{\text{def}}{=} -(-1)^{a+c(a+b+1)}\,{}^cc_{ad}{}^dc_{bc}, \tag{3.3.100}$$

which is known as the *Cartan–Killing matrix*, satisfies eq. (3.3.88). Whenever this matrix is nonsingular the super Lie group therefore belongs to the special class. The converse is not generally true.

Exercises

3.1 Let $x_{\mathbf{X}}(\mathbf{R}_c)$ be the one-parameter Abelian subgroup generated by the real c-type contravariant vector $\mathbf{X}$ at the identity of a finite-dimensional super Lie group G. Show that if $\mathbf{X}$ has vanishing body then for every s in $\mathbf{R}_c$ there exists a nonvanishing infinitesimal ds in $\mathbf{R}_c$ such that $x_{\mathbf{X}}(s+ds)=x_{\mathbf{X}}(s)$. Prove that $x_{\mathbf{X}}(\mathbf{R}_c)$ is a sub-supermanifold of G if and only if $\mathbf{X}$ has nonvanishing body. Conclude that if G has no c-type coordinates then none of its one-parameter Abelian subgroups is a sub-supermanifold.

3.2 Show that eqs. (3.1.87) and (3.1.88) are the integrability conditions for eqs. (3.1.85) and (3.1.86).

3.3 Prove that every matrix representation $D(x)$ of a super Lie group G satisfies

$$\mathbf{G}\cdot\mathbf{X}=D(x)\mathbf{G}\cdot[D_{\mathrm{ad}}(x)^{-1}\mathbf{X}]D(x)^{-1},$$

for all x in G and all $\mathbf{X}$ in $T_e(G)$. (*Hint*: Differentiate the equation $D(xyx^{-1})=D(x)D(y)D(x)^{-1}$ with respect to y^a and then set $y=e$.) This result implies, in particular,

$${}^c c_{bc}=(-1)^{f(e+b)}\,{}^a D_{\mathrm{ad}\;d}(x)\,{}^d c_{ef}\,{}^e D_{\mathrm{ad}}^{-1}{}_b(x)\,{}^f D_{\mathrm{ad}}^{-1}{}_c(x),$$

and hence that the structure constants remain invariant under the transformations of $T_e(G)$ induced by inner automorphisms.

3.4 Prove that if a left-invariant tensor field on a super Lie group is subjected to a right dragging, the result is again a left-invariant tensor field. Prove that the same is true if the words 'right' and 'left' are interchanged.

3.5 Let $\mathbf{X}_{\mathrm{L}}$ be a left-invariant contravariant vector field on a super Lie group and $\mathbf{Y}_{\mathrm{R}}$ a right-invariant contravariant vector field. Prove that $[\mathbf{X}_{\mathrm{L}},\mathbf{Y}_{\mathrm{R}}]=\mathbf{0}$.

3.6 Show that eq. (3.3.8) implies $(-1)^{i(a+1)}(g^{1/2}\,{}^iQ_a)_{,i}=0$.

3.7 Prove that if ∇ is an invariant connection on a coset space Φ of a super Lie group G then the torsion and Riemann tensor fields constructed out of ∇ are invariant. Prove that if g is an invariant metric tensor field on Φ then the Riemannian connection and curvature tensor field constructed out of g are invariant.

3.8 Derive eq. (3.3.70).

3.9 Suppose the adjoint and coadjoint representations of a super Lie group are both equivalent and irreducible. Prove that the matrix η that connects them is necessarily supersymmetric.

3.10 Let Φ be a realization manifold on which a super Lie group G acts, and let $\mathbf{Q_X}$ be the Killing flow on Φ generated by the supervector $\mathbf{X}$ at the identity of G. Let x_Φ be the mapping defined by eq. (3.3.1) and let x'_Φ be the corresponding derivative mapping. Show that

$$x'_\Phi(\mathbf{Q_X}) = \mathbf{Q}_{D_{ad}(x)\mathbf{X}}.$$

3.11 Let G be a super Lie group having equivalent adjoint and coadjoint representations, with a supersymmetric connecting matrix η. Let the coordinate system at the identity be chosen so that η is brought into the canonical form (2.8.10). Let the group supermanifold be endowed with the metric tensor field

$${}_a g_b = {}_a \tilde{L}^{-1c}\, {}_c\eta_d\, {}^d L^{-1}{}_b = {}_a R^{-1c}\, {}_c\eta_d\, {}^d \tilde{R}^{-1}{}_b.$$

Show that the left-invariant field of local frames $\{\mathbf{e}_{La}\}$ then constitutes a global orthosymplectic basis. Show that the same is true of the right-invariant field of local frames $\{\mathbf{e}_{Ra}\}$.

3.12 Prove that the Riemannian connection constructed out of the metric tensor given in the preceding exercise is equal to expressions (3.3.98a, b).

3.13 If the covariant derivative of the curvature tensor field of a Riemannian supermanifold vanishes then the supermanifold is said to be *symmetric*. Prove that the super Lie group endowed with the metric tensor field of exercise **3.11** is a symmetric supermanifold. (*Hint*: Use the left and right invariance of the metric tensor field.)

3.14 Prove that expression (3.3.100) satisfies eq. (3.3.88).

3.15 Extend the classical theory of fibre bundles to supermanifolds. That is, let the base space and fibres now be supermanifolds, and let the associated group be a super Lie group. Develop the theory of connections on *super fibre bundles*.

Comments on chapter 3

The general theory of super Lie groups and their realizations is seen to proceed in complete analogy with the corresponding theory of ordinary Lie groups. The chief new feature is the association, with each super Lie group, of an infinite family of ordinary Lie groups: the skeletons. The skeletons of course are very unobtrusive; they play no direct role in the general theory. On the other hand they will eventually have to be studied. Their potential usefulness in the classification problem of ordinary Lie groups remains completely unexplored.

Another unsolved problem concerns the classification of distinct super Lie groups having the same super Lie algebra. This problem has not yet been approached even for the important super Lie groups described in chapter 4. Among the questions to be asked are: To what extent does a super Lie algebra determine the (algebraic) topology of its associated super Lie groups? Is a super Lie group, regarded as a bundle of soul subspaces over its body, ever a twisted bundle, or is its topology always a trivial extension of that of its body?

The problem of determining all the matrix representations of a given super Lie group, even one that is simple and compact, is more difficult than the corresponding problem for ordinary Lie groups. If a theory of characters exists for representations of compact super Lie groups (it has not yet been worked out), it is bound to be more subtle and complicated than that for ordinary Lie groups. Since the Berezin integral is not a measure-theoretical construct the unitary trick cannot be used, and it is not at all clear whether analogs of the orthogonality relations for characters exist. Another difficulty is the minus sign appearing in the supertrace. This has the consequence that the supertrace of the product of a matrix and its Hermitian conjugate need not have positive body and may even vanish. Also the Cartan–Killing matrix may vanish even when the super Lie group is compact and semisimple.

The author is indebted to G.A. Vilkovisky for the following bit of historical information: the concept of super Lie algebras appears to have been introduced for the first time by G.L. Stawraki (see references at the end of the book). Stawraki called them 'K-algebras' after the first letter of his wife's name.

4

Super Lie groups. Examples

4.1 Construction of super Lie algebras and super Lie groups

Properties of the structure constants

The structure constants of a super Lie group are the components of a rank (1, 2) c-type tensor at the identity. According to the convention (2.4.34) they therefore remain unchanged if the upper index is shifted to the right:

$$^{a}c_{bc} = c^{a}{}_{bc}. \tag{4.1.1}$$

In this chapter we shall use the latter form.

In addition to being of type c, the tensor $c^{a}{}_{bc}$ is real. By virtue of eq. (2.4.21) it follows, in the notation introduced at the beginning of section 1.7, that the components $c^{\mu}{}_{\nu\sigma}$, $c^{\alpha}{}_{\mu\beta}$, $c^{\alpha}{}_{\beta\mu}$ are real c-numbers, the components $c^{\mu}{}_{\alpha\beta}$ are imaginary c-numbers, the components $c^{\alpha}{}_{\mu\nu}$ are real a-numbers, and the components $c^{\mu}{}_{\nu\alpha}$, $c^{\mu}{}_{\alpha\nu}$, $c^{\alpha}{}_{\beta\gamma}$ are imaginary a-numbers. It is not difficult to see that $(c^{\mu}{}_{\nu\sigma})_{B}$ are the structure constants of the body of the super Lie group.

Conventional super Lie groups. Z_2-graded algebras

A super Lie group G will be called *conventional* if a standard basis $\{\mathbf{e}_a\}$ can be introduced in $T_e(G)$ with respect to which the souls of the structure constants all vanish. In such a basis, which will be called a *conventional* basis, the only nonvanishing structure constants are of the type $c^{\mu}{}_{\nu\sigma}$, $c^{\alpha}{}_{\mu\beta}$, $c^{\alpha}{}_{\beta\mu}$ or $c^{\mu}{}_{\alpha\beta}$, those of the first three types being ordinary real numbers and those of the fourth type being ordinary imaginary numbers. Given a conventional basis one can obtain an infinity of others by carrying out transformations of the form (2.4.13) with the matrix elements $^{a}L_{b}$ restricted to have vanishing souls.

A conventional basis in $T_e(G)$ defines a *conventional coordinate system* in a neighborhood of the identity in G, namely the associated canonical coordinate system. Any coordinates differentiably related to the canonical coordinates will also be called *conventional* if their souls vanish whenever the souls of the conventional canonical coordinates vanish.

A super Lie algebra will be called *conventional* if it is the super Lie algebra of some conventional super Lie group. The structure of a conventional super Lie algebra may be fully determined by confining one's attention to real vectors in $T_e(G)$ the components of which in a conventional basis have vanishing souls. If X is such a vector and if it is c-type then its nonvanishing components in the conventional basis are ordinary real numbers and are to be found among the X^μ. If it is a-type then its nonvanishing components in the conventional basis are ordinary imaginary numbers and are to be found among the X^α. Such vectors belong to the subspace $\mathbf{R}^p \oplus (\mathrm{i}\mathbf{R})^q$ of $T_e(G)$.

It is customary among mathematicians to replace the subspace $\mathbf{R}^p \oplus (\mathrm{i}\mathbf{R})^q$ by $\mathbf{R}^p \oplus \mathbf{R}^q$ and to modify the bracket operation by multiplying the structure constants $c^\mu{}_{\alpha\beta}$ by i so that they become real. The resulting algebra, *confined to the real vector space* $\mathbf{R}^p \oplus \mathbf{R}^q$, is called a *$Z_2$-graded algebra.* A Z_2-graded algebra with a bracket operation satisfying the super Jacobi identity is sometimes called a *Lie superalgebra*, to draw attention to its connection with the physics of supersymmetry (see V. G. Kac, *Advances in Mathematics* **26**, 8 (1977)). However, because Lie superalgebras are ordinary vector spaces rather than supervector spaces they cannot be used directly to generate 'supergroups'. One must first replace $\mathbf{R}^p \oplus \mathbf{R}^q$ by a (p, q)-dimensional supervector space and then follow the path outlined in section 3.1 to get the corresponding super Lie group.[†]

Unconventional super Lie groups

If a basis cannot be found in which the souls of its structure constants all vanish, a super Lie group and its associated super Lie algebra will be called *unconventional.* There exists an infinity of unconventional super Lie groups that are not associated with graded algebras at all. Although not as important in practical applications as conventional super Lie groups, they can be highly nontrivial and are not without interest. We give here one example, the simplest: It is of dimension (0, 1), each element having only one a-type coordinate, which may be taken to be both canonical and global. There is only one structure constant c, which may be chosen to be any nonvanishing imaginary a-number. The multiplication mapping

† See also F. A. Berezin & G. I. Kac, *Mat. Sbornik*, **82**, 124 (1970); English translation: *Math. USSR Sbornik*, **11**, 311 (1970).

function F is given (cf. eq. (3.1.104)) by

$$F(x, y) = x + y - \tfrac{1}{2}cxy. \tag{4.1.2}$$

This function is readily verified to satisfy eqs. (3.1.64)–(3.1.66), the coordinates of e and x^{-1} being 0 and $-x$ respectively.

The above group is non-Abelian and simple. (A super Lie group is said to be *simple* if it has no invariant proper subgroup that is a sub-supermanifold of dimension other than (0,0).) None of its skeletons, however, is simple (in the ordinary sense)[†] except the first, S_1. Its skeletons S_1 and S_2, of dimension 1 and 2 respectively, are both free Abelian. Its first non-Abelian skeleton is S_3, of dimension 4. The structure constants of S_3 are obtained by expanding x, y and c in terms of the generators $\zeta^1, \zeta^2, \zeta^3$ of the Grassmann algebra Λ_3:

$$\left.\begin{aligned} x &= x^1\zeta^1 + x^2\zeta^2 + x^3\zeta^3 + \mathrm{i}x^4\zeta^1\zeta^2\zeta^3, \\ y &= y^1\zeta^1 + y^2\zeta^2 + y^3\zeta^3 + \mathrm{i}y^4\zeta^1\zeta^2\zeta^3, \\ c &= -\mathrm{i}(c^1\zeta^1 + c^2\zeta^2 + c^3\zeta^3 + \mathrm{i}c^4\zeta^1\zeta^2\zeta^3), \end{aligned}\right\} \tag{4.1.3}$$

the x^a, y^a, c^a being ordinary real numbers. Regarding the x^a and y^a as coordinates in the skeleton we find, for the multiplication mapping function of S_3,

$$\left.\begin{aligned} &F^1(x, y) = x^1 + y^1, \qquad F^2(x, y) = x^2 + y^2, \qquad F^3(x, y) = x^3 + y^3, \\ &F^4(x, y) = x^4 + y^4 + \tfrac{1}{2}c^1(x^2y^3 - x^3y^2) + \tfrac{1}{2}c^2(x^3y^1 - x^1y^3) \\ &\qquad + \tfrac{1}{2}c^3(x^1y^2 - x^2y^1), \end{aligned}\right\} \tag{4.1.4}$$

whence

$$c^4_{12} = -c^4_{21} = c^3, \qquad c^4_{23} = -c^4_{32} = c^1, \qquad c^4_{31} = -c^4_{13} = c^2, \tag{4.1.5}$$

all other structure constants vanishing.

Mathematicians appear to have devoted no attention to unconventional super Lie groups, so very little is known about them. It is quite possible that their chief interest lies in their skeletons, which seem rarely to be simple or even semisimple. (A Lie group is said to be *semisimple* if it has no invariant Abelian subgroup that is a submanifold of dimension other than 0. An analogous definition of semisimplicity holds for super Lie groups.) Unconventional super Lie groups may provide a new way of classifying infinite families of nonsimple and non-semisimple ordinary Lie groups.

† Its body is, of course, the trivial group having only one element.

Structure of conventional super Lie groups.
The extending representation

Denote by $\mathbf{G}_a$ the matrices $({}^i G_{aj})$ that generate a representation $D(x)$ of a (p, q)-dimensional super Lie group G (see eq. (3.2.92)). If G is conventional then, in a conventional coordinate system, structure constants bearing an odd number of a-type indices vanish, and eq. (3.2.93) may be rewritten in the simpler form

$$[\mathbf{G}_a, \mathbf{G}_b] = \mathbf{G}_c c^c{}_{ab}. \tag{4.1.6}$$

In particular, the generators $\mathbf{c}_a$ of the adjoint representation satisfy

$$[\mathbf{c}_a, \mathbf{c}_b] = \mathbf{c}_c c^c{}_{ab}, \tag{4.1.7}$$

which is an alternative version of eq. (3.1.61).

In a conventional coordinate system the c's have the block structure

$$\mathbf{c}_\mu = \begin{pmatrix} c^\nu{}_{\mu\sigma} & 0 \\ 0 & c^\beta{}_{\mu\gamma} \end{pmatrix}, \quad \mathbf{c}_\alpha = \begin{pmatrix} 0 & c^\nu{}_{\alpha\gamma} \\ c^\beta{}_{\alpha\sigma} & 0 \end{pmatrix}, \tag{4.1.8}$$

and eq. (3.1.61) decomposes into the following four identities:

$$c^\mu{}_{\nu\lambda} c^\lambda{}_{\sigma\tau} + c^\mu{}_{\sigma\lambda} c^\lambda{}_{\tau\nu} + c^\mu{}_{\tau\lambda} c^\lambda{}_{\nu\sigma} = 0, \tag{4.1.9}$$

$$c^\alpha{}_{\mu\gamma} c^\gamma{}_{\nu\beta} + c^\alpha{}_{\nu\gamma} c^\gamma{}_{\beta\mu} + c^\alpha{}_{\beta\sigma} c^\sigma{}_{\mu\nu} = 0, \tag{4.1.10}$$

$$c^\mu{}_{\alpha\gamma} c^\gamma{}_{\nu\beta} - c^\mu{}_{\beta\gamma} c^\gamma{}_{\nu\alpha} + c^\mu{}_{\nu\sigma} c^\sigma{}_{\alpha\beta} = 0, \tag{4.1.11}$$

$$c^\alpha{}_{\beta\mu} c^\mu{}_{\gamma\delta} + c^\alpha{}_{\gamma\mu} c^\mu{}_{\delta\beta} + c^\alpha{}_{\delta\mu} c^\mu{}_{\beta\gamma} = 0, \tag{4.1.12}$$

Equation (4.1.9) shows that the c's bearing only c-type indices are the structure constants of an ordinary p-dimensional Lie group. This group is just the body G_{B} of G. Equations (4.1.9) and (4.1.10) together are equivalent to

$$[\mathbf{c}_\mu, \mathbf{c}_\nu] = \mathbf{c}_\sigma c^\sigma{}_{\mu\nu}, \tag{4.1.13}$$

which shows that the $\mathbf{c}_\mu$ generate a representation of G_{B}, namely the direct sum of the adjoint representation of G_{B} and another representation generated by the $q \times q$ matrices

$$\mathbf{G}_\mu \stackrel{\mathrm{def}}{=} (c^\alpha{}_{\mu\beta}). \tag{4.1.14}$$

The latter representation will be called the *extending representation* because its existence is what makes possible the extension of the Lie group G_{B} to the super Lie group G.

Construction of a class of super Lie algebras

The extending representation cannot be just any representation of G_{B}, for it must also be compatible with the structure equations (4.1.11) and (4.1.12).

Checking this compatibility is the hardest part of the problem of constructing and classifying super Lie groups. We outline here a method that works in many cases or, rather, we outline a method that yields a class of super Lie algebras. (The global problem of the existence of associated super Lie groups will not be considered at this point.)

First, choose G_B to be a Lie group for which there exists a nonsingular symmetric real matrix $\boldsymbol{\eta}_S = (\eta_{\mu\nu})$ such that $\eta_{\mu\tau} c^\tau{}_{\nu\sigma}$ is completely antisymmetric in the indices μ, ν and σ. If G_B is semisimple then one such matrix is the Cartan–Killing matrix[†] $(-c^\sigma{}_{\mu\tau} c^\tau{}_{\nu\sigma})$. However, if G_B is only semisimple but not simple, other possibilities exist. In an appropriate basis $\boldsymbol{\eta}_S$ may be built out of blocks, each block being the Cartan–Killing matrix of one of the simple invariant subgroups of G_B and each block carrying its own scale factor. Flexibility in the choice of the scale factors turns out to be important.

Next, choose for the extending representation a representation of G_B for which there exists an *anti*symmetric imaginary matrix $\boldsymbol{\eta}_A = (\eta_{\alpha\beta})$ such that the matrices $\boldsymbol{\eta}_A \mathbf{G}_\mu = (\eta_{\alpha\gamma} c^\gamma{}_{\mu\beta})$ are all symmetric.[‡] If the matrices $\boldsymbol{\eta}_S$ and $\boldsymbol{\eta}_A$ are combined into a single matrix

$$\boldsymbol{\eta} \overset{\text{def}}{=} \operatorname{diag}(\boldsymbol{\eta}_S, \boldsymbol{\eta}_A) \tag{4.1.15}$$

then the matrices $\boldsymbol{\eta}\mathbf{c}_\mu$ will be antisupersymmetric:

$$(\boldsymbol{\eta}\mathbf{c}_\mu)_{ab} = \eta_{ac} c^c{}_{\mu b} \overset{\text{def}}{=} c_{a\mu b} = -(-1)^{ab} c_{b\mu a}. \tag{4.1.16}$$

Finally, check whether the identity

$$\mathbf{G}_\mu \mathbf{S} \boldsymbol{\eta}_A \mathbf{G}^\mu = -\tfrac{1}{2} \mathbf{G}_\mu \operatorname{tr}(\mathbf{S} \boldsymbol{\eta}_A \mathbf{G}^\mu) \tag{4.1.17}$$

holds, where S is *any* symmetric $q \times q$ matrix and where

$$\mathbf{G}^\mu \overset{\text{def}}{=} \eta^{\mu\nu} \mathbf{G}_\nu, \qquad (\eta^{\mu\nu}) = \boldsymbol{\eta}_S^{-1}. \tag{4.1.18}$$

Equation (4.1.17) is the crucial identity and generally requires a special choice for the scale factors upon which $\boldsymbol{\eta}_S$ depends, if indeed it can be satisfied at all. The exhaustive enumeration of the semisimple Lie groups G_B and extending representations $\mathbf{G}_\mu$ for which the above conditions can be made to hold requires the use of Dynkin weight diagrams and the theory of

[†] Cf. eq. (3.3.100).
[‡] Another way of stating this condition is that the extending representation must leave invariant an antisymmetric bilinear form.

filtrations (see V. G. Kac, *Advances in Mathematics*, **26**, 8 (1977)). We make no attempt to describe such techniques in this book but merely set forth in the following sections the results of the analysis.

Once a Lie group (or, rather, Lie algebra) G_B and an extending representation G_μ have been found satisfying all of the above conditions, then half of the structure constants of the associated super Lie algebra are known, namely $c^\mu{}_{\nu\sigma}$ and $c^\alpha{}_{\mu\beta}$. The remaining structure constants are obtained by defining

$$c^\alpha{}_{\beta\mu} \stackrel{\text{def}}{=} -c^\alpha{}_{\mu\beta}, \tag{4.1.19}$$

$$c^\mu{}_{\alpha\beta} \stackrel{\text{def}}{=} -\eta^{\mu\nu}\,\eta_{\alpha\gamma}\,c^\gamma{}_{\nu\beta} = c^\mu{}_{\beta\alpha}. \tag{4.1.20}$$

It is easy to see that eq. (4.1.11) is then equivalent to (4.1.10), and eq. (4.1.12) is an alternative version of (4.1.17). Moreover, the matrices $\boldsymbol{\eta}\mathbf{c}_a$ are symmetric and $\eta_{ad}\,c^d{}_{bc}$ is completely antisupersymmetric in the indices a, b and c.

In the above construction the matrix $\boldsymbol{\eta}_A$ was not required to be nonsingular. However, if it is singular then the super Lie algebra will generally not be simple, nor even semisimple. (For example, $\boldsymbol{\eta}_A$ could be chosen to vanish, which leads to a trivial extension of G_B.) We shall therefore confine our attention to the case in which $\boldsymbol{\eta}_A$ is nonsingular. The matrix $\boldsymbol{\eta}$ is then also nonsingular, and if the adjoint representation of the super Lie algebra is irreducible the Cartan–Killing matrix is necessarily a multiple of $\boldsymbol{\eta}$:

$$-\operatorname{str}(\mathbf{c}_a\mathbf{c}_b) = \lambda\eta_{ab}, \tag{4.1.21}$$

or, equivalently,

$$\left.\begin{aligned} -c^\sigma{}_{\mu\tau}\,c^\tau{}_{\nu\sigma} + c^\alpha{}_{\mu\beta}\,c^\beta{}_{\nu\alpha} &= \lambda\eta_{\mu\nu}, \\ -c^\mu{}_{\alpha\gamma}\,c^\gamma{}_{\beta\mu} + c^\gamma{}_{\alpha\mu}\,c^\mu{}_{\beta\gamma} &= \lambda\eta_{\alpha\beta}, \end{aligned}\right\} \tag{4.1.22}$$

for some real number λ. Because it is the supertrace 'str' that appears here[†], nonsingularity of the Cartan–Killing matrix is *not* a necessary condition for a super Lie group with irreducible adjoint representation to be simple, as it is for ordinary Lie groups. In the following sections we shall encounter several examples of non-Abelian simple super Lie groups for which the Cartan–Killing matrix vanishes. Among these are two (P(m) and Q(m), see section 4.2) for which the matrix $\boldsymbol{\eta}_A$ does not even exist and which therefore cannot be constructed by the methods outlined here.

[†] For the definition of the supertrace when the matrix is a-type see exercise **1.9**.

Notation

In the following sections we shall need names for specific Lie groups, super Lie groups and their associated algebras. Many authors use capital letters for the groups and lower case or Gothic script for the associated algebras. Since we shall generally not be concerned with the global differences between groups having the same algebra we shall use the same capital-letter symbol for both a group and its algebra. If A and B are two groups then $A \times B$ will denote their direct product, as groups, or their topological product as (super)manifolds. The corresponding algebra is their direct sum as algebras, or as (super) vector spaces, and will be denoted by $A \oplus B$.

We shall normally reserve lower case letters to designate matrix representations of groups and their algebras. The smallest representation that generates all the others by direct products is called the *fundamental representation*. In many cases it is known also as the *defining representation*. If B is a group (or algebra) its fundamental representation will be denoted by b (i.e., by the corresponding lower case symbol) and its adjoint representation by *ad* B. If k and l are representations of a group (algebra) A, their direct product will be denoted by $k \otimes l$. The same symbol may be used also if k and l are representations of different groups (algebras) A and B. $k \otimes l$ is then a representation of the group $A \times B$ (algebra $A \oplus B$), and if k and l are irreducible then $k \otimes l$ is irreducible. If $A = B$ then the symbol $k \otimes l$ may be used to designate a representation either of the group (algebra) A or of the group $A \times A$ (algebra $A \oplus A$). In the latter case the number of group parameters is doubled and $k \otimes l$ is irreducible if k and l are irreducible. In the former case $k \otimes l$ is generally reducible into the direct sum of irreducible representations h_i of $A : k \otimes l = h_1 \oplus h_2 \oplus \ldots$.

The following standard symbols will be used for the classical Lie groups:

GL(n): The group of nonsingular $n \times n$ real matrices.

U(n): The group of $n \times n$ unitary matrices.

SL(n): The group of nonsingular $n \times n$ real matrices with unit determinant.

SU(n): The group of $n \times n$ unitary matrices with unit determinant.

O(n): The group of $n \times n$ real orthogonal matrices.

SO(n): The group of $n \times n$ real orthogonal matrices with unit determinant.

Sp(n): The group of symplectic transformations in an n-dimensional vector space, with n even, i.e., the group of $n \times n$ matrices M that satisfy $MAM^{\sim} = A$, where A is a fixed nonsingular antisymmetric real $n \times n$ matrix. The matrices M are required to be either real or unitary. The symbol USp(n) is often used in the latter case.

As previously remarked, the same symbols will be used for the associated Lie algebras. GL(n) and SL(n) are often denoted by GL(n, **R**) and SL(n, **R**) to distinguish them from their extensions, GL(n, **C**) and SL(n, **C**), which are the corresponding groups of complex matrices. O(n) and SO(n) are also frequently generalized to the groups, O(r, s) and SO(r, s), of $(r+s) \times (r+s)$ matrices M that satisfy $M\,\mathrm{diag}(-1_r, 1_s)M^{\sim} = \mathrm{diag}(-1_r, 1_s)$.

The fundamental representations of the above groups will be denoted by gl(n), u(n), sl(n), su(n), o(n), so(n), sp(n) respectively. In addition to these symbols, and to the symbols for the adjoint representations, we shall use the symbol spin(n) for the spin representation of SO(n) (or rather, of the *covering group* $\underline{\mathrm{SO}}(n)$), and we shall add the prefix 'co' to denote the contragradient representation.

Also in common usage for designating the simple Lie algebras are the Cartan symbols. For the simple classical Lie algebras these are:

A_n; $n = 1, \ldots$: denotes SU($n+1$) or SL($n+1$).
B_n; $n = 2, 3, \ldots$: denotes SO($2n+1$).
C_n; $n = 3, 4 \ldots$: denotes Sp($2n$).
D_n; $n = 4, 5 \ldots$: denotes SO($2n$).

When these symbols are used no distinction is usually made between algebras whose complexifications (i.e., extensions from real to complex vector spaces) are isomorphic, e.g., SU(n) and SL(n), or SO(r, s) and SO($r+s$). In addition to the classical Lie algebras there exist also five exceptional simple Lie algebras, denoted by the Cartan symbols E_6, E_7, E_8, F_4 and G_2. The suffix on each Cartan symbol indicates the rank of the algebra.

4.2 The classical super Lie groups

The group GL(m, n)

This is the group of all nonsingular $(m, n) \times (m, n)$ matrices $\mathbf{x}$ with elements $x^a{}_b$ having the reality and type properties of the components of a real c-type rank (1, 1) tensor; namely, the elements $x^\mu{}_\nu$ and $x^\alpha{}_\beta$ are real c-numbers, the elements $x^\alpha{}_\nu$ are real a-numbers, and the elements $x^\mu{}_\beta$ are imaginary a-numbers. The group is evidently of dimension $(m^2 + n^2, 2mn)$.

If, instead of requiring all the group coordinates to be real, one allows half of the a-number coordinates to be imaginary, one may use the matrix elements $x^a{}_b$ themselves as coordinates in GL(m, n). (One may always convert to all-real coordinates by multiplying the $x^{\mu}{}_{\beta}$ by i.) It is then convenient to use the *pair* of matrix indices, a and b, as coordinate indices instead of trying to map them into a single index. The following index-shifting conventions are then appropriate:

$$ {}^{a}{}_{b}x \overset{\text{def}}{=} (-1)^{b}\, x^{a}{}_{b}, \tag{4.2.1}$$

$$ y^{b}{}_{a}\, {}^{a}{}_{b}x = (-1)^{b}\, y^{b}{}_{a}\, x^{a}{}_{b} = \operatorname{str}(\mathbf{y}\,\mathbf{x}), \tag{4.2.2}$$

$$ y^{b}{}_{a}\, {}^{a}{}_{b}M^{c}{}_{d}\, {}^{d}{}_{c}x = x^{d}{}_{c}\, {}^{c}{}_{d}M^{\sim a}{}_{b}\, {}^{b}{}_{a}y, \tag{4.2.3}$$

$$ {}^{c}{}_{d}M^{\sim a}{}_{b} \overset{\text{def}}{=} (-1)^{(a+b)(c+d)+b+d}\, {}^{a}{}_{b}M^{c}{}_{d}. \tag{4.2.4}$$

Here $\mathbf{x}$ and $\mathbf{y}$ are any two elements of $GL(m,n)$, and the ${}^{a}{}_{b}M^{c}{}_{d}$ may be regarded as elements of an $(m^2+n^2, 2mn)\times(m^2+n^2, 2mn)$ matrix $\mathbf{M}$. Equation (4.2.4) defines the supertranspose $\mathbf{M}^{\sim}$ of $\mathbf{M}$.

If $\mathbf{M}$ is nonsingular then it has an inverse $\mathbf{M}^{-1}$ satisfying

$$ {}^{a}{}_{b}M^{-1\,e}{}_{f}\, {}^{f}{}_{e}M^{c}{}_{d} = {}^{a}{}_{b}M^{e}{}_{f}\, {}^{f}{}_{e}M^{-1\,c}{}_{d} = {}^{a}{}_{b}\delta^{c}{}_{d}, \tag{4.2.5}$$

where the ${}^{a}{}_{b}\delta^{c}{}_{d}$ are the components of the unit matrix $1_{(m^2+n^2, 2mn)}$:

$$ {}^{a}{}_{b}\delta^{c}{}_{d} \overset{\text{def}}{=} {}^{a}\delta_{d}\, {}_{b}\delta^{c}. \tag{4.2.6}$$

The multiplication mapping F of $GL(m,n)$ is defined by

$$ F^{a}{}_{b}(\mathbf{x},\mathbf{y}) = (\mathbf{x}\,\mathbf{y})^{a}{}_{b} = x^{a}{}_{c}\, y^{c}{}_{b}, \tag{4.2.7}$$

and the coordinates of the identity are

$$ e^{a}{}_{b} = \delta^{a}{}_{b} = {}^{a}\delta_{b}. \tag{4.2.8}$$

The left and right auxiliary functions are respectively

$$ {}^{a}{}_{b}L^{c}{}_{d}(\mathbf{x}) = \left[{}^{a}{}_{b}(\mathbf{x}\,\mathbf{y}) \frac{\overleftarrow{\partial}}{\partial^{d}{}_{c}y} \right]_{\mathbf{y}=e} $$

$$ = (-1)^{b+c} \left[(\mathbf{x}\mathbf{y})^{a}{}_{b} \frac{\overleftarrow{\partial}}{\partial y^{d}{}_{c}} \right]_{\mathbf{y}=e} = x^{a}{}_{d}\, {}_{b}\delta^{c}, \tag{4.2.9}$$

$$ {}^{c}{}_{d}R^{a}{}_{b}(\mathbf{x}) = \left[\frac{\overrightarrow{\partial}}{\partial y^{d}{}_{c}} (\mathbf{y}\,\mathbf{x})^{a}{}_{b} \right]_{\mathbf{y}=e} = {}_{d}\delta^{a}\, x^{c}{}_{b}, \tag{4.2.10}$$

whence

$$ {}^{c}{}_{d}L^{\sim a}{}_{b}(\mathbf{x}) = (-1)^{c(a+d)}\, {}_{d}x^{\sim a\,c}\delta_{b}, \tag{4.2.11}$$

$$ {}^{a}{}_{b}R^{\sim c}{}_{d}(\mathbf{x}) = (-1)^{d(b+c)}\, {}^{a}\delta_{d}\, {}_{b}x^{\sim c}, \tag{4.2.12}$$

$$ {}^{a}{}_{b}L^{-1\,c}{}_{d}(\mathbf{x}) = x^{-1\,a}{}_{d}\, {}_{b}\delta^{c}, \tag{4.2.13}$$

$$ {}^{c}{}_{d}R^{-1\,a}{}_{b}(\mathbf{x}) = {}_{d}\delta^{a}\, x^{-1\,c}{}_{b}, \tag{4.2.14}$$

$$\operatorname{sdet} L^{-1}(\mathbf{x}) = \operatorname{sdet} R^{-1}(\mathbf{x}) = (\operatorname{sdet} \mathbf{x})^{-(m-n)}, \tag{4.2.15}$$

$$\begin{aligned} {}^a{}_b D_{\mathrm{ad}}{}^c{}_d(\mathbf{x}) &= {}^a{}_b R^{-1\,e}{}_c(\mathbf{x})\, {}^f{}_e L^c{}_d(\mathbf{x}) \\ &= (-1)^{a(b+c)}\, {}_b x^{\sim -1c}\, x^a{}_d. \end{aligned} \tag{4.2.16}$$

The structure constants are given (cf. eq. (3.1.88)) by

$$\begin{aligned} c^a{}_b{}^c{}_d{}^e{}_f &= (-1)^{b}\, {}^a{}_b c^c{}_d{}^e{}_f \\ &= -(-1)^b \left[(-1)^{e}\, {}^a{}_b L^c{}_d(\mathbf{x}) \frac{\overleftarrow{\partial}}{\partial x^f{}_e} - (-1)^{(c+d)(e+f)+c}\, {}^a{}_b L^e{}_f(\mathbf{x}) \frac{\overleftarrow{\partial}}{\partial x^d{}_c} \right]_{\mathbf{x}=e} \\ &= (-1)^{ac+ce+ea+e}\, \delta^a{}_d\, \delta^c{}_f\, \delta^e{}_b - (-1)^{c+e}\, \delta^a{}_f\, \delta^c{}_b\, \delta^e{}_d, \end{aligned} \tag{4.2.17}$$

and are readily verified to be antisupersymmetric under interchange of index pairs:

$$c^a{}_b{}^c{}_d{}^e{}_f = -(-1)^{(a+b)(c+d)}\, c^c{}_d{}^a{}_b{}^e{}_f = -(-1)^{(c+d)(e+f)}\, c^a{}_b{}^e{}_f{}^c{}_d. \tag{4.2.18}$$

The adjoint and coadjoint representations are therefore identical. The Cartan–Killing matrix (see eq. (3.3.100)) is given by

$$\begin{aligned} &-(-1)^{b+(e+f)(a+b+c+d+1)}\, {}^e{}_f c^a{}_b{}^g{}_h{}^h{}_g c^c{}_d{}^f{}_e \\ &= -2(m-n)\, {}^a{}_b \delta^c{}_d + 2\,(-1)^b\, \delta^a{}_b\, \delta^c{}_d. \end{aligned} \tag{4.2.19}$$

The super Lie algebra associated with $\mathrm{GL}(m,n)$ is the supervector space of all c-type and a-type $(m,n) \times (m,n)$ matrices (see exercise **1.9**) and linear combinations thereof, endowed with a bracket operation given, in the pure case, by

$$\begin{aligned} {}^a[\mathbf{X}, \mathbf{Y}]_b &= (-1)^{b(\mathbf{X}+\mathbf{Y}+1)}\, {}^a{}_b[\mathbf{X}, \mathbf{Y}] \\ &= (-1)^{b(\mathbf{X}+\mathbf{Y}+1)+(e+f)(c+d+\mathbf{X})}\, {}^a{}_b c^c{}_d{}^e{}_f\, {}^d{}_c X^f{}_e Y \\ &= (-1)^{b(\mathbf{X}+\mathbf{Y})+(e+f)(c+d+\mathbf{X})+c(\mathbf{X}+1)}\, c^a{}_b{}^c{}_d{}^e{}_f\, {}^d X_c{}^f Y_e \\ &= {}^a X_c\, {}^c Y_b - (-1)^{\mathbf{X}\mathbf{Y}}\, {}^a Y_c\, {}^c X_b, \end{aligned} \tag{4.2.20}$$

which is just the matrix supercommutator.

It is easy to see that the body of $\mathrm{GL}(m,n)$ is $\mathrm{GL}(m) \times \mathrm{GL}(n)$, having the structure constants

$$\left.\begin{aligned} c^\mu{}_\nu{}^\sigma{}_\tau{}^\rho{}_\lambda &= \delta^\mu{}_\tau\, \delta^\sigma{}_\lambda\, \delta^\rho{}_\nu - \delta^\mu{}_\lambda\, \delta^\sigma{}_\nu\, \delta^\rho{}_\tau, \\ c^\alpha{}_\beta{}^\gamma{}_\delta{}^\varepsilon{}_\zeta &= \delta^\alpha{}_\delta\, \delta^\gamma{}_\zeta\, \delta^\varepsilon{}_\beta - \delta^\alpha{}_\zeta\, \delta^\gamma{}_\beta\, \delta^\varepsilon{}_\delta. \end{aligned}\right\} \tag{4.2.21}$$

The extending representation is generated by the matrices

$$\begin{aligned} \begin{pmatrix} {}^\mu{}_\alpha c^\sigma{}_\tau{}^\beta{}_\nu & 0 \\ 0 & {}^\alpha{}_\mu c^\sigma{}_\tau{}^\nu{}_\beta \end{pmatrix} &= \begin{pmatrix} -c^\mu{}_\alpha{}^\sigma{}_\tau{}^\beta{}_\nu & 0 \\ 0 & c^\alpha{}_\mu{}^\sigma{}_\tau{}^\nu{}_\beta \end{pmatrix} \\ &= \begin{pmatrix} \delta^\mu{}_\tau\, \delta^\sigma{}_\nu\, \delta^\beta{}_\alpha & 0 \\ 0 & -\delta^\alpha{}_\beta\, \delta^\sigma{}_\mu\, \delta^\nu{}_\tau \end{pmatrix} \end{aligned} \tag{4.2.22}$$

and

$$\begin{pmatrix} {}^{\mu}{}_{\alpha}c^{\gamma}{}_{\delta}{}^{\beta}{}_{\nu} & 0 \\ 0 & {}^{\alpha}{}_{\mu}c^{\gamma}{}_{\delta}{}^{\nu}{}_{\beta} \end{pmatrix} = \begin{pmatrix} -c^{\mu}{}_{\alpha}{}^{\gamma}{}_{\delta}{}^{\beta}{}_{\nu} & 0 \\ 0 & c^{\alpha}{}_{\mu}{}^{\gamma}{}_{\delta}{}^{\nu}{}_{\beta} \end{pmatrix}$$
$$= \begin{pmatrix} \delta^{\mu}{}_{\nu}\delta^{\gamma}{}_{\alpha}\delta^{\beta}{}_{\delta} & 0 \\ 0 & -\delta^{\alpha}{}_{\delta}\delta^{\gamma}{}_{\beta}\delta^{\nu}{}_{\mu} \end{pmatrix} \tag{4.2.23}$$

Multiplying these matrices by ${}^{\tau}{}_{\sigma}x$ and ${}^{\delta}{}_{\gamma}x$ respectively, and adding, one obtains the matrix

$$\begin{pmatrix} {}^{\mu}{}_{\nu}x\,\delta^{\beta}{}_{\alpha} + {}^{\beta}{}_{\alpha}x\,\delta^{\mu}{}_{\nu} & 0 \\ 0 & -{}^{\nu}{}_{\mu}x\,\delta^{\alpha}{}_{\beta} - {}^{\alpha}{}_{\beta}x\,\delta^{\nu}{}_{\mu} \end{pmatrix}$$
$$= \begin{pmatrix} x^{\mu}{}_{\nu}\,{}_{\alpha}\delta^{\beta} - \delta^{\mu}{}_{\nu}\,x^{\sim}{}_{\alpha}{}^{\beta} & 0 \\ 0 & x^{\alpha}{}_{\beta}\,\delta^{\nu}{}_{\mu} - \delta^{\alpha}{}_{\beta}\,x^{\sim}{}_{\mu}{}^{\nu} \end{pmatrix} \tag{4.2.24}$$

which, when applied to the column matrix

$$\begin{pmatrix} {}^{\nu}{}_{\beta}y \\ {}^{\beta}{}_{\nu}y \end{pmatrix},$$

yields

$$\begin{pmatrix} x^{\mu}{}_{\nu}\,{}^{\nu}{}_{\alpha}y - x^{\sim}{}_{\alpha}{}^{\beta}\,{}^{\mu}{}_{\beta}y \\ x^{\alpha}{}_{\beta}\,{}^{\beta}{}_{\mu}y - x^{\sim}{}_{\mu}{}^{\nu}\,{}^{\alpha}{}_{\nu}y \end{pmatrix}. \tag{4.2.25}$$

The extending representation is seen to be $[\mathrm{gl}(m)\otimes\mathrm{cogl}(n)]\oplus[\mathrm{cogl}(m)\otimes\mathrm{gl}(n)]$.

The group SL(m, n)

The group GL(m,n) has an invariant Abelian subgroup consisting of all matrices of the form $\lambda 1_{(m,n)}$ where λ is a real c-number with nonvanishing body. This subgroup is a sub-supermanifold of dimension (1, 0), isomorphic to GL(1, 0). GL(m,n) is therefore not semisimple. If $m \neq n$ it may be decomposed into the product SL(m,n) × GL(1, 0), where SL(m,n) is the subgroup consisting of all elements of GL(m,n) that have unit superdeterminant. When $m \neq n$ SL(m,n) is semisimple, although its body $(\mathrm{SL}(m,n))_{\mathrm{B}}$, which is isomorphic to GL(m) × GL(n)/GL(1) or SL(m) × SL(n) × GL(1), *is not*. Its dimension is evidently $(m^2 + n^2 - 1, 2mn)$.

The super Lie algebra associated with SL(m,n) is the supervector space of all c-type and a-type $(m,n) \times (m,n)$ matrices having vanishing supertrace, and linear combinations thereof. When $m \neq n$, one may introduce the projection matrix into this subspace:

$$ {}^{a}{}_{b}\Pi^{c}{}_{d} \overset{\text{def}}{=} {}^{a}{}_{b}\delta^{c}{}_{d} - \frac{1}{m-n}(-1)^{b}\,\delta^{a}{}_{b}\,\delta^{c}{}_{d}. \tag{4.2.26}$$

It is easy to check that this projection matrix, when applied to any of the index pairs of the structure constants (4.2.17) of GL(m,n), leaves these structure constants unchanged. Hence, with the projection understood, one may regard (4.2.17) as giving also the structure constants of SL(m,n). The projection matrix satisfies

$$ {}^a{}_a\Pi^c{}_d = 0, \qquad (-1)^c\,{}^a{}_b\Pi^c{}_c = 0, \tag{4.2.27}$$

$$ {}^a{}_b\Pi^e{}_f\,{}^f{}_e\Pi^c{}_d = {}^a{}_b\Pi^c{}_d, \tag{4.2.28}$$

and may be regarded as the unit matrix in the subspace of supertraceless $(m,n)\times(m,n)$ matrices. The Cartan–Killing matrix is a multiple of it:

$$ -(-1)^{b+(e+f)(a+b+c+d+1)e}\,{}_f c^a{}_b{}^g{}_h\,{}^h{}_g c^c{}_d{}^f{}_e = -2(m-n)\,{}^a{}_b\Pi^c{}_d. \tag{4.2.29}$$

The extending representation of SL(m,n) is generated by all matrices of the form (4.2.24) with $x^\mu{}_\mu = x^\alpha{}_\alpha$. If $x^\mu{}_\nu$ and $x^\alpha{}_\beta$ are replaced by

$$ x^\mu{}_\nu = z^\mu{}_\nu + \frac{\lambda}{m}\delta^\mu{}_\nu, \qquad x^\alpha{}_\beta = z^\alpha{}_\beta + \frac{\lambda}{n}\delta^\alpha{}_\beta, \tag{4.2.30}$$

where $z^\mu{}_\mu = 0$, $z^\alpha{}_\alpha = 0$, then expression (4.2.25) takes the form

$$ \begin{bmatrix} z^\mu{}_\nu{}^\nu{}_\alpha y - z^{\sim\beta}{}_\alpha{}^\mu{}_\beta y - \dfrac{m-n}{mn}\lambda^\mu{}_\alpha y & 0 \\ 0 & z^\alpha{}_\beta{}^\beta{}_\mu y - z^{\sim\nu}{}_\mu{}^\alpha{}_\nu y + \dfrac{m-n}{mn}\lambda^\alpha{}_\mu y \end{bmatrix}, \tag{4.2.31}$$

from which it is easy to see that the extending representation is $[\mathrm{sl}(m)\otimes \mathrm{cosl}(n)\otimes \mathrm{gl}(1)]\oplus[\mathrm{cosl}(m)\otimes \mathrm{sl}(n)\otimes \mathrm{gl}(1)]$.

Since the body $(\mathrm{SL}(m,n))_{\mathrm{B}}$ is not semisimple its Cartan–Killing matrix is singular. There nevertheless exists a nonsingular symmetric matrix $\boldsymbol{\eta}_{\mathrm{S}}$ (see section 4.1) that connects its adjoint and coadjoint representations. It is given by

$$ \begin{pmatrix} \Pi^\mu{}_\nu{}^\sigma{}_\tau & \Pi^\mu{}_\nu{}^\gamma{}_\delta \\ \Pi^\alpha{}_\beta{}^\sigma{}_\tau & \Pi^\alpha{}_\beta{}^\gamma{}_\delta \end{pmatrix} = \begin{pmatrix} {}^\mu{}_\nu\Pi^\sigma{}_\tau & {}^\mu{}_\nu\Pi^\gamma{}_\delta \\ -{}^\alpha{}_\beta\Pi^\sigma{}_\tau & -{}^\alpha{}_\beta\Pi^\gamma{}_\delta \end{pmatrix} $$

$$ = \begin{pmatrix} \pi^\mu{}_\nu{}^\sigma{}_\tau - \dfrac{n}{m(m-n)}\delta^\mu{}_\nu\delta^\sigma{}_\tau & -\dfrac{1}{m-n}\delta^\mu{}_\nu\delta^\gamma{}_\delta \\ -\dfrac{1}{m-n}\delta^\alpha{}_\beta\delta^\sigma{}_\tau & -\pi^\alpha{}_\beta{}^\gamma{}_\delta - \dfrac{m}{n(m-n)}\delta^\alpha{}_\beta\delta^\gamma{}_\delta \end{pmatrix}, \tag{4.2.32}$$

where $\pi^\mu{}_\nu{}^\sigma{}_\tau$ and $\pi^\alpha{}_\beta{}^\gamma{}_\delta$ are the projection matrices onto the subspace of traceless $m\times m$ and $n\times n$ matrices respectively. Using the decomposition (4.2.30) one easily finds

$$ \Pi^\mu{}_\nu{}^\sigma{}_\tau x^\nu{}_\mu x^\tau{}_\sigma - \Pi^\mu{}_\nu{}^\gamma{}_\delta x^\nu{}_\mu x^\delta{}_\gamma - \Pi^\alpha{}_\beta{}^\sigma{}_\tau x^\beta{}_\alpha x^\tau{}_\sigma + \Pi^\alpha{}_\beta{}^\gamma{}_\delta x^\beta{}_\alpha x^\delta{}_\gamma $$

$$ = z^\mu{}_\nu z^\nu{}_\mu - z^\alpha{}_\beta z^\beta{}_\alpha - \frac{m-n}{mn}\lambda^2. \tag{4.2.33}$$

The antisymmetric matrix η_{A} of section 4.1 is given by

$$\begin{pmatrix} \Pi^{\mu}{}_{\alpha}{}^{\nu}{}_{\beta} & \Pi^{\mu}{}_{\alpha}{}^{\beta}{}_{\nu} \\ \Pi^{\alpha}{}_{\mu}{}^{\nu}{}_{\beta} & \Pi^{\alpha}{}_{\mu}{}^{\beta}{}_{\nu} \end{pmatrix} = \begin{pmatrix} 0 & -{}^{\mu}{}_{\alpha}\Pi^{\beta}{}_{\nu} \\ {}^{\alpha}{}_{\mu}\Pi^{\nu}{}_{\beta} & 0 \end{pmatrix} = \begin{pmatrix} 0 & -\delta^{\mu}{}_{\nu}\delta^{\beta}{}_{\alpha} \\ \delta^{\alpha}{}_{\beta}\delta^{\nu}{}_{\mu} & 0 \end{pmatrix} \tag{4.2.34}$$

The group SL(m, m)/GL$(1, 0)$

When $m = n$ the group SL(m, n) is no longer semisimple, for it still has the set of matrices $\lambda 1_{(m,n)}(\lambda_{\mathrm{B}} \neq 0)$ as an invariant Abelian subgroup. (When $m = n$ these matrices themselves have unit superdeterminant.) To get a semisimple super Lie group one must divide by the subgroup: SL(m, m)/GL$(1, 0)$. The resulting group has dimensionality $(2m^2 - 2, 2m^2)$.

The elements **X** of the super Lie algebra associated with SL(m, m)/GL$(1,0)$ may be represented by $(m, m) \times (m, m)$ matrices of the form

$$\mathbf{X} = \begin{pmatrix} A & C \\ D & B \end{pmatrix}, \quad \operatorname{tr} A = 0, \quad \operatorname{tr} B = 0. \tag{4.2.35}$$

However, the bracket operation is no longer the matrix supercommutator but a modification thereof. The matrices (4.2.35) therefore do not generate a genuine matrix representation of SL(m, m)/GL$(1, 0)$ but rather what one may call a *pseudorepresentation*. Using boldface to denote the bracket operation of the super Lie algebra one has, in fact,

$$\boldsymbol{[}\mathbf{X}_1, \mathbf{X}_2\boldsymbol{]} = [\mathbf{X}_1, \mathbf{X}_2] - \frac{1}{m}\operatorname{diag}([\mathbf{X}_1, \mathbf{X}_2]^{\mu}{}_{\mu} 1_m, [\mathbf{X}_1, \mathbf{X}_2]^{\alpha}{}_{\alpha} 1_m), \tag{4.2.36}$$

where the lightface bracket denotes the ordinary matrix supercommutator. It is straightforward to show that this modified supercommutator satisfies the super Jacobi identity provided the **X**'s satisfy the trace constraints indicated in (4.2.35).

The simplest true matrix representation is the adjoint representation, generated by the structure constants. To obtain the structure constants it suffices to compute (4.2.36) for the case in which both $\mathbf{X}_1$ and $\mathbf{X}_2$ are c-type. Denoting by $\mathbf{X}_{12}$ the bracket of $\mathbf{X}_1$ and $\mathbf{X}_2$, and by A_{12}, A_1, A_2, etc. the associated matrix blocks, one finds

$$\left.\begin{aligned} A_{12} &= [A_1, A_2] + C_1 D_2 - C_2 D_1 - \frac{1}{m}\operatorname{tr}(C_1 D_2 - C_2 D_1)\, 1_m, \\ B_{12} &= [B_1, B_2] + D_1 C_2 - D_2 C_1 - \frac{1}{m}\operatorname{tr}(D_1 C_2 - D_2 C_1)\, 1_m, \\ C_{12} &= A_1 C_2 - A_2 C_1 + C_1 B_2 - C_2 B_1, \\ D_{12} &= D_1 A_2 - D_2 A_1 + B_1 D_2 - B_2 D_1. \end{aligned}\right\} \tag{4.2.37}$$

The equation

$$
\begin{aligned}
X_{12}{}^{a}{}_{b} &= (-1)^{(c+d)(e+f)} c^{a}{}_{b}{}^{c}{}_{d}{}^{e}{}_{f}{}^{d}{}_{c} X_{1}{}^{f}{}_{e} X_{2} \\
&= (-1)^{(c+d)(e+f)+c+e} c^{a}{}_{b}{}^{c}{}_{d}{}^{e}{}_{f} X_{1}{}^{d}{}_{c} X_{2}{}^{f}{}_{e}
\end{aligned}
\tag{4.2.38}
$$

then permits one to read off the structure constants directly from (4.2.37). The results are

$$
\left.
\begin{aligned}
&c^{\mu}{}_{\nu}{}^{\sigma}{}_{\tau}{}^{\rho}{}_{\lambda} = \delta^{\mu}{}_{\tau}\delta^{\sigma}{}_{\lambda}\delta^{\rho}{}_{\nu} - \delta^{\mu}{}_{\lambda}\delta^{\sigma}{}_{\nu}\delta^{\rho}{}_{\tau}, \\
&c^{\mu}{}_{\nu}{}^{\alpha}{}_{\sigma}{}^{\tau}{}_{\beta} = c^{\mu}{}_{\nu}{}^{\tau}{}_{\beta}{}^{\alpha}{}_{\sigma} = \delta^{\mu}{}_{\sigma}\delta^{\alpha}{}_{\beta}\delta^{\tau}{}_{\nu} - \frac{1}{m}\delta^{\mu}{}_{\nu}\delta^{\alpha}{}_{\beta}\delta^{\tau}{}_{\sigma}, \\
&c^{\mu}{}_{\alpha}{}^{\beta}{}_{\nu}{}^{\sigma}{}_{\tau} = -c^{\mu}{}_{\alpha}{}^{\sigma}{}_{\tau}{}^{\beta}{}_{\nu} = \delta^{\mu}{}_{\tau}\delta^{\beta}{}_{\alpha}\delta^{\sigma}{}_{\nu} - \frac{1}{m}\delta^{\mu}{}_{\nu}\delta^{\beta}{}_{\alpha}\delta^{\sigma}{}_{\tau}, \\
&c^{\mu}{}_{\alpha}{}^{\beta}{}_{\nu}{}^{\gamma}{}_{\delta} = -c^{\mu}{}_{\alpha}{}^{\gamma}{}_{\delta}{}^{\beta}{}_{\nu} = \delta^{\mu}{}_{\nu}\delta^{\beta}{}_{\delta}\delta^{\gamma}{}_{\alpha} - \frac{1}{m}\delta^{\mu}{}_{\nu}\delta^{\beta}{}_{\alpha}\delta^{\gamma}{}_{\delta}, \\
&c^{\alpha}{}_{\mu}{}^{\nu}{}_{\beta}{}^{\sigma}{}_{\tau} = -c^{\alpha}{}_{\mu}{}^{\sigma}{}_{\tau}{}^{\nu}{}_{\beta} = \delta^{\alpha}{}_{\beta}\delta^{\nu}{}_{\tau}\delta^{\sigma}{}_{\mu} - \frac{1}{m}\delta^{\alpha}{}_{\beta}\delta^{\nu}{}_{\mu}\delta^{\sigma}{}_{\tau}, \\
&c^{\alpha}{}_{\mu}{}^{\nu}{}_{\beta}{}^{\gamma}{}_{\delta} = -c^{\alpha}{}_{\mu}{}^{\gamma}{}_{\delta}{}^{\nu}{}_{\beta} = \delta^{\alpha}{}_{\delta}\delta^{\nu}{}_{\mu}\delta^{\gamma}{}_{\beta} - \frac{1}{m}\delta^{\alpha}{}_{\beta}\delta^{\nu}{}_{\mu}\delta^{\gamma}{}_{\delta}, \\
&c^{\alpha}{}_{\beta}{}^{\gamma}{}_{\mu}{}^{\nu}{}_{\delta} = c^{\alpha}{}_{\beta}{}^{\nu}{}_{\delta}{}^{\gamma}{}_{\mu} = \delta^{\alpha}{}_{\delta}\delta^{\gamma}{}_{\beta}\delta^{\nu}{}_{\mu} - \frac{1}{m}\delta^{\alpha}{}_{\beta}\delta^{\gamma}{}_{\delta}\delta^{\nu}{}_{\mu}, \\
&c^{\alpha}{}_{\beta}{}^{\gamma}{}_{\delta}{}^{\varepsilon}{}_{\zeta} = \delta^{\alpha}{}_{\delta}\delta^{\gamma}{}_{\zeta}\delta^{\varepsilon}{}_{\beta} - \delta^{\alpha}{}_{\zeta}\delta^{\gamma}{}_{\beta}\delta^{\varepsilon}{}_{\delta},
\end{aligned}
\right\}
\tag{4.2.39}
$$

all other structure constants vanishing.

These structure constants may be obtained from the structure constants (4.2.17) by applying the projection matrices $\pi^{\mu}{}_{\nu}{}^{\sigma}{}_{\tau}$ and $\pi^{\alpha}{}_{\beta}{}^{\gamma}{}_{\delta}$ respectively to all index pairs ${}^{\mu}{}_{\nu}$ and ${}^{\alpha}{}_{\beta}$. They are completely antisupersymmetric under interchange of index pairs, and they yield a *vanishing* Cartan–Killing matrix, namely expression (4.2.19) with $m = n$ and the second term suppressed by the projection matrices. The matrices η_S and η_A of section 4.1 are nevertheless well defined:

$$
\eta_S = \begin{pmatrix} \pi^{\mu}{}_{\nu}{}^{\sigma}{}_{\tau} & 0 \\ 0 & \pi^{\alpha}{}_{\beta}{}^{\gamma}{}_{\delta} \end{pmatrix}, \quad \eta_A = \begin{pmatrix} 0 & -\delta^{\mu}{}_{\nu}\delta^{\beta}{}_{\alpha} \\ \delta^{\alpha}{}_{\beta}\delta^{\nu}{}_{\mu} & 0 \end{pmatrix}. \tag{4.2.40}
$$

The body of SL(m, m)/GL$(1,0)$ is obviously SL$(m) \times$ SL(m), and by inserting the structure constants (4.2.39) into expressions (4.2.22) and (4.2.23) one easily checks that the extending representation is $[\mathrm{sl}(m) \otimes \mathrm{cosl}(m)] \oplus [\mathrm{cosl}(m) \otimes \mathrm{sl}(m)]$. Note that when $m = 1$ the group is not simple, being the product of the Abelian group of dimension (0, 1) with itself.

The orthosymplectic group OSp(m,n)

This is the subgroup of GL(m,n) that leaves invariant the quadratic form $X^a{}_a g_b{}^b X$ where **X** is any *c*-type vector and $g\,(\overset{\text{def}}{=}({}_a g_b))$ is a nonsingular supersymmetric matrix whose elements have the reality and type properties of the components of a metric tensor and whose canonical form has no negative diagonal elements. If L is an element of OSp(m,n) then the invariance condition may be expressed in the form

$$L^{\sim} g L = g. \tag{4.2.41}$$

This condition imposes $\frac{1}{2}m(m+1)+\frac{1}{2}n(n-1)$ *c*-type constraints and mn *a*-type constraints on the elements of L. The dimensionality of OSp(m,n) is therefore $(\frac{1}{2}m(m-1)+\frac{1}{2}n(n+1), mn)$. The nonsingularity of g requires that n be an even integer.

Taking the superdeterminant of both sides of eq. (4.2.41) one finds

$$\operatorname{sdet} L = \pm 1. \tag{4.2.42}$$

Hence the connected component of OSp(m,n) containing the identity is actually a subgroup of SL(m,n), and the super Lie algebra OSp(m,n) is a subalgebra of SL(m,n).

The generators of the fundamental representation OSp(m,n) are obtained by considering infinitesimal transformations

$$\delta^a X = {}^a\xi_b{}^b X. \tag{4.2.43}$$

The necessary and sufficient condition that these transformations leave the quadratic form $X^a{}_a g_b{}^b X$ (**X** *c*-type) invariant is that the matrix $({}_a\xi_b)$ be antisupersymmetric, where

$${}_a\xi_b \overset{\text{def}}{=} {}_a g_c{}^c\xi_b. \tag{4.2.44}$$

The transformation (4.2.43) may therefore be rewritten in the form

$$\delta \mathbf{X} = \tfrac{1}{2}\mathbf{G}_{ab}\mathbf{X}\,\xi^{ba}, \tag{4.9.45}$$

where the $\mathbf{G}_{ab}$ are the $(m, n) \times (m, n)$ matrices

$$\mathbf{G}_{ab} \overset{\text{def}}{=} ({}^c G_{ab\,d}) = -(-1)^{ab}\,\mathbf{G}_{ba}, \tag{4.2.46}$$

$${}^c G_{ab\,d} \overset{\text{def}}{=} -\delta^c{}_a g_{bd} + (-1)^{ab}\delta^c{}_b g_{ad}, \tag{4.2.47}$$

$$g_{ab} \overset{\text{def}}{=} (-1)^a{}_a g_b, \tag{4.2.48}$$

and where

$$\xi^{ab} = {}^a\xi^b \overset{\text{def}}{=} {}^a\xi_c g^{cb} = -(-1)^{ab}\xi^{ba}, \tag{4.2.49}$$

(g^{ab}) being the matrix inverse to $({}_ag_b)$.

Without loss of generality, one may introduce a standard basis in the supervector space on which the $\mathbf{G}_{ab}$ act, with respect to which the matrix $({}_ag_b)$ assumes the canonical form (2.8.10). The elements ${}_ag_b$ then vanish when the indices a and b are of opposite type, and it is straightforward to verify that the $\mathbf{G}_{ab}$ satisfy the supercommutation relations

$$\begin{aligned}[\mathbf{G}_{ab}, \mathbf{G}_{cd}] &= (-1)^{bc} g_{ac}\mathbf{G}_{bd} + (-1)^{cd} g_{bd}\mathbf{G}_{ac} - (-1)^{d(b+c)} g_{ad}\mathbf{G}_{bc} - g_{bc}\mathbf{G}_{ad} \\ &= \tfrac{1}{2}\mathbf{G}_{fe}\, c^{ef}{}_{ab\,cd},\end{aligned} \tag{4.2.50}$$

$$\begin{aligned}c^{ef}{}_{ab\,cd} \overset{\text{def}}{=} & -(-1)^{c(a+b)}\delta^e{}_a\delta^f{}_c g_{bd} - (-1)^{b(a+d)}\delta^e{}_b\delta^f{}_d g_{ac} \\ & + (-1)^{ad}\delta^e{}_a\delta^f{}_d g_{bc} + (-1)^{ab+c(a+b)}\delta^e{}_b\delta^f{}_c g_{ad} \\ & + (-1)^{bc}\delta^f{}_a\delta^e{}_c g_{bd} + (-1)^{ab}\delta^f{}_b\delta^e{}_d g_{ac} \\ & + \delta^f{}_a\delta^e{}_d g_{bc} - (-1)^{a(b+c)}\delta^f{}_b\delta^e{}_c g_{ad}.\end{aligned} \tag{4.2.51}$$

The c's are the structure constants of OSp(m, n). Since they have vanishing souls OSp(m, n) is a conventional super Lie group.

If we confine our attention to the connected component of OSp(m, n) containing the identity then the body of OSp(m, n) is SO(m) × Sp(n). Since the canonical form of g is $\operatorname{diag}(1_m, \mathrm{i}\varepsilon)$ with $\varepsilon = \operatorname{diag}\left(\begin{pmatrix} 0 & 1 \\ -1 & 0 \end{pmatrix}, \ldots, \begin{pmatrix} 0 & 1 \\ -1 & 0 \end{pmatrix}\right)$ ($n/2$ blocks), indices from the middle of the Greek alphabet may all be written in the lower position and the structure constants of the body reduce to

$$\begin{aligned}c_{\mu\nu\,\sigma\tau\,\rho\lambda} = & -\delta_{\mu\sigma}\delta_{\nu\rho}\delta_{\tau\lambda} - \delta_{\mu\tau}\delta_{\nu\lambda}\delta_{\sigma\rho} + \delta_{\mu\sigma}\delta_{\nu\lambda}\delta_{\tau\rho} + \delta_{\mu\tau}\delta_{\nu\rho}\delta_{\sigma\lambda} \\ & + \delta_{\nu\sigma}\delta_{\mu\rho}\delta_{\tau\lambda} + \delta_{\nu\tau}\delta_{\mu\lambda}\delta_{\sigma\rho} - \delta_{\nu\sigma}\delta_{\mu\lambda}\delta_{\tau\rho} - \delta_{\nu\tau}\delta_{\mu\rho}\delta_{\sigma\lambda},\end{aligned} \tag{4.2.52}$$

$$\begin{aligned}c^{\alpha\beta}{}_{\gamma\delta\,\varepsilon\tau} = & -\mathrm{i}(\delta^\alpha{}_\gamma\delta^\beta{}_\varepsilon\varepsilon_{\delta\zeta} + \delta^\alpha{}_\delta\delta^\beta{}_\zeta\varepsilon_{\gamma\varepsilon} + \delta^\alpha{}_\gamma\delta^\beta{}_\zeta\varepsilon_{\delta\varepsilon} + \delta^\alpha{}_\delta\delta^\beta{}_\varepsilon\varepsilon_{\gamma\zeta} \\ & + \delta^\beta{}_\gamma\delta^\alpha{}_\varepsilon\varepsilon_{\delta\zeta} + \delta^\beta{}_\delta\delta^\alpha{}_\zeta\varepsilon_{\gamma\varepsilon} + \delta^\beta{}_\gamma\delta^\alpha{}_\zeta\varepsilon_{\delta\varepsilon} + \delta^\beta{}_\delta\delta^\alpha{}_\varepsilon\varepsilon_{\gamma\zeta}),\end{aligned} \tag{4.2.53}$$

where $\varepsilon_{\alpha\beta} = -{}_\alpha\varepsilon_\beta$. Note that the structure constants (4.2.53) are imaginary. This follows from using the ξ^{ab} (eqs. (4.2.45) and (4.2.49)) as the (infinitesimal) coordinates of the group. The $\xi^{\mu\nu}$, $\xi^{\mu\beta}$, $\xi^{\alpha\nu}$ are all real but the $\xi^{\alpha\beta}$ are imaginary. The structure constants (4.2.53) (which are just the structure constants of Sp(n)) would be real if we converted to real coordinates.

The extending representation of OSp (m, n) is generated by the matrices

$$(c_{\mu}{}^{\alpha}{}_{\sigma\tau\ \beta\nu}) = (-(\delta_{\mu\sigma}\delta_{\nu\tau} - \delta_{\mu\tau}\delta_{\nu\sigma})\delta^{\alpha}{}_{\beta}), \tag{4.2.54}$$

$$(c_{\mu}{}^{\alpha}{}_{\gamma\delta\ \beta\gamma}) = (-\mathrm{i}(\delta^{\alpha}{}_{\gamma}\varepsilon_{\delta\beta} + \delta^{\alpha}{}_{\delta}\varepsilon_{\gamma\beta})\delta_{\mu\nu}). \tag{4.2.55}$$

Multiplying these matrices by $\frac{1}{2}\xi_{\tau\sigma}$ and $\frac{1}{2}\xi^{\delta\gamma}$ respectively, and adding, one obtains the matrix $(\xi_{\mu\nu}\delta^{\alpha}{}_{\beta} + \delta_{\mu\nu}\xi^{\alpha}{}_{\beta})$ which, when applied to $(y_{\nu}{}^{\beta})$, yields $(\xi_{\mu\nu}y_{\nu}{}^{\alpha} + \xi^{\alpha}{}_{\beta}y_{\mu}{}^{\beta})$. The extending representation is seen to be simply so(m) $\otimes$sp(n).

The Cartan–Killing matrix is given by

$$-\tfrac{1}{4}(-1)^{a+b+(e+f)(a+b+c+d+1)}c^{ef}{}_{abgh}c^{hg}{}_{cdfe}$$

$$= 2(m-n-2)(-1)^{a+b}[(-1)^{bc}g_{ac}g_{bd} - g_{ad}g_{bd}]. \tag{4.2.56}$$

It vanishes when $m - n = 2$. The symmetric matrix $\boldsymbol{\eta}_{\mathrm{S}}$ (see section 4.1) that connects the adjoint and coadjoint representations of the body is

$$\begin{pmatrix} {}_{\mu\nu}\eta_{\sigma\tau} & {}_{\mu\nu}\eta_{\gamma\delta} \\ {}_{\alpha\beta}\eta_{\sigma\tau} & {}_{\alpha\beta}\eta_{\gamma\delta} \end{pmatrix} = \begin{pmatrix} \delta_{\mu\sigma}\delta_{\nu\tau} - \delta_{\mu\tau}\delta_{\nu\sigma} & 0 \\ 0 & \varepsilon_{\alpha\gamma}\varepsilon_{\beta\delta} + \varepsilon_{\alpha\delta}\varepsilon_{\beta\gamma} \end{pmatrix}. \tag{4.2.57}$$

The antisymmetric matrix $\boldsymbol{\eta}_{\mathrm{A}}$ of section 4.1 is

$$({}_{\mu\alpha}\eta_{\nu\beta}) = (\mathrm{i}\delta_{\mu\nu}{}_{\alpha}\varepsilon_{b}). \tag{4.2.58}$$

The Cartan–Killing matrix is $2(m-n-2)\boldsymbol{\eta}$ where $\boldsymbol{\eta} = \mathrm{diag}(\boldsymbol{\eta}_{\mathrm{S}}, \boldsymbol{\eta}_{\mathrm{A}})$.

If the matrix $\mathbf{L}$ of eq. (4.2.41) is expressed in the block form

$$\mathbf{L} = \begin{pmatrix} A & C \\ D & B \end{pmatrix}, \tag{4.2.59}$$

and if the matrix $\boldsymbol{g}$ is taken in the canonical form $\boldsymbol{g} = \mathrm{diag}(1_m, \mathrm{i}\varepsilon)$, then the invariance condition (4.2.41) decomposes into the equations

$$A^{\sim}A + \mathrm{i}D^{\sim}\varepsilon D = 1_m, \quad A^{\sim}C + \mathrm{i}D^{\sim}\varepsilon B = 0, \quad \mathrm{i}B^{\sim}\varepsilon B - C^{\sim}C = \mathrm{i}\varepsilon. \tag{4.2.60}$$

The general solution of these equations is

$$\begin{aligned} A &= O(1_m - \mathrm{i}D^{\sim}\varepsilon D)^{\frac{1}{2}}, \\ B &= (1_n - \mathrm{i}DD^{\sim}\varepsilon)^{\frac{1}{2}}P, \\ C &= -\mathrm{i}OD^{\sim}\varepsilon P, \end{aligned} \tag{4.2.61}$$

where O and P are arbitrary elements of the fundamental representations of SO(m) and Sp(n) respectively, D is an arbitrary $n \times m$ matrix with real a-number elements, and the square roots are given by the binomial expansion, which terminates after a finite number of terms. The super Lie group 'OSp(m, n) is generally not compact because the component Sp(n) of its body is not compact. However, if P is chosen to be a unitary matrix instead of a real matrix then the body of OSp(m, n) becomes SO$(m) \times$ USp(n) which *is* compact. The two possibilities have isomorphic complex extensions.

OSp(m, n) may also be generalized in another way, by choosing a matrix g that has the canonical form diag$(-1_r, 1_s, i\varepsilon)$, with $r + s = m$. The resulting super Lie group is denoted by OSp$(r, s; n)$ and has SO$(r, s) \times$ Sp(n) as its body. OSp$(r, s; n)$ and OSp(m, n) have isomorphic complex extensions.

The Kac notation

V. G. Kac (*loc. cit.*) has introduced alternative symbols for the graded Lie algebras associated with the simple classical super Lie algebras. Since the classical super Lie algebras are all conventional we shall use the same symbols for the classical super Lie groups themselves. The symbols are the following:

A(m, n);	$m, n = 0, 1, 2 \ldots; m \neq n$:	denotes SL$(m+1, n+1)$.
A(m, m);	$m = 1, 2 \ldots$:	denotes SL$(m+1, m+1)$/GL$(1, 0)$.
B(m, n);	$m = 0, 1, 2 \ldots; n = 1, 2 \ldots$:	denotes OSp$(2m+1, 2n)$.
D(m, n);	$m = 2, 3 \ldots; n = 1, 2 \ldots$:	denotes OSp$(2m, 2n)$.

The notation is evidently based on analogy with the Cartan symbols for the simple ordinary Lie algebras, the integers m, n denoting the respective ranks of the components of the semisimple part of the body. Comparing the above symbols with those of Cartan one sees that, in characterizing the orthosymplectic group, Kac chooses to emphasize the 'ortho' rather than the 'symplectic' component of the body, for the symbol C does not appear above. Only for OSp(m, n) with $m = 2$ does Kac use C:

C(n); $n = 2, 3 \ldots$: denotes OSp$(2, 2n-2)$.

Note in this case that the body is not semisimple and that the integer n does *not* give the rank of its semisimple part.

Another special case is D$(2, 1)$. We shall see in the next section that this super Lie group is the limiting member of an infinite one-parameter family of exceptional super Lie groups. The existence of this family depends, firstly, on the fact that the body of D$(2, 1)$ is SO$(4) \times$ Sp(2), which locally has the same complex extension as SL$(2) \times$ SL$(2) \times$ SL(2), and secondly, on the fact that SL(2) has a special role to play in completing the classification of all the simple super Lie algebras.

Before turning to the exceptional super Lie groups we need to take note of two other families of conventional super Lie groups besides the above groups, which are simple. For these the only names we have are the Kac symbols.

The group P(m)

This is the subset of SL($m+1, m+1$) consisting of $(m+1, m+1) \times (m+1, m+1)$ matrices having the form

$$\exp\begin{pmatrix} A & C \\ D & -A^{\sim} \end{pmatrix} \quad \text{with } \operatorname{tr} A = 0, \quad C^{\sim} = C \quad \text{and} \quad D^{\sim} = -D. \qquad (4.2.62)$$

It is easy to check that the set of matrices $\begin{pmatrix} A & C \\ D & -A^{\sim} \end{pmatrix}$ is closed under the operation of taking the commutator, and hence their exponentials form a group. The super Lie algebra itself is the set of all such matrices, of both c- and a-types, and linear combinations thereof, the bracket operation being the supercommutator.

The dimension of P(m) is $(m(m+2), (m+1)^2)$. Note that this is the first simple conventional super Lie group that we have encountered that can have an odd number of a-type dimensions (when m is even). When m is even P(m) cannot possess an invariant Riemannian metric tensor field. When m is odd it cannot have one, it turns out, that is both right and left invariant.

The structure constants of P(m) may be determined from the bracket operation for c-type matrices:

$$\left.\begin{aligned} A_{12} &= [A_1, A_2] + C_1 D_2 - C_2 D_1, \\ C_{12} &= A_1 C_2 + C_2 A_1^{\sim} - A_2 C_1 - C_1 A_2^{\sim}, \\ D_{12} &= D_1 A_2 + A_2^{\sim} D_1 - D_2 A_1 - A_1^{\sim} D_2. \end{aligned}\right\} \qquad (4.2.63)$$

Designating the components of A, C, D by $A^{\mu}{}_{\nu}$, $C^{\mu\nu}$, $D_{\mu\nu}$ respectively, and using the emplacement of indices to identify the submatrix to which a component belongs,† one finds for the structure constants

$$\begin{aligned} c^{\mu}{}_{\nu}{}^{\sigma}{}_{\tau}{}^{\rho}{}_{\lambda} &= \delta^{\mu}{}_{\tau}\delta^{\sigma}{}_{\lambda}\delta^{\rho}{}_{\nu} - \delta^{\mu}{}_{\lambda}\delta^{\sigma}{}_{\nu}\delta^{\rho}{}_{\tau}, \\ c^{\mu}{}_{\nu\sigma\tau}{}^{\rho\lambda} &= c^{\mu}{}_{\nu}{}^{\rho\lambda}{}_{\sigma\tau} \\ &= \delta^{\mu}{}_{\tau}\delta_{\sigma}{}^{\lambda}\delta^{\rho}{}_{\nu} - \delta^{\mu}{}_{\tau}\delta_{\sigma}{}^{\rho}\delta^{\lambda}{}_{\nu} + \delta^{\mu}{}_{\sigma}\delta_{\tau}{}^{\lambda}\delta^{\rho}{}_{\nu} - \delta^{\mu}{}_{\sigma}\delta_{\tau}{}^{\rho}\delta^{\lambda}{}_{\nu} \end{aligned}$$

† The c-type or a-type character of a component then depends on the emplacement of the indices rather than on what part of the alphabet they come from.

$$\begin{aligned}
c^{\mu\nu\sigma}{}_{\tau\rho\lambda} &= -c^{\mu\nu}{}_{\rho\lambda}{}^{\sigma}{}_{\tau} \\
&= \delta^{\mu}{}_{\tau}\delta^{\sigma}{}_{\lambda}\delta_{\rho}{}^{\nu} + \delta^{\mu}{}_{\tau}\delta^{\sigma}{}_{\rho}\delta_{\lambda}{}^{\nu} + \delta^{\nu}{}_{\tau}\delta^{\sigma}{}_{\lambda}\delta_{\rho}{}^{\mu} + \delta^{\nu}{}_{\tau}\delta^{\sigma}{}_{\rho}\delta_{\lambda}{}^{\mu} \\
&\quad - \frac{2}{m+1}\delta^{\sigma}{}_{\tau}(\delta^{\mu}{}_{\rho}\delta^{\nu}{}_{\lambda} + \delta^{\mu}{}_{\lambda}\delta^{\nu}{}_{\rho}), \\
c_{\mu\nu}{}^{\sigma\tau\rho}{}_{\lambda} &= -c_{\mu\nu}{}^{\rho}{}_{\lambda}{}^{\sigma\tau} \\
&= \delta_{\mu}{}^{\tau}\delta^{\sigma}{}_{\lambda}\delta^{\rho}{}_{\nu} - \delta_{\mu}{}^{\sigma}\delta^{\tau}{}_{\lambda}\delta^{\rho}{}_{\nu} - \delta_{\nu}{}^{\tau}\delta^{\sigma}{}_{\lambda}\delta^{\rho}{}_{\mu} + \delta_{\nu}{}^{\sigma}\delta^{\tau}{}_{\lambda}\delta^{\rho}{}_{\mu} \\
&\quad + \frac{2}{m+1}\delta^{\rho}{}_{\lambda}(\delta_{\mu}{}^{\sigma}\delta_{\nu}{}^{\tau} - \delta_{\mu}{}^{\tau}\delta_{\nu}{}^{\sigma}),
\end{aligned} \tag{4.2.64}$$

all others vanishing.

The body of P(m) is SL(m + 1), with structure constants given by the top line of (4.2.64). The extending representation is generated by the matrices

$$\begin{pmatrix} 0 & c^{\mu\nu\sigma}{}_{\tau\rho\lambda} \\ c_{\mu\nu}{}^{\sigma}{}_{\tau}{}^{\sigma\lambda} & 0 \end{pmatrix}. \tag{4.2.65}$$

Multiplication of these matrices by $x^{\tau}{}_{\sigma}$ (with $x^{\mu}{}_{\mu} = 0$) yields.

$$\begin{pmatrix} 0 & x^{\mu}{}_{\lambda}\delta_{\rho}{}^{\nu} + x^{\mu}{}_{\rho}\delta_{\lambda}{}^{\nu} + x^{\nu}{}_{\lambda}\delta_{\rho}{}^{\mu} + x^{\nu}{}_{\rho}\delta_{\lambda}{}^{\mu} \\ -\delta_{\mu}{}^{\lambda}x^{\rho}{}_{\nu} + \delta_{\mu}{}^{\rho}x^{\lambda}{}_{\nu} + \delta_{\nu}{}^{\lambda}x^{\rho}{}_{\mu} - \delta_{\nu}{}^{\rho}x^{\lambda}{}_{\mu} & 0 \end{pmatrix}, \tag{4.2.66}$$

which, when applied to

$$\begin{pmatrix} \frac{1}{2}y^{\lambda\rho} \\ \frac{1}{2}z_{\lambda\rho} \end{pmatrix}, \quad y^{\lambda\rho} = y^{\rho\lambda}, \quad z_{\lambda\rho} = -z_{\rho\lambda}, \tag{4.2.67}$$

gives $x^{\mu}{}_{\lambda}y^{\lambda\nu} + x^{\nu}{}_{\lambda}y^{\lambda\mu} - z_{\mu\rho}x^{\rho}{}_{\nu} + z_{\nu\rho}x^{\rho}{}_{\mu}$, showing that the extending representation is $(\mathrm{sl}(m+1) \otimes \mathrm{sl}(m+1))_{\mathrm{S}} \oplus (\mathrm{cosl}(m+1) \otimes \mathrm{cosl}(m+1))_{\mathrm{A}}$, where the subscripts S and A denote the symmetric and antisymmetric parts respectively.

The symmetric matrix $\boldsymbol{\eta}_{\mathrm{S}}$ of section 4.2 is just the projection matrix $\pi^{\mu}{}_{\nu}{}^{\sigma}{}_{\tau}$ into the subspace of traceless $(m+1) \times (m+1)$ matrices. The antisymmetric matrix $\boldsymbol{\eta}_{\mathrm{A}}$, on the other hand, does not exist. The group P(m) is exceptional in that it cannot be constructed by the methods outlined in section 4.1. In particular, the identities (4.1.10) and (4.1.11) are now unrelated to one another. A consequence of the nonexistence of $\boldsymbol{\eta}_{\mathrm{A}}$ is that the Cartan–Killing metric must vanish. This may be verified by straightforward computation.

$$\left.\begin{aligned}
&-c^{\rho}{}_{\lambda}{}^{\mu}{}_{\nu}{}^{\iota}{}_{\kappa}c^{\kappa}{}_{\iota}{}^{\sigma}{}_{\tau}{}^{\lambda}{}_{\rho} + \tfrac{1}{4}c_{\rho\lambda}{}^{\mu}{}_{\nu}{}^{\iota\kappa}c_{\kappa\iota}{}^{\sigma}{}_{\tau}{}^{\lambda\rho} + \tfrac{1}{4}c^{\rho\lambda\mu}{}_{\nu\iota\kappa}c^{\kappa\iota\sigma}{}_{\tau\lambda\rho} = 0, \\
&-\tfrac{1}{2}c^{\rho}{}_{\lambda\mu\nu}{}^{\iota\kappa}c_{\kappa\iota}{}^{\sigma\tau\lambda}{}_{\rho} + \tfrac{1}{2}c^{\rho\lambda}{}_{\mu\nu}{}^{\iota}{}_{\kappa}c^{\kappa}{}_{\iota}{}^{\sigma\tau}{}_{\lambda\rho} = 0, \\
&-\tfrac{1}{2}c^{\rho}{}_{\lambda}{}^{\mu\nu}{}_{\iota\kappa}c^{\kappa\iota}{}_{\sigma\tau}{}^{\lambda}{}_{\rho} + \tfrac{1}{2}c_{\rho\lambda}{}^{\mu\nu\iota}{}_{\kappa}c^{\kappa}{}_{\iota\sigma\tau}{}^{\lambda\rho} = 0.
\end{aligned}\right\} \tag{4.2.68}$$

All the P(m)'s are nevertheless simple, except P(1). (See exercise **4.16**.)

The group $\tilde{\mathrm{Q}}(m)$

This is the subset of $\mathrm{SL}(m+1,\ m+1)$ consisting of $(m+1,\ m+1) \times (m+1,\ m+1)$ matrices having the form

$$\exp\begin{pmatrix} A & \mathrm{i}D \\ D & A \end{pmatrix} \quad \text{with} \quad \operatorname{tr} D = 0. \tag{4.2.69}$$

This subset forms a group because, as is easily checked, the set of matrices $\begin{pmatrix} A & \mathrm{i}D \\ D & A \end{pmatrix}$ is closed under the operation of taking the commutator. The dimension of $\tilde{\mathrm{Q}}(m)$ is evidently $((m+1)^2, m(m+2))$.

The structure constants are determined from the bracket operation

$$\left.\begin{aligned} A_{12} &= [A_1, A_2] + \mathrm{i}[D_1, D_2], \\ D_{12} &= [A_1, D_2] + [D_1, A_2]. \end{aligned}\right\} \tag{4.2.70}$$

Designating the components of A, D by $A^{\mu}{}_{\nu}$, $D^{\mu}{}_{\nu}$ respectively, with underlined index pairs identifying a-type components, one finds

$$\left.\begin{aligned} c^{\mu}{}_{\nu}{}^{\sigma}{}_{\tau}{}^{\rho}{}_{\lambda} &= \delta^{\mu}{}_{\tau}\delta^{\sigma}{}_{\lambda}\delta^{\rho}{}_{\nu} - \delta^{\mu}{}_{\lambda}\delta^{\sigma}{}_{\nu}\delta^{\rho}{}_{\tau}, \\ c^{\mu}{}_{\nu}{}^{\sigma}{}_{\underline{\tau}}{}^{\rho}{}_{\underline{\lambda}} &= -\mathrm{i}(\delta^{\mu}{}_{\tau}\delta^{\sigma}{}_{\lambda}\delta^{\rho}{}_{\nu} + \delta^{\mu}{}_{\lambda}\delta^{\sigma}{}_{\nu}\delta^{\rho}{}_{\tau}) \\ &\quad + \frac{2\mathrm{i}}{m+1}(\delta^{\mu}{}_{\lambda}\delta^{\sigma}{}_{\tau}\delta^{\rho}{}_{\nu} + \delta^{\mu}{}_{\tau}\delta^{\sigma}{}_{\nu}\delta^{\rho}{}_{\lambda}) - \frac{2\mathrm{i}}{(m+1)^2}\delta^{\mu}{}_{\nu}\delta^{\sigma}{}_{\tau}\delta^{\rho}{}_{\lambda}, \\ c^{\mu}{}_{\underline{\nu}}{}^{\sigma}{}_{\tau}{}^{\rho}{}_{\underline{\lambda}} &= -c^{\mu}{}_{\underline{\nu}}{}^{\rho}{}_{\underline{\lambda}}{}^{\sigma}{}_{\tau} = \delta^{\mu}{}_{\tau}\delta^{\sigma}{}_{\lambda}\delta^{\rho}{}_{\nu} - \delta^{\mu}{}_{\lambda}\delta^{\sigma}{}_{\nu}\delta^{\rho}{}_{\tau}, \end{aligned}\right\} \tag{4.2.71}$$

all other structure constants vanishing. The body of $\tilde{\mathrm{Q}}(m)$ is $\mathrm{GL}(m+1)$ and the extending representation is $gl(m+1) \otimes \mathrm{cogl}(m+1)$ restricted to the vector space of traceless $(m+1) \times (m+1)$ matrices.

The group $\mathrm{Q}(m)$

The group $\tilde{\mathrm{Q}}(m)$ has an invariant Abelian subgroup consisting of all the matrices $\lambda 1_{(m+1,n+1)}$ where λ is an invertible real c-number. To get a simple group we must divide by this subgroup:

$$\mathrm{Q}(m) \stackrel{\text{def}}{=} \tilde{\mathrm{Q}}(m)/\mathrm{GL}(1,0) \tag{4.2.72}$$

The structure constants of $\mathrm{Q}(m)$ are

$$\left.\begin{aligned} c^{\mu}{}_{\nu}{}^{\sigma}{}_{\tau}{}^{\rho}{}_{\lambda} &= \delta^{\mu}{}_{\tau}\delta^{\sigma}{}_{\lambda}\delta^{\rho}{}_{\nu} - \delta^{\mu}{}_{\lambda}\delta^{\sigma}{}_{\nu}\delta^{\rho}{}_{\tau}, \\ c^{\mu}{}_{\nu}{}^{\sigma}{}_{\underline{\tau}}{}^{\rho}{}_{\underline{\lambda}} &= -\mathrm{i}(\delta^{\mu}{}_{\tau}\delta^{\sigma}{}_{\lambda}\delta^{\rho}{}_{\nu} + \delta^{\mu}{}_{\lambda}\delta^{\sigma}{}_{\nu}\delta^{\rho}{}_{\tau}) \\ &\quad + \frac{2\mathrm{i}}{m+1}(\delta^{\mu}{}_{\nu}\delta^{\sigma}{}_{\lambda}\delta^{\rho}{}_{\tau} + \delta^{\mu}{}_{\lambda}\delta^{\sigma}{}_{\tau}\delta^{\rho}{}_{\nu} + \delta^{\mu}{}_{\tau}\delta^{\sigma}{}_{\nu}\delta^{\rho}{}_{\lambda}) \\ &\quad - \frac{4\mathrm{i}}{(m+1)^2}\delta^{\mu}{}_{\nu}\delta^{\sigma}{}_{\tau}\delta^{\rho}{}_{\lambda}, \\ c^{\mu}{}_{\underline{\nu}}{}^{\sigma}{}_{\tau}{}^{\rho}{}_{\underline{\lambda}} &= -c^{\mu}{}_{\underline{\nu}}{}^{\rho}{}_{\underline{\lambda}}{}^{\sigma}{}_{\tau} = \delta^{\mu}{}_{\tau}\delta^{\sigma}{}_{\lambda}\delta^{\rho}{}_{\nu} - \delta^{\mu}{}_{\lambda}\delta^{\sigma}{}_{\nu}\delta^{\rho}{}_{\tau}, \end{aligned}\right\} \tag{4.2.73}$$

all others vanishing. Every index pair is to be understood as carrying a projection onto the space of traceless $(m+1) \times (m+1)$ matrices. The dimensionality of Q(m) is $(m(m+2), m(m+2))$, its body is SL($m+1$), and the extending representation is *ad* SL($m+1$). When m is odd Q(m) cannot carry a Riemannian metric tensor field. When m is even it cannot carry one that is both left and right invariant.

The matrix η_S of section 4.1 is the projection matrix $\pi^{\mu}{}_{\nu}{}^{\sigma}{}_{\tau}$, and the matrix η_A does not exist. The Cartan–Killing matrix vanishes:

$$\left.\begin{aligned} -c^{\rho}{}_{\lambda}{}^{\mu}{}_{\nu}{}^{\iota}{}_{\kappa} c^{\kappa}{}_{\iota}{}^{\sigma}{}_{\tau}{}^{\lambda}{}_{\rho} + c^{\rho}{}_{\underline{\lambda}}{}^{\mu}{}_{\nu}{}^{\iota}{}_{\underline{\kappa}} c^{\kappa}{}_{\underline{\iota}}{}^{\sigma}{}_{\tau}{}^{\lambda}{}_{\underline{\rho}} &= 0, \\ -c^{\rho}{}_{\lambda}{}^{\mu}{}_{\underline{\nu}}{}^{\iota}{}_{\underline{\kappa}} c^{\kappa}{}_{\underline{\iota}}{}^{\sigma}{}_{\underline{\tau}}{}^{\lambda}{}_{\rho} + c^{\rho}{}_{\underline{\lambda}}{}^{\mu}{}_{\underline{\nu}}{}^{\iota}{}_{\kappa} c^{k}{}_{i}{}^{\sigma}{}_{\underline{\tau}}{}^{\lambda}{}_{\underline{\rho}} &= 0. \end{aligned}\right\} \tag{4.2.74}$$

All the Q(m)'s are nevertheless simple, except Q(1). (See exercise **4.17**).

Note that the set of c- and a-type (as well as impure) matrices of the form $\begin{pmatrix} A & \mathrm{i}D \\ D & A \end{pmatrix}$, with $\operatorname{tr} A = 0$ and $\operatorname{tr} D = 0$, does not form a matrix representation of Q(m). Rather it forms a pseudorepresentation, with the bracket operation defined similarly to that of eq. (4.2.36)

4.3 The exceptional simple super Lie groups

The groups D(2, 1, α)

These groups, which are of dimension (9, 8), can be constructed from the information that they all have identical bodies and identical extending representations, namely, SL(2) × SL(2) × SL(2) and sl(2) ⊗ sl(2) ⊗ sl(2) respectively. Using Latin indices, Greek indices from the middle of the alphabet, and Greek indices from the first of the alphabet, respectively, to characterize the three factors SL(2) of which the body is composed, one may write the structure constants of the body as

$$\left.\begin{aligned} c^{a}{}_{b}{}^{c}{}_{d}{}^{e}{}_{f} &= \delta^{a}{}_{d}\delta^{c}{}_{f}\delta^{e}{}_{b} - \delta^{a}{}_{f}\delta^{c}{}_{b}\delta^{e}{}_{d}, \\ c^{\mu}{}_{\nu}{}^{\sigma}{}_{\tau}{}^{\rho}{}_{\lambda} &= \delta^{\mu}{}_{\tau}\delta^{\sigma}{}_{\lambda}\delta^{\rho}{}_{\nu} - \delta^{\mu}{}_{\lambda}\delta^{\sigma}{}_{\nu}\delta^{\rho}{}_{\tau}, \\ c^{\alpha}{}_{\beta}{}^{\gamma}{}_{\delta}{}^{\varepsilon}{}_{\zeta} &= \delta^{\alpha}{}_{\delta}\delta^{\gamma}{}_{\zeta}\delta^{\varepsilon}{}_{\beta} - \delta^{\alpha}{}_{\zeta}\delta^{\gamma}{}_{\beta}\delta^{\varepsilon}{}_{\delta}, \end{aligned}\right\} \tag{4.3.1}$$

all others vanishing. Every index has two possible values, and every index pair carries a projection into the space of traceless 2 × 2 matrices.

The fundamental representation sl(2) of SL(2) is generated by the traceless 2 × 2 matrices $\mathbf{G}^{a}{}_{b}$ defined by

$$(\mathbf{G}^{a}{}_{b})^{c}{}_{d} \overset{\text{def}}{=} \delta^{a}{}_{d}\delta^{c}{}_{b} - \tfrac{1}{2}\delta^{a}{}_{b}\delta^{c}{}_{d}. \tag{4.3.2}$$

The extending representation of D(2, 1, α) is therefore generated by matrices having the components

$$\left.\begin{aligned} c^{a\mu\alpha\,c}{}_{d\,b\nu\beta} &= (\delta^a{}_d\,\delta^c{}_b - \tfrac{1}{2}\delta^a{}_b\,\delta^c{}_d)\delta^\mu{}_\nu\delta^\alpha{}_\beta, \\ c^{a\mu\alpha\,\sigma}{}_{\tau\,b\nu\beta} &= \delta^a{}_b(\delta^\mu{}_\tau\,\delta^\sigma{}_\nu - \tfrac{1}{2}\delta^\mu{}_\nu\,\delta^\sigma{}_\tau)\delta^\alpha{}_\beta, \\ c^{a\mu\alpha\,\gamma}{}_{\delta\,b\nu\beta} &= \delta^a{}_b\,\delta^\mu{}_\nu(\delta^\alpha{}_\delta\,\delta^\gamma{}_\beta - \tfrac{1}{2}\delta^\alpha{}_\beta\,\delta^\gamma{}_\delta). \end{aligned}\right\} \tag{4.3.3}$$

Up to an overall factor, the matrix η_S of section 4.1 has the nonvanishing components

$$\left.\begin{aligned} \eta^a{}_b{}^c{}_d &= \delta^a{}_d\,\delta^c{}_b - \tfrac{1}{2}\delta^a{}_b\,\delta^c{}_d, \\ \eta^\mu{}_\nu{}^\sigma{}_\tau &= \alpha^{-1}(\delta^\mu{}_\tau\,\delta^\sigma{}_\nu - \tfrac{1}{2}\delta^\mu{}_\nu\,\delta^\sigma{}_\tau), \\ \eta^\alpha{}_\beta{}^\gamma{}_\delta &= \beta^{-1}(\delta^\alpha{}_\delta\,\delta^\gamma{}_\beta - \tfrac{1}{2}\delta^\alpha{}_\beta\,\delta^\gamma{}_\delta), \end{aligned}\right\} \tag{4.3.4}$$

where, by including the adjustable coefficients α^{-1}, β^{-1}, we have allowed for independent scaling of the Cartan–Killing matrices of the factors SL(2). The inverse matrix η_S^{-1} has the nonvanishing components

$$\left.\begin{aligned} \eta^{-1\,a}{}_b{}^c{}_d &= \delta^a{}_d\,\delta^c{}_b - \tfrac{1}{2}\delta^a{}_b\,\delta^c{}_d, \\ \eta^{-1\,\mu}{}_\nu{}^\sigma{}_\tau &= \alpha(\delta^\mu{}_\tau\,\delta^\sigma{}_\nu - \tfrac{1}{2}\delta^\mu{}_\nu\,\delta^\sigma{}_\tau), \\ \eta^{-1\,\alpha}{}_\beta{}^\gamma{}_\delta &= \beta(\delta^\alpha{}_\delta\,\delta^\gamma{}_\beta - \tfrac{1}{2}\delta^\alpha{}_\beta\,\delta^\gamma{}_\delta). \end{aligned}\right\} \tag{4.3.5}$$

The form of the matrix η_A of section 4.1 may be inferred from the remark that every traceless 2×2 matrix is converted into a symmetric matrix through multiplication by ε, where

$$\varepsilon = (\varepsilon_{ab}) \overset{\text{def}}{=} \begin{pmatrix} 0 & 1 \\ -1 & 0 \end{pmatrix}. \tag{4.3.6}$$

The presence of the three factors SL(2) in the body leads unambiguously to

$$\eta_{a\mu\alpha\,b\nu\beta} = \mathrm{i}\varepsilon_{ab}\,\varepsilon_{\mu\nu}\,\varepsilon_{\alpha\beta}. \tag{4.3.7}$$

From eqs. (4.3.1) and (4.3.3) we can now construct the remaining nonvanishing structure constants of D(2, 1, α) (cf. eqs. (4.1.19) and (4.1.20)):

$$\left.\begin{aligned} c^{a\mu\alpha}{}_{b\nu\beta}{}^c{}_d &= -c^{a\mu\alpha\,c}{}_{d\,b\nu\beta}, \\ c^{a\mu\alpha}{}_{b\nu\beta}{}^\sigma{}_\tau &= -c^{a\mu\alpha\,\sigma}{}_{\tau\,b\nu\beta}, \\ c^{a\mu\alpha}{}_{b\nu\beta}{}^\gamma{}_\delta &= -c^{a\mu\alpha\,\gamma}{}_{\delta\,b\nu\beta}, \end{aligned}\right\} \tag{4.3.8}$$

$$\begin{aligned} c^c{}_{d\,a\mu\alpha\,b\nu\beta} &= -\eta^{-1\,c}{}_d{}^e{}_f\,\eta_{a\mu\alpha\,g\sigma\gamma}c^{g\sigma\gamma\,f}{}_{e\,b\nu\beta} \\ &= -\mathrm{i}(\varepsilon_{ad}\,\delta^c{}_b - \tfrac{1}{2}\varepsilon_{ab}\,\delta^c{}_d)\varepsilon_{\mu\nu}\,\varepsilon_{\alpha\beta} \\ &= -\frac{\mathrm{i}}{2}(\varepsilon_{ad}\,\delta^c{}_b + \varepsilon_{bd}\,\delta^c{}_a)\varepsilon_{\mu\nu}\,\varepsilon_{\alpha\beta}, \end{aligned} \tag{4.3.9}$$

and, similarly,

$$c^\sigma{}_{\tau\,a\mu\alpha\,b\nu\beta} = -\frac{\mathrm{i}}{2}\alpha\varepsilon_{ab}(\varepsilon_{\mu\tau}\,\delta^\sigma{}_\nu + \varepsilon_{\nu\tau}\,\delta^\rho{}_\mu)\varepsilon_{\alpha\beta}, \tag{4.3.10}$$

$$c^\gamma{}_{\delta\,a\mu\alpha\,b\nu\beta} = -\frac{\mathrm{i}}{2}\beta\varepsilon_{ab}\,\varepsilon_{\mu\nu}(\varepsilon_{\alpha\delta}\,\delta^\gamma{}_\beta + \varepsilon_{\beta\delta}\,\delta^\gamma{}_\alpha). \tag{4.3.11}$$

There remains only the task of verifying the identity (4.1.12) (or (4.1.17)), which, in the present case, may be written in the form

$$\begin{aligned}0 &= \eta^{e}{}_{f}{}^{g}{}_{h}\, C^{f}{}_{e\,a\mu\alpha\,b\nu\beta}\, C^{h}{}_{g\,c\sigma\gamma\,d\tau\delta} + \eta^{\rho}{}_{\lambda}{}^{i}{}_{\kappa}\, C^{\lambda}{}_{\rho\,a\mu\alpha\,b\nu\beta}\, C^{\kappa}{}_{i\,c\sigma\gamma\,d\tau\delta}\\ &\quad + \eta^{\varepsilon}{}_{\zeta}{}^{\eta}{}_{\theta}\, C^{\zeta}{}_{\varepsilon\,a\mu\alpha\,b\nu\beta}\, C^{\theta}{}_{\eta\,c\sigma\gamma\,d\tau\delta} + \mathrm{cyc}(b\nu\beta, c\sigma\gamma, d\tau\delta)\\ &= \tfrac{1}{2}(\varepsilon_{ac}\,\varepsilon_{bd} + \varepsilon_{ad}\,\varepsilon_{bc})\varepsilon_{\mu\nu}\,\varepsilon_{\sigma\tau}\,\varepsilon_{\alpha\beta}\,\varepsilon_{\gamma\delta}\\ &\quad + \tfrac{1}{2}\alpha\varepsilon_{ab}\,\varepsilon_{cd}(\varepsilon_{\mu\sigma}\,\varepsilon_{\nu\tau} + \varepsilon_{\mu\tau}\,\varepsilon_{\nu\sigma})\varepsilon_{\alpha\beta}\,\varepsilon_{\gamma\delta}\\ &\quad + \tfrac{1}{2}\beta\varepsilon_{ab}\,\varepsilon_{cd}\,\varepsilon_{\mu\nu}\,\varepsilon_{\sigma\tau}(\varepsilon_{\alpha\gamma}\,\varepsilon_{\beta\delta} + \varepsilon_{\alpha\delta}\,\varepsilon_{\beta\gamma}) + \mathrm{cyc}(b\nu\beta, c\sigma\gamma, d\tau\delta). \qquad (4.3.12)\end{aligned}$$

The last expression may be put into an alternative form through use of the identity

$$\varepsilon_{ab}\,\varepsilon_{cd} = \delta_{ac}\,\delta_{bd} - \delta_{ad}\,\delta_{bc}. \qquad (4.3.13)$$

A straightforward but tedious calculation yields

$$\begin{aligned}0 = (1+\alpha+\beta)[\delta_{ac}\,\delta_{bd}(&2\delta_{\mu\sigma}\,\delta_{\nu\tau}\,\delta_{\alpha\gamma}\,\delta_{\beta\delta} + 2\delta_{\mu\tau}\,\delta_{\nu\sigma}\,\delta_{\alpha\beta}\,\delta_{\gamma\delta} + 2\delta_{\mu\nu}\,\delta_{\sigma\tau}\,\delta_{\alpha\delta}\,\delta_{\beta\gamma}\\ &- \delta_{\mu\sigma}\,\delta_{\nu\tau}\,\delta_{\alpha\delta}\,\delta_{\beta\gamma} - \delta_{\mu\sigma}\,\delta_{\nu\tau}\,\delta_{\alpha\beta}\,\delta_{\gamma\delta} - \delta_{\mu\tau}\,\delta_{\nu\sigma}\,\delta_{\alpha\gamma}\,\delta_{\beta\delta}\\ &- \delta_{\mu\tau}\,\delta_{\nu\sigma}\,\delta_{\alpha\delta}\,\delta_{\beta\gamma} - \delta_{\mu\nu}\,\delta_{\sigma\tau}\,\delta_{\alpha\gamma}\,\delta_{\beta\delta})\\ &- \delta_{ad}\,\delta_{bc}\,\delta_{\mu\sigma}\,\delta_{\nu\tau}\,\delta_{\alpha\gamma}\,\delta_{\beta\delta}] + \mathrm{cyc}(b\nu\beta, c\sigma\gamma, d\tau\delta). \qquad (4.3.14)\end{aligned}$$

Evidently we must have

$$\beta = -1-\alpha. \qquad (4.3.15)$$

The real number α itself is arbitrary, except that one must require

$$\alpha \neq 0, -1, \qquad (4.3.16)$$

in order that the matrix (4.3.4) be nonsingular.

Not all values of α lead to distinct groups. $D(2,1,\alpha)$ remains invariant (i.e., up to an isomorphism) under the actions of the following six transformations of the real line (including the point at infinity) into itself:

$$\left.\begin{aligned} e(\alpha) &= \alpha, & a(\alpha) &= -(1+\alpha), & b(\alpha) &= \frac{1}{\alpha},\\ c(\alpha) &= -\frac{1}{1+\alpha}, & d(\alpha) &= -\frac{1+\alpha}{\alpha}, & f(\alpha) &= -\frac{\alpha}{1+\alpha},\end{aligned}\right\} \qquad (4.3.17)$$

The functions e, a, b, c, d, f are the elements of a non-Abelian discrete group. It evidently suffices to restrict α to the range $0 < \alpha \leq 1$. The group $D(2,1)$ (or $OSp(4,2)$) of section 4.2 can be shown to correspond to the limiting case $\alpha = 1$ (or -2, or $-\tfrac{1}{2}$).[†]

[†] One may introduce an alternative parameter $\gamma = -\ln\left|\dfrac{(\alpha-1)(\alpha+\frac{1}{2})(\alpha+2)}{\alpha(\alpha+1)}\right|$ which is invariant under the actions of the discrete group. As α ranges from 0 to 1, γ ranges from $-\infty$ to ∞.

It is straightforward to verify that the Cartan–Killing matrix of D(2, 1, α) vanishes for all values of α:

$$-C^{e}{}_{f}{}^{a}{}_{b}{}^{g}{}_{h}C^{h}{}_{g}{}^{c}{}_{d}{}^{f}{}_{e}+C^{e\mu\alpha\,a}{}_{b\,f\nu\beta}C^{f\nu\beta\,c}{}_{d\,e\mu\alpha}=0, \tag{4.3.18}$$

$$\begin{aligned}
&-C^{d}{}_{e\,a\mu\alpha\,c\sigma\gamma}C^{c\sigma\gamma}{}_{b\nu\beta}{}^{e}{}_{d}-C^{\tau}{}_{\rho\,a\mu\alpha\,c\sigma\gamma}C^{c\sigma\gamma}{}_{b\nu\beta}{}^{\rho}{}_{\tau}\\
&-C^{\delta}{}_{\varepsilon\,a\mu\alpha\,c\sigma\gamma}C^{c\sigma\gamma}{}_{b\nu\beta\,\varepsilon\delta}+C^{c\sigma\gamma}{}_{a\mu\alpha}{}^{e}{}_{d}C^{d}{}_{e\,b\nu\beta\,c\sigma\gamma}\\
&+C^{c\sigma\gamma}{}_{a\mu\alpha}{}^{\rho}{}_{\tau}C^{\tau}{}_{\rho\,b\nu\beta\,c\sigma\gamma}+C^{c\sigma\gamma}{}_{a\mu\alpha}{}^{\varepsilon}{}_{\delta}C^{\delta}{}_{\varepsilon\,\beta\nu\beta\,c\sigma\gamma}\\
&\qquad\qquad=3\mathrm{i}(1+\alpha+\beta)\varepsilon_{ab}\,\varepsilon_{\mu\nu}\,\varepsilon_{\alpha\beta}=0.
\end{aligned} \tag{4.3.19}$$

The group F(4)

The body of this group is $\overline{\mathrm{SO}}(7)\times\mathrm{SL}(2)$, having dimension 24 and rank 4 and the extending representation is spin(7)$\otimes$sl(2). Spin(7) is most easily described in terms of γ-matrices satisfying

$$\{\gamma_\mu,\gamma_\nu\}=2\delta_{\mu\nu}\mathbf{1},\qquad \mu,\nu=1,2,\ldots,7. \tag{4.3.20}$$

There are only two inequivalent irreducible faithful representations for the γ's, each of dimension 8 and each the negative of the other. It does not matter which representation is chosen, for the generators of spin(7) involve the γ's only in the bilinear combinations

$$\mathbf{G}_{\mu\nu}\stackrel{\mathrm{def}}{=}-\tfrac{1}{4}[\gamma_\mu,\gamma_\nu]. \tag{4.3.21}$$

From the dimensionality of the γ-matrices and of sl(2) we infer that F(4) has dimension (24, 16).

It turns out that the γ-matrices for spin(7) can be chosen unitary, antisymmetric and pure imaginary. For example

$$\left.\begin{aligned}
\gamma_1&=\mathbf{1}\times\sigma_3\times\sigma_2,\\
\gamma_2&=\mathbf{1}\times\sigma_1\times\sigma_2,\\
\gamma_3&=\sigma_2\times\mathbf{1}\times\sigma_3,\\
\gamma_4&=\sigma_2\times\mathbf{1}\times\sigma_1,\\
\gamma_5&=\sigma_3\times\sigma_2\times\mathbf{1},\\
\gamma_6&=\sigma_1\times\sigma_2\times\mathbf{1},\\
\gamma_7&=\sigma_2\times\sigma_2\times\sigma_2,
\end{aligned}\right\} \tag{4.3.22}$$

where the σ's are the Pauli matrices. With such a choice, which will be assumed from now on, the 21 matrices $\mathbf{G}_{\mu\nu}$ are antisymmetric and real, while the 35 matrices

$$\mathbf{G}_{\mu\nu\sigma}\stackrel{\mathrm{def}}{=}\tfrac{1}{3}(\gamma_\mu\mathbf{G}_{\nu\sigma}+\gamma_\nu\mathbf{G}_{\sigma\mu}+\gamma_\sigma\mathbf{G}_{\mu\nu}) \tag{4.3.23}$$

are symmetric and imaginary. Any antisymmetric 8×8 matrix $\mathbf{A}$ can be decomposed in the form

$$\mathbf{A} = a_\mu \gamma_\mu + \tfrac{1}{2} a_{\mu\nu} \mathbf{G}_{\nu\mu}, \tag{4.3.24}$$

and any symmetric 8×8 matrix can be expressed as

$$\mathbf{S} = s\mathbf{1} + \tfrac{1}{6} s_{\mu\nu\sigma} \mathbf{G}_{\sigma\nu\mu}, \tag{4.3.25}$$

where

$$a_\mu = \tfrac{1}{8} \operatorname{tr}(\gamma_\mu \mathbf{A}), \quad a_{\mu\nu} = \tfrac{1}{2} \operatorname{tr}(\mathbf{G}_{\mu\nu} \mathbf{A}), \tag{4.3.26}$$
$$s = \tfrac{1}{8} \operatorname{tr} \mathbf{S}, \qquad s_{\mu\nu\sigma} = \tfrac{1}{2} \operatorname{tr}(\mathbf{G}_{\mu\nu\sigma} \mathbf{S}). \tag{4.3.27}$$

We shall need the trace relations

$$\operatorname{tr} \gamma_\mu = 0, \tag{4.3.28}$$
$$\operatorname{tr}(\gamma_\mu \gamma_\nu \gamma_\sigma) = 0, \tag{4.3.29}$$
$$\operatorname{tr}(\gamma_\mu \gamma_\nu \gamma_\sigma \gamma_\tau \gamma_\rho) = 0, \tag{4.3.30}$$
$$\operatorname{tr}(\mathbf{G}_{\mu\nu} \mathbf{G}_{\sigma\tau}) = -2(\delta_{\mu\sigma} \delta_{\nu\tau} - \delta_{\mu\tau} \delta_{\nu\sigma}), \tag{4.3.31}$$

as well as the following identities:

$$\mathbf{G}_{\mu\nu} \mathbf{G}_{\nu\mu} = \tfrac{21}{2} \mathbf{1}, \tag{4.3.32}$$
$$\mathbf{G}_{\mu\nu} \gamma_\sigma \mathbf{G}_{\nu\mu} = \tfrac{9}{2} \gamma_\sigma, \tag{4.3.33}$$
$$\mathbf{G}_{\mu\nu} \mathbf{G}_{\sigma\tau} \mathbf{G}_{\nu\mu} = \tfrac{1}{2} \mathbf{G}_{\sigma\tau}, \tag{4.3.34}$$
$$\mathbf{G}_{\mu\nu} \mathbf{G}_{\rho\sigma\tau} \mathbf{G}_{\nu\mu} = -\tfrac{3}{2} \mathbf{G}_{\rho\sigma\tau}, \tag{4.3.35}$$
$$[\mathbf{G}_{\mu\nu}, \mathbf{G}_{\sigma\tau}] = \delta_{\mu\sigma} \mathbf{G}_{\nu\tau} - \delta_{\mu\tau} \mathbf{G}_{\nu\sigma} + \delta_{\nu\tau} \mathbf{G}_{\mu\sigma} - \delta_{\nu\sigma} \mathbf{G}_{\mu\tau}. \tag{4.3.36}$$

Equation (4.3.36) can be rewritten in the form

$$[\mathbf{G}_{\mu\nu}, \mathbf{G}_{\sigma\tau}] = \tfrac{1}{2} \mathbf{G}_{\rho\lambda} c_{\lambda\rho\,\mu\nu\,\sigma\tau}, \tag{4.3.37}$$

where the c's are the structure constants of $\overline{\mathrm{SO}}(7)$ (cf. eq. (4.2.52)):

$$\begin{aligned} c_{\lambda\rho\,\mu\nu\,\sigma\tau} = {} & \delta_{\lambda\tau} \delta_{\rho\nu} \delta_{\mu\sigma} - \delta_{\rho\tau} \delta_{\lambda\nu} \delta_{\mu\sigma} - \delta_{\lambda\sigma} \delta_{\rho\nu} \delta_{\mu\tau} + \delta_{\rho\sigma} \delta_{\lambda\nu} \delta_{\mu\tau} \\ & + \delta_{\lambda\sigma} \delta_{\rho\mu} \delta_{\nu\tau} - \delta_{\rho\sigma} \delta_{\lambda\mu} \delta_{\nu\tau} - \delta_{\lambda\tau} \delta_{\rho\mu} \delta_{\nu\sigma} + \delta_{\rho\tau} \delta_{\lambda\mu} \delta_{\nu\sigma}. \end{aligned} \tag{4.3.38}$$

These, together with

$$c^{a}{}_{b}{}^{c}{}_{d}{}^{e}{}_{f} = \delta^a{}_d \delta^c{}_f \delta^e{}_b - \delta^a{}_f \delta^c{}_b \delta^e{}_d, \tag{4.3.39}$$

are the nonvanishing structure constants of the body $\overline{\mathrm{SO}}(7) \times \mathrm{SL}(2)$ of $\mathrm{F}(4)$. The extending representation is generated by matrices having the components

$$c_\alpha{}^a{}_{\mu\nu\,\beta b} = (\mathbf{G}_{\mu\nu})_{\alpha\beta} \delta^a{}_b = -\tfrac{1}{4}[\gamma_\mu, \gamma_\nu]_{\alpha\beta} \delta^a{}_b, \tag{4.3.40}$$
$$c_\alpha{}^{ac}{}_{d\beta b} = \delta_{\alpha\beta} (\mathbf{G}^c{}_d)^a{}_b = \delta_{\alpha\beta}(\delta^a{}_d \delta^c{}_d - \tfrac{1}{2}\delta^a{}_b f^c{}_d). \tag{4.3.41}$$

It turns out that the matrix η_S of section 4.1 must be chosen, up to an overall factor, in the form

$$\left.\begin{aligned} \eta_{\mu\nu\sigma\tau} &= -\pi_{\mu\nu\sigma\tau}, \\ \eta^{a}{}_{b}{}^{c}{}_{d} &= \tfrac{2}{3}\pi^{a}{}_{b}{}^{c}{}_{d} \end{aligned}\right\} \tag{4.3.42}$$

(other components vanishing) where $\pi_{\mu\nu\sigma\tau}$ and $\pi^{a}{}_{b}{}^{c}{}_{d}$ are the projection matrices into the spaces of antisymmetric 7×7 matrices and traceless 2×2 matrices respectively:

$$\pi_{\mu\nu\sigma\tau} \overset{\text{def}}{=} \delta_{\mu\tau}\delta_{\nu\sigma} - \delta_{\mu\sigma}\delta_{\nu\tau}, \tag{4.3.43}$$

$$\pi^{a}{}_{b}{}^{c}{}_{d} \overset{\text{def}}{=} \delta^{a}{}_{d}\delta^{c}{}_{b} - \tfrac{1}{2}\delta^{a}{}_{b}\delta^{c}{}_{d}. \tag{4.3.44}$$

They satisfy

$$\tfrac{1}{2}\pi_{\mu\nu\sigma\tau}A_{\tau\sigma} = A_{\mu\nu}, \quad \tfrac{1}{2}\pi_{\mu\nu\rho\lambda}\pi_{\lambda\rho\sigma\tau} = \pi_{\mu\nu\sigma\tau}, \tag{4.3.45}$$

$$\pi^{a}{}_{b}{}^{c}{}_{d}B^{d}{}_{c} = B^{a}{}_{b}, \qquad \pi^{a}{}_{b}{}^{e}{}_{f}\pi^{f}{}_{e}{}^{c}{}_{d} = \pi^{a}{}_{b}{}^{c}{}_{d}, \tag{4.3.46}$$

where $\mathbf{A}^{\sim} = -\mathbf{A}$ and $\operatorname{tr}\mathbf{B} = 0$.

The nonvanishing components of the inverse matrix $\boldsymbol{\eta}_{\mathrm{S}}^{-1}$ are

$$\eta^{-1}{}_{\mu\nu\sigma\tau} = -\pi_{\mu\nu\sigma\tau}, \tag{4.3.47}$$

$$\eta^{-1}{}^{a}{}_{b}{}^{c}{}_{d} = \tfrac{3}{2}\pi^{a}{}_{b}{}^{c}{}_{d}, \tag{4.3.48}$$

and satisfy

$$\tfrac{1}{2}\eta_{\mu\nu\rho\lambda}\eta^{-1}{}_{\lambda\rho\sigma\tau} = \pi_{\mu\nu\sigma\tau}, \tag{4.3.49}$$

$$\eta^{a}{}_{b}{}^{e}{}_{f}\eta^{-1}{}^{f}{}_{e}{}^{c}{}_{d} = \pi^{a}{}_{b}{}^{c}{}_{d}. \tag{4.3.50}$$

The structure of F(4)

Expressions (4.3.38)–(4.3.41) comprise half the structure constants of F(4). The remaining structure constants are obtained by introducing the matrix $\boldsymbol{\eta}_{\mathrm{A}}$ (section 4.1) which, for F(4), takes the form

$$\eta_{\alpha a\,\beta b} = \mathrm{i}\,\delta_{\alpha\beta}\,\varepsilon_{ab}, \tag{4.3.51}$$

ε_{ab} being defined by eq. (4.3.6). This yields

$$\begin{aligned} c_{\mu\nu\,\alpha a\,\beta b} &= -\tfrac{1}{2}\eta^{-1}{}_{\mu\nu\sigma\tau}\eta_{\alpha a\,\gamma c}c_{\gamma}{}^{c}{}_{\tau\sigma\,\beta b} \\ &= \frac{\mathrm{i}}{4}[\gamma_{\mu}, \gamma_{\nu}]_{\alpha\beta}\,\varepsilon_{ab} = c_{\mu\nu\,\beta b\,\alpha a}, \end{aligned} \tag{4.3.52}$$

$$\begin{aligned} c^{c}{}_{d\,\alpha a\,\beta b} &= -\eta^{-1}{}^{c}{}_{d}{}^{e}{}_{f}\,\eta_{\alpha a\,\gamma g}\,c_{\gamma}{}^{g}{}^{f}{}_{e\,\beta b} \\ &= -\tfrac{3}{2}\mathrm{i}\delta_{\alpha\beta}(\varepsilon_{ad}\delta^{c}{}_{b} - \tfrac{1}{2}\varepsilon_{ab}\delta^{c}{}_{d}) \\ &= -\tfrac{3}{4}\mathrm{i}\delta_{\alpha\beta}(\varepsilon_{ad}\delta^{c}{}_{b} + \varepsilon_{bd}\delta^{c}{}_{a}) = c^{c}{}_{d\,\beta b\,\alpha a}, \end{aligned} \tag{4.3.53}$$

as well as

$$c_\alpha{}^a{}_{\beta b\ \mu\nu} = -c_\alpha{}^a{}_{\mu\nu\ \beta b} = \tfrac{1}{4}[\gamma_\mu, \gamma_\nu]_{\alpha\beta}\,\delta^a{}_b, \tag{4.3.54}$$

$$c_\alpha{}^a{}_{\beta b}{}^c{}_d = -c_\alpha{}^{ac}{}_{d\ \beta b} = -\delta_{\alpha\beta}(\delta^a{}_d\delta^c{}_b - \tfrac{1}{2}\delta^a{}_b\delta^c{}_d). \tag{4.3.55}$$

Computation of the Cartan–Killing matrix is now straightforward. One finds that eq. (4.1.21) holds with $\lambda = -6$.

In order to verify the identity (4.1.17) first compute

$$\begin{aligned}
&\tfrac{1}{4}\eta_{\mu\nu\,\sigma\tau}\,c_{\nu\mu\ \alpha a\,\beta b}\,c_{\tau\sigma\ \gamma c\,\delta d} + \eta^e{}_f{}^g{}_h\,c^f{}_{e\ \alpha a\,\beta b}\,c^h{}_{g\ \gamma c\,\delta d}\\
&= \tfrac{1}{2}(\mathbf{G}_{\mu\nu})_{\alpha\beta}(\mathbf{G}_{\nu\mu})_{\gamma\delta}\,\varepsilon_{ab}\,\varepsilon_{cd} + \tfrac{3}{4}\delta_{\alpha\beta}\,\delta_{\gamma\delta}(\varepsilon_{ac}\,\varepsilon_{bd} + \varepsilon_{ad}\,\varepsilon_{bc})\\
&= \tfrac{1}{2}(\mathbf{G}_{\mu\nu})_{\alpha\beta}(\mathbf{G}_{\nu\mu})_{\gamma\delta}(\delta_{ac}\,\delta_{bd} - \delta_{ad}\,\delta_{bc})\\
&\quad - \tfrac{3}{4}\delta_{\alpha\beta}\,\delta_{\gamma\delta}(\delta_{ac}\,\delta_{bd} + \delta_{ad}\,\delta_{bc} - 2\delta_{ab}\,\delta_{cd}).
\end{aligned} \tag{4.3.56}$$

Next note that any 16×16 symmetric matrix $\mathbf{S}$ can be decomposed in the form

$$S_{\alpha a\,\beta b} = \delta_{\alpha\beta}\,S_{ab} + \tfrac{1}{6}(\mathbf{G}_{\mu\nu\sigma})_{\alpha\beta}\,T_{\sigma\nu\mu ab} + U_\mu(\gamma_\mu)_{\alpha\beta}\,\varepsilon_{ab} + \tfrac{1}{2}V_{\mu\nu}(\mathbf{G}_{\nu\mu})_{\alpha\beta}\,\varepsilon_{ab}, \tag{4.3.57}$$

where the S's, T's, U's and V's are appropriate coefficients. A straightforward computation, which makes use of the identities (4.3.28)–(4.3.35), then leads to

$$\begin{aligned}
&(\tfrac{1}{4}\eta_{\mu\nu\,\sigma\tau}\,c_{\nu\mu\ \alpha a\,\gamma c}\,c_{\sigma\tau\ \delta d\,\beta b} + \eta^e{}_f{}^g{}_h\,c^f{}_{e\ \alpha a\,\gamma c}\,c^h{}_{g\ \delta d\,\beta b})S_{\gamma c\,\delta d}\\
&\qquad = 6\delta_{\alpha\beta}(S_{ab} - \delta_{ab}\,S_{cc}) + V_{\mu\nu}(\mathbf{G}_{\nu\mu})_{\alpha\beta}\,\varepsilon_{ab},
\end{aligned} \tag{4.3.58}$$

$$\begin{aligned}
&-\tfrac{1}{2}(\tfrac{1}{4}\eta_{\mu\nu\,\sigma\tau}\,c_{\nu\mu\ \alpha a\,\beta b}\,c_{\sigma\tau\ \gamma c\,\delta d} + \eta^e{}_f{}^g{}_h\,c^f{}_{e\ \alpha a\,\beta b}\,c^h{}_{g\ \gamma c\,\delta d})S_{\gamma c\,\delta d}\\
&\qquad = 6\delta_{\alpha\beta}(S_{ab} - \delta_{ab}\,S_{cc}) + V_{\mu\nu}(\mathbf{G}_{\nu\mu})_{\alpha\beta}\,\varepsilon_{ab}.
\end{aligned} \tag{4.3.59}$$

The equality of the right-hand sides of these equations shows that eq. (4.3.17) is indeed satisfied. The reader who goes through the algebra will discover a delicate cancellation of terms, which comes about only because of the special choice that has been made for the matrix η_S (eqs. (4.3.42)–(4.3.44)).

Pseudorepresentation of F(4)

Every element $\mathbf{X}$ of the super Lie algebra F(4) can be associated with an $(8, 2) \times (8, 2)$ matrix:

$$\mathbf{X} \leftrightarrow \begin{pmatrix} \tfrac{1}{2}A_{\mu\nu}\,\mathbf{G}_{\nu\mu} & \mathrm{i}\mathbf{C}\boldsymbol{\varepsilon} \\ \mathbf{C}^\sim & \mathbf{B} \end{pmatrix}. \tag{4.3.60}$$

$\mathbf{A}(=(A_{\mu\nu}))$ is an antisymmetric 7×7 matrix, $\mathbf{B}$ is a traceless 2×2 matrix and $\mathbf{C}$ is an 8×2 matrix. If $\mathbf{X}$ is c-type then the 24 independent elements of **A** and **B** are the *c*-type components of **X**, and the 16 elements of **C** are the *a*-type components. The obvious relation of (4.3.60) to spin (7)⊗sl(2) suggests that F(4) has an (8, 2) × (8, 2) matrix representation. Unfortunately this is not so. The bracket relation for the super Lie algebra is not faithfully reproduced by the supercommutators of matrices of the form (4.3.60). That is why we are forced to work directly with the structure constants, i.e., with the (24, 16) × (24, 16) adjoint representation, in constructing F(4).

Actually the supercommutator does not fail by much. Denote by $\mathbf{X}_{12}$ the bracket of two *c*-type elements $\mathbf{X}_1$ and $\mathbf{X}_2$ of F(4), and by $\mathbf{A}_{12}$, $\mathbf{A}_1$, $\mathbf{A}_2$, etc., the associated components. Then using the structure constants that have been constructed above, one finds

$$(\mathbf{A}_{12})_{\mu\nu} = [\mathbf{A}_1, \mathbf{A}_2]_{\mu\nu} + \frac{i}{2}\mathrm{tr}[\mathbf{G}_{\mu\nu}(\mathbf{C}_1\,\varepsilon\mathbf{C}_2^{\sim} - \mathbf{C}_2\,\varepsilon\mathbf{C}_1^{\sim})], \tag{4.3.61}$$

$$\mathbf{B}_{12} = [\mathbf{B}_1, \mathbf{B}_2] + \frac{3i}{4}(\mathbf{C}_1^{\sim}\mathbf{C}_2 - \mathbf{C}_2^{\sim}\mathbf{C}_1)\varepsilon, \tag{4.3.62}$$

$$\mathbf{C}_{12} = \tfrac{1}{2}(\mathbf{A}_{1\,\mu\nu}\mathbf{G}_{\nu\mu}\mathbf{C}_2 - \mathbf{A}_{2\,\mu\nu}\mathbf{G}_{\nu\mu}\mathbf{C}_1) - (\mathbf{C}_1\mathbf{B}_2^{\sim} - \mathbf{C}_2\mathbf{B}_1^{\sim}) \tag{4.3.63}$$

By virtue of eqs. (4.3.24) and (4.3.26), eq. (4.3.61) may be rewritten in the form

$$\tfrac{1}{2}(\mathbf{A}_{12})_{\mu\nu}\mathbf{G}_{\nu\mu} = \tfrac{1}{2}[\mathbf{A}_1, \mathbf{A}_2]_{\mu\nu}\mathbf{G}_{\nu\mu} + i(\mathbf{C}_1\,\varepsilon\mathbf{C}_2^{\sim} - \mathbf{C}_2\,\varepsilon\mathbf{C}_1^{\sim}) - \frac{i}{8}\gamma_\mu\,\mathrm{tr}[\gamma_\mu(\mathbf{C}_1\,\varepsilon\mathbf{C}_2^{\sim} - \mathbf{C}_2\,\varepsilon\mathbf{C}_1^{\sim})]. \tag{4.3.64}$$

Equations (4.3.62)–(4.3.64) show that the bracket operation for *c*-type elements of F(4) may be recovered by taking the commutator of matrices of the form (4.3.60) and then subtracting the block matrix

$$\begin{bmatrix} \frac{i}{8}\gamma_\mu\,\mathrm{tr}[\gamma_\mu(\mathbf{C}_1\,\varepsilon\mathbf{C}_2^{\sim} - \mathbf{C}_2\,\varepsilon\mathbf{C}_1^{\sim})] & 0 \\ 0 & \frac{i}{4}(\mathbf{C}_1^{\sim}\mathbf{C}_2 - \mathbf{C}_2^{\sim}\mathbf{C}_1)\varepsilon \end{bmatrix}. \tag{4.3.65}$$

In this way the matrices (4.3.60) may be regarded as providing a *pseudorepresentation* of F(4).

The group G(3)

The body of this group is $G_2 \times SL(2)$, having dimension 17 and rank 3, and the extending representation is $g_2 \times sl(2)$. To construct this group we have first to describe the fundamental representation g_2 of G_2.

Let **X** be an element of the Lie algebra G_2. This algebra has 14 dimensions and hence **X** has 14 real components. In an appropriate basis, eight of these components may be assembled into the real and imaginary parts of the elements of a traceless antihermitian 3×3 matrix **A**, and the remaining six components may be assembled into the real and imaginary parts of a complex 3×1 matrix or 3-vector **V**. If **A** and **V** themselves are assembled into the traceless antihermitian 7×7 matrix

$$D(\mathbf{X}) \stackrel{\text{def}}{=} \begin{bmatrix} \mathbf{A} & \frac{1}{\sqrt{2}}\bar{\varepsilon}\cdot\mathbf{V}^* & \mathbf{V} \\ \frac{1}{\sqrt{2}}\underset{\sim}{\varepsilon}\cdot\mathbf{V} & \mathbf{A}^* & \mathbf{V}^* \\ -\mathbf{V}^\dagger & -\mathbf{V}^\sim & 0 \end{bmatrix} \tag{4.3.66}$$

then the matrices $D(\mathbf{X})$ constitute the fundamental representation g_2, satisfying

$$[D(\mathbf{X}_1), D(\mathbf{X}_2)] = D([\mathbf{X}_1, \mathbf{X}_2]), \tag{4.3.67}$$

where $[\mathbf{X}_1, \mathbf{X}_2]$ is the bracket operation for the Lie algebra G_2. Here $\bar{\varepsilon}$ and $\underset{\sim}{\varepsilon}$ denote the antisymmetric tensors $\varepsilon^{\mu\nu\sigma}$ and $\varepsilon_{\mu\nu\sigma}$ respectively. $\underset{\sim}{\varepsilon}\cdot\mathbf{V}$ and $\bar{\varepsilon}\cdot\mathbf{V}^*$ stand for the 3×3 matrices $(\varepsilon_{\mu\sigma\nu} V^\sigma)$ and $(\varepsilon^{\mu\sigma\nu} V^*_\sigma)$. The asterisk denotes complex conjugation and effects a raising or lowering of indices according to the rules

$$V^{\mu *} = V^*_\mu, \qquad A^\mu{}_\nu{}^* = A^*{}_\mu{}^\nu = -A^\nu{}_\mu. \tag{4.3.68}$$

Denote by $\mathbf{X}_{12}$ the bracket of $\mathbf{X}_1$ and $\mathbf{X}_2$, and by $\mathbf{A}_{12}$, $\mathbf{V}_{12}$, $\mathbf{A}_1$, $\mathbf{V}_1$, etc., the associated components. Then eqs. (4.3.66) and (4.3.67) yield

$$\mathbf{A}_{12} = [\mathbf{A}_1, \mathbf{A}_2] + \tfrac{1}{2}(\bar{\varepsilon}\cdot\mathbf{V}_1^*\,\underset{\sim}{\varepsilon}\cdot\mathbf{V}_2 - \bar{\varepsilon}\cdot\mathbf{V}_2^*\,\underset{\sim}{\varepsilon}\cdot\mathbf{V}_1) - (\mathbf{V}_1\mathbf{V}_2^\dagger - \mathbf{V}_2\mathbf{V}_1^\dagger), \tag{4.3.69}$$

$$\mathbf{V}_{12} = \mathbf{A}_1\mathbf{V}_2 - \mathbf{A}_2\mathbf{V}_1 + \sqrt{2}\,\bar{\varepsilon}\cdot\mathbf{V}_1^*\mathbf{V}_2^*, \tag{4.3.70}$$

$$\begin{aligned} \mathbf{V}_{12}^* &= \mathbf{A}_1^*\mathbf{V}_2^* - \mathbf{A}_2^*\mathbf{V}_1^* + \sqrt{2}\,\underset{\sim}{\varepsilon}\cdot\mathbf{V}_1\mathbf{V}_2 \\ &= -\mathbf{A}_1^\sim\mathbf{V}_2^* + \mathbf{A}_2^\sim\mathbf{V}_1^* + \sqrt{2}\,\underset{\sim}{\varepsilon}\cdot\mathbf{V}_1\mathbf{V}_2. \end{aligned} \tag{4.3.71}$$

Use of the identity

$$\varepsilon^{\mu\nu\rho}\varepsilon_{\rho\sigma\tau} = \delta^\mu{}_\sigma\delta^\nu{}_\tau - \delta^\mu{}_\tau\delta^\nu{}_\sigma$$

permits eq. (4.3.69) to be rewritten in the manifestly traceless form

$$\mathbf{A}_{12} = [\mathbf{A}_1, \mathbf{A}_2] - \tfrac{3}{2}(\mathbf{V}_1\mathbf{V}_2^\dagger - \mathbf{V}_2\mathbf{V}_1^\dagger) - \tfrac{1}{2}\mathbf{1}(\mathbf{V}_1^\dagger\cdot\mathbf{V}_2 - \mathbf{V}_2^\dagger\cdot\mathbf{V}_1). \tag{4.3.72}$$

Multiplication of the identity

$$\begin{aligned} 0 &\equiv \varepsilon_{\mu\sigma\tau}A^\tau{}_\nu - \varepsilon_{\sigma\tau\nu}A^\tau{}_\mu + \varepsilon_{\tau\nu\mu}A^\tau{}_\sigma - \varepsilon_{\nu\mu\sigma}A^\tau{}_\tau \\ &= \varepsilon_{\mu\sigma\tau}A^\tau{}_\nu - A^*{}_\mu{}^\tau\varepsilon_{\tau\sigma\nu} + \varepsilon_{\mu\tau\nu}A^\tau{}_\sigma \end{aligned} \tag{4.3.73}$$

by V^σ yields the relation

$$(\underset{\sim}{\varepsilon}\cdot\mathbf{V})\mathbf{A} - \mathbf{A}^*(\underset{\sim}{\varepsilon}\cdot\mathbf{V}) = -\underset{\sim}{\varepsilon}\cdot(\mathbf{A}\mathbf{V}), \tag{4.3.74}$$

which allows the entry in the second row and first column of $[D(\mathbf{X}_1), D(\mathbf{X}_2)]$ to be expressed in the form

$$\frac{1}{\sqrt{2}}\underset{\sim}{\varepsilon}\cdot(\mathbf{A}_1\mathbf{V}_2 - \mathbf{A}_2\mathbf{V}_1) - (\mathbf{V}_1^*\mathbf{V}_2^\dagger - \mathbf{V}_2^*\mathbf{V}_1^\dagger) = \frac{1}{\sqrt{2}}\underset{\sim}{\varepsilon}\cdot\mathbf{V}_{12}, \tag{4.3.75}$$

as is necessary for consistency with (4.3.66).

The structure of G_2

The structure constants of G_2 are defined by

$$X_{12}{}^A = c^A{}_{BC}\, X_1{}^B\, X_2{}^C \tag{4.3.76}$$

Replacing the X^A by $A^\mu{}_\nu$, V^μ and $V^*{}_\mu$, and keeping only the nonvanishing structure constants, we have

$$A_{12}{}^\mu{}_\nu = c^\mu{}_\nu{}^\sigma{}_\tau{}^\rho{}_\lambda\, A_1{}^\tau{}_\sigma\, A_2{}^\lambda{}_\rho + c^\mu{}_{\nu\sigma}{}^\tau\, V_1{}^\sigma\, V_2{}^*{}_\tau + c^\mu{}_\nu{}^\sigma{}_\tau\, V_1{}^*{}_\sigma\, V_2{}^\tau, \tag{4.3.77}$$

$$V_{12}{}^\mu = c^{\mu\,\nu}{}_{\sigma\tau}\, A_1{}^\sigma{}_\nu\, V_2{}^\tau + c^\mu{}_\nu{}^\sigma{}_\tau\, V_1{}^\nu\, A_2{}^\tau{}_\sigma + c^{\mu\nu\sigma}\, V_1{}^*{}_\nu\, V_2{}^*{}_\sigma, \tag{4.3.78}$$

$$V_{12}{}^*{}_\mu = c_\mu{}^\nu{}_\sigma{}^\tau\, A_1{}^\sigma{}_\nu\, V_2{}^*{}_\tau + c_\mu{}^{\nu\sigma}{}_\tau\, V_1{}^*{}_\nu\, A_2{}^\tau{}_\sigma + c_{\mu\nu\sigma}\, V_1{}^\nu\, V_2{}^\sigma \tag{4.3.79}$$

which, on comparison with eqs. (4.3.70)–(4.3.72), yields

$$c^\mu{}_\nu{}^\sigma{}_\tau{}^\rho{}_\lambda = \delta^\mu{}_\tau\,\delta^\sigma{}_\lambda\,\delta^\rho{}_\nu - \delta^\mu{}_\lambda\,\delta^\sigma{}_\nu\,\delta^\rho{}_\tau, \tag{4.3.80}$$

$$c^\mu{}_{\nu\sigma}{}^\tau = -c^\mu{}_\nu{}^\tau{}_\sigma = -\tfrac{3}{2}(\delta^\mu{}_\sigma\,\delta^\tau{}_\nu - \tfrac{1}{3}\delta^\mu{}_\nu\,\delta^\tau{}_\sigma), \tag{4.3.81}$$

$$c^{\mu\,\nu}{}_{\sigma\tau} = -c^\mu{}_\tau{}^\nu{}_\sigma = \delta^\mu{}_\sigma\,\delta^\nu{}_\tau - \tfrac{1}{3}\delta^\mu{}_\tau\,\delta^\nu{}_\sigma, \tag{4.3.82}$$

$$c_\mu{}^\nu{}_\sigma{}^\tau = -c_\mu{}^{\tau\nu}{}_\sigma = -\delta^\nu{}_\mu\,\delta^\tau{}_\sigma + \tfrac{1}{3}\delta^\tau{}_\mu\,\delta^\nu{}_\sigma, \tag{4.3.83}$$

$$c^{\mu\nu\sigma} = \sqrt{2}\,\varepsilon^{\mu\nu\sigma}, \tag{4.3.84}$$

$$c_{\mu\nu\sigma} = \sqrt{2}\,\varepsilon_{\mu\nu\sigma}. \tag{4.3.85}$$

The terms involving the factor $\frac{1}{3}$ in expressions (4.3.82) and (4.3.83) arise from application of the projection matrix into the space of traceless 3×3 matrices:

$$\pi^\mu{}_\nu{}^\sigma{}_\tau \overset{\text{def}}{=} \delta^\mu{}_\tau\,\delta^\sigma{}_\nu - \tfrac{1}{3}\delta^\mu{}_\nu\,\delta^\sigma{}_\tau. \tag{4.3.86}$$

The nonvanishing components of the Cartan–Killing matrix of G_2 are readily computed:

$$-c^\rho{}_\lambda{}^\mu{}_\nu{}^\iota{}_\kappa\, c^\kappa{}_\iota{}^\sigma{}_\tau{}^\lambda{}_\rho - c^{\rho\,\mu}{}_{\nu\lambda}\, c^{\lambda\,\sigma}{}_{\tau\rho} - c_\rho{}^\mu{}_\nu{}^\lambda\, c_\lambda{}^\sigma{}_\tau{}^\rho = -8\pi^\mu{}_\nu{}^\sigma{}_\tau, \tag{4.3.87}$$

$$-c^\sigma{}_{\tau\mu}{}^\rho\, c_\rho{}^{\nu\tau}{}_\sigma - c^\rho{}_\mu{}^\sigma{}_\tau\, c^\tau{}_\sigma{}^\nu{}_\rho - c_{\sigma\mu\tau}\, c^{\tau\nu\sigma} = 12\delta_\mu^{\ \nu}, \tag{4.3.88}$$

$$-c^\sigma{}_\tau{}^\mu{}_\rho\, c^\rho{}_\nu{}^\tau{}_\sigma - c_\rho{}^{\mu\sigma}{}_\tau\, c^\tau{}_{\sigma\nu}{}^\rho - c^{\sigma\mu\tau}\, c_{\tau\mu\sigma} = 12\delta^\mu{}_\nu. \tag{4.3.89}$$

When this matrix is multiplied by a scale factor expressions (4.3.87), (4.3.88) and (4.3.89) must all scale together.

The structure of G(3)

The nonvanishing structure constants of the body $G_2 \times SL(2)$ of G(3) are given by expressions (4.3.80)–(4.3.85) together with

$$c^{a\ \ c\ \ e}_{\ \ b\ \ d\ \ f} = \delta^a{}_d\,\delta^c{}_f\,\delta^e{}_b - \delta^a{}_f\,\delta^c{}_b\,\delta^e{}_d. \tag{4.3.90}$$

Let the indices $^\mu$, $_\mu$ and $_0$ label respectively the first three rows, the second three rows and the last row of the matrix (4.3.66); and let the indices $_\nu$, $^\nu$ and $_0$ label respectively the first three columns, the second three columns and the last column. We can read off directly from eqs. (4.3.2) and (4.3.66) the nonvanishing matrix elements of the generators of the extending representation $g_2 \otimes sl(2)$ of G(3):

$$c^{\mu a\,\sigma}{}_{\tau\,\nu b} = -c^{\mu a}{}_{\nu b}{}^{\sigma}{}_{t} = (\delta^\mu{}_\tau\,\delta^\sigma{}_\nu - \tfrac{1}{3}\delta^\mu{}_\nu\,\delta^\sigma{}_\tau)\delta^a{}_b, \tag{4.3.91}$$

$$c_\mu{}^{a\,\sigma}{}_\tau{}^\nu{}_b = -c_\mu{}^{a\,\nu}{}_b{}^\sigma{}_\tau = -(\delta_\mu{}^\sigma\,\delta_\tau{}^\nu - \tfrac{1}{3}\delta_\mu{}^\nu\,\delta^\sigma{}_\tau)\delta^a{}_b, \tag{4.3.92}$$

$$c^{\mu a\,\sigma\,\nu}{}_b = -c^{\mu a\,\nu}{}_b{}^\sigma = \frac{1}{\sqrt{2}}\varepsilon^{\mu\sigma\nu}\,\delta^a{}_b, \tag{4.3.93}$$

$$c_\mu{}^a{}_{\sigma\,\nu b} = -c_\mu{}^a{}_{\nu b\,\sigma} = \frac{1}{\sqrt{2}}\varepsilon_{\mu\sigma\nu}\,\delta^a{}_b, \tag{4.3.94}$$

$$c^{\mu a}{}_{\sigma\,0b} = -c^{\mu a}{}_{0b\,\sigma} = \delta^\mu{}_\sigma\,\delta^a{}_b, \tag{4.3.95}$$

$$c_\mu{}^{a\sigma}{}_{0b} = -c_\mu{}^a{}_{0b}{}^\sigma = \delta_\mu{}^\sigma\delta^a{}_b, \tag{4.3.96}$$

$$c_0{}^{a\,\sigma}{}_{\nu b} = -c_0{}^a{}_{\nu b}{}^\sigma = -\delta^\sigma{}_\nu\,\delta^a{}_b, \tag{4.3.97}$$

$$c_0{}^a{}_\sigma{}^\nu{}_b = -c_0{}^{a\,\nu}{}_{b\,\sigma} = -\delta_\sigma{}^\nu\,\delta^a{}_b, \tag{4.3.98}$$

$$c^{\mu a\,c}{}_{d\,b\nu} = -c^{\mu a}{}_{b\nu}{}^c{}_d = \delta^\mu{}_\nu(\delta^a{}_d\,\delta^c{}_b - \tfrac{1}{2}\delta^a{}_b\,\delta^c{}_d), \tag{4.3.99}$$

$$c_\mu{}^{a\,c}{}_{d\,b}{}^\nu = -c_\mu{}^a{}_b{}^{\nu\,c}{}_d = = \delta_\mu{}^\nu(\delta^a{}_d\,\delta^c{}_b - \tfrac{1}{2}\,\delta^a{}_b\,\delta^c{}_d), \tag{4.3.100}$$

$$c_0{}^{a\,c}{}_{d\,0b} = -c_0{}^a{}_{0b}{}^c{}_d = \delta^a{}_d\,\delta^c{}_b - \tfrac{1}{2}\delta^a{}_b\,\delta^c{}_d. \tag{4.3.101}$$

The extending representation has dimension 14 so the dimension of G(3) is (17, 14).

Expressions (4.3.80)–(4.3.85) and (4.3.90)–(4.3.101) constitute the structure constants that were called $c^\mu{}_{\nu\sigma}$, $c^\alpha{}_{\mu\beta}$, $c^\alpha{}_{\beta\mu}$ in section 4.1. To get the remaining structure constants, denoted by $c^\mu{}_{\alpha\beta}$ in section 4.1, we need the matrices η_S and η_A. In the present case it turns out that, up to an overall factor, these must be chosen to be

$$\left.\begin{aligned}\eta^{\mu}{}_{\nu}{}^{\sigma}{}_{\tau} &= -\pi^{\mu}{}_{\nu}{}^{\sigma}{}_{\tau},\\ \eta_{\mu}{}^{\nu} &= \tfrac{3}{2}\delta_{\mu}{}^{\nu},\\ \eta^{\mu}{}_{\nu} &= \tfrac{3}{2}\delta^{\mu}{}_{\nu},\\ \eta^{a}{}_{b}{}^{c}{}_{d} &= \tfrac{3}{4}\pi^{a}{}_{b}{}^{c}{}_{d},\end{aligned}\right\} \tag{4.3.102}$$

and

$$\left.\begin{aligned}\eta^{\mu}{}_{a\,\nu b} &= -\eta_{\nu b}{}^{\mu}{}_{a} = \mathrm{i}\delta^{\mu}{}_{\nu}\varepsilon_{ab},\\ \eta_{0a\,0b} &= \mathrm{i}\varepsilon_{ab}\end{aligned}\right\} \tag{4.3.103}$$

respectively. Using the inverse

$$\left.\begin{aligned}\eta^{-1\,\mu}{}_{\nu}{}^{\sigma}{}_{\tau} &= -\pi^{\mu}{}_{\nu}{}^{\sigma}{}_{\tau},\\ \eta^{-1}{}_{\mu}{}^{\nu} &= \tfrac{2}{3}\delta_{\mu}{}^{\nu},\\ \eta^{-1\,\mu}{}_{\nu} &= \tfrac{2}{3}\delta^{\mu}{}_{\nu},\\ \eta^{-1\,a}{}_{b}{}^{c}{}_{d} &= \tfrac{4}{3}\pi^{a}{}_{b}{}^{c}{}_{d},\end{aligned}\right\} \tag{4.3.104}$$

we obtain

$$\begin{aligned}c^{\mu}{}_{\nu}{}^{\sigma}{}_{a\,\tau b} &= c^{\mu}{}_{\nu\,\tau b}{}^{\sigma}{}_{a}\\ &= -\eta^{-1\,\mu}{}_{\nu}{}^{\rho}{}_{\lambda}\eta^{\sigma}{}_{a\,\kappa c}c^{\kappa c\,\lambda}{}_{\rho\,\tau b} = \mathrm{i}(\delta^{\mu}{}_{\tau}\delta^{\sigma}{}_{\nu} - \tfrac{1}{3}\delta^{\mu}{}_{\nu}\delta^{\sigma}{}_{\tau})\varepsilon_{ab},\end{aligned} \tag{4.3.105}$$

$$c^{\mu}{}_{a}{}^{\nu}{}_{b}{}^{\sigma} = -\eta^{-1\,\mu}{}_{\tau}\eta^{\nu}{}_{a\,\rho c}c^{\rho c\,\tau\sigma}{}_{b} = \mathrm{i}\frac{\sqrt{2}}{3}\varepsilon^{\mu\nu\sigma}\varepsilon_{ab}, \tag{4.3.106}$$

$$c_{\mu\,\nu a\,\sigma b} = -\eta^{-1}{}_{\mu}{}^{\tau}\eta_{\nu a}{}^{\rho}{}_{c}c_{\rho}{}^{c}{}_{\tau\,\sigma b} = \mathrm{i}\frac{\sqrt{2}}{3}\varepsilon_{\mu\nu\sigma}\varepsilon_{ab}, \tag{4.3.107}$$

$$c^{\mu}{}_{\nu a\,0b} = c^{\mu}{}_{0b\,\nu a} = -\eta^{-1\,\mu}{}_{\sigma}\eta_{\nu a}{}^{\tau}{}_{c}c_{\tau}{}^{c\sigma}{}_{0b} = -\mathrm{i}\tfrac{2}{3}\delta^{\mu}{}_{\nu}\varepsilon_{ab}, \tag{4.3.108}$$

$$c_{\mu}{}^{\nu}{}_{a\,0b} = c_{\mu\,0b}{}^{\nu}{}_{a} = -\eta^{-1}{}_{\mu}{}^{\sigma}\eta^{\nu}{}_{a\,\tau c}c^{\tau c}{}_{\sigma\,0b} = -\mathrm{i}\tfrac{4}{3}\delta_{\mu}{}^{\nu}\varepsilon_{ab}, \tag{4.3.109}$$

$$\begin{aligned}c^{a}{}_{b\,c\,d} &= c^{a}{}_{b\,d\,c}\\ &= -\eta^{-1\,a}{}_{b}{}^{e}{}_{f}\eta^{\mu}{}_{c\,\sigma g}c^{\sigma g\,f}{}_{e\,\nu d} = -\mathrm{i}\tfrac{2}{3}\delta^{\mu}{}_{\nu}(\delta^{a}{}_{c}\varepsilon_{db} + \delta^{a}{}_{d}\varepsilon_{cb}),\end{aligned} \tag{4.3.110}$$

$$c^{a}{}_{b\,0c\,0d} = -\eta^{-1\,a}{}_{b}{}^{e}{}_{f}\eta_{0c\,0g}c_{0}{}^{g\,f}{}_{e\,0d} = -\mathrm{i}\tfrac{4}{3}(\delta^{a}{}_{c}\varepsilon_{db} + \delta^{a}{}_{d}\varepsilon_{cb}), \tag{4.3.111}$$

all other structure constants vanishing.

It is now a straightforward computation to check that the identity (4.1.12) holds:

$$\begin{aligned}&\eta^{\rho}{}_{\lambda}{}^{\iota}{}_{\kappa}c^{\lambda}{}_{\rho}{}^{\mu}{}_{a\,\nu b}c^{\kappa}{}_{\iota}{}^{\sigma}{}_{c\,\tau d} + \eta_{\rho}{}^{\lambda}c^{\rho\,\mu}{}_{a\,\nu b}c_{\lambda}{}^{\sigma}{}_{c\,\tau d}\\ &\quad + \eta^{e}{}_{f}{}^{g}{}_{h}c^{f}{}_{e}{}^{\mu}{}_{a\,\nu b}c^{h}{}_{g}{}^{\sigma}{}_{c\,\tau d} + \mathrm{cyc}({}_{\nu b},{}^{\sigma}{}_{c},{}_{\tau d})\\ &= -\tfrac{1}{3}(\delta^{\mu}{}_{\nu}\delta^{\sigma}{}_{\tau} - \delta^{\mu}{}_{\tau}\delta^{\sigma}{}_{\nu})(\varepsilon_{ab}\varepsilon_{cd} + \varepsilon_{ac}\varepsilon_{db} + \varepsilon_{ad}\varepsilon_{bc}) = 0,\end{aligned} \tag{4.3.112}$$

$$\begin{aligned}&\eta_{\sigma}{}^{\tau}c^{\sigma}{}_{\mu a\,0b}c_{\tau}{}^{\nu}{}_{c\,0d} + \eta^{e}{}_{f}{}^{g}{}_{h}c^{f}{}_{e\,\mu a\,0b}c^{h}{}_{g}{}^{\nu}{}_{c\,0d}\\ &\quad + \mathrm{cyc}({}_{0b},{}^{\nu}{}_{c},{}_{0d}) = 0,\end{aligned} \tag{4.3.113}$$

$$\eta_\sigma{}^\tau c^\sigma{}_{0a\,\mu b} c_\tau{}^\nu{}_{c\,0d} + \eta^\sigma{}_\tau c_{\sigma\,0a\,\mu b} c^{\tau\,\nu}{}_{c\,0d} + \eta^e{}_f{}^g{}_h c^f{}_{e\,0a\,\mu b} c^h{}_g{}^\nu{}_{c\,0d} + \mathrm{cyc}({}_{\mu b}, {}^\nu{}_c, {}_{0d}) = 0, \tag{4.3.114}$$

$$\eta^e{}_f{}^g{}_h c^f{}_{e\,0a\,0b} c^h{}_{g\,0c\,0d} + \mathrm{cyc}({}_{0b}, {}_{0c}, {}_{0d}) = 0. \tag{4.3.115}$$

It is also straightforward to verify that eqs. (4.3.22) hold with $\lambda = 4$.

Pseudorepresentation of G(3)

Just as every element of the super Lie algebra F(4) can be associated with an $(8,2) \times (8,2)$ matrix so every element $\mathbf{X}$ of G(3) can be associated with a $(7,2) \times (7,2)$ matrix:

$$\mathbf{X} \leftrightarrow \begin{bmatrix} \mathbf{A} & \frac{1}{\sqrt{2}}\bar{\varepsilon}\cdot\mathbf{V}^* & \mathbf{V} & \mathrm{i}\mathbf{C}\varepsilon \\ \frac{1}{\sqrt{2}}\bar{\varepsilon}\cdot V & \mathbf{A}^* & \mathbf{V}^* & \mathrm{i}\mathbf{C}\varepsilon \\ -\mathbf{V}^\dagger & -\mathbf{V}^\sim & 0 & \mathrm{i}\mathbf{D}\varepsilon \\ \mathbf{C}^\dagger & \mathbf{C}^\sim & \mathbf{D}^\sim & \mathbf{B} \end{bmatrix}. \tag{4.3.116}$$

Here $\mathbf{A}$ and $\mathbf{V}$ are the matrices appearing in (4.3.66), $\mathbf{B}$ is a traceless real 2×2 matrix, $\mathbf{C}$ is a complex 3×2 matrix and $\mathbf{D}$ is a real 1×2 matrix. If $\mathbf{X}$ is *c*-type then the 17 independent real and imaginary parts of the elements of $\mathbf{A}$, $\mathbf{V}$ and $\mathbf{B}$ are the *c*-type components of $\mathbf{X}$, and the 14 independent real and imaginary parts of the elements of $\mathbf{C}$ and $\mathbf{D}$ are the *a*-type components of $\mathbf{X}$.

Using indices ${}^\mu, {}_\mu, {}_0$ and ^{a} to label the rows of the matrix (4.3.116) and indices ${}_\nu, {}^\nu, {}_0$ and ${}_b$ to label the columns, one can read off from the structure constants the bracket relations for G(3):

$$\mathbf{A}_{12} = [\mathbf{A}_1, \mathbf{A}_2] - \tfrac{3}{2}(\mathbf{V}_1\mathbf{V}_2^\dagger - \mathbf{V}_2\mathbf{V}_1^\dagger) - \tfrac{1}{2}\mathbf{1}(\mathbf{V}_1^\dagger\cdot\mathbf{V}_2^\dagger - \mathbf{V}_2\cdot\mathbf{V}_1) + \mathrm{i}(\mathbf{C}_1\varepsilon\mathbf{C}_2^\dagger - \mathbf{C}_2\varepsilon\mathbf{C}_2^\dagger) - \frac{\mathrm{i}}{3}\mathbf{1}\,\mathrm{tr}(\mathbf{C}_1\varepsilon\mathbf{C}_2^\dagger - \mathbf{C}_2\varepsilon\mathbf{C}_1^\dagger), \tag{4.3.117}$$

$$\mathbf{V}_{12} = \mathbf{A}_1\mathbf{V}_2 - \mathbf{A}_2\mathbf{V}_1 + \sqrt{2}\bar{\varepsilon}\cdot\mathbf{V}_1^*\mathbf{V}_2^* + \mathrm{i}\frac{\sqrt{2}}{3}\bar{\varepsilon}:\mathbf{C}_1^*\varepsilon\mathbf{C}_2^\dagger + \mathrm{i}\tfrac{2}{3}(\mathbf{C}_1\varepsilon\mathbf{D}^\sim - \mathbf{C}_2\varepsilon\mathbf{D}_1^\sim), \tag{4.3.118}$$

$$\mathbf{B}_{12} = [\mathbf{B}_1, \mathbf{B}_2] + \mathrm{i}\tfrac{2}{3}(\mathbf{C}_1^\dagger\mathbf{C}_2 - \mathbf{C}_2^\dagger\mathbf{C}_1 + \mathbf{C}_1^\sim\mathbf{C}_2^* - \mathbf{C}_2^\sim\mathbf{C}_1^* + \mathbf{D}_1^\sim\mathbf{D}_2 - \mathbf{D}_2^\sim\mathbf{D}_1)\varepsilon, \tag{4.3.119}$$

$$\mathbf{C}_{12} = \mathbf{A}_1\mathbf{C}_2 - \mathbf{A}_2\mathbf{C}_1 + \frac{1}{\sqrt{2}}(\bar{\varepsilon}\cdot\mathbf{V}_1^*\mathbf{C}_2^* - \bar{\varepsilon}\cdot\mathbf{V}_2^*\mathbf{C}_1) + \mathbf{V}_1\mathbf{D}_2 - \mathbf{V}_2\mathbf{D}_1 - \mathbf{C}_1\mathbf{B}_2^\sim + \mathbf{C}_2\mathbf{B}_1^\sim, \tag{4.3.120}$$

$$\mathbf{D}_{12} = -\mathbf{V}_1^\dagger\mathbf{C}_2 + \mathbf{V}_2^\dagger\mathbf{C}_1 - \mathbf{V}_1^\sim\mathbf{C}_2^* + \mathbf{V}_2^\sim\mathbf{C}_1^* - \mathbf{D}_1\mathbf{B}_2^\sim + \mathbf{D}_2\mathbf{B}_1^\sim. \tag{4.3.121}$$

The double : appearing in eq. (4.3.118) indicates that contractions are to be performed over two pairs of dummy indices. Note that the bracket relations are not given exactly by the matrix commutator. We have here, once again, a pseudorepresentation.

4.4 Super Lie groups of basic importance in physics

The super de Sitter group

This group has body $\overline{\mathrm{SO}}\,(2,3)$ and extending representation spin(2, 3). Its dimension is (10, 4), and its complex extension is the same as that of OSp(1, 4).[†]

Spin(2, 3) is most easily described in terms of γ-matrices satisfying

$$\{\gamma_a, \gamma_b\} = 2\eta_{ab}\,\mathbf{1}, \qquad a, b = -1, 0, 1, 2, 3, \quad (\eta_{ab}) = \mathrm{diag}\,(-1, -1, 1, 1, 1). \tag{4.4.1}$$

There are two inequivalent irreducible faithful representations for the γ's, each of dimension 4 and each the negative of the other. It does not matter which representation is chosen, for the generators of spin(2, 3) involve the γ's only in the bilinear combinations

$$\mathbf{G}_{ab} \overset{\mathrm{def}}{=} -\tfrac{1}{4}[\gamma_a, \gamma_b]. \tag{4.4.2}$$

It turns out that the γ-matrices for spin(2, 3) can be chosen real and orthogonal. For example

$$\left.\begin{aligned} \gamma_{-1} &= \mathrm{i}\sigma_2 \times \sigma_3, \\ \gamma_0 &= \mathrm{i}\mathbf{1} \times \sigma_2, \\ \gamma_1 &= \mathbf{1} \times \sigma_1, \\ \gamma_2 &= \sigma_1 \times \sigma_3, \\ \gamma_3 &= \sigma_3 \times \sigma_3, \end{aligned}\right\} \tag{4.4.3}$$

where the σ's are the Pauli matrices. With such a choice, which will be assumed in this section, there exists a real antisymmetric matrix $\bar{\gamma}$ such that

$$\bar{\gamma}\gamma_a\bar{\gamma}^{-1} = \gamma_a^{\sim}, \quad \text{for all } a. \tag{4.4.4}$$

For example, with the choice (4.4.3) one finds $\bar{\gamma} = \pm\mathrm{i}\sigma_2 \times \sigma_1$.[‡] The matrices $\bar{\gamma}^{-1}$, $\gamma_a\bar{\gamma}^{-1}$, $\mathbf{G}_{ab}\bar{\gamma}^{-1}$ together form a complete linearly independent set,

[†] Since SO(1) is trivial the body of OSp(1,4) is Sp(4), which has the same complex extension as $\overline{\mathrm{SO}}(5)$ or $\overline{\mathrm{SO}}(2, 3)$. The extending representation of OSp(1, 4) is the 4-dimensional fundamental representation sp(4). (See the discussion of OSp(m, n) in section 4.2.)

[‡] More generally $\bar{\gamma} = \pm\gamma_{-1}\gamma_0 = \pm\gamma_1\gamma_2\gamma_3$. The normalization is chosen so that $\bar{\gamma}$ is itself orthogonal and $\det\bar{\gamma} = 1$. Note that the matrices γ_{-1} and γ_0 are antisymmetric and that γ_1, γ_2 and γ_3 are symmetric. Note also that $\gamma_{-1} = \pm\gamma_0\gamma_1\gamma_2\gamma_3$.

the six matrices $\bar{\gamma}^{-1}, \gamma_a\bar{\gamma}^{-1}$ being antisymmetric and the ten matrices $\mathbf{G}_{ab}\bar{\gamma}^{-1}$ being symmetric.

The γ's satisfy the following identities:

$$\left.\begin{aligned}
\operatorname{tr}\gamma_a &= 0,\\
\operatorname{tr}(\gamma_a\gamma_b) &= 4\eta_{ab},\\
\operatorname{tr}\mathbf{G}_{ab} &= 0,\\
\operatorname{tr}(\gamma_a\gamma_b\gamma_c) &= 0,\\
\operatorname{tr}(\mathbf{G}_{ab}\mathbf{G}_{cd}) &= -\eta_{ac}\eta_{bd} + \eta_{ad}\eta_{bc},\\
\gamma_a\gamma^a &= 5\cdot\mathbf{1},\\
\gamma_a\gamma_b\gamma^a &= -3\gamma_b,\\
\gamma_a\mathbf{G}_{bc}\gamma^a &= \mathbf{G}_{bc},\\
\mathbf{G}_{ab}\mathbf{G}^{ba} &= 5\cdot\mathbf{1},\\
\mathbf{G}_{ab}\gamma_c\mathbf{G}^{ba} &= \gamma_c,\\
\mathbf{G}_{ab}\mathbf{G}_{cd}\mathbf{G}^{ba} &= -\mathbf{G}_{cd},
\end{aligned}\right\} \tag{4.4.5}$$

where indices are raised by means of the inverse (η^{ab}) to the matrix (η_{ab}). The generators $\mathbf{G}_{ab}$ themselves satisfy the commutation relations

$$[\mathbf{G}_{ab}, \mathbf{G}_{cd}] = \eta_{ac}\mathbf{G}_{bd} + \eta_{bd}\mathbf{G}_{ac} - \eta_{ad}\mathbf{G}_{bc} - \eta_{bc}\mathbf{G}_{ad}, \tag{4.4.6}$$

which may be rewritten in the form

$$[\mathbf{G}_{ab}, \mathbf{G}_{cd}] = \tfrac{1}{2}\mathbf{G}_{ef}c^{fe}{}_{abcd}, \tag{4.4.7}$$

the c's being the structure constants of $\overline{SO}(2, 3)$:

$$\begin{aligned}
c^{fe}{}_{ab\,cd} = {}& \delta^f{}_d\delta^e{}_b\eta_{ac} - \delta^e{}_d\delta^f{}_b\eta_{ac} - \delta^f{}_c\delta^e{}_b\eta_{ad} + \delta^e{}_c\delta^f{}_b\eta_{ad}\\
&+ \delta^f{}_c\delta^e{}_a\eta_{bd} - \delta^e{}_c\delta^f{}_a\eta_{bd} - \delta^f{}_d\delta^e{}_a\eta_{bc} + \delta^e{}_d\delta^f{}_a\eta_{bc},
\end{aligned} \tag{4.4.8}$$

The nonvanishing structure constants of the super de Sitter group are these c's, together with the matrix elements of the generators G_{ab} and the structure constants obtained from these matrix elements by application of the matrices $\boldsymbol{\eta}_S$ and $\boldsymbol{\eta}_A$ of section 4.1. The latter matrices here take the respective forms

$$\eta_{ab\,cd} = \eta_{ad}\eta_{bc} - \eta_{ac}\eta_{bd}, \tag{4.4.9}$$

$$\eta_{\alpha\beta} = \mathrm{i}\bar{\gamma}_{\alpha\beta}, \tag{4.4.10}$$

yielding

$$c^{\alpha}{}_{ab\,\beta} = -c^{\alpha}{}_{\beta\,ab} = (\mathbf{G}_{ab})^{\alpha}{}_{\beta} = -\tfrac{1}{4}[\gamma_a, \gamma_b]^{\alpha}{}_{\beta}, \tag{4.4.11}$$

$$c^{ab}{}_{\alpha\beta} = -\tfrac{1}{2}\eta^{ab\,cd}\eta_{\alpha\gamma}c^{\gamma}{}_{dc\,\beta} = -\mathrm{i}(\bar{\gamma}\mathbf{G}^{ab})_{\alpha\beta}. \tag{4.4.12}$$

The verification that these structure constants satisfsy the identity (4.1.12) (or, equivalently, (4.1.17)) rests on the verification of the identity

$$\mathbf{G}_{ab}\operatorname{tr}(\mathbf{S}\bar{\gamma}\mathbf{G}^{ba}) + 2\mathbf{G}_{ab}\mathbf{S}\bar{\gamma}\mathbf{G}^{ba} = 0, \tag{4.4.13}$$

where S is any symmetric 4 × 4 matrix. But if S is symmetric it is a linear combination of the matrices $\mathbf{G}_{cd}\bar{\gamma}^{-1}$. It therefore suffices to check (4.4.13) for these matrices. The verification follows immediately from the fifth and eleventh of the identities (4.4.5).

From the structure constants one obtains the commutation relations for the generators G_{ab}, G_α of any representation of the super de Sitter group:

$$[G_{ab}, G_{cd}] = \eta_{ac} G_{bd} + \eta_{bd} G_{ac} - \eta_{ad} G_{bc} - \eta_{bc} G_{ad}, \tag{4.4.14}$$

$$[G_{ab}, G_\beta] = G_\alpha (\mathbf{G}_{ab})^\alpha{}_\beta, \tag{4.4.15}$$

$$[G_\alpha, G_\beta]_S = \{G_\alpha, G_\beta\} = -\frac{i}{2} G_{ab} (\bar{\gamma}\mathbf{G}^{ba})_{\alpha\beta}, \tag{4.4.16}$$

the subscript S in (4.4.16) denoting the supercommutator.

The super Poincaré group

This group is derived from the super de Sitter group by a process known as *contraction.* Let Greek indices from the middle of the alphabet run over the values 0, 1, 2, 3. Then define

$$J_{\mu\nu} \stackrel{\text{def}}{=} iG_{\mu\nu}, \tag{4.4.17}$$

$$P_\mu \stackrel{\text{def}}{=} i\lambda^2 G_{\mu-1}, \tag{4.4.18}$$

$$Q_\alpha \stackrel{\text{def}}{=} i\sqrt{2}\,\lambda G_\alpha, \tag{4.4.19}$$

the G's on the right being the generators of a representation of the super de Sitter group, and λ an arbitrary real number. Equations (4.4.14)–(4.4.16) imply

$$[J_{\mu\nu}, J_{\sigma\tau}] = i(\eta_{\mu\sigma} J_{\nu\tau} + \eta_{\nu\tau} J_{\mu\sigma} - \eta_{\mu\tau} J_{\nu\sigma} - \eta_{\nu\sigma} J_{\mu\tau}), \tag{4.4.20}$$

$$[J_{\mu\nu}, P_\sigma] = i(\eta_{\mu\sigma} P_\nu - \eta_{\nu\sigma} P_\mu), \tag{4.4.21}$$

$$[J_{\mu\nu}, Q_\beta] = iQ_\alpha (\mathbf{G}_{\mu\nu})^\alpha{}_\beta = -\frac{i}{4} Q_\alpha [\gamma_\mu, \gamma_\nu]^\alpha{}_\beta, \tag{4.4.22}$$

$$[P_\mu, P_\nu] = -i\lambda^4 J_{\mu\nu}, \tag{4.4.23}$$

$$[P_\mu, Q_\beta] = \frac{i}{2} \lambda^2 Q_\alpha (\gamma_{-1}\gamma_\mu)^\alpha{}_\beta, \tag{4.4.24}$$

$$\begin{aligned}\{Q_\alpha, Q_\beta\} &= \lambda^2 J_{\mu\nu} (\bar{\gamma}\mathbf{G}^{\nu\mu})_{\alpha\beta} + P_\mu (\bar{\gamma}\gamma_{-1}\gamma^\mu)_{\alpha\beta} \\ &= \tfrac{1}{4}\lambda^2 J_{\mu\nu} (\bar{\gamma}[\gamma^\nu, \gamma^\mu])_{\alpha\beta} + P_\mu (\gamma\gamma^\mu)_{\alpha\beta},\end{aligned} \tag{4.4.25}$$

where

$$\gamma \stackrel{\text{def}}{=} \bar{\gamma}\gamma_{-1}, \qquad \gamma\gamma_\mu\gamma^{-1} = -\gamma_\mu^{\sim}, \qquad \gamma^{\sim} = -\gamma. \tag{4.4.26}$$

Equations (4.4.20)–(4.4.25) define a super Lie group for all values of λ. When $\lambda \neq 0$ the group is the super de Sitter group. When $\lambda = 0$ it 'contracts' to the *super Poincaré group*, with commutation relations

$$\left.\begin{aligned}
&[J_{\mu\nu}, J_{\sigma\tau}] = \mathrm{i}(\eta_{\mu\sigma} J_{\nu\tau} + \eta_{\nu\tau} J_{\mu\sigma} - \eta_{\mu\tau} J_{\nu\sigma} - \eta_{\nu\sigma} J_{\mu\tau}),\\
&[J_{\mu\nu}, P_\sigma] = \mathrm{i}(\eta_{\mu\sigma} P_\nu - \eta_{\nu\sigma} P_\mu),\\
&[J_{\mu\nu}, Q_\beta] = -\frac{\mathrm{i}}{4} Q_\alpha [\gamma_\mu, \gamma_\nu]^\alpha{}_\beta,\\
&[P_\mu, P_\nu] = 0,\\
&[P_\mu, Q_\beta] = 0,\\
&\{Q_\alpha, Q_\beta\} = P_\mu (\gamma\gamma^\mu)_{\alpha\beta}.
\end{aligned}\right\} \tag{4.4.27}$$

The super Poincaré group is not a simple group. The P_μ and Q_α together generate a proper invariant subgroup. The subgroup $\overline{\mathrm{SO}}(1,3)$, generated by the $J_{\mu\nu}$, is a group of automorphisms of this subgroup, and the full group is the semidirect product of the two subgroups. The super Poincaré group is also not semisimple, for it has the group generated by the P_μ as an invariant Abelian subgroup.

The coset space: super Poincaré group/$\overline{\mathrm{SO}}(1,3)$

In the representation generated by the $J_{\mu\nu}$, P_μ and Q_α the representative of an arbitrary element of the super Poincaré group may be expressed in the form

$$D(x, \theta, \omega) = \mathrm{e}^{-\mathrm{i}P_\mu x^\mu - \mathrm{i}\mathbf{Q}\cdot\boldsymbol{\theta}}\, \mathrm{e}^{-(\mathrm{i}/2) J_{\mu\nu}\omega^{\nu\mu}}, \tag{4.4.28}$$

where the x^μ and $\omega^{\mu\nu}$ are real *c*-numbers and $\mathbf{Q}\cdot\boldsymbol{\theta}$ denotes the matrix defined by eq. (3.2.99), the G's in that equation being the Q's and the x's being now *a*-numbers θ^α. Comparing expression (4.4.28) with eq. (3.2.63) one sees that the x^μ and θ^α label the left cosets of the subgroup $\overline{\mathrm{SO}}(1,3)$, and the $\omega^{\mu\nu}$ label the points within each coset. The space of left cosets is a supermanifold of dimension (4, 4), with coordinates x^μ, θ^α.

It is not difficult to compute the action of the super Poincaré group on the coset space. Let $\mathrm{e}^{-\mathrm{i}P_\mu \delta\xi^\mu - \mathrm{i}\mathbf{Q}\cdot\delta\boldsymbol{\xi}}\, \mathrm{e}^{-(\mathrm{i}/2) J_{\mu\nu}\delta\xi^{\nu\mu}}$ be the representative of a group element infinitesimally close to the identity. Then using eqs. (3.1.60) and (3.2.95) to compute the commutator

$$[-\mathrm{i}\mathbf{Q}\cdot\delta\boldsymbol{\xi}, -\mathrm{i}\mathbf{Q}\cdot\boldsymbol{\theta}] = -\{Q_\alpha, Q_\beta\}\theta^\beta\, \delta\xi^\alpha = -P_\mu (\gamma\gamma^\mu)_{\alpha\beta}\, \theta^\beta\, \delta\xi^\alpha, \tag{4.4.29}$$

and using the fact that this commutator commutes with both the P's and Q's, so that

$$\mathrm{e}^{-\mathrm{i}\mathbf{Q}\cdot\delta\boldsymbol{\xi}}\, \mathrm{e}^{-\mathrm{i}\mathbf{Q}\cdot\boldsymbol{\theta}} = \mathrm{e}^{-\mathrm{i}\mathbf{Q}\cdot(\boldsymbol{\theta}+\delta\boldsymbol{\xi}) - \frac{1}{2} P_\mu (\gamma\gamma^\mu)_{\alpha\beta}\, \theta^\beta\, \delta\xi^\alpha}, \tag{4.4.30}$$

one readily verifies that

$$\begin{aligned}&\mathrm{e}^{-\mathrm{i}P_\mu\delta\xi^\mu-\mathrm{i}Q\cdot\delta\xi}\,\mathrm{e}^{-(\mathrm{i}/2)J_{\mu\nu}\delta\xi^{\nu\mu}}\,\mathrm{e}^{-\mathrm{i}P_\sigma x^\sigma-\mathrm{i}Q\cdot\theta}\,\mathrm{e}^{-(\mathrm{i}/2)J_{\sigma\tau}x^{\tau\sigma}}\\&\quad=\mathrm{e}^{-\mathrm{i}P_\mu\bar{x}^\mu-\mathrm{i}Q\cdot\theta}\,\mathrm{e}^{-(\mathrm{i}/2)J_{\mu\nu}\delta\xi^{\nu\mu}}\,\mathrm{e}^{-(\mathrm{i}/2)J_{\sigma\tau}x^{\tau\sigma}},\end{aligned}\tag{4.4.31}$$

where

$$\bar{x}^\mu=x^\mu+\delta\xi^\mu-\frac{\mathrm{i}}{2}(\gamma\gamma^\mu)_{\alpha\beta}\,\theta^\beta\,\delta\xi^\alpha+\delta\xi^\mu{}_\nu\,x^\nu,\tag{4.4.32}$$

$$\bar{\theta}^\alpha=\theta^\alpha+\delta\xi^\alpha-\tfrac{1}{8}[\gamma_\mu,\gamma_\nu]^\alpha{}_\beta\,\theta^\beta\,\delta\xi^{\nu\mu}.\tag{4.4.33}$$

Killing flows and invariant connections

From the infinitesimal transformation laws (4.4.32) and (4.4.33) one immediately obtains the components of the Killing flows on the coset space:

$$\left.\begin{aligned}&{}^\mu Q_\nu=\delta^\mu{}_\nu,\quad &&{}^\mu Q_\alpha=-\frac{\mathrm{i}}{2}(\gamma\gamma^\mu)_{\alpha\beta}\,\theta^\beta,\quad &&{}^\mu Q_{\nu\sigma}=(\delta^\mu{}_\sigma\,\eta_{\nu\tau}-\delta^\mu{}_\nu\,\eta_{\sigma\tau})x^\tau,\\&{}^\alpha Q_\mu=0,&&{}^\alpha Q_\beta=\delta^\alpha{}_\beta,&&{}^\alpha Q_{\mu\nu}=-\tfrac{1}{4}[\gamma_\mu,\gamma_\nu]^\alpha{}_\beta\,\theta^\beta.\end{aligned}\right\}\tag{4.4.34}$$

It is natural to ask whether these Killing flows define canonical geometrical structures on the coset space. We note first that the coset space is reductive, i.e., that equation (3.3.35) is satisfied. This may be seen by simply displaying the nonvanishing structure constants of the super Poincare group:†

$$\begin{aligned}c^{\lambda\rho}{}_{\mu\nu\sigma\tau}=\;&\delta^\lambda{}_\tau\,\delta^\rho{}_\nu\,\eta_{\mu\sigma}-\delta^\rho{}_\tau\,\delta^\lambda{}_\nu\,\eta_{\mu\sigma}-\delta^\lambda{}_\sigma\,\delta^\rho{}_\nu\,\eta_{\mu\tau}+\delta^\rho{}_\sigma\,\delta^\lambda{}_\nu\,\eta_{\mu\tau}\\&+\delta^\lambda{}_\sigma\,\delta^\rho{}_\mu\eta_{\nu\tau}-\delta^\rho{}_\sigma\,\delta^\lambda{}_\mu\,\eta_{\nu\tau}-\delta^\lambda{}_\tau\,\delta^\rho{}_\mu\,\eta_{\nu\sigma}+\delta^\rho{}_\tau\,\delta^\lambda{}_\mu\,\eta_{\nu\sigma},\end{aligned}\tag{4.4.35}$$

$$c^\mu{}_{\sigma\tau\,\nu}=-c^\mu{}_{\nu\,\sigma\tau}=\delta^\mu{}_\tau\,\eta_{\sigma\nu}-\delta^\mu{}_\sigma\,\eta_{\tau\nu},\tag{4.4.36}$$

$$c^\alpha{}_{\mu\nu\,\beta}=-c^\alpha{}_{\beta\,\mu\nu}=-\tfrac{1}{4}[\gamma_\mu,\gamma_\nu]^\alpha{}_\beta,\tag{4.4.37}$$

$$c^\mu{}_{\alpha\beta}=-\mathrm{i}(\gamma\gamma^\mu)_{\alpha\beta}.\tag{4.4.38}$$

From this it follows that the coset space admits a one-parameter family of group-invariant connections. A quick inspection shows that eqs. (3.3.25) and (3.3.26) are satisfied in the coordinates x^μ, θ^α, the fixed point χ (under the action of $\overline{\mathrm{SO}}(1,3)$) being here the origin $x^\mu=0$, $\theta^\alpha=0$. We may therefore use eq. (3.3.40), obtaining, for the connections in question,

$$\Gamma^\mu{}_{\alpha\beta}=-\mathrm{i}\lambda(\gamma\gamma^\mu)_{\alpha\beta}\ (\lambda\text{ an arbitrary real }c\text{-number}),\tag{4.4.39}$$

with all other components vanishing.

Expression (4.4.39) holds not merely at the fixed point but throughout the coset space (in the coordinates x^μ, θ^α), as may be verified by substituting

† The indices A, B in eq. (3.3.35) here correspond to the antisymmetric index pairs $\mu\nu$, $\sigma\tau$, etc., with the indices i,j corresponding to the single indices μ, ν, α, β, etc.

it into eq. (3.3.9) and using the Killing flow components (4.4.34). It is easy to verify that the Riemann tensor field to which the connection (4.4.39) gives rise vanishes for all values of λ. The coset space therefore possesses a one-parameter family of invariant parallelisms at a distance.

Riemannian geometry of the coset space

The connection (4.4.39) is not the only group-invariant connection that the coset space admits. One can also introduce a family of Riemannian connections based on a family of group-invariant metric tensor fields. Note that eq. (3.3.31) is, in the present case, solved by

$$\begin{pmatrix} {}_\mu g_\nu & {}_\mu g_\beta \\ {}_\alpha g_\nu & {}_\alpha g_\beta \end{pmatrix}_{x=0,\theta=0} = \begin{pmatrix} \eta_{\mu\nu} & 0 \\ 0 & \mathrm{i}\lambda\gamma_{\alpha\beta} \end{pmatrix} \tag{4.4.40}$$

where λ is an arbitrary invertible real c-number. It is not difficult to verify that eq. (3.3.8) implies

$$\left.\begin{aligned} &{}_\mu g_{\nu,\sigma} = 0, && {}_\mu g_{\nu,\alpha} = 0, \\ &{}_\mu g_{\alpha,\nu} = 0, && {}_\mu g_{\alpha,\beta} = -\tfrac{\mathrm{i}}{2}\,{}_\mu g_\nu (\gamma\gamma^\nu)_{\alpha\beta}, \\ &{}_\alpha g_{\beta,\mu} = 0, && {}_\alpha g_{\beta,\gamma} = -\tfrac{\mathrm{i}}{2}\,{}_\alpha g_\mu (\gamma\gamma^\mu)_{\beta\gamma} + \tfrac{\mathrm{i}}{2}\,{}_\beta g_\mu (\gamma\gamma^\mu)_{\alpha\gamma}. \end{aligned}\right\} \tag{4.4.41}$$

The solution of these equations, taking account of the symmetry ${}_\mu g_\alpha = -\,{}_\alpha g_\mu$ and the boundary condition (4.4.40), is

$$\begin{pmatrix} {}_\mu g_\nu & {}_\mu g_\beta \\ {}_\alpha g_\nu & {}_\alpha g_\beta \end{pmatrix} = \begin{pmatrix} \eta_{\mu\nu} & -\tfrac{\mathrm{i}}{2}(\gamma\gamma_\mu)_{\beta\delta}\theta^\delta \\ \tfrac{\mathrm{i}}{2}(\gamma\gamma_\nu)_{\alpha\gamma}\theta^\gamma & \mathrm{i}\lambda\gamma_{\alpha\beta} + \tfrac{1}{4}(\gamma\gamma_\sigma)_{\alpha\gamma}\theta^\gamma(\gamma\gamma^\sigma)_{\beta\delta}\theta^\delta \end{pmatrix} \tag{4.4.42}$$

The inverse metric tensor field has the form

$$\begin{pmatrix} g^{\mu\nu} & g^{\mu\beta} \\ g^{\alpha\nu} & g^{\alpha\beta} \end{pmatrix} = \begin{pmatrix} \eta^{\mu\nu}(1 + \tfrac{\mathrm{i}}{4}\lambda^{-1}\gamma_{\gamma\delta}\theta^\delta\theta^\gamma) & -\tfrac{1}{2}\lambda^{-1}(\gamma^\mu)^\beta{}_\delta\theta^\delta \\ -\tfrac{1}{2}\lambda^{-1}(\gamma^\nu)^\alpha{}_\gamma\theta^\gamma & -\mathrm{i}\lambda^{-1}(\gamma^{-1})^{\alpha\beta} \end{pmatrix} \tag{4.4.43}$$

and from eq. (1.6.29f) we may infer that the superdeterminant of the matrix (4.4.42) is $-\lambda^{-4}$. This means that the group-invariant measure function for the coset space is simply a constant, in the coordinates x^μ, θ^α. This also follows from eq. (3.3.10) since $(-1)^{i(a+1)}\,{}^i Q_{a,i} = 0$ in the present case.

It is a straightforward computation to show that (4.4.42) gives rise to the following Riemannian connection:

$$\left.\begin{aligned} &\Gamma^\mu{}_{\nu\sigma} = 0, \qquad \Gamma^\mu{}_{\nu\alpha} = -\tfrac{\mathrm{i}}{4}\lambda^{-1}(\gamma\gamma_\nu\gamma^\mu)_{\alpha\delta}\theta^\delta, \\ &\Gamma^\mu{}_{\alpha\beta} = \tfrac{1}{8}\lambda^{-1}[(\gamma\gamma_\sigma\gamma^\mu)_{\alpha\gamma}(\gamma\gamma^\sigma)_{\beta\delta} - (\gamma\gamma_\sigma\gamma^\mu)_{\beta\gamma}(\gamma\gamma^\sigma)_{\alpha\delta}]\theta^\delta\theta^\gamma, \\ &\Gamma^\alpha{}_{\mu\nu} = 0, \qquad \Gamma^\alpha{}_{\mu\beta} = \tfrac{1}{2}\lambda^{-1}(\gamma_\mu)^\alpha{}_\beta, \\ &\Gamma^\alpha{}_{\beta\gamma} = -\tfrac{\mathrm{i}}{4}\lambda^{-1}[(\gamma_\sigma)^\alpha{}_\beta(\gamma\gamma^\sigma)_{\gamma\delta} - (\gamma_\sigma)^\alpha{}_\gamma(\gamma\gamma^\sigma)_{\beta\delta}]\theta^\delta. \end{aligned}\right\} \tag{4.4.44}$$

From this another computation yields the curvature tensor field:

$$\left.\begin{aligned}
R_{\mu\nu\sigma\tau} &= 0,\\
R_{\mu\nu\sigma\alpha} &= 0,\\
R_{\mu\nu\alpha\beta} &= -\tfrac{i}{4}\lambda^{-1}(\gamma[\gamma_\mu, \gamma_\nu])_{\alpha\beta},\\
R_{\mu\alpha\nu\beta} &= \tfrac{i}{4}\lambda^{-1}(\gamma\gamma_\nu\gamma_\mu)_{\alpha\beta},\\
R_{\mu\alpha\beta\gamma} &= \tfrac{1}{4}\lambda^{-1}[(\gamma[\gamma_\sigma, \gamma_\mu])_{\beta\gamma}(\gamma\gamma^\sigma)_{\alpha\delta} - (\gamma\gamma_\sigma\gamma_\mu)_{\alpha\gamma}(\gamma\gamma^\sigma)_{\beta\delta} - (\gamma\gamma_\sigma\gamma_\mu)_{\alpha\beta}(\gamma\gamma^\sigma)_{\gamma\delta}\\
&\quad - (\gamma\gamma_\sigma\gamma_\mu)_{\alpha\delta}(\gamma\gamma^\sigma)_{\beta\gamma} + \tfrac{1}{2}(\gamma\gamma_\sigma\gamma_\mu)_{\beta\delta}(\gamma\gamma^\sigma)_{\alpha\gamma} + \tfrac{1}{2}(\gamma\gamma_\sigma\gamma_\mu)_{\gamma\delta}(\gamma\gamma^\sigma)_{\alpha\beta}]\theta^\delta,\\
R_{\alpha\beta\gamma\delta} &= \tfrac{1}{4}[2(\gamma\gamma_\sigma)_{\alpha\beta}(\gamma\gamma^\sigma)_{\gamma\delta} - (\gamma\gamma_\sigma)_{\alpha\gamma}(\gamma\gamma^\sigma)_{\beta\delta} - (\gamma\gamma_\sigma)_{\alpha\delta}(\gamma\gamma^\sigma)_{\beta\gamma}]\\
&\quad + \tfrac{1}{32}\lambda^{-1}[(\gamma[\gamma_\sigma, \gamma_\tau])_{\alpha\beta}(\gamma\gamma^\tau)_{\gamma\varepsilon}(\gamma\gamma^\sigma)_{\delta\zeta}\\
&\qquad + (\gamma[\gamma_\sigma, \gamma_\tau])_{\gamma\delta}(\gamma\gamma^\tau)_{\alpha\varepsilon}(\gamma\gamma^\sigma)_{\beta\zeta}\\
&\qquad - (\gamma\gamma_\sigma\gamma_\tau)_{\alpha\gamma}(\gamma\gamma^\tau)_{\beta\varepsilon}(\gamma\gamma^\sigma)_{\delta\zeta}\\
&\qquad - (\gamma\gamma_\sigma\gamma_\tau)_{\beta\delta}(\gamma\gamma^\tau)_{\alpha\varepsilon}(\gamma\gamma^\sigma)_{\gamma\zeta}\\
&\qquad - (\gamma\gamma_\sigma\gamma_\tau)_{\alpha\delta}(\gamma\gamma^\tau)_{\beta\varepsilon}(\gamma\gamma^\sigma)_{\gamma\zeta}\\
&\qquad - (\gamma\gamma_\sigma\gamma_\tau)_{\beta\gamma}(\gamma\gamma^\tau)_{\alpha\varepsilon}(\gamma\gamma^\sigma)_{\delta\zeta}]\theta^\zeta\theta^\varepsilon.
\end{aligned}\right\} \tag{4.4.45}$$

The Ricci tensor and curvature scalar fields are respectively

$$\left.\begin{aligned}
R_{\mu\nu} = \lambda^{-2}\eta_{\mu\nu}, \qquad R_{\mu\alpha} = R_{\alpha\mu} = -\tfrac{i}{2}\lambda^{-2}(\gamma\gamma_\mu)_{\alpha\delta}\theta^\delta,\\
R_{\alpha\beta} = -2i\lambda^{-1}\gamma_{\alpha\beta} - \tfrac{1}{4}\lambda^{-2}(\gamma\gamma_\sigma)_{\alpha\gamma}\theta^\gamma(\gamma\gamma^\sigma)_{\beta\delta}\theta^\delta,
\end{aligned}\right\} \tag{4.4.46}$$

$$R = -4\lambda^{-2}. \tag{4.4.47}$$

The super Lorentz group

This group has body $\overline{\mathrm{SO}}(1,3)$ and extending representation spin(1, 3). It is isomorphic to the complex extension of OSp(1, 2) and has dimension (6, 4).

The generators of spin(1, 3) are

$$G_{\mu\nu} = -\tfrac{1}{4}[\gamma_\mu, \gamma_\nu], \tag{4.4.48}$$

satisfying eqs. (4.4.6), (4.4.7) and (4.4.8), with the indices restricted to range over the values 0, 1, 2, 3. With the matrix γ defined as in eqs. (4.4.26) one easily verifies that the matrices γ^{-1}, $\gamma_\mu\gamma^{-1}$, $\mathbf{G}_{\mu\nu}\gamma^{-1}$, $\gamma_{-1}\gamma_\mu\gamma^{-1}$, $\gamma_{-1}\gamma^{-1}$ form a complete linearly independent set, the ten matrices $\gamma_\mu\gamma^{-1}$, $\mathbf{G}_{\mu\nu}\gamma^{-1}$ being symmetric and the six others antisymmetric. The γ_μ will be assumed to be real.

We shall need the identities

$$\left.\begin{aligned}\gamma_\mu\gamma^\mu&=4\cdot\mathbf{1},\\ \gamma_\mu\gamma_\nu\gamma^\mu&=-2\gamma_\nu,\\ \gamma_\mu\mathbf{G}_{\nu\sigma}\gamma^\mu&=0,\\ \mathbf{G}_{\mu\nu}\gamma_\sigma\mathbf{G}^{\nu\mu}&=0,\\ \mathbf{G}_{\mu\nu}\mathbf{G}_{\sigma\tau}\mathbf{G}^{\nu\mu}&=-\mathbf{G}_{\sigma\tau},\end{aligned}\right\}\tag{4.4.49}$$

as well as

$$\mathrm{tr}(\mathbf{G}_{\mu\nu}\mathbf{G}_{\sigma\tau})=-\eta_{\mu\sigma}\eta_{\nu\tau}+\eta_{\mu\tau}\eta_{\nu\sigma}.\tag{4.4.50}$$

Indices are now raised and lowered by means of the matrices $(\eta^{\mu\nu})$ and $(\eta_{\mu\nu})$.

The generators of the extending representation have the matrix elements

$$c^\alpha{}_{\mu\nu\,\beta}=-c^\alpha{}_{\beta\,\mu\nu}=(\mathbf{G}_{\mu\nu})^\alpha{}_\beta=-\tfrac{1}{4}[\gamma_\mu,\gamma_\nu]^\alpha{}_\beta.\tag{4.4.51}$$

The matrices $\boldsymbol{\eta}_{\mathrm{S}}$ and $\boldsymbol{\eta}_{\mathrm{A}}$ of section 4.1 are given by

$$\eta_{\mu\nu\,\sigma\tau}=\eta_{\mu\tau}\eta_{\nu\sigma}-\eta_{\mu\sigma}\eta_{\nu\tau},\tag{4.4.52}$$

$$\eta_{\alpha\beta}=\mathrm{i}\gamma_{\alpha\beta},\tag{4.4.53}$$

yielding

$$c^{\mu\nu}{}_{\alpha\,\beta}=-\tfrac{1}{2}\eta^{\mu\nu\,\sigma\tau}\eta_{\alpha\gamma}c^\gamma{}_{\tau\sigma\,\beta}=-\mathrm{i}(\gamma\mathbf{G}^{\mu\nu})_{\alpha\beta}.\tag{4.4.54}$$

The verification that these structure constants satisfy the identity (4.1.17) rests on the verification of the identity

$$\mathbf{G}_{\mu\nu}\,\mathrm{tr}(S\gamma\mathbf{G}^{\nu\mu})+2\mathbf{G}_{\mu\nu}S\gamma\mathbf{G}^{\nu\mu}=0,\tag{4.4.55}$$

where S is any symmetric 4×4 matrix. Since a symmetric matrix is a linear combination of $\gamma_\mu\gamma^{-1}$ and $\mathbf{G}_{\mu\nu}\gamma^{-1}$ it suffices to check (4.4.55) for these matrices. The verification is straightforward upon application of the identities (4.4.5), (4.4.49) and (4.4.50).

It is of interest to display a few of the lowest-dimensional representations of the super Lorentz group. We begin with the adjoint representation, which is carried by a *supermultiplet* of the form $(X^{\mu\nu}, X^\alpha)$ where the $X^{\mu\nu}$ and X^α are the components of an antisymmetric tensor and a spinor respectively. The infinitesimal group transformation law for this supermultiplet is

$$\left.\begin{aligned}\delta X^{\mu\nu}&=\delta\xi^\mu{}_\sigma X^{\sigma\nu}+\delta\xi^\nu{}_\sigma X^{\mu\sigma}-\mathrm{i}(\gamma\mathbf{G}^{\mu\nu})_{\alpha\beta}X^\beta\delta\xi^\alpha,\\ \delta X^\alpha&=\tfrac{1}{2}(\mathbf{G}_{\mu\nu})^\alpha{}_\beta X^\beta\delta\xi^{\nu\mu}-\tfrac{1}{2}X^{\nu\mu}(\mathbf{G}_{\mu\nu})^\alpha{}_\beta\delta\xi^\beta,\end{aligned}\right\}\tag{4.4.56}$$

where $\delta\xi^{\mu\nu}$, $\delta\xi^\alpha$ are the infinitesimal group parameters.

Another representation is carried by the supermultiplet (X^μ, X^α, X) where X^μ and X^α are the components of a vector and a spinor respectively and X is a scalar. The transformation law is

$$\left.\begin{aligned}
\delta X^\mu &= \delta\xi^\mu{}_\nu X^\nu + \frac{\mathrm{i}}{2} X^\alpha(\gamma\gamma_{-1}\gamma^\mu)_{\alpha\beta}\,\delta\xi^\beta,\\
\delta X^\alpha &= \tfrac{1}{2}(\mathbf{G}_{\mu\nu})^\alpha{}_\beta X^\beta\,\delta\xi^{\nu\mu} + \tfrac{1}{2}(\gamma_{-1}\gamma_\mu)^\alpha{}_\beta X^\mu\,\delta\xi^\beta + \frac{1}{\sqrt{2}} X\,\delta\xi^\alpha,\\
\delta X &= -\frac{\mathrm{i}}{\sqrt{2}} X^\alpha\gamma_{\alpha\beta}\,\delta\xi^\beta.
\end{aligned}\right| \tag{4.4.57}$$

If (X^μ, X^α, X) and (Y^μ, Y^α, Y) are two supermultiplets of this type then it is straightforward to verify that the following bilinear form is group invariant:

$$X^\mu\eta_{\mu\nu}Y^\nu - \mathrm{i}X^\alpha\gamma_{\alpha\beta}Y^\beta - XY. \tag{4.4.58}$$

Moreover, if $(Z^{\mu\nu}, Z^\alpha)$ is a supermultiplet transforming under the adjoint representation then the following *tri*linear form is invariant:

$$\begin{aligned}
&X^\mu\eta_{\mu\nu}Z^{\nu\sigma}\eta_{\sigma\tau}Y^\tau - \tfrac{1}{2}X^\alpha(\gamma\mathbf{G}_{\mu\nu})_{\alpha\beta}Z^{\nu\mu}Y^\beta\\
&+\tfrac{1}{2}X^\mu Z^\alpha(\gamma\gamma_{-1}\gamma_\mu)_{\alpha\beta}Y^\beta - \tfrac{1}{2}X^\alpha(\gamma\gamma_{-1}\gamma_\mu)_{\alpha\beta}Z^\beta Y^\mu\\
&+\frac{\mathrm{i}}{\sqrt{2}}XZ^\alpha\gamma_{\alpha\beta}Y^\beta - \frac{\mathrm{i}}{\sqrt{2}}X^\alpha\gamma_{\alpha\beta}Z^\beta Y.
\end{aligned} \tag{4.4.59}$$

Still another representation is carried by the supermultiplet (X^α, X_1, X_2) where the X^α are the components of a spinor and X_1 and X_2 are scalars. The transformation law is

$$\left.\begin{aligned}
\delta X^\alpha &= \tfrac{1}{2}(\mathbf{G}_{\mu\nu})^\alpha{}_\beta X^\beta\,\delta\xi^{\nu\mu} + \tfrac{1}{2}X_1\,\delta\xi^\alpha - \tfrac{1}{2}X_2(\gamma_{-1})^\alpha{}_\beta\,\delta\xi^\beta,\\
\delta X_1 &= -\tfrac{1}{2}X^\alpha\gamma_{\alpha\beta}\,\delta\xi^\beta,\\
\delta X_2 &= -\tfrac{1}{2}X^\alpha(\gamma\gamma_{-1})_{\alpha\beta}\,\delta\xi^\beta,
\end{aligned}\right\} \tag{4.4.60}$$

with a bilinear invariant of the form

$$-\mathrm{i}X^\alpha\gamma_{\alpha\beta}Y^\beta - X_1Y_1 + X_2Y_2. \tag{4.4.61}$$

The dimensions of the above representations are (6, 4), (5, 4) and (2, 4) respectively.

4.5 The Cartan super Lie groups

The diffeomorphism group Diff(M)

Let M be an arbitrary connected supermanifold. Denote by Diff(M) the set of all diffeomorphisms $\phi : M \to M$ such that the closure of $\{p \in M | \phi(p) \neq p\}$ is compact. Diff(M) forms a group in an obvious manner. Let the dimension of M be (m, n). If $n \neq 0$ then Diff(M) is a super Lie group; if $m \neq 0$ it is infinite dimensional.

We have seen in section 2.5 that every diffeomorphism defines a derivative mapping of tensor fields, as well as of measure functions and connections (see exercises **2.3** and **2.4**). The super Lie algebra of Diff(M) is defined by the infinitesimal diffeomorphisms. Since each infinitesimal diffeomorphism is an infinitesimal dragging, and since infinitesimal draggings define the Lie derivative, it is not difficult to see that the super Lie algebra of Diff(M) is the set $\mathfrak{X}(M)$ of all vector fields on M, the bracket of the algebra being the super Lie bracket.

Every tensor field (or measure function or connection) on M carries a representation of Diff(M) or, equivalently, of the super Lie algebra of Diff(M). The action of each element of Diff(M) is a derivative mapping and the action of each element of the super Lie algebra is a Lie derivative. For each tensor field (or measure function or connection) the operators $\mathfrak{L}_{\mathbf{X}}$($\mathbf{X}$ c-type) are the generators of the representation, and eq. (3.2.95) takes the form (2.5.28).

The group SDiff(M, μ)

This is the subgroup of Diff(M) consisting of all diffeomorphisms that leave a chosen set of compatible measure functions μ (one in each chart) invariant. It is sometimes called the group of volume-preserving diffeomorphisms. It is generated by the set of vector fields $\mathbf{X}$ satisfying

$$0 = \mathfrak{L}_{\mathbf{X}}\mu = (-1)^i (\mu X^i)_{,i}. \tag{4.5.1}$$

Equation (4.5.1) has the general solution

$$X^i = (-1)^j \mu^{-1} (\mu \mathrm{A}^{ij})_{,j}, \tag{4.5.2}$$

where $\mathbf{A}$ is an arbitrary antisupersymmetric contravariant rank-2 tensor field.

If M can be covered by a single simply connected chart and if M has at least one c-type dimension, then a global coordinate system can be found in which the measure function is everywhere constant. The transformation to such a coordinate system can be induced by a diffeomorphism ϕ in which the coordinates themselves are regarded as dragged. This diffeomorphism is an automorphism connecting SDiff(M, μ) with SDiff(M,const). These two subgroups of Diff(M) are therefore isomorphic in this case, although they are *not* isomorphic in general. Note that SDiff(M,const) is generated by the set of vector fields $\mathbf{X}$ satisfying $(-1)^i X^i{}_{,i} = 0$.

The canonical transformation group Can(M, ω)

This is the subgroup of Diff(M) consisting of all diffeomorphisms that leave invariant a chosen real c-type differentiable 2-form $\boldsymbol{\omega}$ having the properties

$$\mathbf{d}\boldsymbol{\omega} = 0, \tag{4.5.3}$$

$$\text{sdet}({}_i\omega_j) \neq 0 \quad \text{in every chart.} \tag{4.5.4}$$

Equation (4.5.4) can hold only if the number m of c-type dimensions of M is even. In terms of coordinate components eq. (4.5.3) takes the form

$$\omega_{ij,k} + (-1)^{i(j+k)} \omega_{jk,i} + (-1)^{k(i+j)} \omega_{ki,j} = 0. \tag{4.5.5}$$

Can(M, ω) is generated by the set of vector fields $\mathbf{X}$ satisfying

$$\mathfrak{L}_{\mathbf{X}} \boldsymbol{\omega} = 0. \tag{4.5.6}$$

In terms of coordinate components this becomes

$$\begin{aligned} 0 &= (\boldsymbol{\omega} \mathfrak{L}_{\mathbf{X}})_{ij} \\ &= (-1)^{j(i+k+\mathbf{X})} \omega_{kj}{}^k X_{,i} + (-1)^{i\mathbf{X}} \omega_{ik}{}^k X_{,j} + (-1)^{\mathbf{X}(i+j)} \omega_{ij,k}{}^k X \\ &= (-1)^{i\mathbf{X}} (\omega_{ik}{}^k X)_{,j} - (-1)^{j(i+\mathbf{X})} (\omega_{jk}{}^k X)_{,i}, \end{aligned} \tag{4.5.7}$$

in which eq. (4.5.5) has been used in passing to the last line. If M is simply connected eq. (4.5.7) implies the existence of a scalar field X satisfying

$$(-1)^{i\mathbf{X}} \omega_{ik}{}^k X = X_{,i} \quad \text{or} \quad {}_i\omega_k{}^k X = {}_{i,}X \tag{4.5.8}$$

and hence

$${}^i X = \omega^{ij}{}_{j,}X \tag{4.5.9}$$

where (ω^{ij}) is the matrix inverse to $({}_i\omega_j)$:

$$\omega^{ik}{}_k\omega_j = \delta^i{}_j. \tag{4.5.10}$$

If M is not simply connected eqs. (4.5.8) and (4.5.9) can still be used provided we allow for the possibility of X being multi-valued. In this way the elements of the super Lie algebra may be regarded as scalar fields, i.e., as elements of $\mathfrak{F}(M)$, rather than as vector fields.

When the super Lie algebra of $\mathrm{Can}(M, \omega)$ is regarded as the supervector space $\mathfrak{F}(M)$ it is important to determine the bracket relation in $\mathfrak{F}(M)$. Writing

$$ {}^iX = \omega^{ij}{}_{j,}X, \qquad {}^iY = \omega^{ij}{}_{j,}Y, \tag{4.5.11} $$

we have

$$ \begin{aligned} {}^i[X, Y] &= (\omega^{ij}{}_{j,}X)_{,k}\,\omega^{kl}{}_{l,}Y - (-1)^{XY}(\omega^{ij}{}_{j,}Y)_{,l}\,\omega^{lk}{}_{k,}X \\ &= \omega^{ij}{}_{j,}(X_{,k}\,\omega^{kl}{}_{l,}Y) - (-1)^{j(\mathbf{X}+l+1)}\omega^{ij}X_{,k}\,\omega^{kl}{}_{,jl,}Y \\ &\quad - (-1)^{j(\mathbf{X}+l)}\omega^{ij}X_{,k}\,\omega^{kl}{}_{jl,}Y + (-1)^{k(\mathbf{X}+j)}\omega^{ij}{}_{,k\,j,}X\,\omega^{kl}{}_{l,}Y \\ &\quad - (-1)^{\mathbf{XY}+l(Y+j)}\omega^{ij}{}_{,l\,j,}Y\,\omega^{lk}{}_{k,}X - (-1)^{\mathbf{XY}}\omega^{ij}{}_{j,}Y_{,l}\,\omega^{lk}{}_{k,}X. \end{aligned} \tag{4.5.12} $$

By relabeling dummy indices, rearranging the order of factors and indices, and making use of the following corollary of (4.5.5),

$$ \omega^{ij}{}_{,l}\,\omega^{lk} + (-1)^{i(j+k)}\omega^{jk}{}_{,l}\,\omega^{li} + (-1)^{k(i+j)}\omega^{ki}{}_{,l}\,\omega^{lj} = 0, \tag{4.5.13} $$

one may show that all the terms but the first cancel on the right-hand side of eq. (4.5.12). Therefore

$$ {}^i[\mathbf{X}, \mathbf{Y}] = \omega^{ij}{}_{j,}[X, Y], \tag{4.5.14} $$

where

$$ [X, Y] = X_{,i}\,\omega^{ij}{}_{j,}Y. \tag{4.5.15} $$

Equation (4.5.15) defines the bracket in $\mathfrak{F}(M)$.

Suppose M can be covered by a single simply connected chart. Let x^i be the coordinates in that chart and let ${}_i\omega_j(p)$ be the components of ω in these coordinates at a fixed point p. Introduce new global coordinates $\bar{x}^i$ that satisfy the differential equations

$$ {}_{i,}\bar{x}^k{}_k\omega_l(p)\,\bar{x}^l{}_{,j} = {}_i\omega_j \tag{4.5.16} $$

together with the boundary conditions

$$ (\bar{x}^i)_p = 0, \qquad (\bar{x}^i{}_{,j})_p = \delta^i{}_j. \tag{4.5.17} $$

It is not difficult to see that (4.5.5) is the integrability condition for (4.5.16) and hence these equations have a solution. Moreover, since the matrix $({}_i\omega_j)$ is everywhere nonsingular the Jacobian matrix $(\bar{x}^i{}_{,j})$ must be nonsingular, so the solution will be global. In the new coordinates the components of $\boldsymbol{\omega}$ are everywhere constant and equal to ${}_i\omega_j(p)$. This result is known as *Darboux's theorem.*

From a coordinate system in which the ${}_i\omega_j$ are constant it is but a step to pass, by a linear transformation, to a coordinate system in which the

${}_i\omega_j$ take the canonical form

$$({}_i\omega_j) = \zeta \stackrel{\text{def}}{=} \operatorname{diag}\Bigg(\overbrace{\begin{pmatrix} 0 & 1 \\ -1 & 0 \end{pmatrix}, \ldots, \begin{pmatrix} 0 & 1 \\ -1 & 0 \end{pmatrix}}^{m/2 \text{ blocks}}, \overbrace{-\mathrm{i}, \ldots, -\mathrm{i}}^{r\ -\mathrm{i}\text{'s}}, \overbrace{\mathrm{i}, \ldots, \mathrm{i}}^{n-r\ \mathrm{i}\text{'s}}\Bigg). \tag{4.5.18}$$

The proof of this is identical with the proof of the existence of the canonical form (2.8.10) for metric tensors, ω having exactly the opposite symmetry to that of a metric tensor.

The group of contact transformations

Suppose the 2-form ω is the exterior derivative of a 1-form σ:

$$\omega = \mathrm{d}\sigma. \tag{4.5.19}$$

(This will necessarily be the case if, for example, M is diffeomorphic to $\mathbf{R}_c^m \times \mathbf{R}_a^n$.) Introduce the $(m+1, n)$ dimensional supermanifold

$$\bar{M} \stackrel{\text{def}}{=} \mathbf{R}_c \times M, \tag{4.5.20}$$

with points $\bar{p}$ and charts $(\mathbf{R}_c \times \mathscr{U}, \bar{\phi})$, where $(\mathscr{U}, \phi)$ are charts in M and

$$\bar{\phi}(\bar{p}) \stackrel{\text{def}}{=} (x^0, \phi(\pi(\bar{p}))) = (x^0, x^i), \tag{4.5.21}$$

π being the natural projection of $\bar{M}$ onto M, x^0 being a point of $\mathbf{R}_c$, and the x^i being the coordinates of $\pi(\bar{p})$ in $(\mathscr{U}, \phi)$. Introduce in each $(\mathbf{R}_c \times \mathscr{U}, \bar{\phi})$ the following 1-form:

$$\bar{\sigma} \stackrel{\text{def}}{=} \mathrm{d}x^0 + \sigma_i \mathrm{d}x^i, \tag{4.5.22}$$

the σ_i being the components of σ in $(\mathscr{U}, \phi)$. $\bar{\sigma}$ is then a 1-form in $\bar{M}$.

The group of contact transformations is the set of diffeomorphisms $\bar{M} \to \bar{M}$ that induce changes in $\bar{\sigma}$ (by the derivative mapping) of the form

$$\bar{\sigma} \to \bar{\varphi}\bar{\sigma}, \qquad \bar{\varphi} \in \mathfrak{F}(\bar{M}). \tag{4.5.23}$$

The group of contact transformations is generated by the set of contravariant vector fields on $\bar{M}$ that satisfy

$$\mathfrak{L}_{\bar{\mathbf{X}}}\bar{\sigma} = \bar{\psi}\bar{\sigma}, \quad \text{for some } \bar{\psi} \text{ in } \mathfrak{F}(\bar{M}). \tag{4.5.24}$$

Denote by $\Phi(\bar{M})$ the set of all 1-forms in $\bar{M}$ of the form $\bar{\varphi}\bar{\sigma}$ with $\bar{\varphi}$ in $\mathfrak{F}(\bar{M})$. It is easy to see that the group of contact transformations maps $\Phi(\bar{M})$ transitively into itself.

The general solution of eq. (4.5.24) is

$$\left.\begin{aligned} \bar{X}^0 &= \bar{X} - \sigma_i \omega^{ij}{}_{j,}\bar{X}, \\ \bar{X}^i &= \omega^{ij}(-{}_j\sigma\,\bar{X}_{,0} + {}_{j,}\bar{X}), \end{aligned}\right\} \tag{4.5.25}$$

where $\bar{X}$ is an arbitrary element of $\mathfrak{F}(\bar{M})$ and ω^{ij} is the matrix of eq. (4.5.10). We shall not prove this but merely verify that (4.5.25) does indeed satisfy (4.5.24). It suffices to verify this for c-type $\bar{X}$. Using the fact that $\bar{\sigma}_0 = 1$, $\bar{\sigma}_{i,0} = 0$ and $\sigma_{i,j} - (-1)^{ij}\sigma_{j,i} = -\omega_{ij}$ (see eq. (2.6.21)), together with eq. (4.5.10) and the antisupersymmetry of ω^{ij}, we find

$$\begin{aligned} (\mathfrak{L}_{\bar{X}}\bar{\sigma})_0 &= (\bar{\sigma}\mathfrak{L}_{\mathbf{X}})_0 = \bar{X}^0{}_{,0} + \sigma_i \bar{X}^i{}_{,0} \\ &= \bar{X}_{,0} - \sigma_i \omega^{ij}{}_{j,}\bar{X}_{,0} + \sigma_i \omega^{ij}(-{}_j\sigma\,\bar{X}_{,0} + {}_{j,}\bar{X}_{,0}) \\ &= \bar{X}_{,0}, \end{aligned} \tag{4.5.26}$$

$$\begin{aligned} (\mathfrak{L}_{\bar{X}}\bar{\sigma})_i &= (\bar{\sigma}\mathfrak{L}_{\mathbf{X}})_i = \sigma_{i,j}\,{}^j\bar{X} + {}^0\bar{X}_{,i} + \sigma_j\,{}^j\bar{X}_{,i} \\ &= \sigma_{i,j}\omega^{jk}(-{}_k\sigma\,\bar{X}_{,0} + {}_{k,}\bar{X}) + \bar{X}_{,i} - (-1)^{ij}\sigma_{j,i}\omega^{jk}{}_{k,}\bar{X} \\ &\quad - (-1)^{ik}\sigma_j\omega^{jk}{}_{,ik,}\bar{X} - \sigma_j\omega^{jk}{}_{k,}\bar{X}_{,i} \\ &\quad + (-1)^{ik}\sigma_j\omega^{jk}{}_{,i}(-{}_k\sigma\,\bar{X}_{,0} + {}_{k,}\bar{X}) \\ &\quad + \sigma_j\omega^{jk}(-{}_k\sigma_{,i}\,\bar{X}_{,0} + {}_{k,}X_{,i}) \\ &= \sigma_i\,\bar{X}_{,0}, \end{aligned} \tag{4.5.27}$$

whence

$$\mathfrak{L}_{\bar{X}}\bar{\sigma} = \bar{X}_{,0}\,\bar{\sigma}.$$

The solution (4.5.25) allows $\mathfrak{F}(\bar{M})$ to be regarded as the super Lie algebra of the contact transformation group. It is a straightforward but tedious calculation to show that the bracket operation in $\mathfrak{F}(\bar{M})$ is given by

$$[\bar{X}, \bar{Y}] = \bar{X}_{,0}(\bar{Y} - \sigma_i\omega^{ij}{}_{j,}\bar{Y}) - (-1)^{\tilde{X}\tilde{Y}}\,\bar{Y}_{,0}(\bar{X} - \sigma_i\omega^{ij}{}_{j,}\bar{X}) + \bar{X}_{,i}\omega^{ij}{}_{j,}\bar{Y}. \tag{4.5.28}$$

The case $m = 0$

When the number m of c-type dimensions of the supermanifold M is different from zero, all of the above groups are inifinite dimensional. When m vanishes, however, all except the group of contact transformations become finite dimensional super Lie groups. Among these are to be found all the remaining simple super Lie groups that have not already been described in this chapter. We briefly examine each of them, using names that have been assigned by Kac.

The group W(n)

This is the diffeomorphism group for supermanifolds of dimension $(0, n)$. All $(0, n)$-dimensional supermanifolds (for fixed n) are diffeomorphic to one another, and their coordinate systems, being a-number, are global. All diffeomorphisms are equivalent to polynomial coordinate transformations of degree n (or less) with nonsingular super Jacobians. Let $x^\alpha \to \bar{x}^\alpha$ denote such a coordinate transformation. Then each $\bar{x}^\alpha$ may be expanded as a polynomial in the x^α, which terminates at the term of degree n. There will be 2^{n-1} c-type coefficients and 2^{n-1} a-type coefficients in each series. These coefficients may be regarded as the *group* coordinates. Evidently, as a super Lie group, W(n) has dimension $(n \cdot 2^{n-1}, n \cdot 2^{n-1})$. The group W(1), of dimension (1, 1), has an invariant Abelian subgroup of dimension (0, 1) and is therefore not simple. All the other W(n)'s, however, are both conventional and simple. W(2), of dimension (4, 4), may be shown to be isomorphic to SL(2, 1). (See exercise **4.11**.)

The groups S(n) *and* $\tilde{\text{S}}$(n)

These are groups of volume-preserving diffeomorphisms. We have remarked earlier that if a manifold with global coordinates has at least one c-type dimension then, for every measure function μ, a global coordinate system can be introduced in which μ is everywhere constant. This is no longer true when all the coordinates are a-type. Each measure function then falls into one of two classes, those whose polynomial expansions in terms of the coordinates contain nonvanishing terms of nth degree, and those whose expansions don't. The two classes are separately invariant under diffeomorphisms. Indeed, a diffeomorphism leaves the term of nth degree unaffected. This may be seen by noting that an infinitesimal diffeomorphism, corresponding to dragging through an infinitesimal vector $\delta\xi$, changes the functional form of the measure function by an amount

$$\delta\mu = -\mathfrak{L}_{\delta\xi}\mu = (\mu\delta\xi^\alpha)_{,\alpha}. \tag{4.5.29}$$

Since every term in the implicit sum on the right is a derivative, none contains a term of nth degree.

It is not difficult to show that every measure function can be brought either into the form $\mu = 1$ or into the form $\mu = 1 + [\mathrm{i}^{n(n-1)/2}/n!]\beta\varepsilon_{\alpha_1 \ldots \alpha_n} x^{\alpha_n} \ldots x^{\alpha_1}$, where β is a nonvanishing real c-number or a nonvanishing imaginary a-number according as n is even or odd. The group S(n) is the group of diffeomorphisms that leave the first form invariant; $\tilde{\text{S}}$(n) is the group that leaves the second invariant. S(n) is a conventional super Lie group for all n. However, $\tilde{S}(n)$ is unconventional when $\beta_B = 0$ (e.g., when n is odd), for a bodiless β cannot be removed from the group multiplication law. When n is even and β_B is nonvanishing $\tilde{S}(n)$ is conventional, for a rescaling of the coordinates can then be carried out which brings μ into the form $\mu = \text{const} \times \{1 \pm [i^{n(n-1)/2}/n!]\varepsilon_{\alpha_1 \ldots \alpha_n} x^{\alpha_n} \ldots x^{\alpha_1}\}$. Since the 'const' in front can be omitted without loss of generality, the group multiplication law involves only ordinary numbers as coefficients. The two possibilities for $\tilde{S}(n)$, corresponding to the $\pm$ signs, have identical complex extensions. We shall choose the $+$ sign here.

The dimensionality of S(n) and $\tilde{S}(n)$ is easily determined by referring to eq. (4.5.1), which here takes the form

$$0 = \mathfrak{L}_X \mu = -(\mu X^\alpha)_{,\alpha}. \tag{4.5.30}$$

Remembering that the right-hand side of this equation contains no term of nth degree in the x's, we see that when n is even eq. (4.5.30) imposes $2^{n-1} - 1$ c-type conditions and 2^{n-1} a-type conditions, these numbers being reversed when n is odd. Therefore when n is even S(n) and $\tilde{S}(n)$ each have dimension $((n-1)2^{n-1} + 1, (n-1)2^{n-1})$, whereas when n is odd they have dimension $((n-1)2^{n-1}, (n-1)2^{n-1} + 1)$.

S(1) is the Abelian group of dimension (0, 1). $\tilde{S}(1)$ is the (0, 1)-dimensional non-Abelian unconventional super Lie group described in section 4.1. S(2), which has dimension (3, 2), has the Abelian group of dimension (0, 2) as an invariant subgroup and is therefore not simple. However $\tilde{S}(2)$, which has the same dimension, can be shown to be isomorphic to OSp(1, 2), and therefore is simple. (See exercise **4.13**.) All the remaining S(n)'s and $\tilde{S}(n)$'s are simple. S(3), which has dimension (8, 9) can be shown to be isomorphic to P(2). (See exercise **4.14**.)

The groups $\tilde{H}(n)$ *and* H(n)

$\tilde{H}(n)$ is the canonical transformation group on a supermanifold of dimension (0, n). Since the coordinates on such a supermanifold are always global no generality is lost if we restrict our attention to coordinate systems in which the 2-form ω takes the canonical form (4.5.18). We shall consider the case

$$\zeta = i1_n. \tag{4.5.31}$$

The more general case $\zeta = \text{diag}(-i, \ldots, -i, i, \ldots, i)$ leads to groups having complex extensions identical with those of this case.

The super Lie algebra of $\tilde{H}(n)$ consists of all scalar functions X, Y on the $(0, n)$-dimensional supermanifold, the bracket operation being

$$[X, Y] = \mathrm{i}X_{,\alpha\,\alpha,}Y, \tag{4.5.32}$$

with implicit summation over the indices α understood. It will be observed that each term in the sum on the right, being a product of derivatives with respect to one of the coordinates, is independent of that coordinate. If the bracket is expanded as a polynomial in the coordinates, the term of nth degree therefore vanishes. This means that $\tilde{\mathrm{H}}(n)$ is not a simple group. The set of all real c-type scalar functions having no terms of nth degree in their polynomial expansions generates a proper invariant subgroup of $\tilde{\mathrm{H}}(n)$.

Since only derivatives appear in (4.5.32) the terms of zeroth order in the expansions of X and Y are irrelevant, and the corresponding invariant subalgebra may in fact be restricted to scalar functions having *neither* zeroth nor nth order terms in their expansions. These functions generate the group $\mathrm{H}(n)$, the bracket operation being now

$$[X, Y] = \mathrm{i}X_{,\alpha\,\alpha,}Y - \mathrm{i}(X_{,\alpha\,\alpha,}Y)_0, \tag{4.5.33}$$

where $(\quad)_0$ denotes the zeroth order term in the polynomial expansion.

The dimensionality of $\mathrm{H}(n)$ is obtained by counting the number of c-type and a-type coefficients in the polynomial expansion of a real c-type scalar function having no terms of zeroth or nth degree. If n is even then $\mathrm{H}(n)$ has dimension $(2^{n-1} - 2,\ 2^{n-1})$; if n is odd its dimension is $(2^{n-1} - 1,\ 2^{n-1} - 1)$.

$\mathrm{H}(1)$ is the trivial group (dimension $(0,0)$). $\mathrm{H}(2)$ is the Abelian group of dimension $(0,2)$. $\mathrm{H}(3)$, which is of dimension $(3,3)$, has the Abelian group of dimension $(0,3)$ as an invariant subgroup and is therefore not simple. All the other $\mathrm{H}(n)$'s are simple. $\mathrm{H}(4)$, which has dimension $(6,8)$, can be shown to have the same complex extension as $\mathrm{SL}(2,2)/\mathrm{GL}(1,0)$ or $\mathrm{A}(1,1)$.

Exercises

4.1 Construct all super Lie algebras (unconventional as well as conventional) of dimensions $(1,1)$ and $(0,2)$.

4.2 Show that $\mathrm{SL}(m,n)$ is isomorphic to $\mathrm{SL}(n,m)$.

4.3 Show that the super Lie algebras $\mathrm{SL}(1,2)$ and $\mathrm{OSp}(2,2)$ (or $\mathrm{A}(0,1)$ and $\mathrm{C}(2)$) have isomorphic complex extensions.

4.4 Show that the super Lie algebras OSp(4, 2) and D(2, 1, α) have isomorphic complex extensions in the limit $\alpha \to 1$.

4.5 Prove the identities (4.3.28)–(4.3.36) as well as (4.4.5), (4.4.6), (4.4.49) and (4.4.50).

4.6 Verify eqs. (4.4.42)–(4.4.47)

4.7 Verify the invariance of the forms (4.4.58), (4.4.59) and (4.4.61).

4.8 Verify the decomposition laws (4.4.62) and determine the transformation laws for the representation (8, 12). Find a bilinear invariant for this representation, as well as for the representation (6,4). Find as many trilinear invariants as you can, involving the representations (2, 4), (5, 4), (6, 4) and (8, 12).

4.9 Prove that if a supermanifold M bearing a measure function μ can be covered by a single simply connected chart and if M has at least one c-type dimension, then a global coordinate system can be found in which μ is everywhere constant.

4.10 Show that W(1) has an invariant Abelian subgroup of dimension (0, 1).

4.11 Show that W(2) is isomorphic to SL(2, 1). *Hint*: Expand each c-type real vector field (i.e., generator of W(2)) in the form $X^\alpha = \varepsilon^{\alpha\beta} D_\beta + A^\alpha{}_\beta x^\beta + \frac{i}{2} C^\alpha \varepsilon_{\beta\gamma} x^\beta x^\gamma$ and show that the super Lie bracket of pairs of such vector fields is isomorphic to the commutator of pairs of (2, 1) × (2, 1) matrices of the form

$$\begin{pmatrix} A^\alpha{}_\beta & iC^\alpha \\ D_\beta & A^\gamma{}_\gamma \end{pmatrix}.$$

4.12 Show that S(1) is the Abelian group of dimension (0, 1) and that S(2) has the Abelian group of dimension (0, 2) as an invariant subgroup. Show that $\tilde{\text{S}}(1)$ is the non-Abelian unconventional super Lie group of dimension (0, 1) described in section 4.1.

4.13 Show that $\tilde{\text{S}}(2)$ is isomorphic to OSp(1, 2), having body SL(2) (or equivalently Sp(2)) and extending representation sl(2) (or equivalently sp(2)). *Hint*: Show that the generators of $\tilde{\text{S}}(2)$ may be expressed in the form $X^\alpha = D^\alpha + B^\alpha{}_\beta x^\beta + \frac{i}{2} D^\alpha \varepsilon_{\beta\gamma} x^\beta x^\gamma$ with $B^\alpha{}_\alpha = 0$. and show that the super Lie bracket of pairs of such generators is isomorphic to the commutator of pairs of $(1,2)\times(1,2)$ matrices of the form

$$\begin{pmatrix} 0 & \mathrm{i}D^{\sim}\varepsilon \\ D & B \end{pmatrix}, \qquad \operatorname{tr} B = 0,$$

where ε is the matrix $(\varepsilon_{\alpha\beta})$.

4.14 Show that S(3) is isomorphic to P(2). *Hint*: Show that the generators of S(3) may be expressed in the form $X^{\alpha} = \frac{1}{2}\varepsilon^{\alpha\beta\gamma} D_{\beta\gamma} + A^{\alpha}{}_{\beta} x^{\beta} + \frac{\mathrm{i}}{2} C^{\alpha\beta}\varepsilon_{\beta\gamma\delta} x^{\delta} x^{\gamma}$, where $\operatorname{tr} A = 0, C^{\sim} = C$ and $D^{\sim} = -D$. Then show that the super Lie bracket of pairs of such generators is isomorphic to the commutator of pairs of $(3,3)\times(3,3)$ matrices of the form

$$\begin{pmatrix} A & \mathrm{i}C \\ D & -A^{\sim} \end{pmatrix}.$$

4.15 Show that H(2) is the Abelian group of dimension (0, 2). Show that H(3) has the Abelian group of dimension (0, 3) as an invariant subgroup.

4.16 Show that P(1) has an invariant subgroup of dimension (3, 3). (*Hint*: Show that the last line of eqs. (4.2.63) vanishes when $m = 1$.) Show that this subgroup has the same complex extension as H(3).

4.17 Show that Q(1) has the same complex extension as H(3) and is therefore not simple. *Hint*: The bracket relation for Q(m), which replaces eqs. (4.2.70) for $\tilde{\mathrm{Q}}(m)$, is

$$A_{12} = [A_1, A_2] + \mathrm{i}[D_1, D_2] - \frac{2\mathrm{i}}{m+1}\operatorname{tr}(D_1 D_2)\mathbf{1}_{m+1},$$
$$D_{12} = [A_1, D_2] + [D_1, A_2].$$

Show that when $m = 1$ the terms involving D_1 and D_2 in the first of these equations cancel.

Comments on chapter 4

The classification problem for the simple conventional super Lie groups, although solved, is seen to be much more complicated than the corresponding problem for ordinary Lie groups. Several of the simple super Lie groups are analogs of corresponding Lie groups. Others, notably P(m), Q(m), D(2, 1, α), W(n), S(n), $\tilde{\mathrm{S}}(n)$ and $\tilde{\mathrm{H}}(n)$, are without analogs, although the last four are cousins of the infinite-dimensional Cartan groups. It is noteworthy that many have vanishing Cartan–Killing matrices. Several (e.g., P(m) and Q(m)) possess no right- and left-invariant Riemannian geometry.

The classification problem for simple unconventional super Lie groups remains open. It should be stressed that these groups arise naturally (e.g., $\tilde{\mathrm{S}}(n)$ for n odd). There is nothing forced about them, and therefore they deserve study. This study is bound to be complicated. Since each unconventional super Lie group depends on one or more fixed a-numbers for its definition, its detailed properties will depend on the degree of the lowest-order term in the expansion of each of these a-numbers in terms of the generators of the Grassmann algebra Λ_∞.

5

Selected applications of supermanifold theory

5.1 Superclassical dynamical systems

Configuration spaces

Classical mechanics is the study of the motion of dynamical systems either acted upon by external forces or interacting with each other. In simple cases the motion is described by a time-varying real vector in a finite-dimensional ordinary vector space. For example, two interacting particles in Euclidean 3-space are described by a real vector $(\mathbf{r}_1, \mathbf{r}_2)$ in a 6-dimensional ordinary vector space. Here $\mathbf{r}_1$ and $\mathbf{r}_2$ are the position vectors of the respective particles relative to some chosen origin. The vector $(\mathbf{r}_1, \mathbf{r}_2)$ is said to give the *configuration* of the two-particle system and the 6-dimensional vector space is called the *configuration space* of the system. In this example the origin of the vectors $\mathbf{r}_1$ and $\mathbf{r}_2$ is arbitrary and may, in fact, be chosen independently for each particle. This means that the configuration space may equally well be viewed as simply $\mathbf{R}^6$.

The configuration space of a dynamical system need not be of the special form $\mathbf{R}^n$ but may be any (ordinary) differentiable manifold. Examples of systems for which the configuration space is more general than $\mathbf{R}^n$ are (a) a rigid body (free to move and rotate) and (b) a complex molecule at a temperature low enough so that some of its interatomic bonds are frozen. The motion of each of these systems is described by a time-varying point in a nontrivial differentiable manifold. The coordinates of this point, in any chart containing it, are called *dynamical variables* of the system.

As another example of a dynamical system consider a real scalar field φ satisfying the covariant Klein–Gordon equation in a curved spacetime, i.e., in a (pseudo)Riemannian manifold M possessing a metric tensor field with canonical form diag$(-1, 1, 1, 1)$. Assume that there exists a global time coordinate t which foliates M into topologically identical spacelike sections Σ. Each Σ inherits a positive-definite metric tensor field g from M. The dynamical variables of the scalar field are the collection of values that φ assumes over each Σ. These variables, which change with time (i.e., from one

Σ to another), may be regarded as the coordinates of a point in an *infinite-dimensional* differentiable manifold C.

That C, which is the configuration space of the scalar field, has the structure of a differentiable manifold may be inferred as follows. First recall that the distinguishing properties of a differentiable manifold are (1) a topology that is locally Euclidean and (2) the existence in it of a notion of differentiation. In the case of C a topology can be introduced that is not merely locally Euclidean but *metric*. Let φ_1 and φ_2 be any two real scalar fields over a spacelike section Σ. The *distance* between them may be defined to be

$$s(\varphi_1, \varphi_2) \stackrel{\text{def}}{=} \left[\int_\Sigma g^{\frac{1}{2}}(\varphi_1 - \varphi_2)^2 \, \mathrm{d}^3 x \right]^{\frac{1}{2}}, \tag{5.1.1}$$

where $g^{\frac{1}{2}}$ is the measure function defined in each chart of Σ by the metric tensor field $\boldsymbol{g}$ (see eqs (2.9.7) and (2.9.8)). The atlas chosen for each Σ may be assumed to be obtained by projection along the vector field $\partial/\partial t$ from a given atlas on a given Σ at a fiducial instant of time. Since $\boldsymbol{g}$ may depend on t the distance concept may change quantitatively from one Σ to another, but this does not change the locally Euclidean character of the topology of C. Note that the integral (5.1.1) will not generally converge for all pairs φ_1, φ_2 unless C is restricted to the set of square integrable scalar fields on Σ, with measure function $g^{\frac{1}{2}}$. However, this restriction is often a convenient one to impose physically.†

Differentiation in C may be introduced as follows. Let F be a real scalar field on C, i.e., a mapping from C to $\mathbf{R}$. Such a mapping is often called a *functional*, and its value at a point φ of C is denoted by $F[\varphi]$. Let $\delta\varphi$ be a C^∞ infinitesimal variation in φ and let $\delta F[\varphi]$ denote the corresponding change in $F[\varphi]$. If, for all φ in C and all C^∞ variations $\delta\varphi$, $\delta F[\varphi]$ can be written in the form

$$\delta F[\varphi] = \int_\Sigma \frac{\delta F[\varphi]}{\delta\varphi(x)} \delta\varphi(x) \, \mathrm{d}^3 x, \tag{5.1.2}$$

where the coefficient $\delta F[\varphi]/\delta\varphi(x)$ in the integrand is either the value at x of a scalar density on Σ or else the formal value at x of a distribution on Σ, and is independent of $\delta\varphi$, then F is called a *differentiable functional.* For every x in Σ, $\delta F[\varphi]/\delta\varphi(x)$ may be regarded as a *component* (in a coordinate basis) of the *differential* $\mathbf{d}F$ of F at the point φ of C. The differential $\mathbf{d}F$ itself is often called the *variational derivative* or *functional*

† Under this restriction a countable basis can always be introduced in the tangent space at each point of C.

derivative of F. By an abuse of notation one often uses the symbol $\delta F[\varphi]/\delta\varphi(x)$ to represent the functional derivative.

It is often possible to subject F to repeated variations:

$$\delta^n F[\varphi] = \int_\Sigma d^3x_1 \ldots \int_\Sigma d^3x_n \frac{\delta^n F[\varphi]}{\delta\varphi(x_1)\ldots\delta\varphi(x_n)} \delta\varphi(x_1)\ldots\delta\varphi(x_n) \quad (5.1.3)$$

where $\delta^n F[\varphi]/\delta\varphi(x_1)\ldots\delta\varphi(x_n)$ is the formal value, at the point $(x_1, \ldots, x_n)$, of a $\delta\varphi$-independent distribution on Σ^n. If such distributions exist for all positive integers n and for all φ in C then F may be called a C^∞ functional on C.

Analogous constructions may be introduced for any set of scalar, vector, or tensor fields on M. The components of such a set at any point x in any chart of a spatial section Σ of M may be assembled into a single collection $\{\varphi^i(x)\}$ where i is a generic index. The distance between two such sets of fields, φ_1 and φ_2, may be defined by

$$s(\varphi_1, \varphi_2) \stackrel{\text{def}}{=} \left[\int_\Sigma g^{\frac{1}{2}} G_{ij}(\varphi_1^i - \varphi_2^i)(\varphi_1^j - \varphi_2^j)\, d^3x\right]^{\frac{1}{2}}, \quad (5.1.4)$$

where (G_{ij}) is any positive-definite x-dependent and chart-dependent symmetric real matrix that satisfies the following conditions: (1) G_{ij} in any chart is obtained, via the vector $\partial/\partial t$, from its value in the corresponding chart of the fiducial Σ. (2) In overlapping regions of overlapping charts the corresponding G_{ij} are determined by the transformation laws of the scalar, vector, and tensor fields in question, in such a way that expression (5.1.4) is chart independent.

By such a procedure one may accommodate all combinations of boson fields, including gauge fields and the gravitational field. In every case the configuration space is an infinite-dimensional differentiable manifold. Note that the distance concept defined by expression (5.1.4) is chart independent even in the case of gauge fields. Although a given gauge potential is generally representable only by patching several Lie-algebra-valued 1-forms together, the difference between two neighboring gauge potentials is always a global 1-form.

Supermanifolds as configuration spaces

In quantum field theory one encounters fermion fields in addition to boson fields. Fermion fields do not possess a classical limit. It is nevertheless convenient for many purposes to treat them formally *as if* they possessed a classical limit or, rather, a *superclassical* limit.

The superclassical limit of a collection of boson and fermion fields is obtained as follows. Let $(\mathscr{U}, \phi)$ be a chart of the spacetime manifold M and let $\{e_\alpha\}$ be a field of orthonormal frames defined over $\mathscr{U}$. Denote by $\{\varphi^i(x)\}$ the collection of components, at the point x in the chart $(\mathscr{U}, \phi)$ and relative to the frame field $\{e_\alpha\}$, of the boson and fermion fields in question. If fermion fields are present and the gravitational field is dynamical then the $\varphi^i(x)$ must include the sixteen components of the frame field. By choosing a Majorana representation for the fermion fields one may assume, without loss of generality, that all the $\varphi^i(x)$ are real (self-adjoint).

Suppose m of the $\varphi^i(x)$ are components of boson fields and n of them are components of fermion fields. In the quantum theory the collection of fields over $\mathscr{U}$ may be regarded as a mapping from $\mathscr{U}$ to $\mathscr{O}^{m+n}$, where $\mathscr{O}$ is the space of self-adjoint linear operators on an appropriate Hilbert space or Fock space. (If the gravitational field is quantized then the frame field components themselves become nontrivial quantum operators.) The superclassical limit is obtained by making the replacement $\mathscr{O}^{m+n} \to \mathbf{R}_c^m \times \mathbf{R}_a^n$ and leaving the dynamical equations formally intact. Each $\varphi^i(x)$ then becomes either a real c-number or a real a-number according as the corresponding quantum field component is that of a boson field or a fermion field. The resulting dynamical system is said to be *superclassical*.

The configuration space C of a superclassical dynamical system is obviously a supermanifold. If the dynamical variables are field variables the supermanifold is infinite dimensional, but there exist models (e.g., the fermi oscillator to be discussed in section 5.5) for which it is finite dimensional. In the field case the topology of C is again defined by eq. (5.1.4) where the matrix (G_{ij}) is now a nonsingular supersymmetric matrix with a positive-definite boson sector, and the square root with positive body is understood. Each point φ of C is contained in an open neighborhood $\mathscr{U}_\varphi$ defined by

$$\mathscr{U}_\varphi \stackrel{\text{def}}{=} \{\varphi' \varepsilon C \,|\, [s(\varphi, \varphi')]_{\mathrm{B}} < \varepsilon\} \tag{5.1.5}$$

where ε is a sufficiently small positive number.

Space of histories

In relativistic field theories it is convenient to let the symbol x include also the time t so that 'x' refers not merely to a point of a spatial cross-section Σ but to a point of spacetime M itself. Spacetime will always be assumed to be topologically $\mathbf{R} \times \Sigma$, with each spatial section being a global Cauchy hypersurface for the dynamical equations. However, it is generally convenient to regard the '$\mathbf{R}$' in '$\mathbf{R} \times \Sigma$' as representing an open interval of the time axis rather than the whole axis, so that M is metrically incomplete. This is done for two reasons: (1) so that the action functional (see below) may be imagined as ranging over well-defined values, with the dynamical variables satisfying specified boundary conditions at the time limits (as well as at spatial infinity when Σ is noncompact); (2) so as to avoid prejudicing in advance the issue of the onset of spacetime singularities, which are known to occur frequently in the classical theory when the gravitational field itself is dynamical.

When x includes the time the functions $\varphi^i(x)$, over all the charts of some atlas of M, represent the *history* of the combined fields. The set of all possible histories (whether dynamically allowed or not) is called the *space of histories* and will be denoted by Φ. Like the configuration space C, Φ is a differentiable supermanifold. Unlike C, it is always infinite dimensional.

In developing the general formalism of field theory it is convenient to lump the symbol x with the generic index i and to make the latter do double duty as a discrete label for the field components and as a continuous label for the points of spacetime. The symbol φ^i may then be regarded as representing the coordinates of the point φ in a chart of Φ. The index i on φ^i will be called c-type or a-type according as the corresponding field component is bosonic or fermionic.

Let F be a supernumber-valued scalar field on Φ, i.e., a mapping from Φ to Λ_∞. The field F will be called a *functional* on Φ, and its value at the point φ will be denoted by $F[\varphi]$. When fermion fields are included among the φ^i one must distinguish right functional differentiation from left functional differentiation. If, for each i, $\delta\varphi^i$ is a C^∞ infinitesimal variation in φ^i, then the nth variation of F is

$$\begin{aligned} \delta^n F[\varphi] &= \delta\varphi^{i_1}\ldots\delta\varphi^{i_n}{}_{i_n\ldots i_1,}F[\varphi] \\ &= F_{,i_1\ldots i_n}[\varphi]\delta\varphi^{i_n}\ldots\delta\varphi^{i_1}, \end{aligned} \tag{5.1.6}$$

where

$$\left.\begin{aligned} {}_{i_n\ldots i_1,}F[\varphi] &\overset{\text{def}}{=} \frac{\vec{\delta}}{\delta\varphi^{i_n}}\cdots\frac{\vec{\delta}}{\delta\varphi^{i_1}}F[\varphi], \\ F_{,i_1\ldots i_n}[\varphi] &\overset{\text{def}}{=} F[\varphi]\frac{\overleftarrow{\delta}}{\delta\varphi^{i_1}}\cdots\frac{\overleftarrow{\delta}}{\delta\varphi^{i_n}}, \end{aligned}\right\} \tag{5.1.7}$$

and where the summation convention over repeated indices now includes (by virtue of their role as continuous labels) integration over M. With these notational conventions indices can be manipulated much as in chapter 1. In particular, if F is a pure scalar field (i.e., either c-number-valued or a-number-valued) the same index-shifting rules hold.

The action functional and the dynamical equations

The nature and dynamical properties of a physical system are completely determined by specifying an *action functional* S for it. An action functional is a differentiable real-valued c-type scalar field on Φ, i.e. a C^∞ mapping $S: \Phi \to \mathbf{R}_c$. The choice of action functional (and even of Φ) for a given system is generally not unique but depends on (1) the choice of dynamical variables used to describe the system (e.g., Lagrangian variables vs. Hamiltonian variables) and (2) the kind of boundary conditions imposed at the time limits (and at spatial infinity if Σ is noncompact). However, all the possible action functionals for a given system must yield equivalent families of dynamical histories. Each such family is defined as the set of points φ of Φ that satisfy

$$_{i,}S[\varphi] \equiv (-1)^i S_{,i}[\varphi] = 0. \tag{5.1.8}$$

Equations (5.1.8) are called the *dynamical equations* of the system. They will be assumed to be *local* in time, i.e., involving not more than a finite number of time derivatives. In a relativistic field theory this implies that they must also be local in space. In this chapter it will be assumed that no local gauge group is present and that solving eqs. (5.1.8) with a given set of initial data constitutes a well-posed Cauchy problem with no gauge freedom. It will also be assumed that the initial data may be specified on any spacelike hypersurface Σ (i.e., that every Σ is a *complete Cauchy hypersurface*) and that no disturbance travels faster than light.

Infinitesimal disturbances and Green's functions

Let A be a differentiable real-valued c-type scalar field on Φ and let ε be an infinitesimal real c-number. Let the action functional S suffer the infinitesimal change

$$S \to S \to \varepsilon A. \tag{5.1.9}$$

This change induces a displacement of the family of dynamical histories. Each dynamical history φ for the original action S gets shifted to an infinitesimally nearby history $\varphi + \delta\varphi$. The shift $\delta\varphi$ may be regarded as a disturbance in the solution φ of (5.1.8) resulting from the change (5.1.9). The disturbed history $\varphi + \delta\varphi$ satisfies the equation

$$\begin{aligned} 0 &= {}_{i,}S[\varphi + \delta\varphi] + \varepsilon\, {}_{i,}A[\varphi + \delta\varphi] \\ &= {}_{i,}S[\varphi] + {}_{i,}S_{,j}[\varphi]\delta\varphi^j + \varepsilon\, {}_{i,}A[\varphi], \end{aligned} \tag{5.1.10}$$

correct to first infinitesimal order. In view of (5.1.8) this may be rewritten in the form

$$ {}_{i,}S_{,j}\,\delta\varphi^j = -\varepsilon\,{}_{i,}A, \tag{5.1.11}$$

in which the arguments have been suppressed.

The second functional derivative ${}_{i,}S_{,j}$ appearing in eq. (5.1.11) is effectively a linear differential operator. (Examples will be displayed in later sections.) The equation itself can be solved with the aid of Green's functions. We shall consider only retarded and advanced boundary conditions. Let the associated Green's functions be denoted by G^{-ij} and G^{+ij} respectively. Then†

$$ {}_{i,}S_{,k}\,G^{\pm kj} = -{}_i\delta^j, \tag{5.1.12}$$

with

$$\left.\begin{array}{lll} G^{-ij}=0 & \text{if} & i<j, \\ G^{+ij}=0 & \text{if} & i>j, \end{array}\right\} \tag{5.1.13}$$

where '$i<j$' means 'the time associated with the index i lies to the past of the time associated with the index j' and '$i>j$' means '$j<i$'. In relativistic field theories ${}_{i,}S_{,j}$ is a differential operator of hyperbolic type, and 'past' and 'future' can be defined with respect to any foliation of spacetime into spacelike hypersurfaces. Consequently the boundary conditions (5.1.13) imply that $G^{-ij}(G^{+ij})$ is nonvanishing only when the spacetime point associated with i lies on or inside the future (past) light cone emanating from the spacetime point associated with j.

Let the retarded and advanced solutions of equation (5.1.11) be denoted by $\delta^-\varphi$ and $\delta^+\varphi$ respectively. Then

$$\delta^\pm\varphi^i = \varepsilon G^{\pm ij}{}_{j,}A. \tag{5.1.14}$$

Let B be any differentiable scalar function on Φ. The associated disturbances in B caused by the change (5.1.9) are

$$\delta^\pm B = B_{,i}\,\delta^\pm\varphi^i = \varepsilon B_{,i}\,G^{\pm ij}{}_{j,}A. \tag{5.1.15}$$

It will be useful to define

$$D_A B \overset{\text{def}}{=} \lim_{\varepsilon\to 0}\frac{1}{\varepsilon}\delta^- B = B_{,i}\,G^{-ij}{}_{j,}A, \tag{5.1.16}$$

where the body of ε is understood to vanish only in the limit $\varepsilon\to 0$.

A cautionary remark is in order regarding the use of expressions such as $B_{,i}G^{\pm ij}{}_{j,}A$ in which there is an implicit summation–integration over *two* dummy labels. The integrals may not always converge or, if they do, the result may depend on which integral is performed first. The order of integration is

† The ${}_i\delta^j$ appearing in eq. (5.1.12) is a combined delta function–Kronecker delta.

generally immaterial if the supports (in spacetime) of the functions $B_{,i}$ and ${}_{j,}A$ are compact (i.e., if A and B are constructed out of dynamical variables taken from compact regions of spacetime). Another example is the expression $X^i{}_{i,}S_{,j}{}^jY$ in which, because ${}_{i,}S_{,j}$ is effectively a differential operator, a change of integration order corresponds to an integration by parts. In this case the order is immaterial if the *intersection* of the supports of X^i and ${}^jY$ is compact.

Reciprocity relations

Consider the expression $G^{-ki}{}_{k,}S_{,l}G^{+lj}$. Because of the boundary conditions (5.1.13) the intersection of the supports of the functions G^{-ki} and G^{+lj} (i and j fixed) is compact. Therefore it makes no difference whether the k integration or the l integration is performed first. Because S is c-type one may write

$$ {}_{k,}S_{,l} = (-1)^{k+l+kl}{}_{l,}S_{,k}. \tag{5.1.17}$$

Note that both ${}_{k,}S_{,l}$ and $G^{\pm kl}$ are c-type when k and l have the same type and a-type when k and l are of opposite types. From these facts obtain

$$\begin{aligned} 0 &= (-1)^{ik}G^{-ki}[{}_{k,}S_{,l} - (-1)^{k+l+kl}{}_{l,}S_{,k}]G^{+lj} \\ &= -(-1)^{ij}G^{-ji} - (-1)^{ik+k+l+kl+(k+i)(l+k)}{}_{l,}S_{,k}G^{-ki}G^{+lj} \\ &= -(-1)^{ij}G^{-ji} - (-1)^{l(i+1)}{}_{l,}S_{,k}G^{-ki}G^{+lj} \\ &= -(-1)^{ij}G^{-ji} + G^{+ij}, \end{aligned}$$

whence one may infer the following important reciprocity relations satisfied by the retarded and advanced Green's functions:

$$G^{\pm ij} = (-1)^{ij}G^{\pm ji}. \tag{5.1.18}$$

These reciprocity relations may be used to show that the $G^{\pm ij}$ are not only *right* Green's functions of ${}_{i,}S_{,j}$, as in eq. (5.1.12), but *left* Green's functions as well:†

$$\begin{aligned} -{}^j\delta_i &= -(-1)^{i(j+1)}{}_i\delta^j = (-1)^{i(j+1)}{}_{i,}S_{,k}G^{\pm kj} \\ &= (-1)^{i(j+1)+i+k+ik+(i+k)(k+j)}G^{\pm kj}{}_{k,}S_{,i} \\ &= (-1)^{jk}G^{\pm kj}{}_{k,}S_{,i} = G^{\mp jk}{}_{k,}S_{,j}. \end{aligned} \tag{5.1.19}$$

† The identity of the left and right Green's functions holds for any hyperbolic differential operator ${}_iF_j$ whether or not it has the symmetry (5.1.17) and hence regardless of whether the reciprocity relations are satisfied. Thus let $G_R^{\pm ij}$ and $G_L^{\pm ij}$ be respectively right and left Green's functions of ${}_iF_j$ and let ${}^iX$ be arbitrary functions of compact support in spacetime. Let ${}^iY = (G_R^{-ij} - G_L^{-ij}){}_jF_k{}^kX$. Obviously, ${}_iF_j{}^jY = 0$. But ${}^iY = 0$ when $i \lessdot \mathrm{supp}({}^kX)$, which implies that the Cauchy data for ${}^iY$ vanish when $i \lessdot \mathrm{supp}({}^kX)$. This in turn implies that the ${}^iY$ vanish everywhere which, because the ${}^kX$ are arbitrary, implies $0 = (G_R^{-ij} - G_L^{-ij}){}_jF_k = G_R^{-ij}{}_jF_k + {}^i\delta_k$, whence $G_R^{-ij} = G_L^{-ij}$. Similarly $G_R^{+ij} = G_L^{+ij}$.

The Peierls bracket

Let A and B be differentiable real c-type scalar fields on Φ. Their *Peierls bracket* (cf. R. E. Peierls, *Proc. Roy. Soc.* (*London*) **A214**, 143 (1952)) is defined by

$$(A, B) \overset{\text{def}}{=} D_A B - D_B A = B_{,i} G^{-ij}{}_{j,} A - A_{,i} G^{-ij}{}_{j,} B, \qquad (5.1.20)$$

eq. (5.1.16) being used in passing to the final expression. By virtue of the reciprocity relations (5.1.18) the Peierls bracket may be expressed in the alternative form

$$(A, B) = A_{,j} G^{+ji}{}_{i,} B - A_{,i} G^{-ij}{}_{j,} B = A_{,i} \tilde{G}^{ij}{}_{j,} B, \qquad (5.1.21)$$

where

$$\tilde{G}^{ij} \overset{\text{def}}{=} G^{+ij} - G^{-ij}. \qquad (5.1.22)$$

The function $\tilde{G}^{ij}$ is called the *supercommutator function* for the dynamical system. It satisfies the homogeneous equations

$${}_{i,}S_{,k} \tilde{G}^{kj} = 0, \qquad \tilde{G}^{ik}{}_{k,}S_{,j} = 0, \qquad (5.1.23)$$

and, by virtue of the reciprocity relations (5.1.18), is antisupersymmetric:

$$\tilde{G}^{ij} = -(-1)^{ij} \tilde{G}^{ji}. \qquad (5.1.24)$$

In eqs. (5.1.20) and (5.1.21) the unwritten argument φ is to be understood as being a solution of the dynamical equations (5.1.8). It does not matter whether the functional derivatives appearing in (5.1.20) and (5.1.21) are evaluated before or after this constraint is imposed. The functionals A and B, when constrained, are defined only *modulo* ${}_{i,}S$, and if they are replaced by

$$\bar{A} \equiv A + a^i{}_{i,}S, \qquad \bar{B} \equiv B + b^i{}_{i,}S \qquad (5.1.25)$$

(a^i and b^i arbitrary), it is easy, using eqs. (5.1.23), to verify that the Peierls bracket remains unaffected:

$$(\bar{A}, \bar{B}) = (A, B). \qquad (5.1.26)$$

Peierls bracket identities

When all the dynamical variables φ^i are c-type the Peierls bracket can be shown to be identical to the conventional Poisson bracket defined in terms of canonical variables (see section 6.1). The above definitions generalize the Poisson bracket to cases in which a-type dynamical variables (e.g., fermion fields) are present. The definition (5.1.20) is valid only if A and B are both c-type. Equation (5.1.21), on the other hand, may be used as the definition of the Peierls bracket when A and B are of arbitrary type, or even impure.

This generalized Peierls bracket has the same symmetry as a supercommutator:

$$(A,B) = -(-1)^{AB}(B,A). \tag{5.1.27}$$

It also satisfies the super Jacobi identity

$$(A(B,C)) + (-1)^{A(B+C)}(B(C,A)) + (-1)^{C(A+B)}(C,(A,B)) = 0. \tag{5.1.28}$$

This is most easily proved by choosing A, B, C to be any three members of an indexed set $\{A^\alpha\}$ of pure functionals A^α. Equation (5.1.28) is then equivalent to

$$\varepsilon_{\alpha\beta\gamma}(A^\gamma,(A^\beta,A^\alpha)) = 0, \tag{5.1.29}$$

where $\varepsilon_{\alpha\beta\gamma}$ is any three-index quantity having the same anti-supersymmetry as the components of a 3-form. Applying eq. (5.1.21) to the left-hand side of (5.1.29) one obtains

$$\varepsilon_{\alpha\beta\gamma}(A^\gamma,(A^\beta,A^\alpha)) = \varepsilon_{\alpha\beta\gamma}\, A^\gamma{}_{,k}\, \tilde{G}^{kc}{}_{c,}(A^\beta{}_{,j}\, \tilde{G}^{ji}{}_{i,}A^\alpha), \tag{5.1.30}$$

which shows that one needs to compute functional derivatives of the Green's functions. A more general result will first be derived.

Suppose the action functional S suffers an infinitesimal variation δS. The corresponding variation in eq. (5.1.12) yields

$${}_{i,}S_{,k}\,\delta G^{\pm kj} = -(\delta\,{}_{i,}S_{,k})G^{\pm kj}. \tag{5.1.31}$$

The solution of this equation, having the required boundary conditions, is

$$\delta G^{\pm ij} = G^{\pm ik}(\delta\,{}_{k,}S_{,l})G^{\pm lj}. \tag{5.1.32}$$

Noting that differentiation is just a special case of variation, we infer

$$\begin{aligned} {}_{a,}G^{\pm ij} &= (-1)^{a(i+k)}\, G^{\pm ik}{}_{ak,}S_{,l}\, G^{\pm lj}. \\ &= (-1)^{a(i+j)+jk}\, G^{\pm ik}\, G^{\mp jl}{}_{lka,}S, \end{aligned} \tag{5.1.33}$$

and hence

$${}_{a,}\tilde{G}^{ij} = (-1)^{a(i+j)+jk}(G^{+ik}G^{-jl} - G^{-ik}G^{+jl})_{lka,}S. \tag{5.1.34}$$

Applying this to eq. (5.1.30) one now has

$$\begin{aligned} \varepsilon_{\alpha\beta\gamma}(A^\gamma,(A^\beta,A^\alpha)) = \varepsilon_{\alpha\beta\gamma}[&(-1)^{\beta c}2A^\gamma{}_{,k}\,\tilde{G}^{kc}\,A^\beta{}_{,j}\,\tilde{G}^{ji}{}_{ic,}A^\alpha \\ &+(-1)^{a(b+c)+\beta c}A^\gamma{}_{,k}(G^{+kc} - G^{-kc}) \\ &\times(A^\beta{}_{,j}G^{+jb}A^\alpha{}_{,i}G^{-ia} - A^\beta{}_{,j}G^{-jb}A^\alpha{}_{,i}G^{+ia})_{abc,}S] \end{aligned} \tag{5.1.35}$$

The first term inside the square brackets is supersymmetric in γ and β. The second term, when multiplied out, yields four terms, each of which is supersymmetric in either γ and β or γ and α. Since $\varepsilon_{\alpha\beta\gamma}$ is completely antisupersymmetric in its indices every term makes a vanishing total contribution, and eq. (5.1.29) thereby follows.

The existence of the identities (5.1.27) and (5.1.28) is what makes possible the transition from the (super)classical theory to the full quantum theory, in which Peierls brackets are replaced by (super)commutators. As is well known, the transition is not always unique; ambiguities in factor ordering can occur. Less well known is the fact that the use of a-number-valued external sources, which is a common device when fermion fields are present, lifts one out of the standard framework of conventional Hilbert-space (or Fock-space) theory. A generalization of the conventional theory is needed.

5.2 Super Hilbert spaces

Definition

A super Hilbert space $\mathscr{H}$ is a supervector space (see section 1.4) for which the notion of a real (or imaginary) supervector is undefined and for which the complex conjugation mapping is replaced by an *inner product*, i.e., by a one-to-one mapping $*: \mathscr{H} \to \mathscr{H}^*$, from $\mathscr{H}$ to its dual $\mathscr{H}^*$, which satisfies the following axioms:[†]

(0) $$*(|\varphi\rangle) \overset{\text{def}}{=} |\varphi\rangle^* \overset{\text{def}}{=} \langle\varphi|, \tag{5.2.1}$$

$$\langle\varphi|\cdot|\chi\rangle \overset{\text{def}}{=} \langle\varphi|\chi\rangle. \tag{5.2.2}$$

(1) $*$ maps c-type (a-type) supervectors in $\mathscr{H}$ into c-type (a-type) dual supervectors in $\mathscr{H}^*$.

(2) For every $|\varphi\rangle$ and $|\chi\rangle$ in $\mathscr{H}$ and every α in Λ_∞,

$$(\alpha|\varphi\rangle)^* = \langle\varphi|\alpha^*, \tag{5.2.3}$$

$$(|\varphi\rangle + |\chi\rangle)^* = \langle\varphi| + \langle\chi|, \tag{5.2.4}$$

where α^* denotes the usual complex conjugate of α.

(3) For every $|\varphi\rangle$ in $\mathscr{H}$ and every $\langle\chi|$ in $\mathscr{H}^*$

$$\langle\chi|\varphi\rangle^* = \langle\varphi|\chi\rangle, \tag{5.2.5}$$

where $\langle\chi|\varphi\rangle^*$ denotes the complex conjugate of the supernumber $\langle\chi|\varphi\rangle$. Evidently $\langle\varphi|\varphi\rangle$ is real for all $|\varphi\rangle$ in $\mathscr{H}$.

(4) $|\varphi\rangle$ has nonvanishing body if and only if $\langle\varphi|\varphi\rangle_B$ is positive. $\langle\varphi|\varphi\rangle_B$ is never negative.

As has already been remarked, super Hilbert spaces are generalizations of ordinary Hilbert spaces, designed so as to enable one to consider

[†] Here the Dirac notation is used. Elements of $\mathscr{H}$ are denoted by $|\varphi\rangle$, $|\chi\rangle$, $|\psi\rangle$ etc. $\langle\varphi|$ is an element of $\mathscr{H}^*$ and $\langle\varphi|\chi\rangle$ is the inner product of $\langle\varphi|$ and $|\chi\rangle$.

quantum systems with supernumber-valued parameters (e.g., *a*-type external sources) which are themselves introduced in order to present, in a compact algebraic way, certain relationships between real physical amplitudes. *Real* physics is restricted to the ordinary Hilbert space that sits inside the super Hilbert space. It will nevertheless be convenient to abuse terminology somewhat by calling an element $|\varphi\rangle$ of $\mathscr{H}$ *physical* if it merely has nonvanishing body. Physical elements of $\mathscr{H}$ are also called *state vectors*. It is easy to show that if $|\varphi\rangle$ is physical it can be normalized (i.e. multiplied by an appropriate supernumber with nonvanishing body) so that $\langle\varphi|\varphi\rangle = 1$. In practice, one often extends both $\mathscr{H}$ and its dual in such a way as to allow delta-function orthonormalization of state vectors.

It is sometimes convenient to append the following definition:

$$\langle\varphi|^* \stackrel{\text{def}}{=} *^{-1}(\langle\varphi|) = |\varphi\rangle. \tag{5.2.6}$$

Linear operators

A mapping $A: \mathscr{H} \to \mathscr{H}$ is said to be a *linear operator* if and only if, for all $|\varphi\rangle$ and $|\chi\rangle$ in $\mathscr{H}$ and all α in Λ_∞,

$$A(|\varphi\rangle\alpha) = (A|\varphi\rangle)\alpha \stackrel{\text{def}}{=} A|\varphi\rangle\alpha, \tag{5.2.7}$$

$$A(|\varphi\rangle + |\chi\rangle) = A|\varphi\rangle + A|\chi\rangle, \tag{5.2.8}$$

where $A|\varphi\rangle$ is shorthand for $A(|\varphi\rangle)$. A linear operator acting on $\mathscr{H}$ may equally well be regarded as acting on $\mathscr{H}^*$ through the rule

$$(\langle\chi|A)|\varphi\rangle \stackrel{\text{def}}{=} \langle\chi|(A|\varphi\rangle) \stackrel{\text{def}}{=} \langle\chi|A|\varphi\rangle, \tag{5.2.9}$$

for all $|\varphi\rangle$ in $\mathscr{H}$ and all $\langle\chi|$ in $\mathscr{H}^*$. By virtue of eqs. (1.5.49) and (1.5.51) this implies

$$(\alpha\langle\chi|)A|\varphi\rangle = \alpha\langle\chi|A|\varphi\rangle, \tag{5.2.10}$$

$$\begin{aligned}(\langle\chi| + \langle\psi|)A|\varphi\rangle &= \langle\chi|A|\varphi\rangle + \langle\psi|A|\varphi\rangle \\ &= (\langle\chi|A + \langle\psi|A)|\varphi\rangle,\end{aligned} \tag{5.2.11}$$

for all α in Λ_∞, all $|\varphi\rangle$ in $\mathscr{H}$, and all $\langle\chi|$ and $\langle\psi|$ in $\mathscr{H}^*$. Evidently A is also a linear operator when acting on $\mathscr{H}^*$:

$$(\alpha\langle\chi|)A = \alpha(\langle\chi|A) \stackrel{\text{def}}{=} \alpha\langle\chi|A, \tag{5.2.12}$$

$$(\langle\chi| + \langle\psi|)A = \langle\chi|A + \langle\psi|A. \tag{5.2.13}$$

Linear operators may be combined with each other and with supernumbers through the definitions

$$(A+B)|\varphi\rangle \overset{\text{def}}{=} A|\varphi\rangle + B|\varphi\rangle, \tag{5.2.14}$$

$$(AB)|\varphi\rangle \overset{\text{def}}{=} A(B|\varphi\rangle) \overset{\text{def}}{=} AB|\varphi\rangle, \tag{5.2.15}$$

$$(\alpha A)|\varphi\rangle \overset{\text{def}}{=} \alpha(A|\varphi\rangle) \overset{\text{def}}{=} \alpha A|\varphi\rangle, \quad (A\alpha)|\varphi\rangle \overset{\text{def}}{=} A(\alpha|\varphi\rangle) \overset{\text{def}}{=} A\alpha|\varphi\rangle, \tag{5.2.16}$$

for all α in Λ_∞, all $|\varphi\rangle$ in $\mathscr{H}$ and all linear operators A and B. It is straightforward to verify that $A+B$, AB, αA and $A\alpha$ are themselves linear operators and furthermore that

$$\langle\varphi|(A+B) = \langle\varphi|A + \langle\varphi|B, \tag{5.2.17}$$

$$\langle\varphi|(AB) = (\langle\varphi|A)B \overset{\text{def}}{=} \langle\varphi|AB, \tag{5.2.18}$$

$$\langle\varphi|(\alpha A) = (\langle\varphi\alpha)A \overset{\text{def}}{=} \langle\varphi|\alpha A, \quad \langle\varphi|(A\alpha) = (\langle\varphi|A)\alpha \overset{\text{def}}{=} \langle\varphi|A\alpha, \tag{5.2.19}$$

for all $\langle\varphi|$ in $\mathscr{H}^*$. The set of all linear operators constitutes what may be called a *superalgebra*.

A linear operator A is said to be pure if it maps pure supervectors into pure supervectors, c-type if it leaves the types unchanged and a-type if it changes the types. If $|\varphi\rangle, \langle\chi|$, α and A are pure then

$$\begin{aligned}\langle\chi|\alpha A|\varphi\rangle &= (-1)^{\alpha\chi}\alpha\langle\chi|A|\varphi\rangle \\ &= (-1)^{\alpha(A+\varphi)}\langle\chi|A|\varphi\rangle\alpha = (-1)^{\alpha A}\langle\chi|A\alpha|\varphi\rangle,\end{aligned}$$

whence

$$\alpha A = (-1)^{\alpha A} A\alpha. \tag{5.2.20}$$

The *adjoint* A^* of a linear operator A is defined by

$$A^*|\varphi\rangle = (\langle\varphi|A)^*, \tag{5.2.21}$$

for all $|\varphi\rangle$ in $\mathscr{H}$. It is straightforward to verify that A^* is also a linear operator. The operator A is said to be *self-adjoint* if and only if $A^* = A$.

Physical observables

A linear operator A, whether c-type, a-type or impure, will be called a *physical observable* if and only if (1) it is self-adjoint; (2) all its eigenvalues are c-numbers; (3) for every eigenvalue there is at least one corresponding physical eigenvector; and (4) the set of physical eigenvectors that correspond to *soulless* eigenvalues contains a complete basis. The soulless eigenvalues will be called *physical eigenvalues*.

Theorem

All the eigenvalues of a physical observable are real.

Proof: Denote by $|A'\rangle$ a physical eigenvector corresponding to the eigenvalue A' of the physical observable A. Since A' is a c-number it commutes with everything, and hence

$$A|A'\rangle = A'|A'\rangle = |A'\rangle A'. \tag{5.2.22}$$

Since A is self-adjoint we have

$$\begin{aligned}\langle A'|A &= (A^*|A'\rangle)^* = (A|A'\rangle)^* \\ &= (|A'\rangle A')^* = A'^*\langle A'|,\end{aligned} \tag{5.2.23}$$

whence

$$\langle A'|A'\rangle A' = \langle A'|A|A'\rangle = A'^*\langle A'|A'\rangle = \langle A'|A'\rangle A'^*. \tag{5.2.24}$$

Since $|A'\rangle$ is physical $\langle A'|A'\rangle_B$ is positive and hence $\langle A'|A'\rangle$ is invertible. Therefore

$$A'^* = A'. \tag{5.2.25}$$

Theorem

Eigenvectors corresponding to different physical eigenvalues of a physical observable are orthogonal.

Proof: Let A' and A'' be two distinct physical eigenvalues of the physical observable A, and let $|A'\rangle$ and $|A''\rangle$ be corresponding eigenvectors. Since A' and A'' are distinct ordinary real numbers, $A' - A''$ is invertible. Making use of eqs. (5.2.22), (5.2.23) and (5.2.25) we have

$$\langle A''|A'\rangle A' = \langle A''|A|A'\rangle = A''\langle A''|A'\rangle = \langle A''|A'\rangle A''$$

and hence

$$\langle A''|A'\rangle(A' - A'') = 0.$$

Since $A' - A''$ is invertible it follows that

$$\langle A''|A'\rangle = 0 \quad \text{when} \quad A' \neq A'' \text{ and } A'_S = A''_S = 0. \tag{5.2.26}$$

5.3 Quantum systems

Transition to the quantum theory

In practice one begins with a superclassical dynamical system and then obtains from it a corresponding quantum system. Each dynamical variable φ^i is replaced by a self-adjoint linear operator (denoted here by the same symbol), and these linear operators are assumed to satisfy differential equations identical in form with those of the superclassical theory, with a particular choice of factor ordering.† The super Hilbert space on which these operators act is not given *a priori* but is constructed in such a way as to yield a representation of the operator superalgebra generated by the dynamical variables. The operator superalgebra itself is determined by the quantization rule

$$[\varphi^i, \varphi^j] = \mathrm{i}(\varphi^i, \varphi^j) = \mathrm{i}\tilde{G}^{ij}, \tag{5.3.1}$$

which identifies, up to a factor i, each Peierls bracket with a supercommutator.‡

When the dynamical equations are nonlinear the supercommutator function $\tilde{G}^{ij}$ itself depends on the φ^i and becomes a linear operator. It is usually difficult to give a simple factor-ordering prescription for passing from the superclassical $\tilde{G}^{ij}$ to the corresponding quantum $\tilde{G}^{ij}$. It is, however, possible to start with equal-time supercommutators for Cauchy data (which are usually uniquely defined by (5.3.1) up to possible renormalization factors) and then determine $\tilde{G}^{ij}$ by the dynamical equations.

A similar difficulty arises with the more general quantization rule

$$[A, B] = \mathrm{i}(A, B) = \mathrm{i}A_{,i}\tilde{G}^{ij}{}_{j,}B \tag{5.3.2}$$

for arbitrary functionals A and B of the φ^i. Even though simple operator-ordering prescriptions may be adopted in defining A and B there will often be no simple prescription for passing from the superclassical $A_{,i}\tilde{G}^{ij}{}_{j,}B$ to its quantum analog. One must build up the analog, starting from (5.3.1) and making repeated use of the supercommutator identity (2.3.3) which, incidentally, has the Peierls-bracket analog:

$$(A, BC) = (A, B)C + (-1)^{AB}B(A, C). \tag{5.3.3}$$

Once the operator superalgebra is set up the super Hilbert space is constructed in such a way that the following two conditions are satisfied: (1) Each dynamical variable becomes a physical observable (for definition see section 5.2) of the same type as the corresponding superclassical variable. (2) The representation of the operator superalgebra that the super Hilbert space yields is the minimal representation compatible with condition (1). In practice one also requires certain functionals of the φ^i, such as the energy operator, the angular momentum operator, etc., to satisfy the conditions for being physical observables. If the above conditions cannot be met, then the quantum theory of the system is either inconsistent or unphysical.

† In order to maintain consistency one sometimes has to add extra terms to these equations, which do not appear in the superclassical theory from which one starts.

‡ Units with $\hbar = 1$ are implied here.

The Schwinger variational principle

Instead of making direct use of the operator dynamical equations one may express the dynamical content of the quantum theory in an alternative form which is often more useful in applications. Let A and B be any two physical observables of a given system which satisfy

$$\operatorname{supp}(A_{,i}) \gg \operatorname{supp}(B_{,i}). \tag{5.3.4}$$

That is, A is constructed out of φ^i's taken from a region of spacetime that lies to the future of the region from which the φ^i's making up B are taken. We shall be interested in the inner product $\langle A'|B'\rangle$ where $|A'\rangle$ and $|B'\rangle$ are physical eigenvectors of A and B respectively, corresponding to the eigenvalues A' and B'.

Suppose the action functional suffers an infinitesimal change δS. This produces a change in the dynamical equations and hence a change in their solutions φ^i. Suppose the forms of A and B as functionals of the φ's remain unchanged. As operators, A and B will nevertheless be changed because the φ's have changed. Denote these changes by δA and δB respectively. The eigenvectors $|A'\rangle$ and $|B'\rangle$ too will suffer changes $\delta|A'\rangle$ and $\delta|B'\rangle$. The precise nature of these changes will depend on boundary conditions.

Suppose δS satisfies the condition

$$\operatorname{supp}(A_{,i}) \gg \operatorname{supp}(\delta S_{,i}) \gg \operatorname{supp}(B_{,i}). \tag{5.3.5}$$

That is, suppose δS is constructed out of φ^i's taken from a region of spacetime that lies to the past of the region associated with A and to the future of the region associated with B. Suppose furthermore that retarded boundary conditions are adopted. Then, at times to the past of the region associated with δS, the dynamical variables φ^i will remain unchanged. This means that

$$\delta B = 0. \tag{5.3.6}$$

The observable A, on the other hand, suffers the change

$$\delta A = D_{\delta S} A, \tag{5.3.7}$$

where the notation of eq. (5.1.16) has been borrowed. If one were working with the superclassical theory instead of the full quantum theory one could also write[†]

$$D_A \delta S = 0, \tag{5.3.8}$$

$$\delta A = D_{\delta S} A = D_{\delta S} A - D_A \delta S = (\delta S, A), \tag{5.3.9}$$

[†] In section 5.1 the quantity $D_A \delta S$ is not defined unless A is c-type. It can be generalized to the case in which A is a-type by defining $D_{A\zeta}\delta S = D_A \delta S \zeta$ where ζ is an arbitrary imaginary a-number. Since δS is c-type the Peierls bracket $(\delta S, A)$ is still given by $D_{\delta S} A - D_A \delta S$.

which it is tempting to translate into the quantum theory as

$$\delta A = -\mathrm{i}[\delta S, A]. \tag{5.3.10}$$

If an arbitrary operator ordering is chosen for the operator dynamical equations it is not generally possible to show that eq. (5.3.10) holds. However, one may turn the problem around and require that the operator dynamical equations *be* such that eq. (5.3.10) *does* hold. Note that if it holds for $A = \varphi^i$, for all i whose associated spacetime points lie to the future of the region associated with δS, then it also holds for all A satisfying condition (5.3.5), because (5.3.10) is simply a (super)unitary transformation:

$$\begin{aligned} A[\varphi] + \delta A[\varphi] &= u^{-1}[\varphi]A[\varphi]u[\varphi] \\ &= A[u^{-1}[\varphi]\varphi u[\varphi]] = A[\varphi + \delta\varphi], \end{aligned} \tag{5.3.11}$$

where

$$u[\varphi] = 1 + \mathrm{i}\delta S[\varphi]. \tag{5.3.12}$$

Note also that the requirement that eq. (5.3.10) hold makes the operator dynamical equations depend on a particular choice of factor ordering for δS, and hence for S.

It is worth remarking that if one starts from a linear theory, for which the dynamical equations are unambiguous, one can construct a wide class of nonlinear theories simply by adding sequentially, from the future to the past, small pieces to the action, which, in the case of field theories, take the form

$$\delta_n S = \int_{\Sigma_n}^{\Sigma_{n+1}} L_{\mathrm{int}}(\varphi, \partial\varphi/\partial x, \ldots)\,\mathrm{d}^4x, \tag{5.3.13}$$

where L_{int} is a local addition to the Lagrangian. One demands that each increment $\delta_n S$ induce changes of the form (5.3.10) in all observables A for which $\mathrm{supp}(A_{,i}) > \Sigma_{n+1}$ and then passes to the limit in which all the spacelike hypersurfaces Σ_n become infinitely close together. If one denotes by φ_0^i the dynamical variables of the initial linear theory and by φ^i the dynamical variables of the final nonlinear theory then

$$\varphi^i = U^{-1}[\Sigma]\varphi_0^i U[\Sigma], \tag{5.3.14}$$

$$U[\Sigma] = \mathrm{T}(\exp \mathrm{i}\int_{\Sigma_{in}}^{\Sigma} L_{int}(\varphi_0, \partial\varphi_0/\partial x, \ldots)\,\mathrm{d}^4x), \tag{5.3.15}$$

where T is the chronological ordering operator (see below), Σ is any spacelike hypersurface containing the spacetime point associated with the index i, and Σ_{in} is a hypersurface marking the time boundary on which initial conditions for the dynamics are set.

Suppose now that eq. (5.3.10) does hold. Then it is easy to see that the change $\delta|A'\rangle$ in the eigenvector $|A'\rangle$ resulting from this change in A is given by

$$|A'\rangle + \delta|A'\rangle = u^{-1}|A'\rangle \tag{5.3.16}$$

or

$$\delta|A'\rangle = -\mathrm{i}\delta S|A'\rangle, \tag{5.3.17}$$

modulo an infinitesimal phase change $\mathrm{i}\delta\phi|A'\rangle$ where $\delta\phi$ is a real c-number. Equation (5.3.6), on the other hand, implies

$$\delta|B'\rangle = 0, \tag{5.3.18}$$

whence

$$\delta\langle A'|B'\rangle = \mathrm{i}\langle A'|\delta S|B'\rangle. \tag{5.3.19}$$

Equation (5.3.19) is known as the *Schwinger variational principle*. It is particularly useful because it is a direct statement about *transition amplitudes* $\langle A'|B'\rangle$, and it has a wide variety of applications. Although it was derived here through the imposition of retarded boundary conditions it is actually independent of boundary conditions. For example, if advanced boundary conditions are imposed and use is made of the reciprocity relations (5.1.18) then eqs. (5.3.6) and (5.3.9) get replaced by

$$\delta A = 0, \tag{5.3.20}$$

$$\delta B = D_B\delta S = D_B\delta S - D_{\delta S}B = (B, \delta S) \to -\mathrm{i}[B, \delta S], \tag{5.3.21}$$

which imply

$$A[\varphi + \delta\varphi] = A[\varphi], \qquad B[\varphi + \delta\varphi] = u[\varphi]B[\varphi]u^{-1}[\varphi], \tag{5.3.22}$$

$$\delta|A'\rangle = 0, \qquad \delta|B'\rangle = \mathrm{i}\delta S|B'\rangle, \tag{5.3.23}$$

once again leading to (5.3.19). A change from retarded to advanced boundary conditions simply corresponds to an overall unitary transformation, generated by the operator $u[\varphi]$ of (5.3.12), which leaves the Schwinger variational principle unchanged.†

It will be noted that the particular choice of physical observables A and B in the above derivations is irrelevant. Only the condition (5.3.5) is important. Since the eigenvalues of more than one observable usually have to be specified in order to determine a quantum state uniquely, it

† Whether one imposes retarded or advanced boundary conditions, or something in between, the following statements are always true: (1) The unperturbed dynamical equations $S_{,i} = 0$ continue to hold in the regions to the past and to the future of supp($\delta S_{,i}$). (2) The $\varphi^i + \delta\varphi^i$ in these regions differ from the unperturbed φ^i by unitary transformations. This agrees with a theorem (not proved here) that says that any two solutions of the operator dynamical equations are related by either a unitary or (in the time-reversal-invariant case) an antiunitary transformation.

will be convenient to replace eq. (5.3.19) from now on by the more general statement

$$\delta\langle \text{out}|\text{in}\rangle = \mathrm{i}\langle \text{out}|\delta S|\text{in}\rangle, \tag{5.3.24}$$

where $|\text{in}\rangle$ and $|\text{out}\rangle$ are state vectors determined by some unspecified conditions on the values of the dynamical variables φ^i in regions respectively to the past and to the future of the region in which one may wish to vary the action.

External sources

A particularly convenient way to vary the action is to append to it a term of the form $J_i\varphi^i$ where the J_i are pure supernumber-valued functions over spacetime, real when i is c-type and imaginary when i is a-type. The J_i are called *external sources.*

Let the external sources suffer variations δJ_i whose supports are confined to the spacetime region lying, in time, between the regions associated with the state vectors $|\text{in}\rangle$ and $|\text{out}\rangle$. Then the amplitude $\langle \text{out}|\text{in}\rangle$ suffers the change

$$\delta\langle \text{out}|\text{in}\rangle = \mathrm{i}\langle \text{out}|\delta J_i\varphi^i|\text{in}\rangle \tag{5.3.25}$$

Evidently $\langle \text{out}|\text{in}\rangle$ is a functional of the sources. Let us assume that the state vector $|\text{out}\rangle$ is c-type. Then

$$\frac{\vec{\delta}}{i\delta J_j}\langle \text{out}|\text{in}\rangle = \langle \text{out}|\varphi^i|\text{in}\rangle = \sum\langle \text{out}|\varphi'\rangle\varphi'^j\langle\varphi'|\text{in}\rangle, \tag{5.3.26}$$

where the summation in the final expression is over a complete set of normalized physical eigenvectors $|\varphi'\rangle$ of φ^j, corresponding to the eigenvalues φ'^j. (Since the φ^j are assumed to be physical observables such eigenvectors exist.)

Now let δJ_i be a second variation in the sources. Suppose $j \lessdot \text{supp}(\delta J_i)$. Then the amplitudes $\langle\varphi'|\text{in}\rangle$ in eq. (5.3.26) remain unchanged, and

$$\begin{aligned}\delta\langle \text{out}|\varphi^j|\text{in}\rangle &= \mathrm{i}\sum\langle \text{out}|\delta J_i\varphi^i|\varphi'\rangle\varphi'^j\langle\varphi'|\text{in}\rangle\\ &= \mathrm{i}\langle \text{out}|\delta J_i\varphi^i\varphi^j|\text{in}\rangle,\end{aligned} \tag{5.3.27}$$

whence

$$\frac{\vec{\delta}}{\mathrm{i}\delta J_i}\langle \text{out}|\varphi^j|\text{in}\rangle = \langle \text{out}|\varphi^i\varphi^j|\text{in}\rangle. \tag{5.3.28}$$

If, on the other hand, $\text{supp}(\delta J_i) \lessdot \text{j}$, then

$$\delta\langle \text{out}|\varphi^j|\text{in}\rangle = \mathrm{i}\langle \text{out}|\varphi^j\delta J_i\varphi^i|\text{in}\rangle \tag{5.3.29}$$

and

$$\frac{\vec{\delta}}{\mathrm{i}\delta J_i}\langle \mathrm{out}|\varphi^j|\mathrm{in}\rangle = (-1)^{ij}\langle \mathrm{out}|\varphi^j\varphi^i|\mathrm{in}\rangle. \tag{5.3.30}$$

Evidently

$$\frac{\vec{\delta}}{\mathrm{i}\delta J_i}\frac{\vec{\delta}}{\mathrm{i}\delta J_j}\langle \mathrm{out}|\mathrm{in}\rangle = \langle \mathrm{out}|T(\varphi^i\varphi^j)|\mathrm{in}\rangle \tag{5.3.31}$$

and, more generally,

$$\frac{\vec{\delta}}{\mathrm{i}\delta J_{i_1}}\cdots\frac{\vec{\delta}}{\mathrm{i}\delta J_{i_n}}\langle \mathrm{out}|\mathrm{in}\rangle = \langle \mathrm{out}|T(\varphi^{i_1}\ldots\varphi^{i_n})|\mathrm{in}\rangle, \tag{5.3.32}$$

where T is the *chronological ordering operator* which rearranges the factors $\varphi^{i_1}\ldots\varphi^{i_n}$ so that the times associated with the indices appear in chronological sequence, increasing from right to left, and which inserts an additional factor -1 for each interchange of a pair of a-type indices that occurs in carrying out this rearrangement. When the points associated with an index pair i,j are separated by a spacelike interval there is no ambiguity in the T-operation because the supercommutator function $\tilde{G}^{ij}$ then vanishes. The only ambiguities that can occur arise in the limit when two points coincide, for expression (5.3.32) is, in quantum field theory, a distribution that becomes singular when two points coincide. It turns out that such ambiguities can be eliminated, and the distribution uniquely defined, by demanding that the T-operation commute with differentiation with respect to the spacetime coordinates.

Equation (5.3.32) may be further generalized to

$$\langle \mathrm{out}|T(A[\varphi])|\mathrm{in}\rangle = A\left[\frac{\vec{\delta}}{\mathrm{i}\delta J}\right]\langle \mathrm{out}|\mathrm{in}\rangle, \tag{5.3.33}$$

where $A[\varphi]$ is any functional of the φ^i for which $\mathrm{supp}(A_{,i})$ lies between the 'in' and 'out' regions and which, in the superclassical theory, possesses a functional Taylor expansion with finite radius of convergence about some point φ_0 of Φ. Equation (5.3.33) finds a wide variety of applications in quantum field theory.

Chronologically ordered form of the operator dynamical equations

It is often convenient to write the quantum dynamical variables in the form

$$\varphi^i = \varphi^i_c + \phi^i, \tag{5.3.34}$$

where the φ^i_c are supernumber-valued variables having the same reality and type properties as the φ^i. Equation (5.3.34) amounts to a redefinition of the zero point for the quantum variables. The φ^i are replaced by the ϕ^i which, because the φ_c^i are supernumbers, obey identical supercommutation relations:

$$[\phi^i, \phi^j] = \mathrm{i}\tilde{G}^{ij}[\varphi]. \tag{5.3.35}$$

The φ_c^i are called *background* variables and are often, though not always, chosen to satisfy the superclassical dynamical equations.

Suppose external sources are present so that the total action functional takes the form $S + J_i\varphi^i$. Then the operator dynamical equations may be written in the form

$$-J_i = S_{,i}[\varphi] = S_{,i}[\varphi_c] + S_{,ij}[\varphi_c]\phi^j + \tfrac{1}{2}S_{,ijk}[\varphi_c]\phi^k\phi^j + \cdots. \tag{5.3.36}$$

The final expression is meant to represent a functional Taylor expansion about the point φ_c of Φ. Unfortunately, factor-ordering ambiguities prevent one from establishing any *a priori* rule (independent of the particular dynamical system under consideration) for the forms of the cubic and higher-order terms of the series, and therefore they have been left unwritten. However, eq. (5.3.36) will suffice for present purposes.

It will be convenient to introduce the step function θ_Σ corresponding to an arbitrary foliation of spacetime into spacelike hypersurfaces Σ:

$$\theta_\Sigma(i,j) \overset{\text{def}}{=} \begin{cases} 1 & i \succ j, \\ 0 & i \prec j, \end{cases} \tag{5.3.37}$$

the symbol '$\succ$' being relative to the foliation. Using this step function, together with eq. (5.3.35), one may rewrite the quadratic term of (5.3.36) in the form

$$\begin{aligned} &\tfrac{1}{2}S_{,ijk}[\varphi_c]\{[\theta_\Sigma(k,j) + \theta_\Sigma(j,k)]\phi^k\phi^j \\ &\qquad - [\theta_\Sigma(k,j)\phi^k\phi^j + (-1)^{jk}\theta_\Sigma(j,k)\phi^j\phi^k] + T(\phi^k\phi^j)\} \\ &= \tfrac{1}{2}S_{,ijk}[\varphi_c]\{T(\phi^k\phi^j) + \theta_\Sigma(j,k)[\phi^k, \phi^j]\} \\ &= \tfrac{1}{2}S_{,ijk}[\varphi_c]\{T(\phi^k\phi^j) + \mathrm{i}\theta_\Sigma(j,k)\tilde{G}^{kj}[\varphi]\} \\ &= \tfrac{1}{2}S_{,ijk}[\varphi_c]\{T(\phi^k\phi^j + \mathrm{i}G^{+kj}[\varphi]) + \cdots\}, \end{aligned} \tag{5.3.38}$$

where '$+\cdots$' stands for the difference between $\mathrm{i}\theta_\Sigma(j,k)\tilde{G}^{kj}[\varphi]$ (which in the superclassical theory is just $\mathrm{i}G^{+kj}[\varphi]$) and the result of chronological ordering $\mathrm{i}G^{+kj}[\varphi]$ with the superclassical φ's replaced by quantum φ's.

If the unwritten terms of (5.3.36) are imagined as reordered in a similar way then one may rewrite the operator dynamical equations in the form

$$-J_i = T(S_{,i}[\varphi]) + \frac{\mathrm{i}}{2}T(S_{,ijk}[\varphi]G^{+kj}[\varphi]) + \cdots, \tag{5.3.39}$$

where each term is to be understood as obtained by taking the corresponding superclassical quantity, replacing the superclassical φ's by quantum φ's and performing the chronological ordering operation. In most cases encountered in practice a well-defined unitary theory is obtained by keeping only the terms shown explicitly in eq. (5.3.39). (See, for example, section 6.4.) The '$+\cdots$' will therefore be omitted from now on.

Using the variational law (1.6.8) for the superdeterminant and remembering that ${}_{i,}S_{,j}$ is the negative of the inverse of G^{+ij}, one may express the second term of (5.3.39) in the form

$$\frac{i}{2}T(G^{+kj}[\varphi]_{kj,}S_{,i}[\varphi]) = \frac{i}{2}T(\{\ln \operatorname{sdet} G^{+}[\varphi]\}_{,i}) \tag{5.3.40}$$

and hence finally

$$-J_i = T\left(\left\{S[\varphi] + \frac{i}{2}\ln \operatorname{sdet} G^{+}[\varphi]\right\}\frac{\overleftarrow{\delta}}{\delta\varphi^i}\right). \tag{5.3.41}$$

This will be regarded as the correct form for the operator dynamical equations. The superdeterminant of $G^{+}[\varphi]$ is, of course, a purely formal expression. In practice it must be replaced by a ratio, $\operatorname{sdet} G^{+}[\varphi]/\operatorname{sdet} G_0^{+}$, where G_0^{+} is a fiducial Green's function independent of φ, and, in field theory, regularizations and renormalizations must be performed to exorcise surviving infinities.

The Feynman functional integral

Consider now the amplitude $\langle \text{out}|\text{in}\rangle$. It is a functional of the sources J_i. Let us try to express it as a functional Fourier integral:

$$\langle \text{out}|\text{in}\rangle = \int F[\varphi]e^{iJ\varphi}\,d\varphi = \int e^{iJ\varphi}F[\varphi]\,d\varphi. \tag{5.3.42}$$

Here the φ^i are once again superclassical variables, as in section 5.1. In the present context, however, they are variables of integration rather than dynamical variables. Integration over the a-type φ's is to be understood as an infinite limit of a multiple Berezin integral, obeying the rules set forth in section 1.3. The volume element for the combined integration, c-type as well as a-type, is denoted by $d\varphi$. The integral (5.3.42) is, of course, purely a formal expression. How it may be given precise meaning in specific cases will be shown in the following sections.

Proceeding purely formally, assuming the validity of integrating by parts, and making use of eqs. (5.3.33) and (5.3.41), one may now write

$$
\begin{aligned}
&\int F[\varphi]\frac{\overleftarrow{\delta}}{\mathrm{i}\delta\varphi^i}\mathrm{e}^{\mathrm{i}J\varphi}\,\mathrm{d}\varphi \\
&= -(-1)^i\int F[\varphi]\frac{\overrightarrow{\delta}}{\mathrm{i}\delta\varphi^i}\mathrm{e}^{\mathrm{i}J\varphi}\,\mathrm{d}\varphi \\
&= -\int F[\varphi]J_i\mathrm{e}^{\mathrm{i}J\varphi}\,\mathrm{d}\varphi = -(-1)^{iF}J_i\langle \mathrm{out}|\mathrm{in}\rangle \\
&=(-1)^{iF}\langle\mathrm{out}|\, T\left(\left\{S[\varphi]+\frac{\mathrm{i}}{2}\ln\operatorname{sdet}G^+[\varphi]\right\}\frac{\overleftarrow{\delta}}{\delta\varphi^i}\right)|\mathrm{in}\rangle \\
&=(-1)^{iF}\left[\left(S[\varphi]+\frac{\mathrm{i}}{2}\ln\operatorname{sdet}G^+[\varphi]\right)\frac{\overleftarrow{\delta}}{\delta\varphi^i}\right]_{\varphi\to\overrightarrow{\delta}/\mathrm{i}\delta J}\langle\mathrm{out}|\mathrm{in}\rangle \\
&=(-1)^{iF}\int\left(S[\varphi]+\frac{\mathrm{i}}{2}\ln\operatorname{sdet}G^+[\varphi]\right)\frac{\overleftarrow{\delta}}{\delta\varphi^i}F[\varphi]\mathrm{e}^{\mathrm{i}J\varphi}\,\mathrm{d}\varphi. \qquad (5.3.43)
\end{aligned}
$$

The functional Fourier transform of this equation is simply

$$F[\varphi]\frac{\overleftarrow{\delta}}{\mathrm{i}\delta\varphi^i}=(-1)^{iF}\left(S[\varphi]+\frac{\mathrm{i}}{2}\ln\operatorname{sdet}G^+[\varphi]\right)\frac{\overleftarrow{\delta}}{\delta\varphi^i}F[\varphi], \qquad (5.3.44)$$

which implies

$$
\begin{aligned}
F[\varphi] &= Z\mathrm{e}^{\mathrm{i}(S[\varphi]+(\mathrm{i}/2)\ln\operatorname{sdet}G^+[\varphi])} \\
&= Z\mathrm{e}^{\mathrm{i}S[\varphi]}(\operatorname{sdet}G^+[\varphi])^{-\frac{1}{2}}, \qquad (5.3.45)
\end{aligned}
$$

where Z is a constant of integration. Hence finally

$$\langle\mathrm{out}|\mathrm{in}\rangle = Z\int \mathrm{e}^{\mathrm{i}(S[\varphi]+J\varphi)}(\operatorname{sdet}G^+[\varphi])^{-\frac{1}{2}}\,\mathrm{d}\varphi. \qquad (5.3.46)$$

Expression (5.3.46) is known as *Feynman's functional integral* or *sum over histories*. Two remarks must be made in regard to its use:

(1) The constant of integration Z is not determined by the above analysis. It obviously depends in part on the normalization adopted for the supervectors $|\mathrm{in}\rangle$ and $|\mathrm{out}\rangle$. In specific cases its logarithm is found to diverge and hence it can actually be computed only formally. Fortunately one does not generally have to know Z. Only the ratio of $\langle\mathrm{out}|\mathrm{in}\rangle$ to some fiducial amplitude is needed in practice, and then the Z's cancel out.

(2) The choice of action function $S[\varphi]$ appearing in (5.3.46) must be adapted to the boundary conditions appropriate to the supervectors $|\mathrm{in}\rangle$ and $|\mathrm{out}\rangle$. How this choice is made in specific cases will appear in the following sections.

5.4 A simple Fermi system

Action functional and Green's functions

Perhaps the simplest of all dynamical systems is that which is described by the action functional

$$S \equiv \mathrm{i} \int (\tfrac{1}{2} x\dot{x} + \eta x)\,\mathrm{d}t. \tag{5.4.1}$$

Here η is a given real *a*-number-valued function of the time t (i.e., an external source), x is a real *a*-type dynamical variable, and the dot denotes differentiation with respect to t. At the superclassical level x is *a*-number-valued; in the quantum theory it is a self-adjoint *a*-type linear operator. It will be assumed to be a physical observable.

The action functional (5.4.1) yields the dynamical equation

$$0 = \frac{\overrightarrow{\delta}}{\delta x(t)} S \equiv \mathrm{i}(\dot{x} - \eta), \tag{5.4.2}$$

which is linear inhomogeneous. The second functional derivative of S is

$$\frac{\overrightarrow{\partial}}{\partial x(t)} S \frac{\overleftarrow{\delta}}{\delta x(t')} \equiv \mathrm{i} \frac{\partial}{\partial t} \delta(t, t'), \tag{5.4.3}$$

and the Green's functions for small disturbances are simply the Green's functions of the operator $\mathrm{i}\partial/\partial t$. One readily verifies that they are given by

$$G^-(t, t') = \mathrm{i}\theta(t, t'), \tag{5.4.4}$$

$$G^+(t, t') = -\mathrm{i}\theta(t', t), \tag{5.4.5}$$

yielding an almost trivial supercommutator function:

$$\tilde{G}(t, t') = G^+(t, t') - G^-(t, t') = -\mathrm{i}. \tag{5.4.6}$$

The quantization rule (5.3.1) takes the form

$$[x(t), x(t')] = \mathrm{i}\tilde{G}(t, t') = 1, \tag{5.4.7}$$

where the supercommutator is here an anticommutator.

Eigenvectors of x

Setting $t' = t$ in eq. (5.4.7) one obtains, for all t,

$$x^2 = \tfrac{1}{2}, \tag{5.4.8}$$

from which it may be inferred that the eigenvalues of x are $\pm 1/\sqrt{2}$. When the external source η vanishes the system becomes completely trivial, because x is then a constant operator, and if the state vector is an eigenvector of $x(t)$ at one instant of time it is an eigenvector forever. When η is nonvanishing the system is nonphysical but less trivial.

Let $|\pm 1/\sqrt{2}, t\rangle$ be physical eigenvectors of x at the time t:

$$x(t)|\pm 1/\sqrt{2}, t\rangle = \pm \frac{1}{\sqrt{2}}|\pm 1/\sqrt{2}, t\rangle. \qquad (5.4.9)$$

Assume them to be normalized so that

$$\langle \pm 1/\sqrt{2}, t | \pm 1/\sqrt{2}, t\rangle = 1, \quad \langle \pm 1/\sqrt{2}, t | \mp 1/\sqrt{2}, t\rangle = 0. \qquad (5.4.10)$$

Note that they are necessarily impure, for if they were pure eq. (5.4.9) would be an identity between supervectors of opposite type. On the other hand they suffice to form a complete basis for a super Hilbert space, yielding a minimal representation for the operator algebra (5.4.7). The minimal super Hilbert space has total dimension 2, which is finite. It therefore has a pure basis (see section 1.4). Since normalized impure supervectors exist, its dimension is necessarily (1, 1).

The energy

Before constructing a pure basis it will be convenient to introduce the energy or Hamiltonian:

$$H = L\frac{\overleftarrow{\partial}}{\partial \dot{x}}\dot{x} - L = -\mathrm{i}\eta x, \qquad (5.4.11)$$

where the Lagrangian L is just expression (5.4.1) with the symbols $\int$ and $\mathrm{d}t$ removed. The energy is c-type but, owing to the presence of the a-number source η, unphysical. That it generates time displacements follows from the equal-time commutator

$$-\mathrm{i}[x, H] = -x\eta x + \eta x^2 = 2\eta x^2 = \eta = \dot{x}. \qquad (5.4.12)$$

Evidently

$$x(t + \delta t) = [1 + \mathrm{i}H(t)\delta t]x(t)[1 - \mathrm{i}H(t)\,\delta t], \qquad (5.4.13)$$

whence, with a natural choice of phase for the eigenvectors $|\pm 1/\sqrt{2}, t\rangle$ at different times,

$$|\pm 1/\sqrt{2}, t + \delta t\rangle = [1 + \mathrm{i}H(t)\,\delta t]|\pm 1/\sqrt{2}, t\rangle, \qquad (5.4.14)$$

$$\begin{aligned}\frac{\partial}{\partial t}|\pm 1/\sqrt{2}, t\rangle &= \mathrm{i}H(t)|\pm 1/\sqrt{2}, t\rangle \\ &= \eta(t)x(t)|\pm 1/\sqrt{2}, t\rangle \\ &= \pm\frac{1}{\sqrt{2}}\eta(t)|\pm 1/\sqrt{2}, t\rangle. \end{aligned} \qquad (5.4.15)$$

A pure basis

Since H is c-type it leaves supervector types unchanged. This means that if the supervector

$$|c,t\rangle \stackrel{\text{def}}{=} |1/\sqrt{2},t\rangle\alpha + |-1/\sqrt{2},t\rangle\beta, \tag{5.4.16}$$

where α and β are constant supernumbers, is of a given type at one instant of time then it is of the same type for all t. Suppose it is c-type. Let $|a,t\rangle$ be the a-type supervector into which $|c,t\rangle$ is transformed by application of the operator $\sqrt{2}x(t)$:

$$|a,t\rangle \stackrel{\text{def}}{=} |1/\sqrt{2},t\rangle\alpha - |-1/\sqrt{2},t\rangle\beta. \tag{5.4.17}$$

Then $|c,t\rangle$ and $|a,t\rangle$ will together form an orthonormal pure basis satisfying

$$\langle c,t|c,t\rangle = \langle a,t|a,t\rangle = 1, \qquad \langle c,t|a,t\rangle = 0, \tag{5.4.18}$$

provided

$$\alpha^*\alpha = \beta^*\beta = \tfrac{1}{2}. \tag{5.4.19}$$

The general solution of eq. (5.4.19) is

$$\alpha = \frac{1}{\sqrt{2}}e^{i\varphi_1}, \qquad \beta = \frac{1}{\sqrt{2}}e^{i\varphi_2}, \tag{5.4.20}$$

where φ_1 and φ_2 are arbitrary real c-numbers. The simplest choice is $\varphi_1 = \varphi_2 = 0$ which yields

$$|c,t\rangle = \frac{1}{\sqrt{2}}|1/\sqrt{2},t\rangle + \frac{1}{\sqrt{2}}|-1/\sqrt{2},t\rangle, \tag{5.4.21}$$

$$|a,t\rangle = \frac{1}{\sqrt{2}}|1/\sqrt{2},t\rangle - \frac{1}{\sqrt{2}}|-1/\sqrt{2},t\rangle, \tag{5.4.22}$$

but it will be observed that one could equally well interchange the right-hand sides of these equations. This means that there is an arbitrariness in choosing the supervectors that one calls c-type and the supervectors that one calls a-type. This arbitrariness is characteristic of Fermi systems and will be encountered again in the case of the Fermi oscillator. Note that with either of the above choices one has

$$\frac{\partial}{\partial t}|c,t\rangle = \frac{1}{\sqrt{2}}\eta(t)|a,t\rangle, \qquad \frac{\partial}{\partial t}|a,t\rangle = \frac{1}{\sqrt{2}}\eta(t)|c,t\rangle. \tag{5.4.23}$$

An alternative representation

Let x' be a variable real a-number. The differential operator $\frac{1}{2}x' + \vec{\partial}/\partial x'$ satisfies the same algebra as the super Hilbert space operator $x(t)$:

$$\left(\tfrac{1}{2}x' + \frac{\partial}{\partial x'}\right)^2 = \tfrac{1}{2}. \tag{5.4.24}$$

Therefore a class of supervectors $|x', t\rangle$ may be introduced satisfying

$$\langle x', t|x(t) = \left(\tfrac{1}{2}x' + \frac{\vec{\partial}}{\partial x'}\right)\langle x', t|. \tag{5.4.25}$$

Note that $\frac{1}{2}x' + \vec{\partial}/\partial x'$ is not the only differential operator satisfying (5.4.24); any operator $\alpha x' + \beta\vec{\partial}/\partial x'$ where α and β are c-numbers satisfying $\alpha\beta = \frac{1}{2}$ would do as well. However, it turns out that the present choice enables one to make easy contact with the Feynman functional integral for the system.

It is not difficult to verify that, *modulo* phase factors, eq. (5.4.25) has two distinct solutions satisfying

$$\langle x', t|x', t\rangle = 1, \tag{5.4.26}$$

namely

$$\langle x', t| = \langle c, t| + \frac{1}{\sqrt{2}}x'\langle a, t| \tag{5.4.27}$$

and the same expression with $\langle c, t|$ and $\langle a, t|$ interchanged. We shall choose the solution (5.4.27), which has the property that the $\langle x', t|$ are then all c-type. Note that this choice may also be written

$$\langle x', t| = \langle c, t| - \frac{1}{\sqrt{2}}\langle a, t|x'. \tag{5.4.28}$$

The supervectors $|x', t\rangle$ do *not* form a complete set. However, they do satisfy an important integral identity. Using eqs. (1.3.5), (1.3.6) and (1.3.15) one has

$$\begin{aligned}\left(\frac{\mathrm{i}}{\pi}\right)^{\frac{1}{2}}&\int |x', t\rangle\langle x', t|\,\mathrm{d}x' \\ &= \left(\frac{\mathrm{i}}{\pi}\right)^{\frac{1}{2}}\int\left(|c, t\rangle + \frac{1}{\sqrt{2}}|a, t\rangle x'\right)\left(\langle c, t| - \frac{1}{\sqrt{2}}\langle a, t|x'\right)\mathrm{d}x' \\ &= P,\end{aligned} \tag{5.4.29}$$

where

$$P \stackrel{\text{def}}{=} \mathrm{i}(|a, t\rangle\langle c, t| - |c, t\rangle\langle a, t|). \tag{5.4.30}$$

The operator P a self-adjoint and a-type, satisfies

$$P^2 = |c,t\rangle\langle c,t| + |a,t\rangle\langle a,t| = 1, \tag{5.4.31}$$

$$P|c,t\rangle = \mathrm{i}|a,t\rangle, \qquad P|a,t\rangle = -\mathrm{i}|c,t\rangle, \tag{5.4.32}$$

and supercommutes (anticommutes) with both $x(t)$ and $\eta(t)$. It therefore supercommutes (commutes) with the Hamiltonian and is t-independent.

Two other useful relations satisfied by the $|x',t\rangle$ are

$$\langle x'',t|x',t\rangle = 1 + \tfrac{1}{2}x''x', \tag{5.4.33}$$

and

$$\begin{aligned}\langle x'',t|P|x',t\rangle &= \left(\langle c,t| + \frac{1}{\sqrt{2}}x''\langle a,t|\right)P\left(|c,t\rangle + \frac{1}{\sqrt{2}}|a,t\rangle x'\right)\\ &= \mathrm{i}\left(\langle c,t| + \frac{1}{\sqrt{2}}x''\langle a,t|\right)\left(|a,t\rangle - \frac{1}{\sqrt{2}}|c,t\rangle x'\right)\\ &= \frac{\mathrm{i}}{\sqrt{2}}(x''-x') = -\left(\frac{\pi}{\mathrm{i}}\right)^{\frac{1}{2}}\delta(x''-x'),\end{aligned} \tag{5.4.34}$$

where the δ-function in the final expression is the function defined by eqs. (1.3.10) and (1.3.15).

The functional integral representation of $\langle x'',t''|x',t'\rangle$

The inner product $\langle x'',t''|x',t'\rangle$ has a remarkable representation as a functional integral, which can be obtained by first computing $\langle x'',t''|x',t'\rangle$ for the case in which t' and t'' are infinitesimally close together. Using eqs. (5.4.23), (5.4.27) and (5.4.33) one has

$$\begin{aligned}&\langle x'',t+\delta t|x',t\rangle\\ &= \langle x'',t|x',t\rangle + \delta t\left[\frac{\partial}{\partial t}\left(\langle c,t| - \frac{1}{\sqrt{2}}\langle a,t|x''\right)\right]\left(|c,t\rangle + \frac{1}{\sqrt{2}}|a,t\rangle x'\right)\\ &= 1 + \tfrac{1}{2}x''x' + \delta t\frac{1}{\sqrt{2}}\left(\langle a,t|\eta(t) - \frac{1}{\sqrt{2}}\langle c,t|\eta(t)x''\right)\left(|c,t\rangle + \frac{1}{\sqrt{2}}|a,t\rangle x'\right)\\ &= 1 + \tfrac{1}{2}x''x' - \tfrac{1}{2}\eta(t)(x''+x')\,\delta t\\ &= \exp[\tfrac{1}{2}x''x' - \tfrac{1}{2}\eta(t)(x''+x')\,\delta t]\end{aligned} \tag{5.4.35}$$

and hence, for infinitesimal $t''-t'$,

$$\langle x'',t''|x',t'\rangle = \exp[\tfrac{1}{2}x''x' - \tfrac{1}{2}\eta(\tfrac{1}{2}(t''+t'))(x''+x')(t''-t')]. \tag{5.4.36}$$

Now let $x(t)$ be a straight-line path between the a-numbers x' and x''

at the times t' and t'' respectively:

$$x(t) = x' + (t - t')\frac{x'' - x'}{t'' - t'}. \qquad (5.4.37)$$

It is easy to verify that

$$\int_{t'}^{t''} (\tfrac{1}{2}x\dot{x} + \eta x)\,\mathrm{d}t = -\tfrac{1}{2}x''x' + \tfrac{1}{2}\eta(\tfrac{1}{2}(t'' + t'))(x'' + x')(t'' - t'), \qquad (5.4.38)$$

correct to first order in $t'' - t'$. Therefore eq. (5.4.36) may be rewritten in the form

$$\langle x'', t''|x', t'\rangle = \exp\left[-\int_{x',t'}^{x'',t''} (\tfrac{1}{2}x\dot{x} + \eta x)\,\mathrm{d}t\right], \qquad (5.4.39)$$

provided it is understood that the integral in the exponent is to be evaluated for the straight-line path. The exponent is seen to be just i times the action functional (5.4.1) for this path.

When $t'' - t'$ is finite one may proceed as follows. Let N be a large *even* integer and let $t_0, t_1, t_2, \ldots, t_N, t_{N+1}$ be a sequence of times satisfying

$$t'' = t_{N+1} > t_N > \cdots > t_2 > t_1 > t_0 = t'. \qquad (5.4.40)$$

Set

$$x_{N+1} = x'', \qquad x_0 = x'. \qquad (5.4.41)$$

Then, making use of eq. (5.4.29) and noting that the symbol $\mathrm{d}x'$ appearing in this equation anticommutes with P, write

$$\int \langle x_{N+1}, t_{N+1}|x_N, t_N\rangle\langle x_N, t_N|x_{N-1}, t_{N-1}\rangle \cdots$$

$$\cdots\langle x_2, t_2|x_1, t_1\rangle\langle x_1, t_1|x_0, t_0\rangle\left(-\frac{\mathrm{i}}{\pi}\right)^{N/2} \mathrm{d}x_1\,\mathrm{d}x_2\ldots\mathrm{d}x_N$$

$$= \langle x_{N+1}, t_{N+1}|P^N|x_0, t_0\rangle = \langle x_{N+1}, t_{N+1}|x_0, t_0\rangle$$

$$= \langle x'', t''|x', t'\rangle. \qquad (5.4.42)$$

Finally, pass to the limit $N \to \infty$, $t_n - t_{n-1} \to 0$ for all n, and make use of (5.4.39), obtaining

$$\langle x'', t''|x', t'\rangle = Z\int \exp\left[-\int_{x',t'}^{x'',t''} (\tfrac{1}{2}x\dot{x} + \eta x)\,\mathrm{d}t\right]\mathrm{d}x, \qquad (5.4.43)$$

where, formally,

$$Z = \left(-\frac{\mathrm{i}}{\pi}\right)^{(t''-t')/2\mathrm{d}t - \frac{1}{2}}, \qquad (5.4.44)$$

$$\mathrm{d}x = \prod_{t'<t<t''} \mathrm{d}x(t). \qquad (5.4.45)$$

Equation (5.4.43) is a special case of (5.3.46), in which $(\mathrm{sdet}\,G^{+})^{-\frac{1}{2}}$, which is here just a constant, has been absorbed into the factor Z. The method by which (5.4.43) has been derived serves to make precise the meaning of the Feynman functional integral in the present context. Several remarks are in order: (1) The functional integral (5.4.43) is not a sum over paths joining eigenvalues of the operator x. The supervectors $|x',t\rangle$ are not eigenvectors of $x(t)$ and the a-numbers x' are not its eigenvalues. The operator $x(t)$, being a physical observable, has only c-numbers as eigenvalues. The paths in question are paths for an associated nonphysical superclassical system. (2) Since the dynamical equation is a first-order equation there is, at each instant of time, only a single Cauchy datum for the superclassical motion, namely the value of x at that time. Therefore, except in the special case $x'' = x' + \int_{t'}^{t''} \eta(t)\,dt$, the path that makes the exponent stationary in (5.4.43) cannot be a smooth path satisfying eq. (5.4.2). (3) The paths that are summed over in the functional integral are *broken* paths consisting always of an *odd* number of straight-line segments, the number tending to infinity in the limit.

Evaluation of the functional integral

Insertion of (5.4.36) into the left-hand side of eq. (5.4.42) leads to the following approximation for the exponent appearing in (5.4.43):

$$\begin{aligned}
-\int_{x',t'}^{x'',t''} &(\tfrac{1}{2}x\dot{x} + \eta x)\mathrm{d}t \\
&\approx -\tfrac{1}{2}\eta_{N+\frac{1}{2}} x_{N+1}\,\Delta t + \tfrac{1}{2}x_{N+1}x_N - \tfrac{1}{2}\eta_{N+\frac{1}{2}} x_N \Delta t \\
&\quad -\tfrac{1}{2}\eta_{N-\frac{1}{2}} x_N\,\Delta t + \tfrac{1}{2}x_N x_{N-1} - \tfrac{1}{2}\eta_{N-\frac{1}{2}} x_{N-1}\,\Delta t \\
&\quad - \vdots \\
&\quad -\tfrac{1}{2}\eta_{\frac{3}{2}} x_2\,\Delta t + \tfrac{1}{2}x_2 x_1 - \tfrac{1}{2}\eta_{\frac{3}{2}} x_1\,\Delta t \\
&\quad -\tfrac{1}{2}\eta_{\frac{1}{2}} x_1\,\Delta t + \tfrac{1}{2}x_1 x_0 - \tfrac{1}{2}\eta_{\frac{1}{2}} x_0\,\Delta t.
\end{aligned} \tag{5.4.46}$$

Here all the time intervals have been chosen equal to

$$\Delta t = \frac{t'' - t'}{N+1} \tag{5.4.47}$$

and the η's are defined by

$$\eta_{n+\frac{1}{2}} \stackrel{\text{def}}{=} \eta(t' + (n+\tfrac{1}{2})\Delta t). \tag{5.4.48}$$

Since none of the x's appears more than quadratically in (5.4.46), the functional integral (5.4.43) is Gaussian.

It is easy to show (see exercise **5.2**) that the value of a Gaussian integral is just the integrand evaluated at the stationary point of the exponent, times the usual Gaussian factor (1.7.64) which includes the inverse square root of a superdeterminant.† In the present case the Gaussian factor is independent of the external source and can be absorbed into the normalizing constant Z. One therefore needs only to evaluate the integrand at the stationary point.

The stationary point is determined by equating to zero the derivatives of expression (5.4.46) with respect to $x_1, x_2, \ldots, x_N$:

$$\left.\begin{aligned} &\tfrac{1}{2}(x_{N+1} - x_{N-1}) - \tfrac{1}{2}(\eta_{N+\frac{1}{2}} + \eta_{N-\frac{1}{2}})\Delta t = 0, \\ &\tfrac{1}{2}(x_N - x_{N-2}) - \tfrac{1}{2}(\eta_{N-\frac{1}{2}} + \eta_{N-\frac{3}{2}})\Delta t = 0, \\ &\vdots \\ &\tfrac{1}{2}(x_3 - x_1) - \tfrac{1}{2}(\eta_{\frac{5}{2}} + \eta_{\frac{3}{2}})\Delta t = 0, \\ &\tfrac{1}{2}(x_2 - x_0) - \tfrac{1}{2}(\eta_{\frac{3}{2}} + \eta_{\frac{1}{2}})\Delta t = 0. \end{aligned}\right\} \tag{5.4.49}$$

In the limit $N \to \infty$ these equations imply

$$x_n = \begin{cases} x_{N+1} - \displaystyle\int_{t'+n\Delta t}^{t''} \eta(t)\mathrm{d}t, & n \text{ odd}, \\[2ex] x_0 + \displaystyle\int_{t'}^{t'+n\Delta t} \eta(t)\mathrm{d}t, & n \text{ even}. \end{cases} \tag{5.4.50}$$

The stationary path is seen to be a broken-line path that *oscillates* infinitely rapidly between two solutions of eq. (5.4.2), namely the solution determined by the Cauchy datum $x(t') = x_0 = x'$ and that determined by the Cauchy datum $x(t'') = x_{N+1} = x''$. These two solutions may be referred to as the *retarded* and *advanced superclassical trajectories* respectively.

Inserting expressions (5.4.50) into (5.4.46) one obtains.

$$\begin{aligned} &-\tfrac{1}{2}\eta_{N+\frac{1}{2}} x'' \Delta t + \tfrac{1}{2}x''\left[x' + \int_{t'}^{t'+N\Delta t} \eta(t)\mathrm{d}t\right] \\ &-\tfrac{1}{2}\eta_{N+\frac{1}{2}}\left[x' + \int_{t'}^{t'+N\Delta t} \eta(t)\mathrm{d}t\right]\Delta t - \tfrac{1}{2}\eta_{N-\frac{1}{2}}\left[x' + \int_{t'}^{t'+N\Delta t} \eta(t)\mathrm{d}t\right]\Delta t \\ &+\frac{1}{2}\left[x' + \int_{t'}^{t'+N\Delta t} \eta(t)\mathrm{d}t\right]\left[x'' - \int_{t'+(N-1)\Delta t}^{t''} \eta(t)\mathrm{d}t\right] \\ &-\tfrac{1}{2}\eta_{N-\frac{3}{2}}\left[x'' - \int_{t'+(N-1)\Delta t}^{t''} \eta(t)\mathrm{d}t\right]\Delta t - \cdots - \tfrac{1}{2}\eta_{\frac{3}{2}}\left[x' + \int_{t'}^{t'+2\Delta t} \eta(t)\mathrm{d}t\right]\Delta t \end{aligned}$$

† Note that the superdeterminant in the present case is nonsingular only if the integer N is even.

$$+\frac{1}{2}\left[x'+\int_{t'}^{t'+2\Delta t}\eta(t)\mathrm{d}t\right]\left[x''-\int_{t'+\Delta t}^{t''}\eta(t)\mathrm{d}t\right]$$

$$-\tfrac{1}{2}\eta_{\frac{3}{2}}\left[x''-\int_{t'+\Delta t}^{t''}\eta(t)\mathrm{d}t\right]\Delta t-\tfrac{1}{2}\eta_{\frac{1}{2}}\left[x''-\int_{t'+\Delta t}^{t''}\eta(t)\mathrm{d}t\right]\Delta t$$

$$+\frac{1}{2}\left[x''-\int_{t'+\Delta t}^{t''}\eta(t)\mathrm{d}t\right]x'-\tfrac{1}{2}\eta_{\frac{1}{2}}x'\,\Delta t.$$

When all the factors in this expression are multiplied out seven distinct types of terms are encountered: (1) Terms of the forms $-\frac{1}{2}\eta_{n+\frac{1}{2}}x''\,\Delta t$ and $-\frac{1}{2}\eta_{n+\frac{1}{2}}x'\,\Delta t$. In the limit $N\to\infty$ these all add up to $\frac{1}{2}(x''+x')\times\int_{t'}^{t''}\eta(t)\mathrm{d}t$. (2) Terms of the forms $\frac{1}{2}x''x'$ and $\frac{1}{2}x'x''$. There are $N/2+1$ of the former and $N/2$ of the latter. Their total contribution is therefore $\frac{1}{2}x''x'$. (3) Terms of the forms $\frac{1}{2}x''\int_{t'}^{t'+n\Delta t}\eta(t)\mathrm{d}t$ and $\frac{1}{2}\int_{t'}^{t'+n\Delta t}\eta(t)\mathrm{d}t\,x''$, with n even. Because of the anticommutativity of x'' and η these terms mutually cancel. (4) Terms of the forms $-\frac{1}{2}x'\int_{t'+n\Delta t}^{t''}\eta(t)\mathrm{d}t$ and $-\frac{1}{2}\int_{t'+n\Delta t}^{t''}\eta(t)\mathrm{d}t\,x'$, with n odd. These too mutually cancel. (5) Terms of the forms $-\frac{1}{2}\eta_{n+\frac{1}{2}}\int_{t'}^{t'+n\Delta t}\eta(t)\mathrm{d}t\,\Delta t$ and $-\frac{1}{2}\eta_{n-\frac{1}{2}}\int_{t'}^{t'+n\Delta t}\eta(t)\,\mathrm{d}t\,\Delta t$, with n even. In the limit $N\to\infty$ these all add up to $-\frac{1}{2}\int_{t'}^{t''}\mathrm{d}s\int_{t'}^{s}\mathrm{d}t\,\eta(s)\eta(t)$. (6) Terms of the forms $\frac{1}{2}\eta_{n+\frac{1}{2}}\int_{t'+n\Delta t}^{t''}\eta(t)\mathrm{d}t\,\Delta t$ and $\frac{1}{2}\eta_{n-\frac{1}{2}}\int_{t'+n\Delta t}^{t''}\eta(t)\mathrm{d}t\,\Delta t$, with n odd. In the limit $N\to\infty$ these all add up to $\frac{1}{2}\int_{t'}^{t''}\mathrm{d}s\int_{s}^{t''}\mathrm{d}t\,\eta(s)\eta(t)$. (7) Terms of the forms $-\frac{1}{2}\int_{t'}^{t'+n\Delta t}\eta(t)\mathrm{d}t\int_{t'+(n-1)\Delta t}^{t''}\eta(t)\mathrm{d}t$ and $-\frac{1}{2}\int_{t'+(n-1)\Delta t}^{t''}\eta(t)\mathrm{d}t\int_{t'}^{t'+(n-2)\Delta t}\eta(t)\mathrm{d}t$, with n even. Because of the anticommutativity of the η's these terms, pairwise, give $-\frac{1}{2}\int_{t'+(n-2)\Delta t}^{t'+n\Delta t}\eta(t)\mathrm{d}t\times\int_{t'+(n-1)\Delta t}^{t''}\eta(t)\mathrm{d}t$ and, in the limit $N\to\infty$, all add up to $-\frac{1}{2}\int_{t'}^{t''}\mathrm{d}s\int_{s}^{t''}\mathrm{d}t\times\eta(s)\eta(t)$, thus cancelling the contribution from the type-6 terms.

The whole analysis yields

$$-\int_{x',t'}^{x'',t''}(\tfrac{1}{2}x\dot{x}+\eta x)_{\text{stationary}}\mathrm{d}t$$

$$=\tfrac{1}{2}x''x'+\tfrac{1}{2}(x''+x')\int_{t'}^{t''}\eta(t)\mathrm{d}t-\tfrac{1}{2}\int_{t'}^{t''}\mathrm{d}s\int_{t'}^{s}\mathrm{d}t\,\eta(s)\eta(t),\qquad(5.4.51)$$

and hence, in view of the boundary condition (5.4.36),

$$\langle x'',t''|x',t'\rangle$$

$$=\exp\left[\tfrac{1}{2}x''x'+\tfrac{1}{2}(x''+x')\int_{t'}^{t''}\eta(t)\mathrm{d}t-\tfrac{1}{2}\int_{t'}^{t''}\mathrm{d}s\int_{t'}^{s}\mathrm{d}t\,\eta(s)\eta(t)\right]$$

$$=\left[1+\tfrac{1}{2}x''x'+\tfrac{1}{2}(x''+x')\int_{t'}^{t''}\eta(t)\mathrm{d}t\right]\exp\left[-\tfrac{1}{2}\int_{t'}^{t''}\mathrm{d}s\int_{t'}^{s}\mathrm{d}t\,\eta(s)\eta(t)\right],$$

(5.4.52)

a result that is not so easy to obtain by other means. Although it was obtained under the assumption that $t'' > t'$, complex conjugation and reordering of factors show that it holds also for $t'' < t'$.

Expression (5.4.52) can be used to obtain other amplitudes. For example,

$$\begin{aligned}\langle c,t''|c,t'\rangle &= -\frac{i}{\pi}\int \langle c,t''|x'',t''\rangle\langle x'',t''|x',t'\rangle\langle x',t'|c,t'\rangle \mathrm{d}x'\,\mathrm{d}x'' \\ &= -\frac{i}{\pi}\int\left[1+\tfrac{1}{2}x''x' + \tfrac{1}{2}(x''+x')\int_{t'}^{t''}\eta(t)\mathrm{d}t\right]\mathrm{d}x'\,\mathrm{d}x'' \\ &\quad\times \exp\left[-\frac{1}{2}\int_{t'}^{t''}\mathrm{d}s\int_{t'}^{s}\mathrm{d}t\,\eta(s)\eta(t)\right] \\ &= \exp\left[-\frac{1}{2}\int_{t'}^{t''}\mathrm{d}s\int_{t'}^{s}\mathrm{d}t\,\eta(s)\eta(t)\right],\end{aligned} \tag{5.4.53}$$

$$\begin{aligned}\langle a,t''|c,t'\rangle &= -\frac{i}{\pi}\int \langle a,t''|x'',t''\rangle\langle x'',t''|x',t'\rangle\langle x',t'|c,t'\rangle \mathrm{d}x'\,\mathrm{d}x'' \\ &= -\frac{i}{\pi}\int \frac{1}{\sqrt{2}}x''\left[1+\tfrac{1}{2}x''x' + \tfrac{1}{2}(x''+x')\int_{t'}^{t''}\eta(t)\mathrm{d}t\right]\mathrm{d}x'\,\mathrm{d}x'' \\ &\quad\times \exp\left[-\frac{1}{2}\int_{t'}^{t''}\mathrm{d}s\int_{t'}^{s}\mathrm{d}t\,\eta(s)\eta(t)\right] \\ &= \frac{1}{\sqrt{2}}\int_{t'}^{t''}\eta(t)\mathrm{d}t\,\exp\left[-\frac{1}{2}\int_{t'}^{t''}\mathrm{d}s\int_{t'}^{s}\mathrm{d}t\,\eta(s)\eta(t)\right]\end{aligned} \tag{5.4.54}$$

and, similarly,

$$\langle c,t''|a,t'\rangle = -\frac{1}{\sqrt{2}}\int_{t'}^{t''}\eta(t)\mathrm{d}t\,\exp\left[-\frac{1}{2}\int_{t'}^{t''}\mathrm{d}s\int_{t'}^{s}\mathrm{d}t\,\eta(s)\eta(t)\right] \tag{5.4.55}$$

$$\langle a,t''|a,t'\rangle = \exp\left[-\frac{1}{2}\int_{t'}^{t''}\mathrm{d}s\int_{t'}^{s}\mathrm{d}t\,\eta(s)\eta(t)\right]. \tag{5.4.56}$$

The consistency of these results with eqs. (5.4.23) is readily checked.

The average superclassical trajectory

Another check on expression (5.4.52) may be obtained by functionally differentiating it with respect to $\eta(t)$ and using the Schwinger variational principle. One easily finds

$$\begin{aligned}\langle x'',t''|x(t)|x',t'\rangle &= -\frac{\vec{\delta}}{\delta\eta(t)}\langle x'',t''|x',t'\rangle \\ &= x_{\mathrm{av}}(t)\langle x'',t''|x',t'\rangle,\end{aligned} \tag{5.4.57}$$

where

$$x_{\text{av}}(t) \stackrel{\text{def}}{=} \tfrac{1}{2}(x'' + x') + \frac{1}{2}\int_{t'}^{t} \eta(s)\text{d}s - \frac{1}{2}\int_{t}^{t''} \eta(s)\text{d}s. \tag{5.4.58}$$

The function $x_{\text{av}}(t)$ is the *average* of the retarded and advanced superclassical trajectories.

That it should be the average trajectory that appears here becomes clear when one notes that in every row of the sum (5.4.46) the source $\eta_{n+\frac{1}{2}}$ is coupled equally to x_{n+1} and to x_n. Functional differentiation with respect to $\eta_{n+\frac{1}{2}}$ therefore brings down a factor $\frac{1}{2}(x_{n+1} + x_n)$ in the integrand of the Gaussian integral (5.4.43). It is well known that the resulting integral is just the original Gaussian integral times $\frac{1}{2}(x_{n+1} + x_n)$ evaluated at the stationary point. Equation (5.4.50) shows that the stationary values of x_n and x_{n+1} always lie at opposite ends of one of the broken-line segments of which the stationary path is composed, and hence their average lies half way between the retarded and advanced superclassical trajectories.

Propagator for $x_{\text{av}}(t)$

If the source η suffers an infinitesimal change $\delta\eta$ then the change in the average superclassical trajectory can be expressed as a Green's function integral:

$$\delta x_{\text{av}}(t) = -\int G(t,s)\text{i}\,\delta\eta(s)\text{d}s, \tag{5.4.59}$$

where

$$\begin{aligned} G(t,s) &= -x_{\text{av}}(t)\frac{\overleftarrow{\delta}}{\text{i}\,\delta\eta(s)} \\ &= \text{i}[\ln\langle x'',t''|x',t'\rangle]\frac{\overleftarrow{\delta}}{\delta\eta(t)}\frac{\overleftarrow{\delta}}{\delta\eta(s)} \\ &= \frac{\text{i}}{2}\theta(t,s) - \frac{\text{i}}{2}\theta(s,t) \\ &= \tfrac{1}{2}[G^{+}(t,s) + G^{-}(t,s)]. \end{aligned} \tag{5.4.60}$$

This function is called the *propagator for the average trajectory.* It is antisymmetric in s and t and satisfies the boundary conditions

$$\left.\begin{aligned} \tfrac{1}{2}[G(t',s) + G(t'',s)] = 0 \\ \tfrac{1}{2}[G(t,t') + G(t,t'')] = 0 \end{aligned}\right\} \quad t'' > t, s > t', \tag{5.4.61}$$

corresponding to the fact that the average trajectory is fixed not by specifying individual endpoint conditions x' and x'' but by specifying the average of these conditions.

5.5 The Fermi oscillator

Action functional and Green's functions

In contrast to the system considered in the preceding section the Fermi oscillator is both nontrivial and physically important. The configuration space of the previous system was $\mathbf{R}_a$; that of the Fermi oscillator is $\mathbf{R}_a^2$. The latter is described by *two* real *a*-type dynamical variables, which will be denoted by x_1 and x_2 respectively. It is convenient to assemble these variables into a column array

$$x \overset{\text{def}}{=} \begin{pmatrix} x_1 \\ x_2 \end{pmatrix} \tag{5.5.1}$$

and to introduce the matrix

$$M \overset{\text{def}}{=} \begin{pmatrix} 0 & 1 \\ -1 & 0 \end{pmatrix}. \tag{5.5.2}$$

The action functional is then given by

$$S = \frac{\mathrm{i}}{2}\int (x^{\sim}\dot{x} + \omega x^{\sim} M x)\mathrm{d}t = \frac{\mathrm{i}}{2}\int (x_1\dot{x}_1 + x_2\dot{x}_2 + \omega x_1 x_2 - \omega x_2 x_1)\mathrm{d}t, \tag{5.5.3}$$

where $x^{\sim}$ denotes the transpose of x and ω is a positive ordinary real number.

The dynamical equation is

$$0 = \frac{\vec{\delta}}{\delta x} S \equiv \mathrm{i}(\dot{x} + \omega M x), \tag{5.5.4}$$

which, because $M^2 = -\mathbf{1}_2$, implies also

$$\ddot{x} + \omega^2 x = 0. \tag{5.5.5}$$

The second functional derivative is

$$\frac{\vec{\delta}}{\delta x(t)} S \frac{\overleftarrow{\delta}}{\delta x(t')} \equiv \mathrm{i}\left(\mathbf{1}_2 \frac{\partial}{\partial t} + \omega M\right)\delta(t,t'), \tag{5.5.6}$$

and the retarded and advanced Green's functions of the operator $\mathrm{i}(\mathbf{1}_2\,\partial/\partial t + \omega M)$ are readily verified to be

$$G^{-}(t,t') = \mathrm{i}\theta(t,t')\begin{pmatrix} \cos\omega(t-t') & -\sin\omega(t-t') \\ \sin\omega(t-t') & \cos\omega(t-t') \end{pmatrix}, \tag{5.5.7}$$

$$G^+(t,t') = -\mathrm{i}\theta(t',t)\begin{pmatrix} \cos\omega(t-t') & -\sin\omega(t-t') \\ \sin\omega(t-t') & \cos\omega(t-t') \end{pmatrix}. \tag{5.5.8}$$

The supercommutator function is

$$\tilde{G}(t,t') = G^+(t,t') - G^-(t,t') = -\mathrm{i}\begin{pmatrix} \cos\omega(t-t') & -\sin\omega(t-t') \\ \sin\omega(t-t') & \cos\omega(t-t') \end{pmatrix}. \tag{5.5.9}$$

and the operator superalgebra of the quantum theory is defined by

$$[x(t), x^{\sim}(t')] = \mathrm{i}\tilde{G}(t,t') = \begin{pmatrix} \cos\omega(t-t') & -\sin\omega(t-t') \\ \sin\omega(t-t') & \cos\omega(t-t') \end{pmatrix}, \tag{5.5.10}$$

the supercommutator being here, as in the preceding section, an anticommutator.

Setting $t = t'$ in eq. (5.5.10) one finds

$$(x_1)^2 = (x_2)^2 = \tfrac{1}{2}, \quad \text{for all } t, \tag{5.5.11}$$

and hence the eigenvalues of both x_1 and x_2 are $\pm 1/\sqrt{2}$. The values of x_1 and x_2 cannot be specified simultaneously, however, because the two operators anticommute rather than commute with one another at equal times.

Mode functions and Hamiltonian

The general solution of the dynamical equation (5.5.4) is

$$x(t) = au(t) + a^*u^*(t) \tag{5.5.12}$$

where

$$u(t) \overset{\text{def}}{=} \begin{pmatrix} 1/\sqrt{2} \\ \mathrm{i}/\sqrt{2} \end{pmatrix} \mathrm{e}^{-\mathrm{i}\omega t}, \tag{5.5.13}$$

satisfying

$$u^{\sim}(t)u(t) = 0, \qquad u^{\dagger}(t)u(t) = 1, \tag{5.5.14}$$

'†' denoting the transpose conjugate. The functions $u(t)$ and $u^*(t)$ are known as *mode functions*. The supercommutator function has a simple expression in terms of them:

$$\tilde{G}(t,t') = G^{(+)}(t,t') + G^{(-)}(t,t'), \tag{5.5.15}$$

$$G^{(+)}(t,t') \overset{\text{def}}{=} -\mathrm{i}u(t)u^{\dagger}(t'), \tag{5.5.16}$$

$$G^{(-)}(t,t') \overset{\text{def}}{=} -\mathrm{i}u^*(t)u^{\sim}(t') = -G^{(+)}(t,t')^*. \tag{5.5.17}$$

The functions $G^{(+)}$ and $G^{(-)}$ are known as *positive* and *negative frequency functions* respectively.

In the superclassical theory the coefficient a in eq. (5.5.12) is an arbitrary complex a-number. In the quantum theory it becomes a non-self-adjoint a-type operator. By virtue of eqs. (5.5.14) it and its adjoint may be expressed in the forms

$$a = u^{\dagger}(t)x(t), \qquad a^* = x^{\sim}(t)u(t), \tag{5.5.18}$$

leading to the supercommutators

$$[a, a] = u^{\dagger}(t)[x(t), x^{\sim}(t)]u^*(t) = u^{\dagger}(t)\mathbf{1}_2 u^*(t) = 0, \tag{5.5.19}$$

$$[a, a^*] = u^{\dagger}(t)[x(t), x^{\sim}(t)]u(t) = u^{\dagger}(t)\mathbf{1}_2 u(t) = 1. \tag{5.5.20}$$

The reader is again reminded that these supercommutators are anticommutators. Hence

$$a^2 = 0 \tag{5.5.21}$$

and similarly

$$a^{*2} = 0. \tag{5.5.22}$$

Since all dynamical quantities can be expressed in terms of a and a^*, eqs. (5.5.20)–(5.5.22) may be used in place of eq. (5.5.10) to define the operator superalgebra of the theory.

The Lagrangian for the action (5.5.3) yields the following Hamiltonian or energy operator:

$$\begin{aligned} H &= L\frac{\overleftarrow{\partial}}{\partial \dot{x}}\dot{x} - L = -\frac{i}{2}\omega x^{\sim} M x = -\frac{i}{2}\omega(x_1 x_2 - x_2 x_1) \\ &= -i\omega x_1 x_2 = -\frac{i}{2}\omega(a e^{-i\omega t} + a^* e^{i\omega t})(i a e^{-i\omega t} - i a^* e^{i\omega t}) \\ &= \tfrac{1}{2}\omega(a^* a - a a^*) = \omega a^* a - \tfrac{1}{2}\omega. \end{aligned} \tag{5.5.23}$$

It is readily verified that

$$\dot{x} = -i[x, H]. \tag{5.5.24}$$

Basis supervectors

If the energy is to be a physical observable then the self-adjoint operator a^*a must be a physical observable. Denote the eigenvalues of a^*a by n and its corresponding physical eigenvectors by $|n\rangle$.

$$a^* a|n\rangle = n|n\rangle = |n\rangle n. \tag{5.5.25}$$

Assume the normalization

$$\langle n|n\rangle = 1. \tag{5.5.26}$$

Suppose there exists a nonvanishing n. Since

$$\langle n|a^*a|n\rangle = n,$$

$a|n\rangle$ cannot be the zero supervector. Furthermore, since

$$0 = a^*a^2|n\rangle = (a^*a)a|n\rangle$$

it follows that $a|n\rangle$ is an eigenvector of a^*a corresponding to the eigenvalue 0. Evidently, if a^*a has any eigenvalues at all, 0 is one of them. There is then also a corresponding physical eigenvector $|0\rangle$, which may be assumed to be normalized: $\langle 0|0\rangle = 1$.

Next note that

$$\langle 0|aa^*|0\rangle = \langle 0|[a, a^*]|0\rangle = \langle 0|0\rangle = 1, \tag{5.5.27}$$

which implies that $a^*|0\rangle$ is not the zero supervector. Moreover,

$$(a^*a)a^*|0\rangle = a^*[a, a^*]|0\rangle = a^*|0\rangle,$$

which, together with (5.5.27), implies (*modulo* an irrelevant phase factor) the identification

$$a^*|0\rangle = |1\rangle. \tag{5.5.28}$$

This, in turn, leads to

$$\left.\begin{aligned} a|1\rangle = aa^*|0\rangle = (1 - a^*a)|0\rangle = |0\rangle, \\ a|0\rangle = a^2|1\rangle = 0, \qquad a^*|1\rangle = a^{*2}|0\rangle = 0. \end{aligned}\right\} \tag{5.5.29}$$

The operator a^*a evidently has the two eigenvalues 0 and 1. Suppose there were another eigenvalue n. Then $a|n\rangle$ would be a nonvanishing eigenvector of a^*a corresponding to the eigenvalue 0. The simplest representation of the operator superalgebra is achieved by assuming $a|n\rangle$ to be proportional to $|0\rangle$:

$$a|n\rangle = |0\rangle Z,$$

where Z is some nonvanishing supernumber. Suppose n has nonvanishing body. Then since

$$n = \langle n|a^*a|n\rangle = Z^*\langle 0|0\rangle Z = Z^*Z$$

it follows that Z has nonvanishing body. But then

$$|n\rangle = a^*a|n\rangle\frac{1}{n} = a^*|0\rangle\frac{Z}{n} = |1\rangle\frac{Z}{n}, \tag{5.5.30}$$

implying that $n = 1$, which is a contradiction. Therefore n must have vanishing body. Since

$$\langle n|aa^*|n\rangle = \langle n|(1 - a^*a)|n\rangle = 1 - n$$

it follows that $a^*|n\rangle$ cannot be the zero supervector. It is easy to see, moreover, that $a^*|n\rangle$ is an eigenvector of a^*a corresponding to the eigenvalue 1. *Modulo* an irrelevant phase factor one may make the identification

$$|1\rangle = a^*|n\rangle(1-n)^{-\frac{1}{2}},$$

whence

$$|0\rangle = a|1\rangle = aa^*|n\rangle(1-n)^{-\frac{1}{2}} = (1-a^*a)|n\rangle(1-n)^{-\frac{1}{2}}$$
$$= |n\rangle(1-n)^{\frac{1}{2}}, \qquad (5.5.31)$$

implying that $n=0$, which is again a contradiction.

The simplest representation of the operator superalgebra is evidently two dimensional, with a^*a having only the eigenvalues 0 and 1, and with $|0\rangle$ and $|1\rangle$ chosen as basis supervectors. In spite of the fact that the configuration space of the Fermi oscillator is bigger than that of the previous example, the super Hilbert space is the same size: Total dimension $=2$.

Eigenvectors of x_1 and x_2. Choice of pure basis

Denote by $|1, 1/\sqrt{2}, t\rangle$ and $|1, -1/\sqrt{2}, t\rangle$ the normalized eigenvectors of $x_1(t)$ corresponding to the eigenvalues $1/\sqrt{2}$ and $-1/\sqrt{2}$ respectively. Denote by $|2, 1/\sqrt{2}, t\rangle$ and $|2, -1/\sqrt{2}, t\rangle$ the corresponding eigenvectors of $x_2(t)$. Using

$$x_1(t) = \frac{1}{\sqrt{2}}(ae^{-i\omega t} + a^*e^{i\omega t}), \quad x_2(t) = \frac{i}{\sqrt{2}}(ae^{-i\omega t} - a^*e^{-i\omega t}), \qquad (5.5.32)$$

it is easy to verify that, up to inessential phase factors, these eigenvectors are given in terms of the basis supervectors $|0\rangle$ and $|1\rangle$ by

$$\left.\begin{aligned} |1, \pm 1/\sqrt{2}, t\rangle &= \frac{1}{\sqrt{2}}(|0\rangle \pm e^{i\omega t}|1\rangle), \\ |2, \pm 1/\sqrt{2}, t\rangle &= \frac{1}{\sqrt{2}}(|0\rangle \mp ie^{i\omega t}|1\rangle), \end{aligned}\right\} \qquad (5.5.33)$$

and, inversely,

$$\left.\begin{aligned} |0\rangle &= \frac{1}{\sqrt{2}}(|1, 1/\sqrt{2}, t\rangle + |1, -1/\sqrt{2}, t\rangle) \\ &= \frac{1}{\sqrt{2}}(|2, 1/\sqrt{2}, t\rangle + |2, -1/\sqrt{2}, t\rangle), \\ |1\rangle &= \frac{1}{\sqrt{2}}e^{-i\omega t}(|1, 1/\sqrt{2}, t\rangle - |1, -1/\sqrt{2}, t\rangle) \\ &= \frac{i}{\sqrt{2}}e^{-i\omega t}(|2, 1/\sqrt{2}, t\rangle - |2, -1/\sqrt{2}, t\rangle). \end{aligned}\right| \qquad (5.5.34)$$

By arguments familiar from the preceding section one knows that the supervectors $|1, \pm 1/\sqrt{2}, t\rangle$ and $|2, \pm 1/\sqrt{2}, t\rangle$ are necessarily impure. By arguments identical to those following eq. (5.4.10) it is clear that $|0\rangle$ and $|1\rangle$ may be regarded as constituting a pure basis and that each may be assigned to be *either* c-type *or* a-type, the other necessarily having the opposite type and the super Hilbert space having dimension (1, 1).

For definiteness in the remainder of this section the supervectors $|0\rangle$ and $|1\rangle$ will be assumed to be c-type and a-type respectively. But it must be stressed that this is an arbitrary convention. The operator superalgebra of the theory is invariant under interchange of a and a^*, and the operator aa^* has the same spectrum as a^*a. It is sometimes convenient to adopt the 'hole theory' point of view and make opposite assignments for $|0\rangle$ and $|1\rangle$. Note that in any case the energy spectrum is $\pm\frac{1}{2}\omega$.

Coherent states

Introduce the time-dependent operators

$$\left.\begin{aligned} a(t) &\overset{\text{def}}{=} a\mathrm{e}^{-\mathrm{i}\omega t} = \frac{1}{\sqrt{2}}[x_1(t) - \mathrm{i}x_2(t)], \\ a^*(t) &\overset{\text{def}}{=} a^*\mathrm{e}^{\mathrm{i}\omega t} = \frac{1}{\sqrt{2}}[x_1(t) + \mathrm{i}x_2(t)]. \end{aligned}\right\} \tag{5.5.35}$$

Evidently

$$x_1(t) = \frac{1}{\sqrt{2}}[a(t) + a^*(t)], \qquad x_2(t) = \frac{\mathrm{i}}{\sqrt{2}}[a(t) - a^*(t)], \tag{5.5.36}$$

whence

$$\left.\begin{aligned} x_1(t)|0\rangle &= \frac{1}{\sqrt{2}}\mathrm{e}^{\mathrm{i}\omega t}|1\rangle, & x_1(t)|1\rangle &= \frac{1}{\sqrt{2}}\mathrm{e}^{-\mathrm{i}\omega t}|0\rangle, \\ x_2(t)|0\rangle &= -\frac{\mathrm{i}}{\sqrt{2}}\mathrm{e}^{\mathrm{i}\omega t}|1\rangle, & x_2(t)|1\rangle &= \frac{\mathrm{i}}{\sqrt{2}}\mathrm{e}^{-\mathrm{i}\omega t}|0\rangle. \end{aligned}\right\} \tag{5.5.37}$$

Let a' be an arbitrary complex a-number. Write

$$a' = \frac{1}{\sqrt{2}}(x_1' - \mathrm{i}x_2') \tag{5.5.38}$$

where x_1' and x_2' are real a-numbers. Introduce the c-type supervector

$$\begin{aligned} |a', t\rangle &\overset{\text{def}}{=} \mathrm{e}^{-\frac{1}{2}a'^*a'}\mathrm{e}^{\mathrm{i}Ht}(|0\rangle - a'|1\rangle) \\ &= (1 - \tfrac{1}{2}a'^*a')(\mathrm{e}^{-(\mathrm{i}/2)\omega t}|0\rangle - a'\mathrm{e}^{(\mathrm{i}/2)\omega t}|1\rangle) \\ &= \left(1 + \frac{\mathrm{i}}{2}x_1'x_2'\right)\mathrm{e}^{-(\mathrm{i}/2)\omega t}|0\rangle - \frac{1}{\sqrt{2}}\mathrm{e}^{(\mathrm{i}/2)\omega t}(x_1' - \mathrm{i}x_2')|1\rangle. \end{aligned} \tag{5.5.39}$$

It is easy to verify that $|a',t\rangle$ is a right eigenvector of $a(t)$ corresponding to the eigenvalue a':

$$a(t)|a',t\rangle = a'|a',t\rangle = |a',t\rangle a'. \tag{5.5.40}$$

$|a',t\rangle$ is said to be the supervector corresponding to a *coherent state*. Its dual under the mapping * is a left eigenvector of $a^*(t)$ corresponding to the eigenvalue a'^*. It will be convenient here to depart from the convention (5.2.1) and to denote the dual by $\langle a'^*,t|$ so that we may write

$$\langle a'^*,t|a^*(t) = \langle a'^*,t|a'^* = a'^*\langle a'^*,t|. \tag{5.5.41}$$

In terms of the supervectors $\langle 0|$ and $\langle 1|$ the dual $\langle a'^*,t|$ is given by

$$\begin{aligned}\langle a'^*,t| &= \mathrm{e}^{-\frac{1}{2}a'^*a'}(\langle 0| + a'^*\langle 1|)\mathrm{e}^{-iHt} \\ &= \left(1+\frac{\mathrm{i}}{2}x_1'x_2'\right)\mathrm{e}^{(\mathrm{i}/2)\omega t}\langle 0| + \frac{1}{\sqrt{2}}(x_1'+\mathrm{i}x_2')\mathrm{e}^{-(\mathrm{i}/2)\omega t}\langle 1|.\end{aligned} \tag{5.5.42}$$

and it is easy to check that

$$\langle a'^*,t|a',t\rangle = 1. \tag{5.5.43}$$

With the aid of eqs. (5.5.37) and (5.5.42) it is straightforward to verify that $\langle a'^*,t|$ satisfies the differential equations (cf. eq. (5.4.25))

$$\left.\begin{aligned}\left(\tfrac{1}{2}x_1' + \frac{\vec{\partial}}{\partial x_1'}\right)\langle a'^*,t| &= \langle a'^*,t|x_1(t), \\ \left(\tfrac{1}{2}x_2' + \frac{\vec{\partial}}{\partial x_2'}\right)\langle a'^*,t| &= \langle a'^*,t|x_2(t).\end{aligned}\right\} \tag{5.5.44}$$

The operators $\frac{1}{2}x_1' + \vec{\partial}/\partial x_1'$ and $\frac{1}{2}x_2' + \vec{\partial}/\partial x_2'$ appearing in these equations obviously satisfy the same algebra as $x_1(t)$ and $x_2(t)$. An equivalent pair of equations is†

$$\left.\begin{aligned}\left(\tfrac{1}{2}a'^* + \frac{\vec{\partial}}{\partial a'}\right)\langle a'^*,t| &= \langle a'^*,t|a^*(t), \\ \left(\tfrac{1}{2}a' + \frac{\vec{\partial}}{\partial a'^*}\right)\langle a'^*,t| &= \langle a'^*,t|a(t).\end{aligned}\right\} \tag{5.5.45}$$

The coherent-state supervectors do not form a complete set. (For example, the supervector $|1\rangle$ cannot be constructed out of them.) They nevertheless satisfy an integral identity that has the formal appearance of

† With the normalization (5.5.43) the supervector $\langle a'^*,t|$ depends on *both* a' and a'^*, as does $|a',t\rangle$.

a completeness relation:

$$\frac{1}{2\pi i}\int |a',t\rangle\langle a'^*,t|\mathrm{d}a'^*\,\mathrm{d}a'$$
$$=\frac{1}{2\pi}\int\left\{\mathrm{e}^{-(i/2)\omega t}|0\rangle\left(1+\frac{i}{2}x_1'x_2'\right)+\frac{1}{\sqrt{2}}\mathrm{e}^{(i/2)\omega t}|1\rangle(x_1'+x_2')\right\}$$
$$\times\left\{\left(1+\frac{i}{2}x_1'x_2'\right)\mathrm{e}^{(i/2)\omega t}\langle 0|+\frac{1}{\sqrt{2}}(x_1'+ix_2')\mathrm{e}^{-(i/2)\omega t}\langle 1|\right\}\mathrm{d}x_1'\,\mathrm{d}x_2'$$
$$=\frac{i}{2\pi}\int(|0\rangle\langle 0|+|1\rangle\langle 1|)x_1'x_2'\,\mathrm{d}x_1'\,\mathrm{d}x_2'$$
$$=|0\rangle\langle 0|+|1\rangle\langle 1|=1. \tag{5.5.46}$$

In the change from the integration variables a'^*, a' to x_1', x_2' the super Jacobian (1.7.24) has been included.

The functional integral representation of $\langle a''^*,t''|a',t'\rangle$

The amplitude $\langle a''^*,t''|a',t'\rangle$ is easily computed:

$$\langle a''^*,t''|a',t'\rangle=\mathrm{e}^{-\frac{1}{2}a''^*a''-\frac{1}{2}a'^*a'}(\mathrm{e}^{(i/2)\omega t''}\langle 0|+\mathrm{e}^{-(i/2)\omega t''}a''^*\langle 1|)$$
$$\times(\mathrm{e}^{-(i/2)\omega t'}|0\rangle+\mathrm{e}^{(i/2)\omega t'}|1\rangle a')$$
$$=\mathrm{e}^{-\frac{1}{2}a''^*a''-\frac{1}{2}a'^*a'+(i/2)\omega(t''-t')}[1+a''^*\mathrm{e}^{-i\omega(t''-t')}a']$$
$$=\exp\left[-\tfrac{1}{2}a''^*a''+a''^*\mathrm{e}^{-i\omega(t''-t')}a'-\tfrac{1}{2}a'^*a'+\frac{i}{2}\omega(t''-t')\right]. \tag{5.5.47}$$

When $t''-t'$ is infinitesimal the exponent in the final expression may be written

$$i\left[\frac{i}{2}a''^*(a''-a')-\frac{i}{2}(a''^*-a'^*)a'-\omega a''^*a'(t''-t')+\tfrac{1}{2}\omega(t''-t')\right]$$
$$=i\int_{a',t'}^{a''^*,t''}\left[\frac{i}{2}(a^*\dot{a}-\dot{a}^*a)-\omega a^*a+\tfrac{1}{2}\omega\right]\mathrm{d}t, \tag{5.5.48}$$

where a on the right-hand side is a complex a-number-valued function of t passing through the endpoint values indicated. The integrand on the right is just the superclassical Lagrangian of eq. (5.5.3) with the x's replaced by a a and a^* according to eqs. (5.5.36), and with a constant $\frac{1}{2}\omega$ added.†

† This constant accounts for the quantum energy $-\frac{1}{2}\omega$ in the state $|0\rangle$.

It is to be emphasized that the left-hand side of eq. (5.5.48) constitutes a very special representation of the integral as a sum of differences. In the rest of this section the superclassical action is to be understood as *defined* to be the limit of a sum of differences of exactly this form.

When $t'' - t'$ is finite, use of eq. (5.5.46) permits one to write

$$\langle a''^*, t'' | a', t' \rangle = \int \langle a^*_{N+1}, t_{n+1} | a_N, t_N \rangle \langle a^*_N, t_N | a_{N-1}, t_{N-1} \rangle \ldots$$

$$\ldots \langle a^*_2, t_2 | a_1, t_1 \rangle \langle a^*_1, t_1 | a_0, t_0 \rangle \left(\frac{1}{2\pi \mathrm{i}} \right)^N \mathrm{d}a^*_1 \, \mathrm{d}a_1 \ldots \mathrm{d}a^*_N \, \mathrm{d}a_N, \quad (5.5.49)$$

where

$$a^*_{N+1} = a''^*, \qquad a_0 = a'. \quad (5.5.50)$$

and where the t's satisfy (5.4.40). (Note that the integer N does not have to be even here.) One may then pass to the limit $N \to \infty$, $t_n - t_{n-1} \to 0$ for all n, obtaining, by virtue of (5.5.4),

$$\langle a''^*, t'' | a', t' \rangle = Z \int \exp\left\{ \mathrm{i} \int_{a',t'}^{a''^*,t''} \left[\frac{\mathrm{i}}{2}(a^* \dot{a} - \dot{a}^* a) - \omega a^* a + \tfrac{1}{2}\omega \right] \mathrm{d}t \right\} \mathrm{d}x, \quad (5.5.51)$$

where, formally,

$$\left. \begin{aligned} Z &= \left(\frac{1}{2\pi} \right)^{(t''-t')/\mathrm{d}t - 1}, \\ \mathrm{d}x &= \prod_{t'<t<t''} \mathrm{d}x_1(t) \mathrm{d}x_2(t). \end{aligned} \right\} \quad (5.5.52)$$

Equation (5.5.51), like eq. (5.4.43), is a special case of (5.3.46), in which $(\mathrm{sdet}\, G^+)^{-\frac{1}{2}}$, which is again just a constant, has been absorbed into the factor Z.

Direct evaluation of the functional integral

The functional integral (5.5.51), like (5.4.43), is a Gaussian integral. Its evaluation involves finding a stationary point which, when found, gives additional insight into the meaning of the integral itself. Inserting (5.5.47) into the right-hand side of (5.5.49) and remembering that short-time contributions to the action integral are to be understood in the sense of eq. (5.5.48), one obtains the following *differenced* form of the exponent appearing in (5.5.51):

$$\mathrm{i}\int_{a',t'}^{a''*,t''}\left[\frac{\mathrm{i}}{2}(a^*\dot{a}-\dot{a}^*a)-\omega a^*a+\tfrac{1}{2}\omega\right]\mathrm{d}t$$

$$= -\tfrac{1}{2}a^*_{N+1}a_{N+1}+a^*_{N+1}\mathrm{e}^{-\mathrm{i}\omega\Delta t}a_N-\tfrac{1}{2}a^*_Na_N+\frac{\mathrm{i}}{2}\omega\Delta t$$

$$-\tfrac{1}{2}a^*_Na_N+a^*_N\mathrm{e}^{-\mathrm{i}\omega\Delta t}a_{N-1}-\tfrac{1}{2}a^*_{N-1}a_{N-1}+\frac{\mathrm{i}}{2}\omega\Delta t$$

$$\vdots$$

$$-\tfrac{1}{2}a^*_2a_2+a^*_2\mathrm{e}^{-\mathrm{i}\omega\Delta t}a_1-\tfrac{1}{2}a^*_1a_1+\frac{\mathrm{i}}{2}\omega\Delta t$$

$$-\tfrac{1}{2}a^*_1a_1+a^*_1\mathrm{e}^{-\mathrm{i}\omega\Delta t}a_0-\tfrac{1}{2}a^*_0a_0+\frac{\mathrm{i}}{2}\omega\Delta t, \tag{5.5.53}$$

Δt being given by eq. (5.4.47).

The stationary trajectory is the solution of the following equations, obtained by differentiating (5.5.53) with respect to $a_1, a_2, \ldots, a_N$ and $a^*_1, a^*_2, \ldots, a^*_N$:

$$\left.\begin{aligned} a^*_{N+1}\mathrm{e}^{-\mathrm{i}\omega\Delta t}-a^*_N&=0,\\ a^*_N\mathrm{e}^{-\mathrm{i}\omega\Delta t}-a^*_{N-1}&=0,\\ &\vdots\\ a^*_2\mathrm{e}^{-\mathrm{i}\omega\Delta t}-a^*_1&=0,\end{aligned}\right\} \tag{5.5.54}$$

$$\left.\begin{aligned} -a_N+\mathrm{e}^{-\mathrm{i}\omega\Delta t}a_{N-1}&=0,\\ &\vdots\\ -a_2+\mathrm{e}^{-\mathrm{i}\omega\Delta t}a_1&=0,\\ -a_1+\mathrm{e}^{-\mathrm{i}\omega\Delta t}a_0&=0.\end{aligned}\right\} \tag{5.5.55}$$

Evidently

$$a^*_n=a^*_{N+1}\mathrm{e}^{-\mathrm{i}\omega(N+1-n)\Delta t}, \qquad a_n=\mathrm{e}^{-\mathrm{i}\omega n\Delta t}a_0. \tag{5.5.56}$$

Introduce the subscript 'st' to denote the stationary trajectory in the limit $N\to\infty$. Then

$$a^*_{\mathrm{st}}(t)=\mathrm{e}^{-\mathrm{i}\omega(t''-t)}a''^*, \qquad a_{\mathrm{st}}(t)=\mathrm{e}^{-\mathrm{i}\omega(t-t')}a', \tag{5.5.57}$$

and, by virtue of eqs. (5.5.36),

$$x_{\mathrm{st}}(t)=\frac{1}{\sqrt{2}}\begin{pmatrix}\mathrm{e}^{-\mathrm{i}\omega(t-t')}a'+\mathrm{e}^{-\mathrm{i}\omega(t''-t)}a''^*\\ \mathrm{i}\mathrm{e}^{-\mathrm{i}\omega(t-t')}a'-\mathrm{i}\mathrm{e}^{-\mathrm{i}\omega(t''-t)}a''^*\end{pmatrix}. \tag{5.5.58}$$

Note that, except when $a'=a''=0$, the functions $a_{\mathrm{st}}(t)$ and $a^*{}_{\mathrm{st}}(t)$ are *not* complex conjugates of one another, and the components of $x_{\mathrm{st}}(t)$ are *not* real. This means that the stationary trajectory lies outside the configuration space $\mathbf{R}_a^2$ of the superclassical system, in its complex extension $\mathbf{C}_a^2$.

It is straightforward to evaluate expression (5.5.53) at the stationary point. The terms in $\frac{i}{2}\omega\Delta t$ add up to $\frac{i}{2}\omega(t''-t')$. The identity

$$a_{st}^*(t)a_{st}(t) = a''^* e^{-i\omega(t''-t')}a', \tag{5.5.59}$$

implies that the terms in $-\frac{1}{2}a_n^*a_n$ $(n = 1, 2, \ldots, N)$ all add up to $-Na''^* \times e^{-i\omega(t''-t')}a'$. The terms in $a_n^* e^{-i\omega\Delta t}a_{n-1}(n = 1, 2, \ldots, N+1)$, on the other hand, add up to $(N+1)a''^* e^{-i\omega(t''-t')}a'$. In view of eqs. (5.5.50), therefore, the total sum (5.5.53) is precisely equal to the final exponent in (5.5.47). Equation (5.5.47) itself follows from the functional integral by virtue of the normalization condition (5.5.43).

The importance of endpoint contributions

Expressions (5.5.57) and (5.5.58) for the stationary trajectory satisfy the dynamical equation (5.5.4) or, equivalently,

$$\dot{a}_{st}(t) = -i\omega a_{st}(t), \qquad \dot{a}_{st}^*(t) = i\omega a_{st}^*(t). \tag{5.5.60}$$

If one were naively to insert (5.5.58) into the action (5.5.3), or, equivalently, expressions (5.5.57) into the left-hand side of eq. (5.5.53), one would obtain zero, in apparent contradiction to the nonvanishing value that has just been obtained for (5.5.53). The contradiction is resolved when account is taken of *endpoint contributions* in (5.5.53). The stationary trajectory (5.5.58) does not join smoothly to its endpoint values

$$\begin{pmatrix} x_1' \\ x_2' \end{pmatrix} \text{ at } t = t' \quad \text{and} \quad \begin{pmatrix} x_1'' \\ x_2'' \end{pmatrix} \text{ at } t = t''.$$

The real stationary trajectory involves step functions at the end points. Equations (5.5.57), for example, should really be replaced by

$$\left.\begin{aligned} a_{st}^*(t) &= \lim_{\varepsilon\to 0}\,[\theta(t, t'+\varepsilon)e^{-i\omega(t''-t)}a''^* + \theta(t'+\varepsilon, t)a'^*], \\ a_{st}(t) &= \lim_{\varepsilon\to 0}\,[\theta(t''-\varepsilon, t)e^{-i\omega(t-t')}a' + \theta(t, t''-\varepsilon)a''], \end{aligned}\right\} \tag{5.5.61}$$

which yield

$$\begin{aligned} \dot{a}_{st}^*(t) &= \lim_{\varepsilon\to 0}\,\{i\omega\theta(t, t'+\varepsilon)e^{-i\omega(t''-t)}a''^* + \delta(t, t'+\varepsilon)[e^{-i\omega(t''-t)}a''^* - a'^*]\}, \\ \dot{a}_{st}(t) &= \lim_{\varepsilon\to 0}\,\{-i\omega\theta(t''-\varepsilon, t)e^{-i\omega(t-t')}a' + \delta(t, t''-\varepsilon)[a'' - e^{-i\omega(t-t')}a']\}. \end{aligned}$$

$$\tag{5.5.62}$$

Insertion of (5.5.61) and (5.5.62) into the left-hand side of (5.5.53) immediately reproduces the expression previously found for the value at the stationary point.

The stationary trajectory as a matrix element

If an external source term were added to the action functional (5.5.3) then differentiation of the amplitude $\langle a''^*, t''|a', t'\rangle$ with respect to the source would yield the matrix element $\langle a''^*, t''|x(t)|a', t'\rangle$. Differentiation of the functional integral, on the other hand, would yield $x_{st}(t)\ \langle a''^*, t''|a', t'\rangle$. The identity of the two expressions when the source vanishes is easily shown:

$$\begin{aligned}\langle a''^*, t''|a(t)|a', t'\rangle &= \langle a''^*, t''|e^{-i\omega(t-t')}a(t')|a', t'\rangle \\ &= e^{-i\omega(t-t')}a'\langle a''^*, t''|a', t'\rangle = a_{st}(t)\langle a''^*, t''|a', t'\rangle, \qquad (5.5.63)\\ \langle a''^*, t''|a^*(t)|a', t'\rangle &= \langle a''^*, t''|a^*(t'')e^{-i\omega(t''-t)}|a', t'\rangle \\ &= e^{-i\omega(t''-t)}a''^*\langle a''^*, t''|a', t'\rangle = a^*_{st}(t)\langle a''^*, t''|a', t'\rangle, \qquad (5.5.64)\end{aligned}$$

and hence

$$\langle a''^*, t''|x(t)|a', t'\rangle = x_{st}(t)\langle a''^*, t''|a', t'\rangle, \qquad t' < t < t''. \qquad (5.5.65)$$

The Feynman propagator

In the case of the Fermi oscillator the Green's function for the stationary trajectory is known as the *Feynman propagator*. It may be obtained by differentiating $\ln\langle a''^*, t''|a', t'\rangle$ twice with respect to the source (cf. eq. (5.4.60)) or, alternatively, from

$$\begin{aligned}G(s,t)\langle a''^*, t''|a', t'\rangle = {} & i\langle a''^*, t''|T(x(s)\tilde{x}(t))|a', t'\rangle \\ & - i x_{st}(s)\tilde{x}_{st}(t)\langle a''^*, t''|a', t'\rangle. \qquad (5.5.66)\end{aligned}$$

Using

$$a(s)a(t) = 0, \qquad a^*(s)a^*(t) = 0, \qquad [a(s), a^*(t)] = e^{-i\omega(s-t)} \qquad (5.5.67)$$

one readily obtains

$$\begin{aligned}&\langle a''^*, t''|x(s)\tilde{x}(t)|a', t'\rangle \\ &= \frac{i}{2}\langle a''^*, t''|\begin{pmatrix} a(s) + a^*(s) \\ ia(s) - ia^*(s)\end{pmatrix}(a(t) + a^*(t) \quad ia(t) - ia^*(t))|a', t'\rangle \\ &= \frac{i}{2}\left[\begin{pmatrix} a^*_{st}(s)a_{st}(t) - a^*_{st}(t)a_{st}(s) & ia^*_{st}(s)a_{st}(t) + ia^*_{st}(t)a_{st}(s) \\ -ia^*_{st}(s)a_{st}(t) - ia^*_{st}(t)a_{st}(s) & a^*_{st}(s)a_{st}(t) - a^*_{st}(t)a_{st}(s)\end{pmatrix}\right. \\ &\quad \left. + \begin{pmatrix} 1 & -i \\ i & 1\end{pmatrix} e^{-i\omega(s-t)}\right]\langle a''^*, t''|a', t'\rangle \\ &= i[x_{st}(s)\tilde{x}_{st}(t) + u(s)u^\dagger(t)]\langle a''^*, t''|a', t'\rangle, \qquad (5.5.68)\end{aligned}$$

$$\left.\begin{aligned} G(s,t) &= \mathrm{i}[\theta(s,t)u(s)u^{\dagger}(t) - \theta(t,s)u^{*}(s)u^{\sim}(t)] \\ &= -\theta(s,t)G^{(+)}(s,t) + \theta(t,s)G^{(-)}(s,t) \\ &= G^{-}(s,t) + G^{(-)}(s,t) = G^{+}(s,t) - G^{(+)}(s,t). \end{aligned}\right\} \tag{5.5.69}$$

When $s > t$ the Feynman propagator behaves as $\mathrm{e}^{-\mathrm{i}\omega s}$ as a function of s and as $\mathrm{e}^{\mathrm{i}\omega t}$ as a function of t. The same statement holds true with s and t interchanged. This property is sometimes expressed by saying that $G(s,t)$ propagates *positive* frequencies to the future and *negative* frequencies to the past.

When $t'' > s > t'$ the Feynman propagator satisfies the boundary conditions

$$\left.\begin{aligned} (1 \quad -\mathrm{i})G(t',s) &= 0, \qquad G(s,t')\begin{pmatrix} 1 \\ -\mathrm{i} \end{pmatrix} = 0, \\ (1 \quad \mathrm{i})G(t'',s) &= 0, \qquad G(s,t'')\begin{pmatrix} 1 \\ \mathrm{i} \end{pmatrix} = 0, \end{aligned}\right\} \tag{5.5.70}$$

corresponding to the fact that the stationary trajectory is fixed by specifying the following boundary conditions:

$$\left.\begin{aligned} x_{\mathrm{st}\,1}(t') - \mathrm{i}x_{\mathrm{st}\,2}(t') &= x_1' - \mathrm{i}x_2', \\ x_{\mathrm{st}\,2}(t'') + \mathrm{i}x_{\mathrm{st}\,2}(t'') &= x_1'' + \mathrm{i}x_2''. \end{aligned}\right\} \tag{5.5.71}$$

5.6 The Bose oscillator

Action functional and Green's functions

The configuration space of the Bose oscillator is $\mathbf{R}_c$. Denote by x the real c-type dynamical variable. The action functional is

$$S = \tfrac{1}{2}\int(\dot{x}^2 - \omega^2 x^2)\,\mathrm{d}t, \tag{5.6.1}$$

where ω is a positive ordinary real number. The dynamical equation is

$$0 = \frac{\delta S}{\delta x} \equiv -(\ddot{x} + \omega^2 x), \tag{5.6.2}$$

and the second functional derivative of the action is

$$\frac{\delta^2 S}{\delta x(t)\delta x(t')} = -\left(\frac{\partial^2}{\partial t^2} + \omega^2\right)\delta(t,t'). \tag{5.6.3}$$

The retarded and advanced Green's function of the operator

$-(\partial^2/\partial t^2+\omega^2)$ are readily verified to be

$$G^-(t,t')=\omega^{-1}\theta(t,t')\sin\omega(t-t'), \tag{5.6.4}$$

$$G^+(t,t')=-\omega^{-1}\theta(t',t)\sin\omega(t-t'), \tag{5.6.5}$$

yielding the supercommutator function

$$\tilde{G}(t,t')=-\omega^{-1}\sin\omega(t-t'). \tag{5.6.6}$$

The operator superalgebra of the quantum theory is defined by

$$[x(t),x(t')]=\mathrm{i}\tilde{G}(t,t')=-\mathrm{i}\omega^{-1}\sin\omega(t-t'), \tag{5.6.7}$$

the supercommutator being here an ordinary commutator. The quantum $x(t)$ is assumed to be a self-adjoint c-type operator as well as a physical observable. The corollary of eq. (5.6.7),

$$[x(t),\dot{x}(t)]=\mathrm{i}, \tag{5.6.8}$$

implies, by well-known arguments, that the eigenvalues of both $x(t)$ and $\dot{x}(t)$ range over the whole of $\mathbf{R}_c$. Thus, if $|x',t\rangle$ is an eigenvector of $x(t)$ corresponding to the eigenvalue x' in $\mathbf{R}_c$ then $\mathrm{e}^{-\mathrm{i}\xi\dot{x}(t)}|x',t\rangle$ is, for all ξ in $\mathbf{R}_c$, an eigenvector of $x(t)$ corresponding to the eigenvalue $x'+\xi$. Similarly, if $|p',t\rangle$ is an eigenvector of $\dot{x}(t)$ corresponding to the eigenvalue p' in $\mathbf{R}_c$ then $\mathrm{e}^{\mathrm{i}\pi x(t)}|p',t\rangle$ is, for all π in $\mathbf{R}_c$, an eigenvector of $\dot{x}(t)$ corresponding to the eigenvalue $p'+\pi$.

Because the spectra of $x(t)$ and $\dot{x}(t)$ are continuous the super Hilbert space of the theory is infinite dimensional and it is convenient to impose delta-function normalization on the supervectors $|x',t\rangle$ and $|p',t\rangle$:

$$\langle x'',t|x',t\rangle=\delta(x'',x'),\quad \langle p'',t|p',t\rangle=\delta(p'',p'). \tag{5.6.9}$$

The $|x',t\rangle$ with $x'_S=0$ (or, alternatively, the $|p',t\rangle$ with $p'_S=0$) may be chosen as a complete set of basis supervectors for the super Hilbert space. Since all of the operators of the theory are c-type, no generality is lost by assuming that these basis supervectors are all c-type. The normalizations (5.6.9), together with the completeness conditions, imply

$$\int_{-\infty}^{\infty}|x',t\rangle\langle x',t|\,\mathrm{d}x'=1,\quad \int_{-\infty}^{\infty}|p',t\rangle\langle p',t|\,\mathrm{d}p'=1. \tag{5.6.10}$$

These equations hold for arbitrary contours of integration in $\mathbf{R}_c$ provided only the contours begin at $-\infty$ and end at ∞.

The normalizations (5.6.9) also imply

$$\langle x',t|p',t\rangle=(2\pi)^{-\frac{1}{2}}\mathrm{e}^{ip'x'} \tag{5.6.11}$$

up to an arbitrary x'-dependent (or, alternatively, p'-dependent) phase factor. Thus, with a natural choice of phase,

$$|x'+\delta x',t\rangle = [1 - \mathrm{i}\delta x'\dot{x}(t)]|x',t\rangle, \quad \delta x' \text{ infinitesimal,} \tag{5.6.12}$$

whence

$$\frac{\partial}{\partial x'}|x',t\rangle = -\mathrm{i}\dot{x}(t)|x',t\rangle, \tag{5.6.13}$$

$$-\mathrm{i}\frac{\partial}{\partial x'}\langle x',t|p',t\rangle = \langle x',t|\dot{x}(t)|p',t\rangle = p'\langle x',t|p',t\rangle, \tag{5.6.14}$$

of which (5.6.11) is the solution, the constant of integration being determined by Fourier transform theory in $\mathbf{R}_c$ (see section 1.2).

Mode functions and Hamiltonian

The general solution of the dynamical equation (4.1.2) is

$$x(t) = au(t) + a^*u^*(t), \tag{5.6.15}$$

where

$$u(t) \stackrel{\text{def}}{=} (2\omega)^{-\frac{1}{2}}\mathrm{e}^{-\mathrm{i}\omega t} \tag{5.6.16}$$

satisfying

$$\mathrm{i}u(t)\frac{\overleftrightarrow{\mathrm{d}}}{\mathrm{d}t}u(t) = 0, \qquad \mathrm{i}u^*(t)\frac{\overleftrightarrow{\mathrm{d}}}{\mathrm{d}t}u(t) = 1, \tag{5.6.17}$$

$$\frac{\overleftrightarrow{\mathrm{d}}}{\mathrm{d}t} \stackrel{\text{def}}{=} \frac{\overrightarrow{\mathrm{d}}}{\mathrm{d}t} - \frac{\overleftarrow{\mathrm{d}}}{\mathrm{d}t}. \tag{5.6.18}$$

The functions $u(t)$ and $u^*(t)$ are the mode functions of the Bose oscillator. In terms of them the supercommutator function may be written

$$\tilde{G}(t,t') = G^{(+)}(t,t') + G^{(-)}(t,t'), \tag{5.6.19}$$

$$G^{(+)}(t,t') \stackrel{\text{def}}{=} -\mathrm{i}u(t)u^*(t'), \tag{5.6.20}$$

$$G^{(-)}(t,t') \stackrel{\text{def}}{=} \mathrm{i}u^*(t)u(t') = G^{(+)}(t,t')^*. \tag{5.6.21}$$

In the quantum theory the coefficients a and a^* in eq. (5.6.15) are non-self-adjoint c-type operators. By virtue of eqs. (5.6.17) they may be expressed in the forms

$$a = \mathrm{i}u^*(t)\frac{\overleftrightarrow{\mathrm{d}}}{\mathrm{d}t}x(t) = (2\omega)^{-\frac{1}{2}}[\mathrm{i}\dot{x}(t) + \omega x(t)]\,\mathrm{e}^{\mathrm{i}\omega t}, \tag{5.6.22}$$

$$a^* = (2\omega)^{-\frac{1}{2}}[-\mathrm{i}\dot{x}(t) + \omega x(t)]\,\mathrm{e}^{-\mathrm{i}\omega t}, \tag{5.6.23}$$

leading, with the aid of (5.6.8), to the (super)commutator

$$[a, a^*] = (2\omega)^{-1}[\mathrm{i}\dot{x}(t) + \omega x(t), -\mathrm{i}\dot{x}(t) + \omega x(t)] = 1. \tag{5.6.24}$$

Equation (5.6.24) can be used in place of (5.6.8) to define the operator algebra of the theory.

The Hamiltonian or energy of the Bose oscillator is given by

$$H=\frac{\partial L}{\partial \dot{x}}\dot{x}-L=\tfrac{1}{2}(\dot{x}^2+\omega^2 x^2)=\tfrac{1}{2}(\dot{x}^2-x\ddot{x})$$
$$=-\tfrac{1}{2}x\frac{\overleftrightarrow{\mathrm{d}}}{\mathrm{d}t}\dot{x}=\frac{\mathrm{i}\omega}{2}[au(t)+a^*u^*(t)]\frac{\overleftrightarrow{\mathrm{d}}}{\mathrm{d}t}[au(t)-a^*u^*(t)]$$
$$=\tfrac{1}{2}\omega(aa^*+a^*a)=\omega a^*a+\tfrac{1}{2}\omega. \tag{5.6.25}$$

It is readily verified that

$$\dot{x}=-\mathrm{i}[x,H]. \tag{5.6.26}$$

Energy eigenvectors

As in the case of the Fermi oscillator, here too the operator a^*a must be a physical observable. Denote its eigenvalues and corresponding physical eigenvectors by n and $|n\rangle$ respectively. Assume the normalization

$$\langle n|n\rangle=1. \tag{5.6.27}$$

Suppose there exists a nonvanishing n. Because

$$\langle n|a^*a|n\rangle=n, \tag{5.6.28}$$

it follows that $a|n\rangle$ cannot be the zero supervector. It is, in fact, an eigenvector of a^*a corresponding to the eigenvalue $n-1$:

$$(a^*a)a|n\rangle=(aa^*a-[a,a^*]a)|n\rangle=(n-1)a|n\rangle. \tag{5.6.29}$$

Evidently

$$a|n\rangle=|n-1\rangle Z, \tag{5.6.30}$$

$$n=\langle n|a^*a|n\rangle=Z^*\langle n-1|n-1\rangle Z=Z^*Z, \tag{5.6.31}$$

where Z is some nonvanishing supernumber.

It is clear that if n is an eigenvalue, so also are $n-1, n-2,\ldots$ so long as the corresponding Z does not vanish. Eventually, however, one must reach a Z that vanishes, for otherwise an eigenvalue with negative body would be reached, which is impossible because

$$0\le\langle n|a^*a|n\rangle_{\mathrm{B}}=n_{\mathrm{B}}. \tag{5.6.32}$$

Let n_0 be the eigenvalue in the descending series for which the corresponding Z vanishes. Then

$$a|n_0\rangle=0, \tag{5.6.33}$$

which implies

$$n_0|n_0\rangle=a^*a|n_0\rangle=0, \tag{5.6.34}$$

whence $n_0 = 0$ and

$$a|0\rangle = 0. \tag{5.6.35}$$

Evidently the existence of a nonvanishing eigenvalue n implies the existence of a vanishing eigenvalue. Moreover, since the series n, $n-1$, $n-2,\ldots$ must terminate in zero, n must be a positive integer. Conversely, the existence of a zero eigenvalue implies that all the positive integers are eigenvalues. The eigenvectors (*modulo* arbitrary phase factors) are given by

$$|n\rangle = (n!)^{-\frac{1}{2}} a^{*n}|0\rangle, \qquad \langle n| = (n!)^{-\frac{1}{2}}\langle 0|a^n. \tag{5.6.36}$$

Thus

$$\begin{aligned} a^*a|n\rangle &= (n!)^{-\frac{1}{2}} a^*aa^{*n}|0\rangle = (n!)^{-\frac{1}{2}} a^*[a, a^{*n}]|0\rangle \\ &= n(n!)^{-\frac{1}{2}} a^{*n}|0\rangle = n|n\rangle, \end{aligned} \tag{5.6.37}$$

$$\begin{aligned} \langle n|n\rangle &= (n!)^{-1}\langle 0|a^n a^{*n}|0\rangle = (n!)^{-1}\langle 0|a^{n-1}[a, a^{*n}]|0\rangle \\ &= [(n-1)!]^{-1}\langle 0|a^{n-1}a^{*n-1}|0\rangle = \cdots = \langle 0|0\rangle = 1. \end{aligned} \tag{5.6.38}$$

The operator a^*a has the nonnegative integers as eigenvalues, and no others. The simplest representation of the operator superalgebra is obtained by assuming the eigenvectors $|n\rangle$, $n = 0, 1, 2, \ldots$, to constitute a complete basis. The normalization

$$\langle n|n'\rangle = \delta_{nn'} \tag{5.6.39}$$

implies the completeness relation

$$\sum_{n=0}^{\infty} |n\rangle\langle n| = 1. \tag{5.6.40}$$

For simplicity the eigenvector $|0\rangle$ may be assumed to be c-type. Since a and a^* are c-type it then follows that all the $|n\rangle$ are c-type, and the super Hilbert space of the theory has dimension $(\infty, 0)$.

The $|n\rangle$ are also eigenvectors of the energy operator (5.6.25), corresponding to the eigenvalues $(n + \frac{1}{2})\omega$. The assumption that the $|n\rangle$ form a complete basis implies that the spectrum of the energy is nondegenerate.

Coherent states

Introduce the time-dependent operators

$$\left.\begin{aligned} a(t) &\overset{\text{def}}{=} a\,e^{-i\omega t} = (2\omega)^{-\frac{1}{2}}[i\dot{x}(t) + \omega x(t)], \\ a^*(t) &= a^*\,e^{i\omega t} = (2\omega)^{-\frac{1}{2}}[-i\dot{x}(t) + \omega x(t)]. \end{aligned}\right\} \tag{5.6.41}$$

Let a' be an arbitrary complex c-number. Introduce the c-type supervector

$$|a',t\rangle \stackrel{\text{def}}{=} e^{-\frac{1}{2}a'^*a'} e^{iHt} \sum_{n=0}^{\infty} (n!)^{-\frac{1}{2}} a'^n |n\rangle$$

$$= e^{-\frac{1}{2}a'^*a'} \sum_{n=0}^{\infty} (n!)^{-\frac{1}{2}} a'^n e^{i(n+\frac{1}{2})\omega t} |n\rangle. \qquad (5.6.42)$$

With the aid of the corollaries

$$a|n\rangle = n^{\frac{1}{2}}|n-1\rangle, \qquad a^*|n\rangle = (n+1)^{\frac{1}{2}}|n+1\rangle, \qquad (5.6.43)$$

of eqs. (5.6.36) it is straightforward to verify that $|a', t\rangle$ is a right eigenvector of $a(t)$ corresponding to the eigenvalue a':

$$a(t)|a',t\rangle = e^{-\frac{1}{2}a'^*a'} \sum_{n=0}^{\infty} (n!)^{-\frac{1}{2}} a'^n e^{i[(n-1)+\frac{1}{2}]\omega t} a|n\rangle$$

$$= a' e^{-\frac{1}{2}a'^*a'} \sum_{n=1}^{\infty} [(n-1)!]^{-\frac{1}{2}} a'^{n-1} e^{i[(n-1)+\frac{1}{2}]\omega t} |n-1\rangle$$

$$= a'|a',t\rangle \qquad (5.6.44)$$

The state associated with the supervector $|a',t\rangle$ is called a *coherent state.* The dual to $|a',t\rangle$, which will be denoted by $\langle a'^*,t|$, is a left eigenvector of $a^*(t)$ corresponding to the eigenvalue a'^*:

$$\langle a'^*,t| = e^{-\frac{1}{2}a'^*a'} \sum_{n=0}^{\infty} (n!)^{-\frac{1}{2}} a'^{*n} \langle n| e^{-iHt}$$

$$= e^{-\frac{1}{2}a'^*a'} \sum_{n=0}^{\infty} (n!)^{-\frac{1}{2}} a'^{*n} e^{-i(n+\frac{1}{2})\omega t} \langle n|, \qquad (5.6.45)$$

$$\langle a'^*,t|a^*(t) = \langle a'^*,t|a'^* = a'^*\langle a'^*,t|. \qquad (5.6.46)$$

Using eqs. (5.6.42) and (5.6.45) it is straightforward to compute $\langle a''^*,t''|a',t'\rangle$

$$= e^{-\frac{1}{2}a''^*a''-\frac{1}{2}a'^*a'} \sum_{n=0}^{\infty} (n!)^{-1} a''^{*n} e^{-i(n+\frac{1}{2})\omega t''} a'^n e^{i(n+\frac{1}{2}\omega)t'}$$

$$= e^{-\frac{1}{2}a''^*a''-\frac{1}{2}a'^*a'-i/2\omega(t''-t')} \sum_{n=0}^{\infty} (n!)^{-1} [a''^* e^{-i\omega(t''-t')} a']^n$$

$$= \exp\left[-\tfrac{1}{2}a''^*a'' + a''^* e^{-i\omega(t''-t')} a' - \tfrac{1}{2}a'^*a' - \frac{i}{2}\omega(t''-t')\right] \qquad (5.6.47)$$

which, except for the sign of the last term in the exponent, is formally identical to eq. (5.5.46) for the Fermi oscillator. Note the normalization

$$\langle a'^*,t|a',t\rangle = 1. \qquad (5.6.48)$$

Unlike the coherent-state supervectors for the Fermi oscillator, the supervectors (5.6.42) *do* form a complete set. In fact they constitute an *overcomplete* set. Their completeness can be expressed by a relation that is formally identical to eq. (5.5.45). Writing.

$$a' = re^{i\theta}, \qquad a'^* = re^{-i\theta}, \tag{5.6.49}$$

where r and θ are real c-numbers, and taking note of the Jacobian relation

$$da'^*\, da' = \frac{\partial(a'^*, a')}{\partial(r,\theta)} dr\, d\theta = 2ir\, dr\, d\theta, \tag{5.6.50}$$

one obtains

$$\begin{aligned}
&\frac{1}{2\pi i}\int |a',t\rangle\langle a'^*,t|\, da'^*\, da' \\
&= \frac{1}{\pi}\sum_{m,n=0}^{\infty}\int_0^\infty r\, dr \int_0^{2\pi} d\theta (m!n!)^{-\frac{1}{2}} e^{i(m-n)(\theta+\omega t)} r^{m+n} e^{-r^2} |m\rangle\langle n| \\
&= \sum_{n=0}^{\infty}\int_0^\infty (n!)^{-1} z^n e^{-z}\, dz |n\rangle\langle n| \qquad (z = r^2) \\
&= \sum_{n=0}^{\infty} |n\rangle\langle n| = 1.
\end{aligned} \tag{5.6.51}$$

Hamilton–Jacobi theory

Equations (5.6.47) and (5.6.51) can form a point of departure for constructing a functional integral representation of the amplitude $\langle a''^*, t''|a', t'\rangle$ that is formally identical (except for the sign of the $(i/2)\omega t$ terms) to that of the corresponding amplitude for the Fermi oscillator. The action integral (5.5.47) in the present case (with $\frac{1}{2}\omega$ replaced by $-\frac{1}{2}\omega$) is just an alternative version of the so-called Hamiltonian form of the action, which is often used in discussing Bose systems. It will be left to the reader to carry out the construction. Here attention will be focussed on the Lagrangian form (5.6.1) of the action and the functional integral to which *it* leads. It is necessary first to assemble a few results that may be regarded as quantum extensions of classical Hamilton–Jacobi theory. In particular, an expression for the amplitude $\langle x'', t''|a', t'\rangle$ is needed.

Using eqs. (5.6.13) and (5.6.41) one obtains the following differential equation:

$$\begin{aligned}
a'\langle x'',t''|a',t'\rangle &= \langle x'',t''|a(t')|a',t'\rangle = e^{i\omega(t''-t')}\langle x'',t''|a(t'')|a',t'\rangle \\
&= (2\omega)^{-\frac{1}{2}} e^{i\omega(t''-t')}\langle x'',t''|[i\dot{x}(t'') + \omega x(t'')]|a',t'\rangle \\
&= (2\omega)^{-\frac{1}{2}} e^{i\omega(t''-t')}\left(\frac{\partial}{\partial x''} + \omega x''\right)\langle x'',t''|a',t'\rangle,
\end{aligned} \tag{5.6.52}$$

of which the general solution is

$$\langle x'', t''|a', t'\rangle = Z(a', t'', t')\exp\{-\tfrac{1}{2}\omega[x'' - (2/\omega)^{\frac{1}{2}} e^{-i\omega(t''-t')}a']^2\}. \tag{5.6.53}$$

The function $Z(a', t'', t')$ is fixed by eqs. (5.6.10) and (5.6.47):

$$\begin{aligned}\langle a''^*, t''|a', t'\rangle &= \int_{-\infty}^{\infty} \langle a''^*, t''|x'', t''\rangle\langle x'', t''|a', t'\rangle \, dx'' \\ &= Z^*(a'', t'', t'')Z(a', t'', t') \\ &\quad \times \int_{-\infty}^{\infty} \exp(-\omega\{x'' - (2\omega)^{-\frac{1}{2}}[a''^* + e^{-i\omega(t''-t')}a']\}^2 \\ &\qquad - \tfrac{1}{4}[a''^* - e^{-i\omega(t''-t')}a']^2)\, dx'' \\ &= (\pi/\omega)^{\frac{1}{2}} Z^*(a'', t'', t'')Z(a', t'', t') \\ &\quad \times \exp\{-\tfrac{1}{4}[a''^* - e^{-i\omega(t''-t')}a']^2\}.\end{aligned} \tag{5.6.54}$$

Comparison with (5.6.47) shows that $Z(a', t'', t')$ may be chosen to be

$$Z(a', t'', t') = (\omega/\pi)^{\frac{1}{4}} \exp\left\{\tfrac{1}{2} e^{-2i\omega(t''-t')}a'^2 - \tfrac{1}{2}a'^*a' - \frac{i}{2}\omega(t'' - t')\right\}, \tag{5.6.55}$$

whence

$$\langle x'', t''|a', t'\rangle = (2\pi)^{-\frac{1}{4}}\, i^{\frac{1}{2}} D^{\frac{1}{2}}(t'', t')\, e^{iS(x'', t''|a', t') - \frac{1}{2}a'^*a'}, \tag{5.6.56}$$

where

$$S(x'', t''|a', t') \overset{\text{def}}{=} \frac{i}{2}\omega[x'' - (2/\omega)^{\frac{1}{2}} e^{-i\omega(t''-t')}a']^2 - \frac{i}{2}e^{-2i\omega(t''-t')}a'^2, \tag{5.6.57}$$

$$D(t'', t') \overset{\text{def}}{=} \partial^2 S/\partial x''\partial a' = -i(2\omega)^{\frac{1}{2}} e^{-i\omega(t''-t')} \tag{5.6.58}$$

The function $S(x'', t''|a', t')$ is readily verified to satisfy the Hamilton–Jacobi equation for the Bose oscillator:

$$\begin{aligned}&\partial S/\partial t'' + H(x'', \partial S/\partial x'') \qquad (H(x, p) = \tfrac{1}{2}(p^2 + \omega^2 x^2)) \\ &\quad = \partial S/\partial t'' + \tfrac{1}{2}[(\partial S/\partial x'')^2 + \omega^2 x''^2] = 0.\end{aligned} \tag{5.6.59}$$

It generates the canonical transformation from the 'canonical' variables a', a'^* at time t' to the canonical variables x'', p'' at time t'':

$$p'' = \partial S/\partial x'' = i\omega[x'' - (2\omega)^{\frac{1}{2}} e^{-i\omega(t''-t')}a'] \tag{5.6.60}$$

or

$$a' = (2\omega)^{-\frac{1}{2}} e^{i\omega(t''-t')}(ip'' + \omega x'') \tag{5.6.61}$$

and

$$a'^* = i\partial S/\partial a' = -e^{-2i\omega(t''-t')}a' + (2\omega)^{\frac{1}{2}}e^{-i\omega(t''-t')}x''$$
$$= (2\omega)^{-\frac{1}{2}}e^{-i\omega(t''-t')}(-ip'' + \omega x'') \tag{5.6.62}$$

(cf. eqs. (5.6.41)). The function $D(t'',t')$, which in the case of Bose theories that have more than one degree of freedom becomes a determinant, is known as the *Van Vleck–Morette determinant* for the generator S.

The generator S also satisfies a Hamilton–Jacobi equation in terms of the variables a' and a'^*:

$$-\partial S/\partial t' + \bar{H}(a', i\partial S/\partial a') \qquad (\bar{H}(a,a^*) = \omega a^* a)$$
$$= -\partial S/\partial t' + i\omega(\partial S/\partial a')a' = 0. \tag{5.6.63}$$

The amplitude $\langle x'',t''|x',t'\rangle$ and its functional integral representation

From eqs. (5.6.51) and (5.6.56) a straightforward but tedious Gaussian integral leads to the amplitude $\langle x'',t''|x',t'\rangle$:

$$\langle x'',t''|x',t'\rangle = \frac{1}{2\pi i}\int \langle x'',t''|a',t'\rangle\langle a'^*,t'|x',t'\rangle\, da'^*\, da'$$
$$= (2\pi i)^{-\frac{1}{2}}D^{\frac{1}{2}}(t''|t')e^{iS(x'',t''|x',t')}, \tag{5.6.64}$$

where

$$S(x'',t''|x',t') \stackrel{\text{def}}{=} \frac{\omega}{2\sin\omega(t''-t')}[(x''^2 + x'^2)\cos\omega(t''-t') - 2x''x'], \tag{5.6.65}$$

$$D(t''|t') \stackrel{\text{def}}{=} -\frac{\partial^2}{\partial x''\partial x'}S(x'',t''|x',t') = \frac{\omega}{\sin\omega(t''-t')}. \tag{5.6.66}$$

The function $S(x'',t''|x',t')$ is the generator of the canonical transformation from the canonical variables x', p' at time t' to the canonical variables x'', p'' at time t''. It satisfies the Hamilton–Jacobi equations

$$\left.\begin{aligned} \partial S/\partial t'' + H(x'', \partial S/\partial x'') &= 0, \\ -\partial S/\partial t' + H(x', -\partial S/\partial x') &= 0. \end{aligned}\right\} \tag{5.6.67}$$

When $t'' - t'$ is infinitesimal, expression (5.6.65) becomes, correct to first order in $t'' - t'$,

$$S(x'',t''|x',t') = \frac{1}{2}\frac{(x''-x')^2}{t''-t'} - \tfrac{1}{6}\omega^2(x''^2 + x''x' + x'^2)(t''-t')$$
$$= \frac{1}{2}\int_{x',t'}^{x'',t''}(\dot{x}^2 - \omega^2 x^2)\,dt, \tag{5.6.68}$$

where the final integral is to be understood as evaluated for the straight-line path between x', t' and x'', t''. When $t'' - t'$ is finite, one may write

$$\langle x'', t'' | x', t' \rangle = \int \langle x_{N+1}, t_{N+1} | x_N, t_N \rangle \langle x_N, t_N | x_{N-1}, t_{N-1} \rangle \\ \ldots \langle x_2, t_2 | x_1, t_1 \rangle \langle x_1, t_1 | x_0, t_0 \rangle \, dx_1 \, dx_2 \ldots dx_N, \qquad (5.6.69)$$

where the x's and t's satisfy eqs. (5.4.40) and (5.4.41) and the integration with respect to each x is over an arbitrary contour in $\mathbf{R}_c$ between $-\infty$ and ∞. One may then pass to the limit $N \to \infty, t_n - t_{n-1} \to 0, D(t_n | t_{n-1}) \to (t_n - t_{n-1})^{-1}$ for all n, obtaining

$$\langle x'', t'' | x', t' \rangle = Z \int \exp\left[\frac{i}{2} \int_{x',t'}^{x'',t''} (\dot{x}^2 - \omega^2 x^2) \, dt \right] \mathrm{d}x \qquad (5.6.70)$$

where, formally,

$$Z = (2\pi i \, dt)^{-(t''-t')/2dt}, \qquad (5.6.71)$$

$$\mathrm{d}x = \prod_{t'<t<t''} dx(t). \qquad (5.6.72)$$

Equation (5.6.70), like eqs. (5.4.43) and (5.5.50), is another special case of (5.3.46). Again the factor $(\operatorname{sdet} G^+)^{-\frac{1}{2}}$ is a constant which can be absorbed into the coefficient Z. The paths for which the action integral in the exponent of the functional integrand is to be evaluated are broken-straight-line paths. The functional integral itself is a Gaussian integral which may be evaluated by determining the stationary point, i.e., the superclassical path between x', t' and x'', t''. The value assumed by the action functional at the superclassical path is well known to be given by expression (5.6.65), which is called the (super) classical action *function*. The functional integral therefore leads back again to eq. (5.6.64).

The functional-integral representation of $\langle a''^*, t'' | a', t' \rangle$

Equations (5.6.56) and (5.6.70) together yield a functional integral representation for $\langle a''^*, t'' | a', t' \rangle$ in which the x's rather than the a's and a^*'s appear as integration variables:

$$\begin{aligned} \langle a''^*, t'' | a', t' \rangle &= \int dx'' \int dx' \langle a''^*, t'' | x'', t'' \rangle \langle x'', t'' | x', t' \rangle \langle x', t' | a', t' \rangle \\ &= (\omega/\pi)^{\frac{1}{2}} (2\pi i \, dt)^{-(t''-t')/2dt} \, e^{-\frac{1}{2}(a''^* a'' + a'^* a')} \\ &\quad \times \int \exp\Big[iS(a''^*, t'' | x'', t'') + \frac{i}{2} \int_{x',t'}^{x'',t''} (\dot{x}^2 - \omega^2 x^2) \, dt \\ &\qquad + iS(x', t' | a', t') \Big] \, \mathrm{d}x, \end{aligned} \qquad (5.6.73)$$

where

$$S(a''^*, t''|x', t') \stackrel{\text{def}}{=} -S(x', t'|a'', t'')^* \tag{5.6.74}$$

and

$$\mathrm{d}x \stackrel{\text{def}}{=} \prod_{t' \leqslant t \leqslant t''} \mathrm{d}x(t). \tag{5.6.75}$$

The stationary path between coherent states

Equation (5.6.73) shows that the action functional appropriate to coherent-state boundary conditions is not the simple integral (5.6.1) but rather

$$S(a''^*, t''|x'', t'') + \frac{1}{2}\int_{x',t'}^{x'',t''} (\dot{x}^2 - \omega^2 x^2)\,\mathrm{d}t + S(x', t'|a', t'), \tag{5.6.76}$$

in which the Hamilton–Jacobi functions $S(x', t'|a', t')$ and $S(a''^*, t''|x'', t'')$ have been added as endpoint contributions. The stationary path is no longer determined by x' and x'', for these are now freely variable, but rather by a' and a''^*.

To determine the stationary path it suffices to 'break into' the action integral at an arbitrary time t and to solve the equation

$$\left\{\frac{\partial}{\partial x'}[S(a''^*, t''|x', t) + S(x', t|a', t')]\right\}_{x' = x_{st}(t)} = 0. \tag{5.6.77}$$

A simple calculation, using (5.6.57) and (5.6.74), leads to

$$x_{st}(t) = (2\omega)^{-\frac{1}{2}}[\mathrm{e}^{-\mathrm{i}\omega(t''-t)}a''^* + \mathrm{e}^{-\mathrm{i}\omega(t-t')}a']. \tag{5.6.78}$$

The function x_{st} is seen to be generally complex valued, and hence the stationary path lies outside the superclassical configuration space $\mathbf{R}_c$, in its complex extension $\mathbf{C}_c$.

Since the functional integral (5.6.73) is Gaussian it follows immediately that

$$\langle a''^*, t''|x(t)|a', t'\rangle = x_{st}(t)\langle a''^*, t''|a', t'\rangle, \tag{5.6.79}$$

a result that can be obtained also directly from eqs. (5.6.41) and their corollary

$$x(t) = (2\omega)^{-\frac{1}{2}}[a^*(t) + a(t)]. \tag{5.6.80}$$

The Feynman propagator

The Feynman propagator for the Bose oscillator, like that for the Fermi oscillator, is the Green's function for the stationary path between coherent states. It may be defined by

$$G(s,t)\langle a''^*,t''|a',t'\rangle = \mathrm{i}\langle a''^*,t''|T(x(s)x(t))|a',t'\rangle - \mathrm{i}x_{\mathrm{st}}(s)x_{\mathrm{st}}(t)\langle a''^*,t''|a',t'\rangle. \tag{5.6.81}$$

(cf. eq. (5.5.66)). Making use of eq. (5.6.80) together with

$$[a(s), a^*(t)] = \mathrm{e}^{-\mathrm{i}\omega(s-t)}, \tag{5.6.82}$$

$$a_{\mathrm{st}}(s) = \mathrm{e}^{-\mathrm{i}\omega(s-t')}a', \qquad a^*_{\mathrm{st}}(s) = \mathrm{e}^{-\mathrm{i}\omega(t''-s)}a''^*, \tag{5.6.83}$$

one readily obtains

$$\begin{aligned}\langle a''^*,t''|x(s)x(t)|a',t'\rangle &= (2\omega)^{-1}\langle a''^*,t''|[a^*(s)+a(s)][a^*(t)+a(t)]|a',t'\rangle \\ &= (2\omega)^{-1}\{[a^*_{\mathrm{st}}(s)+a_{\mathrm{st}}(s)][a^*_{\mathrm{st}}(t)+a_{\mathrm{st}}(t)] + \mathrm{e}^{-\mathrm{i}\omega(s-t)}\} \\ &\quad\times\langle a''^*,t''|a',t'\rangle,\end{aligned} \tag{5.6.84}$$

whence

$$\left.\begin{aligned}G(s,t) &= \mathrm{i}(2\omega)^{-1}\,\mathrm{e}^{-\mathrm{i}\omega|s-t|} \\ &= \mathrm{i}[\theta(s,t)u(s)u^*(t) + \theta(t,s)u^*(s)u(t)] \\ &= -\theta(s,t)G^{(+)}(s,t) + \theta(t,s)G^{(-)}(s,t) \\ &= G^-(s,t) + G^{(-)}(s,t) = G^+(s,t) - G^{(+)}(s,t).\end{aligned}\right\} \tag{5.6.85}$$

When $t'' > s > t'$ the Feynman propagator satisfies the boundary conditions

$$\left.\begin{aligned}&\left(\mathrm{i}\frac{\partial}{\partial t'} + \omega\right)G(t',s) = 0, \qquad \left(\mathrm{i}\frac{\partial}{\partial t'} + \omega\right)G(s,t') = 0, \\ &\left(-\mathrm{i}\frac{\partial}{\partial t''} + \omega\right)G(t'',s) = 0, \quad \left(-\mathrm{i}\frac{\partial}{\partial t''} + \omega\right)G(s,t'') = 0,\end{aligned}\right\} \tag{5.6.86}$$

corresponding to the fact that the stationary path is fixed by specifying the boundary conditions

$$\left.\begin{aligned}&(2\omega)^{-\frac{1}{2}}\left(\mathrm{i}\frac{\partial}{\partial t'} + \omega\right)x_{\mathrm{st}}(t') = a', \\ &(2\omega)^{-\frac{1}{2}}\left(-\mathrm{i}\frac{\partial}{\partial t''} + \omega\right)x_{\mathrm{st}}(t'') = a''^*.\end{aligned}\right\} \tag{5.6.87}$$

Energy eigenfunctions

A final example showing the utility of the coherent-state supervectors is the computation of the eigenfunctions of the energy operator. Using eqs. (5.6.45), (5.6.51) and (5.6.56)–(5.6.58), one has

$$
\begin{aligned}
\langle x',t|n\rangle &= (2\pi i)^{-1}\int \langle x',t|a',t\rangle\langle a'^*,t|n\rangle\, da'^*\, da' \\
&= (2\pi i)^{-1}\int (\omega/\pi)^{\frac{1}{4}} \exp\{-\tfrac{1}{2}\omega[x' - (2/\omega)^{\frac{1}{2}}a']^2 \\
&\quad + \tfrac{1}{2}a'^2 - a'^*a' - i(n+\tfrac{1}{2})\omega t\}(n!)^{-\frac{1}{2}} a'^{*n}\, da'^*\, da' \\
&= \pi^{-1}(\omega/\pi)^{\frac{1}{4}}(n!)^{-\frac{1}{2}} e^{-i(n+\frac{1}{2})\omega t}\int_0^\infty r\, dr \int_0^{2\pi} d\theta \\
&\quad \times \exp\{-\tfrac{1}{2}\omega[x' - (2/\omega)^{\frac{1}{2}} r e^{i\theta}]^2 + \tfrac{1}{2}r^2 e^{2i\theta} - r^2\} r^n e^{-in\theta} \\
&= \frac{1}{\pi i}\left(\frac{\omega}{\pi}\right)^{\frac{1}{4}}(n!)^{-\frac{1}{2}} e^{-\frac{1}{2}\omega x'^2 - i(n+\frac{1}{2})\omega t}\int_0^\infty r^{n+1} e^{-r^2}\, dr \\
&\quad \times \oint \frac{dz}{z^{n+1}} \exp[-\tfrac{1}{2}r^2 z^2 + (2\omega)^{\frac{1}{2}} x' r z] \\
&= 2\left(\frac{\omega}{\pi}\right)^{\frac{1}{4}}(n!)^{-\frac{3}{2}} e^{\frac{1}{2}\omega x'^2 - i(n+\frac{1}{2})\omega t}\int_0^\infty r^{n+1} e^{-r^2}\, dr \\
&\quad \times \left(\frac{d^n}{dz^n}\exp\{-\tfrac{1}{2}[rz - (2\omega)^{\frac{1}{2}}x']^2\}\right)_{z=0} \\
&= 2\left(\frac{\omega}{\pi}\right)^{\frac{1}{4}}(n!)^{-\frac{3}{2}} e^{\frac{1}{2}\omega x'^2 - i(n+\frac{1}{2})\omega t}\int_0^\infty r^{2n+1} e^{-r^2}\, dr \\
&\quad \times \left[\frac{d^n}{d\xi^n} e^{-\frac{1}{2}\xi^2}\right]_{\xi = -(2\omega)^{\frac{1}{2}}x'} \\
&= \left(\frac{\omega}{\pi}\right)^{\frac{1}{4}}(n!)^{-\frac{1}{2}} e^{-\frac{1}{2}\omega x'^2 - i(n+\frac{1}{2})\omega t}\left[e^{\frac{1}{2}\xi^2}\frac{d^n}{d\xi^n} e^{-\frac{1}{2}\xi^2}\right]_{\xi = -(2\omega)^{\frac{1}{2}}x'}.
\end{aligned}
\tag{5.6.88}
$$

5.7 Bose–Fermi supersymmetry

The simplest model

When Bose and Fermi systems are combined into a single system new kinds of symmetries and conservation laws can occur. The simplest model that illustrates this possibility is obtained by combining the systems considered in the preceding two sections. The total action functional for the model is just the sum of the action functionals (5.5.3) and (5.6.1). The configuration space of the combined superclassical system is $\mathbf{R}_c \times \mathbf{R}_a^2$ and the super Hilbert space of the combined quantum system is $\mathfrak{H}_B \otimes \mathfrak{H}_F$ where $\mathfrak{H}_B$ and $\mathfrak{H}_F$ are the super Hilbert spaces for the Bose and Fermi oscillators respectively. Since $\mathfrak{H}_B$ has dimension $(\infty, 0)$ and $\mathfrak{H}_F$ has dimension $(1, 1)$, the dimension of $\mathfrak{H}_B \otimes \mathfrak{H}_F$ is (∞, ∞).

In order to avoid notational confusion the dynamical variables x_1, x_2 of section 5.5 will in this section be denoted by y_1, y_2 and the operators a, a^* will be denoted by b, b^*. The notation of section 5.6, on the other hand, will be left intact. Other notation will be obvious. The total action functional then takes the form

$$S = \int \left[\tfrac{1}{2}(\dot{x}^2 - \omega^2 x^2) + \frac{i}{2}(y^\sim \dot{y} + \omega y^\sim M y) \right] dt. \tag{5.7.1}$$

Note that the frequency ω is chosen to be the same in both terms of the integrand of (5.7.1). In principle a different frequency could be chosen for each term, but *only when the frequencies are identical* does the new *Bose–Fermi supersymmetry* occur.

This new symmetry will be displayed in a moment, but first some basic results of the preceding two sections will be restated in the revised notation:

$$\left.\begin{aligned} x(t) &= (2\omega)^{-\frac{1}{2}} (a e^{-i\omega t} + a^* e^{i\omega t}), \\ y(t) &= b \begin{pmatrix} 1/\sqrt{2} \\ i/\sqrt{2} \end{pmatrix} e^{-i\omega t} + b^* \begin{pmatrix} 1/\sqrt{2} \\ -i/\sqrt{2} \end{pmatrix} e^{i\omega t}, \end{aligned}\right\} \tag{5.7.2}$$

$$\left.\begin{aligned} [x(t), x(t')] &= -i\omega^{-1} \sin\omega(t-t'), \\ [y(t), y^\sim(t')] &= \begin{pmatrix} \cos\omega(t-t') & -\sin\omega(t-t') \\ \sin\omega(t-t') & \cos\omega(t-t') \end{pmatrix}, \end{aligned}\right\} \tag{5.7.3}$$

$$[a, a^*] = 1, \qquad b^2 = 0, \qquad b^{*2} = 0, \qquad [b, b^*] = 1. \tag{5.7.4}$$

Equations (5.7.3) and (5.7.4) determine the operator superalgebras for the Bose and Fermi oscillators separately. The operator superalgebra for the combined system is obtained by appending to these also the following bracket relations:

$$[x(t), y(t')] = 0, \tag{5.7.5}$$

$$[a, b] = 0, \qquad [a, b^*] = 0. \tag{5.7.6}$$

All the above brackets are supercommutators. The reader must remember that they are equivalent to the ordinary anticommutator when both entries are *a*-type. In the above equations x, a and a^* are *c*-type operators; y, b and b^* are *a*-type.

The energy operator for the combined system is

$$\left.\begin{aligned} H &= \tfrac{1}{2}(\dot{x}^2 + \omega^2 x^2) - \frac{i}{2}\omega y^\sim M y \\ &= \omega(a^* a + b^* b). \end{aligned}\right\} \tag{5.7.7}$$

Note that the terms $\pm\frac{1}{2}\omega$ in the Hamiltonians (5.5.23) and (5.6.25) cancel when they are added together. The energy eigenvalues are therefore integral multiples of ω. All except the lowest (zero) are doubly degenerate.

New conserved quantities

Let $\delta\alpha$ be a 2×1 matrix (i.e., column vector) having infinitesimal real a-numbers as elements, and let these a-numbers be arbitrary functions of the time t, of compact support. Let the dynamical variables x and y suffer the infinitesimal changes

$$\left.\begin{aligned}\delta x &= \mathrm{i}y^{\sim}\delta\alpha,\\ \delta y &= (\dot{x}\mathbf{1}_2 - \omega x M)\delta\alpha.\end{aligned}\right\} \tag{5.7.8}$$

These changes induce the following change in the action functional:

$$\begin{aligned}\delta S &\equiv \int\left(\frac{\delta S}{\delta x}\delta x + S\frac{\overleftarrow{\delta}}{\delta y}\delta y\right)\mathrm{d}t\\ &\equiv \int[-\mathrm{i}(\ddot{x}+\omega^2 x)y^{\sim}\delta\alpha - \mathrm{i}(\dot{y}^{\sim}-\omega y^{\sim}M)(\dot{x}\mathbf{1}_2-\omega x M)\delta\alpha]\,\mathrm{d}t\\ &\equiv -\mathrm{i}\int\dot{Q}^{\sim}\delta\alpha\,\mathrm{d}t,\end{aligned} \tag{5.7.9}$$

where

$$Q^{\sim} \stackrel{\text{def}}{=} y^{\sim}(\dot{x}\mathbf{1}_2-\omega x M), \qquad Q = (\dot{x}\mathbf{1}_2+\omega x M)y. \tag{5.7.10}$$

The variation δS obviously vanishes when the dynamical equations are satisfied. Since $\delta\alpha$ is arbitrary it follows that the quantity Q is conserved:

$$\dot{Q} = 0. \tag{5.7.11}$$

Being conserved, Q may be expected to be the generator of the very transformations (5.7.8) that gave rise to it. This, in fact, follows at once from the supercommutation relations (5.7.3) and (5.7.5.):

$$\left.\begin{aligned}[x, Q^{\sim}\delta\alpha] &= [x, y^{\sim}(\dot{x}\mathbf{1}_2-\omega x M)]\,\delta\alpha = \mathrm{i}y^{\sim}\delta\alpha,\\ [y, Q^{\sim}\delta\alpha] &= [y, y^{\sim}(\dot{x}\mathbf{1}_2-\omega x M)]\,\delta\alpha = (\dot{x}\mathbf{1}_2-\omega x M)\,\delta\alpha.\end{aligned}\right\} \tag{5.7.12}$$

Since Q is conserved it suffices to evaluate these supercommutators at equal times.

Equations (5.7.12) also follow from more general arguments, which are worth presenting because they apply to a wide variety of systems. Let the symbols x, y be replaced by φ^i, $i = 0, 1, 2$, where $\varphi^0 = x$, $\varphi^1 = y_1$, $\varphi^2 = y_2$.

The change (5.7.9) in the action functional can then be represented as

$$\delta S \equiv \int S \frac{\overleftarrow{\delta}}{\delta\varphi^i} \delta\varphi^i \, \mathrm{d}t, \tag{5.7.13}$$

where the $\delta\varphi^i$ are given by (5.7.8). Let this change be regarded as a change in the explicit functional form of the action. Such a change induces retarded and advanced changes in the dynamical variables given (see eq. (5.1.14)) by

$$\begin{aligned}\delta^{\pm}\varphi^i(t) &= \int G^{\pm ij}(t,t') \frac{\overrightarrow{\delta}}{\delta\varphi^j(t')} \delta S \, \mathrm{d}t' \\ &= \int \mathrm{d}t' \int \mathrm{d}t'' G^{\pm ij}(t,t') \frac{\overrightarrow{\delta}}{\delta\varphi^j(t')} \left[S \frac{\overleftarrow{\delta}}{\delta\varphi^k(t'')} \delta\varphi^k(t'') \right] \\ &= \int \mathrm{d}t' \int \mathrm{d}t'' G^{\pm ij}(t,t') \left[\frac{\overrightarrow{\delta}}{\delta\varphi^j(t')} S \frac{\overleftarrow{\delta}}{\delta\varphi^k(t'')} \right] \delta\varphi^k(t'') \\ &= -\,\delta\varphi^i(t),\end{aligned} \tag{5.7.14}$$

the dynamical equations having been used in passing to the third line. The final result follows from the fact that the support of $\delta\varphi^i$ is compact, so that the order of integration is immaterial, and from the fact that the $G^{\pm ij}(t,t')$ are left as well as right Green's functions. The retarded and advanced changes in the φ^i are seen to be identical. Among other things this implies that the Peierls bracket (or supercommutator) of δS with anything vanishes, which one also knows because δS vanishes by virtue of the dynamical equations.

Now use the fact that δS has the explicit form (5.7.9), or, equivalently,

$$\delta S \equiv \mathrm{i} \int Q^{\sim} \delta\dot{\alpha} \, \mathrm{d}t. \tag{5.7.15}$$

Let $\delta\alpha$ be chosen to be

$$\delta\alpha(t) \equiv \theta(t'', t)\theta(t, t')\delta\beta, \tag{5.7.16}$$

where $t'' > t'$ and $\delta\beta$ is time independent. Then

$$\delta S \equiv -\mathrm{i}[Q^{\sim}(t'') - Q^{\sim}(t')]\,\delta\beta, \tag{5.7.17}$$

which, of course, vanishes when the dynamical equations are satisfied. Let A be an arbitrary function of the φ^i such that $t' < \mathrm{supp}(A\overleftarrow{\delta}/\delta\varphi^i) < t''$. Then the change in A produced by the transformation (5.7.8) may be expressed in the form

$$\begin{aligned}\delta A &= \int A \frac{\overleftarrow{\delta}}{\delta\varphi^i(t)} \delta\varphi^i(t) \, \mathrm{d}t \\ &= -\int A \frac{\overleftarrow{\delta}}{\delta\varphi^i(t)} \delta^{\pm}\varphi^i(t) \, \mathrm{d}t\end{aligned}$$

$$= -\int dt \int dt'' A \frac{\overleftarrow{\delta}}{\delta\varphi^i(t)} G^{\pm ij}(t,t'') \frac{\overrightarrow{\delta}}{\delta\varphi^j(t'')} \delta S$$

$$= \begin{cases} \mathrm{i}\int dt \int dt'' A \dfrac{\overleftarrow{\delta}}{\delta\varphi^i(t)} G^{+ij}(t,t'') \dfrac{\overrightarrow{\delta}}{\delta\varphi^j(t'')} Q^{\sim}(t'')\,\delta\beta \\ -\mathrm{i}\int dt \int dt'' A \dfrac{\overleftarrow{\delta}}{\delta\varphi^i(t)} G^{-ij}(t,t'') \dfrac{\overrightarrow{\delta}}{\delta\varphi^j(t')} Q^{\sim}(t')\,\delta\beta \end{cases}$$

$$= \begin{cases} \mathrm{i}\int dt \int dt'' A \dfrac{\overleftarrow{\delta}}{\delta\varphi^i(t)} \tilde{G}^{ij}(t,t'') \dfrac{\overrightarrow{\delta}}{\delta\varphi^j(t'')} Q^{\sim}(t'')\,\delta\beta \\ \mathrm{i}\int dt \int dt'' A \dfrac{\overleftarrow{\delta}}{\delta\varphi^i(t)} \tilde{G}^{ij}(t,t'') \dfrac{\overrightarrow{\delta}}{\delta\varphi^j(t'')} Q^{\sim}(t')\delta\beta. \end{cases} \tag{5.7.18}$$

The expressions after the third line follow from the support condition assumed for $A\overleftarrow{\delta}/\delta\varphi^i$ and from the fact that Q is a *local* operator, constructed out of the dynamical variables and a *finite* number of their (time) derivatives. Since t' and t'' are arbitrary and since Q is constant by virtue of the dynamical equations, eq. (5.7.18) implies quite generally

$$\delta A = \mathrm{i}(A, Q^{\sim}\delta\beta) = [A, Q^{\sim}\delta\beta], \tag{5.7.19}$$

Equations (5.7.12) are just special cases of this.

The Bose–Fermi supersymmetry group

Applying (5.7.19) to Q itself, taking careful account of the order of factors, and making use of the dynamical equations as well as the corollary

$$yy^{\sim} = \tfrac{1}{2}[\mathbf{1}_2 + (y^{\sim}My)M] = \tfrac{1}{2}[\mathbf{1}_2 + M(y^{\sim}My)] \tag{5.7.20}$$

of the equal-time supercommutator $[y, y^{\sim}] = \mathbf{1}_2$, one finds

$$\begin{aligned} [Q, Q^{\sim}\delta\alpha] = \delta Q &= (\delta\dot{x}\mathbf{1}_2 + \omega\delta x M)y + (\dot{x}\mathbf{1}_2 + \omega x M)\,\delta y \\ &= \mathrm{i}(\dot{y}^{\sim}\delta\alpha)y + \mathrm{i}\omega(y^{\sim}\delta\alpha)My + (\dot{x}\mathbf{1}_2 + \omega x M)(\dot{x}\mathbf{1}_2 - \omega x M)\,\delta\alpha \\ &= \mathrm{i}\omega(yy^{\sim} - \mathbf{1}_2)M\delta\alpha + \mathrm{i}\omega M(yy^{\sim} - \mathbf{1}_2)\,\delta\alpha \\ &\quad + (\dot{x}^2\mathbf{1}_2 + \omega^2 x^2\mathbf{1}_2 + \mathrm{i}\omega M)\,\delta\alpha \\ &= (-\mathrm{i}\omega y^{\sim}My + \dot{x}^2 + \omega^2 x^2)\,\delta\alpha = 2H\delta\alpha, \end{aligned} \tag{5.7.21}$$

whence

$$[Q, Q^{\sim}] = 2H\mathbf{1}_2. \tag{5.7.22}$$

Combining this with the supercommutator

$$[Q, H] = 0, \tag{5.7.23}$$

one sees that H, Q_1 and Q_2 are the generators of a super Lie algebra (and hence of a super Lie group) of dimension (1, 2). The group elements may be expressed in the form $e^{iHt+Q\cdot\theta}$ where t, θ_1 and θ_2 are canonical coordinates. The elements e^{iHt} constitute an invariant Abelian subgroup and hence the group is neither simple nor semisimple. It is a kind of poor man's analog of the super Poincaré group of section 4.3. It is called a *Bose–Fermi supersymmetry group.*

Eigenvectors of Q_1 and Q_2

Equation (5.7.22) implies

$$Q_1^2 = Q_2^2 = H, \qquad Q_1Q_2 + Q_2Q_1 = 0, \tag{5.7.24}$$

from which one may infer at once that the eigenvalues of Q_1 and Q_2 are the square roots of the eigenvalues of H. Denote by $|n, m\rangle$ the simultaneous eigenvector of a^*a and b^*b corresponding to the eigenvalues n and m respectively. The quantum number m can assume the value 0 or 1 whereas n ranges over the nonnegative integers: $n = 0, 1, 2, \ldots$. From eq. (5.7.7) one sees that

$$H|n, m\rangle = (n+m)\omega|n, m\rangle. \tag{5.7.25}$$

The $|n, m\rangle$ may be assumed to constitute a complete orthonormal pure basis. With the conventions of the preceding two sections the supervectors $|n, 0\rangle$ are c-type and the supervectors $|n, 1\rangle$ are a-type.

The $|n, m\rangle$, being pure, are not eigenvectors of the a-type operators Q_1, Q_2. In order to construct the latter it is convenient first to express Q_1 and Q_2 in terms of the operators a, a^*, b, b^*. Using the decompositions (5.7.2) one readily obtains

$$\begin{aligned} Q &= (\dot{x}1_2 + \omega x M)y \\ &= -i\omega(2\omega)^{-\frac{1}{2}}(ae^{-i\omega t} - a^*e^{i\omega t})\left[b\begin{pmatrix} 1/\sqrt{2} \\ i/\sqrt{2}\end{pmatrix}e^{-i\omega t} + b^*\begin{pmatrix} 1/\sqrt{2} \\ -i/\sqrt{2}\end{pmatrix}e^{i\omega t}\right] \\ &\quad + \omega(2\omega)^{-\frac{1}{2}}(ae^{-i\omega t} + a^*e^{i\omega t})\left[b\begin{pmatrix} i/\sqrt{2} \\ -1/\sqrt{2}\end{pmatrix}e^{-i\omega t} + b^*\begin{pmatrix} -i/\sqrt{2} \\ -i/\sqrt{2}\end{pmatrix}e^{i\omega t}\right] \\ &= \omega^{\frac{1}{2}}\begin{pmatrix} -i \\ -1\end{pmatrix}ab^* + \omega^{\frac{1}{2}}\begin{pmatrix} i \\ -1\end{pmatrix}a^*b, \end{aligned} \tag{5.7.26}$$

whence

$$Q_1 = -i\omega^{\frac{1}{2}}(ab^* - a^*b), \qquad Q_2 = -\omega^{\frac{1}{2}}(ab^* + a^*b). \tag{5.7.27}$$

From this it is easy to obtain eqs. (5.7.24).

The eigenvalues of Q_1 and Q_2 are $\pm(n\omega)^{\frac{1}{2}}$, $n = 0, 1, 2, \ldots$. Denote the corresponding normalized eigenvectors by $|1, \pm(n\omega)^{\frac{1}{2}}\rangle$ and $|2, \pm(n\omega)^{\frac{1}{2}}\rangle$.

Using eqs. (5.5.28), (5.5.29) and (5.6.43) it is then straightforward to verify that

$$\left.\begin{aligned} &|1, \pm(n\omega)^{\frac{1}{2}}\rangle = |2, \pm(n,\omega)^{\frac{1}{2}}\rangle = |0,0\rangle \quad \text{when } n=0,\\ &\left.\begin{aligned}|1, \pm(n\omega)^{\frac{1}{2}}\rangle &= 2^{-\frac{1}{2}}(|n,0\rangle \mp \mathrm{i}|n-1,1\rangle)\\ |2, \pm(n\omega)^{\frac{1}{2}}\rangle &= 2^{-\frac{1}{2}}(|n,0\rangle \mp |n-1,1\rangle)\end{aligned}\right\} \text{when } n=1,2,\ldots,\end{aligned}\right\} \tag{5.7.28}$$

up to irrelevant phase factors. Note that the values of Q_1 and Q_2 cannot be fixed simultaneously because Q_1 and Q_2 do not commute. The supervectors $|1, \pm(n\omega)^{\frac{1}{2}}\rangle$ (or, alternatively, the supervectors $|2, \pm(n\omega)^{\frac{1}{2}}\rangle$), like the $|n,m\rangle$, are energy eigenvectors and constitute by themselves a complete orthonormal basis, albeit an impure one.

The supersymmetry group as a transformation group

The transformations (5.7.8), with the infinitesimal *a*-number column vector $\delta\alpha$ chosen to be constant in time, may be regarded as an action of the supersymmetry group on the dynamical variables x, y. Because of eqs. (5.7.12) this action is a time-independent unitary transformation and hence the transformed variables satisfy the same dynamical equations ($\ddot{x}+\omega^2 x=0$ and $\dot{y}+\omega M y=0$) as the original variables. The existence of these transformations is remarkable, for they mix the Bose and Fermi variables together. The system is said to possess a *global* Bose–Fermi supersymmetry, the word 'global' distinguishing this type of symmetry from invariance under gauge transformations (or supergauge transformations), which is sometimes called local supersymmetry. States or supervectors are also said to possess supersymmetry if they remain invariant under the actions of the group. Thus in the present example the lowest-energy eigenvector or *ground state* $|0,0\rangle$ is supersymmetric because it remains unchanged when the operator $e^{iHt+Q^{\sim}\theta}$ is applied to it.

Equations (5.7.8) do not fully represent the actions of the supersymmetry group because they involve only two *a*-number parameters, the components of $\delta\alpha$. The full transformation laws are the following:

$$\left.\begin{aligned}\delta x &= \dot{x}\,\delta t + \mathrm{i}y^{\sim}\delta\alpha,\\ \delta y &= \dot{y}\,\delta t + (\dot{x}\mathbf{1}_2 - \omega x M)\,\delta\alpha,\end{aligned}\right\} \tag{5.7.29}$$

where δt is a constant infinitesimal real *c*-number. (If δt and $\delta\alpha$ are allowed to be switched on and off so as to have compact support, the corresponding change in the action is

$$\delta S \equiv -\int (\dot{H}\,\delta t + \mathrm{i}\dot{Q}^{\sim}\delta\alpha)\,\mathrm{d}t \tag{5.7.30}$$

which, of course, vanishes by virtue of the dynamical equations.) Using the dynamical equations one may readily show that the commutator of two such transformations, with parameters δt_1, $\delta\alpha_1$ and δt_2, $\delta\alpha_2$ respectively, is again a transformation of the same type, with parameters δt_{12}, $\delta\alpha_{12}$ given by

$$\delta t_{12} = -2\mathrm{i}\delta\tilde{\alpha}_1\,\delta\alpha_2, \qquad \delta\alpha_{12} = 0. \tag{5.7.31}$$

Auxiliary variable

The above *closure* property of the transformations holds only if x and y satisfy the dynamical equations. It does not hold when x and y are arbitrary functions of the time. However, it can be made to hold in the latter case if a single additional real c-type variable (function of t) is introduced. Denote this *auxiliary* variable by z. Then replace eqs. (5.7.29) by

$$\left.\begin{aligned} \delta x &= (\dot{x}+z)\delta t + \mathrm{i}y^{\sim}\delta\alpha, \\ \delta y &= -\omega M y\,\delta t + (\dot{x}\mathbf{1}_2 - \omega x M + z\mathbf{1}_2)\delta\alpha, \\ \delta z &= -(\ddot{x} + \omega^2 x + \dot{z})\delta t - \mathrm{i}(\dot{y}^{\sim} - \omega y^{\sim} M)\delta\alpha. \end{aligned}\right\} \tag{5.7.32}$$

It is straightforward to verify that the commutator of two of *these* transformations is again a transformation of the same type, with parameters given, as before, by eqs. (5.7.31).

Equations (5.7.32) can, in fact, be made completely equivalent to eqs. (5.7.29) simply by elevating z to the status of a dynamical variable and replacing the action (5.7.1) by

$$S = \int \left[\tfrac{1}{2}(\dot{x}^2 - \omega^2 x^2) + \frac{\mathrm{i}}{2}(y^{\sim}\dot{y} + \omega y^{\sim} M y) - \tfrac{1}{2}z^2 \right] \mathrm{d}t. \tag{5.7.33}$$

The dynamical equations for x and y are the same as before, whereas the dynamical equation for z renders z completely innocuous:

$$0 = \frac{\delta S}{\delta z} \equiv -z. \tag{5.7.34}$$

When the new as well as old dynamical equations are satisfied eqs. (5.7.32) reduce to eqs. (5.7.29). Moreover, eq. (5.7.30) remains intact.

Nonlinear Bose–Fermi supersymmetry

The system described by the action (5.7.1) is linear. Bose–Fermi supersymmetry can exist even for nonlinear systems. An example is provided by the action

$$S = \tfrac{1}{2}\int\{\dot{x}^2 - [V'(x)]^2 + \mathrm{i}y^{\sim}\dot{y} + \mathrm{i}V''(x)y^{\sim}My\}\mathrm{d}t, \tag{5.7.35}$$

which is a simple generalization of (5.7.1). The action (5.7.1) is a special case of (5.7.35), with the function $V(x)$ chosen to be $\frac{1}{2}\omega x^2$. A more interesting, nonlinear case is obtained by choosing

$$\left.\begin{aligned} V(x) &= \tfrac{1}{3}\lambda x^3 - \mu^2 x, & V'(x) &= \lambda x^2 - \mu^2, \\ V''(x) &= 2\lambda x, & [V'(x)]^2 &= \lambda^2 x^4 - 2\lambda\mu^2 x^2 + \mu^4. \end{aligned}\right\} \tag{5.7.36}$$

In general $V(x)$ will be assumed to have vanishing soul when x has vanishing soul.

The supersymmetry transformation associated with (5.7.35) is the following generalization of eqs. (5.7.29):

$$\left.\begin{aligned} \delta x &= \dot{x}\,\delta t + \mathrm{i}y^{\sim}\delta\alpha, \\ \delta y &= \dot{y}\,\delta t + [\dot{x}\mathbf{1}_2 - V'(x)M]\delta\alpha. \end{aligned}\right\} \tag{5.7.37}$$

To compute the generators H and Q of these transformations it will be convenient first to work with the superclassical theory so as to avoid for the moment any factor-ordering problems that may arise in the quantum theory. The dynamical equations are

$$0 = \frac{\vec{\delta}}{\delta x}S \equiv -\ddot{x} - V'(x)V''(x) + \tfrac{1}{2}V'''(x)y^{\sim}My, \tag{5.7.38}$$

$$0 = \frac{\vec{\delta}}{\delta y}S \equiv \mathrm{i}[\dot{y} + V''(x)My]. \tag{5.7.39}$$

If δt and $\delta\alpha$ are chosen to have compact support as functions of time then the transformation (5.7.37) induces the following change in the action:

$$\begin{aligned} \delta S &\equiv \int(-[\ddot{x} + V'(x)V''(x) - \tfrac{1}{2}V'''(x)y^{\sim}My][\dot{x}\,\delta t + \mathrm{i}y^{\sim}\delta\alpha] \\ &\qquad - \mathrm{i}[\dot{y}^{\sim} - V''(x)y^{\sim}M]\{\dot{y}\,\delta t + [\dot{x}\mathbf{1}_2 - V'(x)M]\delta\alpha\})\mathrm{d}t \\ &\equiv -\int(\dot{H}\,\delta t + \mathrm{i}\dot{Q}^{\sim}\delta\alpha)\mathrm{d}t, \end{aligned} \tag{5.7.40}$$

where

$$H \stackrel{\mathrm{def}}{=} \tfrac{1}{2}\dot{x}^2 + \tfrac{1}{2}[V'(x)]^2 - \tfrac{1}{2}V''(x)y^{\sim}My, \tag{5.7.41}$$

$$Q^{\sim} \stackrel{\mathrm{def}}{=} y^{\sim}[\dot{x}\mathbf{1}_2 - V'(x)M], \qquad Q = [\dot{x}\mathbf{1}_2 + V'(x)M]y. \tag{5.7.42}$$

In obtaining (5.7.40) use is made of the fact that in the superclassical theory $\dot{y}^{\sim}\dot{y} = 0$, and a product of three y's vanishes.

As a first step to the quantum theory let us examine some simple properties of the Green's functions for infinitesimal disturbances. First compute

$$\left.\begin{aligned}
\frac{\vec{\delta}}{\delta x(t)} S \frac{\overleftarrow{\delta}}{\delta x(t')} &\equiv \left\{ -\frac{\partial^2}{\partial t^2} - [V''(x)]^2 - V'(x)V'''(x) + \tfrac{1}{2}V''''(x) y^{\sim} M y \right\} \delta(t,t'),\\
\frac{\vec{\delta}}{\delta x(t)} S \frac{\overleftarrow{\delta}}{\delta y(t')} &\equiv \mathrm{i}\, V'''(x) y^{\sim} M\, \delta(t,t'),\\
\frac{\vec{\delta}}{\delta y(t)} S \frac{\overleftarrow{\delta}}{\delta x(t')} &\equiv \mathrm{i}\, V'''(x) M y\, \delta(t,t'),\\
\frac{\vec{\delta}}{\delta y(t)} S \frac{\overleftarrow{\delta}}{\delta y(t')} &\equiv \mathrm{i} \left[\mathbf{1}_2 \frac{\partial}{\partial t} + V''(x) M \right] \delta(t,t').
\end{aligned}\right\} \tag{5.7.43}$$

If the function $V(x)$ is chosen as in eqs. (5.7.36) then the differential equation satisfied by the Green's functions is

$$\begin{pmatrix} -\partial^2/\partial t^2 - 6\lambda^2 x^2 + 2\lambda\mu^2 & 2\mathrm{i}\lambda y^{\sim} M \\ 2\mathrm{i}\lambda M y & \mathrm{i}(\mathbf{1}_2 \partial/\partial t + 2\lambda x M) \end{pmatrix} \begin{pmatrix} G^{\pm xx}(t,t') & G^{\pm xy}(t,t') \\ G^{\pm yx}(t,t') & G^{\pm yy}(t,t') \end{pmatrix}$$

$$= -\begin{pmatrix} \delta(t,t') & 0 \\ 0 & \mathbf{1}_2 \delta(t,t') \end{pmatrix}. \tag{5.7.44}$$

It is straightforward to verify that the behaviour of these Green's functions when $t - t'$ is small is given by

$$\left.\begin{aligned}
G^{-xx}(t,t') = G^{+xx}(t',t) &= \theta(t,t')[t - t' + O(t-t')^3],\\
G^{\pm xy}(t,t') = O(t-t')^2, &\qquad G^{\pm yx}(t,t') = O(t-t')^2,\\
G^{-yy}(t,t') = -G^{+yy}(t',t) &= \mathrm{i}\theta(t,t')[\mathbf{1}_2 + O(t-t')],
\end{aligned}\right\} \tag{5.7.45}$$

and the same behaviour holds even when $V(x)$ is not given by (5.7.36).

From this the short-time behavior of the supercommutator function immediately follows:

$$\left.\begin{aligned}
\tilde{G}^{xx}(t,t') &= -(t-t') + O(t-t')^3, &\qquad \tilde{G}^{xy}(t,t') &= O(t-t')^2,\\
\tilde{G}^{yx}(t,t') &= O(t-t')^2, &\qquad \tilde{G}^{yy}(t,t') &= -\mathrm{i}\mathbf{1}_2 + O(t-t').
\end{aligned}\right\} \tag{5.7.46}$$

In the quantum theory the supercommutator function takes its values in the space of linear operators of the super Hilbert space of the theory. However, its short-time behavior continues to be given by (5.7.46), and this suffices to yield the following equal-time supercommutators:

$$\left.\begin{array}{lll}[x,\dot{x}]=\mathrm{i}, & [y,y^{\sim}]=\mathbf{1}_2, & \\ {[x,y]=0,} & [x,\dot{y}]=0, & [\dot{x},y]=0.\end{array}\right\} \tag{5.7.47}$$

When these supercommutation relations are satisfied there is no factor-ordering ambiguity in the dynamical equations (5.7.38) and (5.7.39), and none in the expressions for H and Q, eqs. (5.7.41) and (5.7.42).

The supersymmetry group

With the aid of the dynamical equations and the supercommutation relations (5.7.47), together with their corollaries

$$\tfrac{1}{2}y^{\sim}Myy=y_1y_2y=-\tfrac{1}{2}My, \tag{5.7.48}$$

and eq. (5.7.20), it is straightforward to verify the time independence of H and Q:

$$\begin{aligned}\dot{H}&=\tfrac{1}{2}\{\dot{x},\ddot{x}+V'(x)V''(x)-\frac{\mathrm{i}}{2}V'''(x)y^{\sim}My\}-\frac{\mathrm{i}}{2}V''(x)(\dot{y}^{\sim}My+y^{\sim}M\dot{y})\\ &=-\frac{\mathrm{i}}{2}[V''(x)]^2y^{\sim}(M^2-M^2)y=0,\end{aligned} \tag{5.7.49}$$

$$\begin{aligned}\dot{Q}&=[\ddot{x}\mathbf{1}_2+\tfrac{1}{2}\{\dot{x},V''(x)\}M]y+[\dot{x}\mathbf{1}_2+V'(x)M]\dot{y}\\ &=\left[-V'(x)V''(x)\mathbf{1}_2+\frac{\mathrm{i}}{2}V'''(x)y^{\sim}My\mathbf{1}_2+\tfrac{1}{2}\{\dot{x},V''(x)\}M\right]y\\ &\quad-[\dot{x}\mathbf{1}_2+V'(x)M]V''(x)My\\ &=-\frac{\mathrm{i}}{2}V'''(x)My-\tfrac{1}{2}[\dot{x},V''(x)]My=0.\end{aligned} \tag{5.7.50}$$

Since H and Q are time independent it suffices to use equal-time supercommutators in computing their action on the dynamical variables. One readily verifies

$$[x,H]=\mathrm{i}\dot{x},\qquad [y,H]=\mathrm{i}\dot{y}, \tag{5.7.51}$$

$$[x,Q^{\sim}\delta\alpha]=\mathrm{i}y^{\sim}\delta\alpha,\qquad [y,Q^{\sim}\delta\alpha]=[\dot{x}\mathbf{1}_2-V'(x)M]\delta\alpha, \tag{5.7.52}$$

(cf. eqs. (5.7.37)). These yield, in turn,

$$\begin{aligned}[Q,Q^{\sim}\delta\alpha]&=\mathrm{i}(\dot{y}^{\sim}\delta\alpha)y+\mathrm{i}V''(x)(y^{\sim}\delta\alpha)My\\ &\quad+[\dot{x}\mathbf{1}_2+V'(x)M][\dot{x}\mathbf{1}_2-V'(x)M]\,\delta\alpha\\ &=\mathrm{i}V''(x)(yy^{\sim}-\mathbf{1}_2)M\,\delta\alpha+\mathrm{i}V''(x)M(yy^{\sim}-\mathbf{1}_2)\delta\alpha\\ &\quad+\{\dot{x}^2\mathbf{1}_2+[V'(x)]^2+\mathrm{i}V''(x)M\}\delta\alpha\\ &=\{\dot{x}^2\mathbf{1}_2+[V'(x)]^2-\mathrm{i}V''(x)y^{\sim}My\}\delta\alpha=2H\,\delta\alpha,\end{aligned} \tag{5.7.53}$$

whence

$$[Q, Q^{\sim}] = 2H\mathbf{1}_2, \qquad [Q, H] = 0 \tag{5.7.54}$$

(cf. eqs. (5.7.22) and (5.7.23)). The supersymmetry group, as an abstract group, is seen to be the same as in the linear case.

A pure basis

Introduce the non-self-adjoint *a*-type variables,

$$b \overset{\text{def}}{=} 2^{-\frac{1}{2}}(y_1 - \mathrm{i}y_2), \qquad b^* = 2^{-\frac{1}{2}}(y_1 + \mathrm{i}y_2). \tag{5.7.55}$$

These variables satisfy the equal-time supercommutation relations

$$[b, b^*] = 1, \qquad b^2 = 0, \qquad [b, x] = 0. \tag{5.7.56}$$

Although they are time dependent their product b^*b is not. Using the dynamical equation (5.7.39) one finds

$$\left.\begin{aligned} &\dot{b} = 2^{-\frac{1}{2}}(\dot{y}_1 - \mathrm{i}\dot{y}_2) = 2^{-\frac{1}{2}}V''(x)(-y_2 - \mathrm{i}y_1) = -\mathrm{i}V''(x)b, \\ &\dot{b}^* = \mathrm{i}V''(x)b^*, \end{aligned}\right\} \tag{5.7.57}$$

whence

$$d(b^*b)/\mathrm{d}t = 0. \tag{5.7.58}$$

Introduce simultaneous eigenvectors $|x', n, t\rangle$ of $x(t)$ and b^*b:

$$x(t)|x', n, t\rangle = x'|x', n, t\rangle, \qquad b^*b|x', n, t\rangle = n|x', n, t\rangle. \tag{5.7.59}$$

The arguments of sections 5.5 and 5.6 show that n can take only the values 0 and 1 whereas x' ranges over $\mathbf{R}_c$. One may assume the normalization

$$\langle x'', m, t | x', n, t\rangle = \delta(x'', x')\delta_{mn}, \tag{5.7.60}$$

and one also has

$$\langle x'', 0, t'' | x', 1, t'\rangle = 0 \tag{5.7.61}$$

even when $t'' \neq t'$.

The supervectors $|x', 0, t\rangle$ and $|x', 1, t\rangle$ may be assumed to be related by

$$|x', 1, t\rangle = b^*(t)|x', 0, t\rangle, \qquad |x', 0, t\rangle = b(t)|x', 1, t\rangle, \tag{5.7.62}$$

and by virtue of eqs. (5.7.57), the equations

$$b(t)|x', 0, t'\rangle = 0, \qquad b^*(t)|x', 1, t'\rangle = 0, \tag{5.7.63}$$

hold even when $t \neq t'$. It is clear from the arguments of section 5.5. that the supervectors $|x',0,t\rangle$ may be assumed to be c-type. The supervectors $|x',1,t\rangle$ are then a-type and the $|x',n,t\rangle$ constitute a complete orthonormal pure basis for the (∞,∞)-dimensional super Hilbert space of the system.

The energy spectrum

Because the system is nonlinear it is not generally possible to obtain exact expressions for the eigenvalues of the energy operator. However, one can infer a number of important features of its spectrum. Since b^*b is time independent it commutes with the energy operator and one may introduce simultaneous eigenvectors of H and b^*b:

$$H|E,n\rangle = E|E,n\rangle, \qquad b^*b|E,n\rangle = n|E,n\rangle. \tag{5.7.64}$$

These eigenvectors are time independent and satisfy the relations

$$\left.\begin{aligned} b|E,0\rangle &= b(1-b^*b)|E,0\rangle = b^2b^*|E,0\rangle = 0,\\ b^*|E,1\rangle &= b^*(b^*b)|E,1\rangle = b^{*2}b|E,1\rangle = 0. \end{aligned}\right\} \tag{5.7.65}$$

From the theorems of section 5.2 they also satisfy

$$\langle x',0,t|E,1\rangle = 0, \qquad \langle x',1,t|E,0\rangle = 0. \tag{5.7.66}$$

The function $V(x)$ will be assumed to satisfy

$$|V'(\pm\infty)| = \infty. \tag{5.7.67}$$

The energy spectrum is then discrete and the normalization

$$\langle E,m|E',n\rangle = \delta_{EE'}\delta_{mn} \tag{5.7.68}$$

may be imposed.

Using

$$y_1 = 2^{-\frac{1}{2}}(b+b^*), \qquad y_2 = \mathrm{i}2^{-\frac{1}{2}}(b-b^*), \tag{5.7.69}$$

it is straightforward to compute

$$y^{\sim}My = 2\mathrm{i}(b^*b-\tfrac{1}{2}), \tag{5.7.70}$$

and hence to rewrite the energy operator in the form

$$H = \tfrac{1}{2}\dot{x}^2 + \tfrac{1}{2}[V'(x)]^2 + V''(x)(b^*b-\tfrac{1}{2}). \tag{5.7.71}$$

From this, with the aid of eq. (5.6.13), it follows that

$$\begin{aligned} E\langle x',n,t|E,n\rangle &= \langle x',n,t|H|E,n\rangle \\ &= \left\{-\frac{1}{2}\frac{\partial^2}{\partial x'^2} + \tfrac{1}{2}[V'(x')]^2 + (n-\tfrac{1}{2})V''(x')\right\}\langle x',n,t|E,n\rangle. \end{aligned} \tag{5.7.72}$$

The energy eigenvalues are seen to be the eigenvalues of the differential operators $-\frac{1}{2}\partial^2/\partial x'^2 + \frac{1}{2}[V'(x')]^2 \pm \frac{1}{2}V''(x')$. Since $V(x)$ is assumed to have vanishing soul when x does, all these eigenvalues must be ordinary real numbers.

In sorting out the energy eigenvalues it will be convenient to make use of the following operators:

$$\left.\begin{aligned} \mathfrak{Q} &\stackrel{\text{def}}{=} 2^{-\frac{1}{2}}(Q_1 - iQ_2) = 2^{-\frac{1}{2}}[\dot{x}(y_1 - iy_2) + V'(x)(y_2 + iy_1)] \\ &= [\dot{x} + iV'(x)]b, \\ \mathfrak{Q}^* &= 2^{-\frac{1}{2}}(Q_1 + iQ_2) = [\dot{x} - iV'(x)]b^*. \end{aligned}\right\} \tag{5.7.73}$$

These a-type operators satisfy

$$[\mathfrak{Q}, \mathfrak{Q}^*] = 2H, \qquad \mathfrak{Q}^2 = 0, \qquad \mathfrak{Q}^{*2} = 0, \tag{5.7.74}$$

$$\mathfrak{Q}|E,0\rangle = 0, \qquad \mathfrak{Q}^*|E,1\rangle = 0. \tag{5.7.75}$$

Since the energy operator is the square of a self-adjoint operator ($H = Q_1^2 = Q_2^2$) none of its eigenvalues can be negative. Suppose it has a vanishing eigenvalue, and suppose the eigenvalue 0 for b^*b is compatible with this. Then there exists a normalized eigenvector $|0,0\rangle$:

$$H|0,0\rangle = 0, \qquad b|0,0\rangle = 0. \tag{5.7.76}$$

The eigenvalues of Q_1 and Q_2 are the square roots of those of H, and hence

$$Q_1|0,0\rangle = 0, \qquad Q_2|0,0\rangle = 0, \tag{5.7.77}$$

which implies

$$\mathfrak{Q}|0,0\rangle = 0, \qquad \mathfrak{Q}^*|0,0\rangle = 0. \tag{5.7.78}$$

The first of eqs. (5.7.78) is trivial by virtue of the second of eqs. (5.7.76). The second of eqs. (5.7.78), however, is not trivial and implies

$$\begin{aligned} 0 &= \langle x', 1, t|[\dot{x}(t) - iV'(x(t))]b^*(t)|0,0\rangle \\ &= -i[\partial/\partial x' + V'(x')]\langle x', 0, t|0,0\rangle, \end{aligned} \tag{5.7.79}$$

which, in turn, implies

$$\left.\begin{aligned} &\langle x', 0, t|0,0\rangle = Z\exp[-V(x')], \\ &|Z| = \left\{\int_{-\infty}^{\infty} \exp[-2V(x')]dx'\right\}^{-\frac{1}{2}}. \end{aligned}\right\} \tag{5.7.80}$$

In order that the function (5.7.80) be normalizable and the coefficient Z exist, the function $V(x')$ must satisfy

$$V(\pm\infty) = \infty. \tag{5.7.81}$$

This condition holds in the case of the linear Bose–Fermi system but *not*, for example, when $V(x)$ is given by (5.7.36). Let us suppose, for a moment, that condition (5.7.81) does hold. Then H *has* a vanishing eigenvalue and, moreover, this eigenvalue is nondegenerate. The lowest eigenvalue of each of the operators $-\frac{1}{2}\partial^2/\partial x'^2 + \frac{1}{2}[V'(x')]^2 \pm \frac{1}{2}V''(x')$ is always nondegenerate, and the only way H could have a degenerate vanishing eigenvalue would be if both of these operators had zero eigenvalues, i.e., if there were a normalizable eigenvector $|0,1\rangle$ in addition to $|0,0\rangle$. But this would imply

$$\mathfrak{Q}|0,1\rangle = 0, \qquad \mathfrak{Q}^*|0,1\rangle = 0,$$

in addition to eqs. (5.7.78). The second of these equations is trivial, but the first implies

$$\begin{aligned} 0 &= \langle x',0,t|[\dot{x}(t) + \mathrm{i}V'(x(t))]b(t)|0,1\rangle \\ &= -\mathrm{i}[\partial/\partial x' - V'(x')]\langle x',1,t|0,1\rangle \end{aligned}$$

and hence

$$\langle x',1,t|0,1\rangle = Z\exp V(x') \tag{5.7.82}$$

which is *not* normalizable.

A vanishing eigenvalue, if it exists, is therefore isolated and nondegenerate. All other eigenvalues, however, are doubly degenerate. Thus, let $|E,0\rangle$ be an eigenvector of H and b^*b corresponding to the eigenvalues E and 0 respectively, with $E \neq 0$. Then $\mathfrak{Q}|E,0\rangle = 0$, but

$$|E,1\rangle = (2E)^{-\frac{1}{2}}\mathfrak{Q}^*|E,0\rangle \tag{5.7.83}$$

modulo an irrelevant phase factor. Thus

$$\begin{aligned} H\mathfrak{Q}^*|E,0\rangle &= \tfrac{1}{2}[\mathfrak{Q},\mathfrak{Q}^*]\mathfrak{Q}^*|E,0\rangle = \tfrac{1}{2}\mathfrak{Q}^*\mathfrak{Q}\mathfrak{Q}^*|E,0\rangle \\ &= \tfrac{1}{2}\mathfrak{Q}^*[\mathfrak{Q},\mathfrak{Q}^*]|E,0\rangle = \mathfrak{Q}^*H|E,0\rangle = E\mathfrak{Q}^*|E,0\rangle \end{aligned}$$

and

$$\begin{aligned} \langle E,0|\mathfrak{Q}\mathfrak{Q}^*|E,0\rangle &= \langle E,0|[\mathfrak{Q},\mathfrak{Q}^*]|E,0\rangle \\ &= 2\langle E,0|H|E,0\rangle = 2E, \end{aligned}$$

in which eqs. (5.7.74) have been used. Similarly

$$|E,0\rangle = (2E)^{-\frac{1}{2}}\mathfrak{Q}|E,1\rangle. \tag{5.7.84}$$

Note that $|E,0\rangle$ and $|E,1\rangle$, although corresponding to the same energy, are opposite in type.

Spontaneously broken supersymmetry

When H has a vanishing eigenvalue the corresponding eigenvector $|0,0\rangle$ is invariant under the actions of the supersymmetry group, i.e., under application of the operator $e^{iHt+\bar{Q}\theta}$. When H has no vanishing eigenvalue, for example when condition (5.7.81) is replaced by

$$V(-\infty) = -\infty, \qquad V(\infty) = \infty, \tag{5.7.85}$$

then the lowest eigenvalue is doubly degenerate and there is no linear combination of the corresponding eigenvectors that is invariant under application of $e^{iHt+Q\cdot\theta}$. In fact, there is no state that is invariant under the actions of the group, and the supersymmetry is said to be *spontaneously broken.*

When the supersymmetry is spontaneously broken the lowest energy eigenvalue is necessarily positive, although it can be arbitrarily small. A generic case in which the lowest eigenvalue is exponentially small compared to the others arises when the function $V(x')$ has just one local maximum, at x_1 say, and one local minimum, at a point x_0, and when the derivative $V'(x')$ has a single local extremum, at a point $x_{\frac{1}{2}}$ in between:

$$V'(x_{1,0}) = 0, \qquad V''(x_{\frac{1}{2}}) = 0. \tag{5.7.86}$$

In addition, the derivatives must satisfy

$$|V''(x_{1,0})| \ll \tfrac{1}{2}[V'(x_{\frac{1}{2}})]^2. \tag{5.7.87}$$

With the boundary conditions (5.7.85) it follows generically, that

$$x_1 < x_{\frac{1}{2}} < x_0, \quad V'(x_{\frac{1}{2}}) < 0, \quad V''(x_1) < 0, \quad V''(x_0) > 0, \quad V'''(x_{\frac{1}{2}}) > 0. \tag{5.7.88}$$

In the case of the function (5.7.36) condition (5.7.87) requires $\lambda^{\frac{1}{2}}\mu^{-3} \ll \frac{1}{4}$.

The operators $-\frac{1}{2}\partial^2/\partial x'^2 + \frac{1}{2}[V'(x')]^2 \pm \frac{1}{2}V''(x')$, whose eigenvalues are the energy eigenvalues, may be viewed as energy operators for particles moving in double-well potentials $\frac{1}{2}[V'(x')]^2 \pm \frac{1}{2}V''(x')$. When condition (5.7.87) is satisfied the two wells, as will be seen, are separated by a significant barrier. This will permit the lowest-energy eigenvalue to be estimated by introducing approximate eigenfunctions having significant amplitudes in only the lower well in each case. The separation ΔE of the lowest energy level from the first excited level may be estimated by approximating either well by a Bose oscillator potential (see section 5.6). Since the wells are located approximately at x_0 and x_1 one obtains

$$\Delta E \sim \left\{\frac{d^2}{dx'^2}\tfrac{1}{2}[V'(x')]^2\right\}^{1/2}_{x'=x_{0,1}} = |V''(x_{1,0})|. \tag{5.7.89}$$

Condition (5.7.87) insures that ΔE is small compared to the barrier height, which implies that the phenomenon of tunneling will make only slight corrections to this estimate.

Combining the approximate relation

$$|V''(x_{1,0})| \sim \frac{2|V'(x_{\frac{1}{2}})|}{x_0 - x_1} \qquad (5.7.90)$$

with condition (5.7.87) one obtains the following additional inequalities:

$$|V'(x_{\frac{1}{2}})| \gg \frac{4}{x_0 - x_1}, \qquad |V''(x_{1,0})| \gg \frac{8}{(x_0 - x_1)^2}, \qquad (5.7.91)$$

$$V(x_1) - V(x_0) \sim |V'(x_{\frac{1}{2}})|(x_0 - x_1) \gg 4. \qquad (5.7.92)$$

Condition (5.7.91), combined with the relation

$$V'''(x_{1,0}) \sim \frac{V''(x_0) - V''(x_1)}{x_0 - x_1} \sim \frac{2|V''(x_{1,0})|}{x_0 - x_1}, \qquad (5.7.93)$$

permits one to verify that the bottoms of the wells of the potentials $\frac{1}{2}[V'(x')]^2 \pm \frac{1}{2}V''(x')$ are located very nearly at x_0 and x_1. The actual bottoms are located at the solutions of the equations

$$V'(x')V''(x') \pm V'''(x') = 0. \qquad (5.7.94)$$

Writing the solutions as $x_1 \mp \Delta x_1$ and $x_0 \mp \Delta x_0$ and working to first order in $\Delta x_{0,1}$, one obtains

$$[V''(x_{1,0})]^2 \Delta x_{1,0} = V'''(x_{1,0}), \qquad (5.7.95)$$

whence

$$\Delta x_{1,0} = \frac{V'''(x_{1,0})}{[V''(x_{1,0})]^2} \sim \frac{2}{|V''(x_{1,0})|(x_0 - x_1)} \ll \tfrac{1}{4}(x_0 - x_1). \qquad (5.7.96)$$

From now on the $\Delta x_{0,1}$ will be neglected.

The approximate ground-state eigenfunctions that will be used are based on eqs. (5.7.80) and (5.7.82). When condition (5.7.85) holds the functions appearing in these equations blow up exponentially at $-\infty$ or ∞ and hence cannot be normalized. The approximate eigenfunctions are obtained by simply cutting off these functions, with the aid of step functions, at the points where they turn around and start blowing up:

$$\langle x', 0, 0|E, 0\rangle \approx \left(\frac{|V''(x_0)|}{\pi}\right)^{\frac{1}{4}} \theta(x', x_1) e^{-V(x') + V(x_0)}, \qquad (5.7.97)$$

$$\langle x', 1, 0|E, 1\rangle \approx -i\left(\frac{|V''(x_1)|}{\pi}\right)^{\frac{1}{4}} \theta(x_0, x') e^{V(x') - V(x_1)}. \qquad (5.7.98)$$

The normalization factors standing in front are obtained by treating the exponentials (which are sharply peaked when condition (5.7.87) holds) as

Gaussians, i.e., by approximating the exponents themselves thus:

$$\left.\begin{aligned} -V(x')+V(x_0) &\approx -\tfrac{1}{2}V''(x_0)(x'-x_0)^2, \\ V(x')-V(x_1) &\approx \tfrac{1}{2}V''(x_1)(x'-x_1)^2. \end{aligned}\right\} \tag{5.7.99}$$

The factor $-\mathrm{i}$ in (5.7.98) is included so that these equations will (as presently seen) be compatible with eqs. (5.7.83) and (5.7.84).

The latter equations provide the key to determining the energy of E of the doubly degenerate ground state. Multiplying them on the left by $\langle x',1,0|$ and $\langle x',0,0|$ respectively one obtains

$$\begin{aligned} \langle x',1,0|E,1\rangle &= (2E)^{-\frac{1}{2}}\langle x',1,0|[\dot{x}(0)-\mathrm{i}V'(x(0))]b^*(0)|E,0\rangle \\ &= -\mathrm{i}(2E)^{-\frac{1}{2}}[\partial/\partial x'+V'(x')]\langle x',0,0|E,0\rangle, \end{aligned} \tag{5.7.100}$$

$$\begin{aligned} \langle x',0,0|E,0\rangle &= (2E)^{-\frac{1}{2}}\langle x',0,0|[\dot{x}(0)+\mathrm{i}V'(x(0))]b(0)|E,1\rangle \\ &= -\mathrm{i}(2E)^{-\frac{1}{2}}[\partial/\partial x'-V'(x')]\langle x',1,0|E,1\rangle. \end{aligned} \tag{5.7.101}$$

Each of these equations expresses one of the ground-state eigenfunctions in terms of the other. One may ask how consistent these exact equations are with the approximations (5.7.97), (5.7.98). To test for consistency it will not be a good procedure simply to insert the approximations into the right-hand sides, for two reasons: (1) Derivative operators tend to magnify errors. (2) Since E is exponentially small, multiplication by $(2E)^{-\frac{1}{2}}$ will further exaggerate the errors. However, one can convert (5.7.100) and (5.7.101) into integral equations, with the aid of Green's functions $G^{\pm}(x',x'')$ satisfying

$$[\partial/\partial x' \mp V(x')]G^{\pm}(x',x'') = -\delta(x',x'').$$

One has

$$\langle x',0,0|E,0\rangle = -\mathrm{i}(2E)^{\frac{1}{2}}\int_{-\infty}^{\infty} G^{-}(x',x'')\langle x'',1,0|E,1\rangle\,\mathrm{d}x'', \tag{5.7.102}$$

$$\langle x',1,0|E,1\rangle = -\mathrm{i}(2E)^{\frac{1}{2}}\int_{-\infty}^{\infty} G^{+}(x',x'')\langle x'',0,0|E,0\rangle\,\mathrm{d}x''. \tag{5.7.103}$$

It is easy to verify that, with the boundary conditions (5.7.85), the Green's functions are given by

$$G^{-}(x',x'') = -\theta(x',x'')\mathrm{e}^{-V(x')+V(x'')}, \tag{5.7.104}$$

$$G^{+}(x',x'') = \theta(x'',x')\mathrm{e}^{V(x')-V(x'')}. \tag{5.7.105}$$

Inserting expressions (5.7.97) and (5.7.98) into the right-hand sides of the integral equations, one readily finds

$$\langle x',0,0|E,0\rangle$$

$$\approx (2E)^{\frac{1}{2}}\left(\frac{|V''(x_1)|}{\pi}\right)^{\frac{1}{4}}\mathrm{e}^{\Delta V}\mathrm{e}^{-V(x')+V(x_0)}\int_{-\infty}^{\min(x',x_0)}\mathrm{e}^{2[V(x'')-V(x_1)]}\mathrm{d}x'', \tag{5.7.106}$$

$\langle x', 1, 0|E, 1\rangle$

$$\approx -\mathrm{i}(2E)^{\frac{1}{2}}\left(\frac{|V''(x_0)|}{\pi}\right)^{\frac{1}{4}} \mathrm{e}^{\Delta V} \mathrm{e}^{V(x')-V(x_1)} \int_{\max(x',x_1)}^{\infty} \mathrm{e}^{-2[V(x'')-V(x_0)]} \mathrm{d}x'', \tag{5.7.107}$$

where

$$\Delta V \overset{\text{def}}{=} V(x_1) - V(x_0). \tag{5.7.108}$$

The integrands of eqs. (5.7.106) and (5.7.107) are sharply peaked around x_1 and x_0 respectively. Hence the integrals fall off sharply when $x' < x_1$ and $x' > x_0$ respectively. They are effectively proportional to $\theta(x', x_1)$ and $\theta(x_0, x')$. The factors of proportionality may be determined by evaluating the integrals in the Gaussian approximation. One finally gets

$$\langle x', 0, 0|E, 0\rangle \approx (2E)^{\frac{1}{2}}\left(\frac{\pi}{|V''(x_1)|}\right)^{\frac{1}{4}} \mathrm{e}^{\Delta V} \theta(x', x_1) \mathrm{e}^{-V(x')+V(x_0)}, \tag{5.7.109}$$

$$\langle x', 1, 0|E, 1\rangle \approx -\mathrm{i}(2E)^{\frac{1}{2}}\left(\frac{\pi}{|V''(x_0)|}\right)^{\frac{1}{4}} \mathrm{e}^{\Delta V} \theta(x_0, x') \mathrm{e}^{V(x')-V(x_1)}. \tag{5.7.110}$$

Consistency with eqs. (5.7.97) and (5.7.98) is immediate provided

$$E = (\mathrm{e}^{-2\Delta V}/2\pi)|V''(x_1)V''(x_0)|^{\frac{1}{2}}. \tag{5.7.111}$$

Since $2\Delta V \gg 8$ (see (5.7.92)) the ground-state energy is, as promised, exponentially small compared to the spacing (5.7.89) between it and the next level. It is, moreover, exponentially small compared to the difference between the bottoms of the two wells in the potentials $\frac{1}{2}[V'(x')]^2 \pm \frac{1}{2}V''(x')$, thus insuring that the ground-state eigenfunctions do indeed have significant amplitudes in only the lower well in each case.

Exercises

5.1 Let the supervectors $|\text{in}\rangle$ and $|\text{out}\rangle$ of eq. (5.3.25) be coherent-state supervectors for either the Fermi oscillator or the Bose oscillator coupled to an added external source having compact support in time. Integrate this equation in each case to obtain $\langle\text{out}|\text{in}\rangle$ as a functional of the external source. Define $W = -\mathrm{i}\ln\langle\text{out}|\text{in}\rangle$. Show that the Feynman propagator is given in each case by $G^{ij} = \vec{\delta}/\delta J_i \vec{\delta}/\delta J_j W$.

5.2 Consider the following modification of the Gaussian integral (1.7.52) of Chapter 1:

$$I_J = \int \exp\left(\frac{i}{2}x^i{}_iM_jx^j + iJ_ix^i\right)d^{m,n}x$$

Show that its value is just expression (1.7.64) multiplied by the value the integrand assumes at the stationary point of the exponent, i.e., at the value of x^i for which

$$x^j{}_jM_i + J_i = 0.$$

Let $\langle x^i \rangle$ be the solution of this equation. Show that

$$\int x^i \exp\left(\frac{i}{2}x^j{}_jM_kx^k + iJ_jx^j\right)d^{m,n}x = \langle x^i \rangle I_J.$$

5.3 Let $x_i(t)$, $i = 1, 2, 3$, be three real *a*-number-valued functions of time and let $(i/4)\int x_i\dot{x}_i\,dt$ be the action functional for a superclassical system having configuration space $\mathbf{R}_a^3$. Quantize this system. Show that the quantized x_i are time independent and satisfy the supercommutation relation $[x_i, x_j] = 2\delta_{ij}$. Introduce the self-adjoint *c*-type operators

$$J_i \overset{\text{def}}{=} -\frac{i}{4}\varepsilon_{ijk}x_jx_k$$

where ε_{ijk} is the antisymmetric permutation symbol. Show that the J_i satisfy the supercommutation relations

$$[x_i, J_j] = i\varepsilon_{ijk}x_k, \quad [J_i, J_j] = i\varepsilon_{ijk}J_k,$$

as well as

$$J_1^2 = J_2^2 = J_3^2 = \tfrac{1}{4}.$$

Show that the minimal super Hilbert space that is needed for representing the operator algebra defined by these supercommutation relations, and for respecting the type properties of the operators, has dimension (2, 2). Introduce simultaneous eigenvectors $|x_3', J_3'\rangle$ of x_3 and J_3. These form a complete orthonormal basis, with the x_3' and J_3' assuming the values ± 1 and $\pm\frac{1}{2}$ respectively. This is necessarily an impure basis. Construct a pure basis, and show that the usual ambiguity about type assignments exists for this system as for all Fermi systems. Show that phase relations may be chosen in such a way that the matrix representation of the operators x_i, J_i in the basis $|x_3', J_3'\rangle$ takes the form

$$x_i = \sigma_i \times \sigma_3, \qquad J_i = \tfrac{1}{2}\sigma_i \times 1_2,$$

where the σ_i are the Pauli matrices. Obtain the corresponding matrices in the pure basis.

5.4 The supervector $|a',t\rangle$ defined by eq. (5.5.39) is a normalized right eigenvector of the a-type operator a corresponding to the a-number eigenvalue a'. Show that one could equally well introduce normalized right eigenvectors of the operator a^*. (These are *not* the images of the supervectors (5.5.42) under the mapping $*^{-1}$.) Show that, with the conventions of section 5.5, these eigenvectors are necessarily a-type supervectors, and hence a special convention must be adopted for writing the eigenvector relation. Choose the convention

$$a^*(t)|a'^*,t\rangle = |a'^*,t\rangle a'^* = -a'^*|a'^*,t\rangle$$

and construct the $|a'^*,t\rangle$ in terms of $|0\rangle, |1\rangle$, a' and a'^*. Obtain analogs of as many equations of section 5.5 as possible in terms of the $|a'^*,t\rangle$ and their adjoints. In particular, obtain an analog for eq. (5.5.45) as well as a functional integral representation for $\langle a'',t''|a'^*,t'\rangle$. How are the new and old Feynman propagators related?

5.5 Show that analogously to expression (5.6.57) there exists a Hamilton–Jacobi function $S(x'',t''|a'^*,t')$ that generates the canonical transformation from the variables a', a'^* at time t' to the variables x'', p'' at time t'', and that satisfies the Hamilton–Jacobian equation (5.6.59) as well as the following alternative to equation (5.6.63):

$$-\partial S/\partial t' + \bar{H}(-\mathrm{i}\,\partial S/\partial a'^*, a'^*).$$

Show, however, that it can*not* be inserted into an expression like (5.6.56) to define a right eigenvector of the operator a^*, because the resulting eigenfunction would be nonnormalizable. Show that a^*, in fact, does not possess *any* right eigenvectors. (*Hint*: Suppose that it did: $a^*(t)|a'^*,t\rangle = a'^*|a'^*,t\rangle$. Let $\langle a',t|$ be the image of $|a'^*,t\rangle$ under the mapping $*$. Show that the body of $\langle a',t|[a^*(t)-a'^*][a(t)-a']|a'^*,t\rangle$ would have to be negative, in contradiction to the basic axioms of super Hilbert-space theory.) Why does a^* have right eigenvectors in the Fermi-oscillator case but not in the case of the Bose oscillator?

5.6 Derive expressions (5.6.65) and (5.6.66) by carrying out the integral (5.6.64).

5.7 Derive expression (5.6.78) from eq. (5.6.77).

5.8 For the case of the nonlinear supersymmetric Bose–Fermi system construct normalized eigenvectors of the a-type self-adjoint operators Q_1 and Q_2 out of the supervectors $|E,n\rangle$, $n=0,1$.

5.9 Let the linear supersymmetric Bose–Fermi system of section 5.7 be at a finite temperature T. Show that the partition function for the system is given by

$$\operatorname{tr}(e^{-H/T}) = \operatorname{ctnh}(\tfrac{1}{2}\omega/T).$$

Show that the mean values at temperature T of the Bose and Fermi components of the energy H are given respectively by

$$\langle H_B \rangle = \tfrac{1}{2}\omega \operatorname{ctnh}(\tfrac{1}{2}\omega/T), \qquad \langle H_F \rangle = -\tfrac{1}{2}\omega \tanh(\tfrac{1}{2}\omega/T)$$

and hence

$$\langle H_B \rangle \langle H_F \rangle = -\tfrac{1}{4}\omega^2$$

at all temperatures. Note that it is the ordinary matrix trace and not the supertrace that appears in the definition of the partition function and other thermal averages. Develop the formalism of quantum statistical mechanics in such a way that the supertrace has a natural role to play.

5.10 Let A be any c-type operator constructed out of dynamical variables of the Fermi oscillator of section 5.5. Let $|a', t\rangle$, $\langle a'^*, t|$ be the coherent-state supervectors defined in that section. Show that

$$\frac{1}{2\pi i}\int \langle a'^*, t|A|a', t\rangle \, da'^* da' = \langle 0|A|0\rangle - \langle 1|A|1\rangle = \operatorname{str} A.$$

Comments on chapter 5

The aim of this chapter is almost exclusively pedagogical. Only the very simplest examples are chosen in order to illustrate basic applications of supermanifold theory, functional Berezin integration and super Hilbert space theory. The latter theory has never previously been spelled out axiomatically, and yet it is needed as soon as one introduces a-number external sources or a-number background fields. The distinction between a-type and c-type state supervectors and the type ambiguity that exists for fermion systems are worthy of note.

It is hoped that the explicit constructions of the functional Berezin integrals for simple Fermi-type systems will make these integrals, which in field theory are generally encountered purely formally, a little more real to the average student. The section on nonlinear Bose–Fermi supersymmetry (eq. (31.35) ff.) stems from work of E. Witten as expounded by Salomonson and van Holten (see references at the end of the book) and is included to show that Bose–Fermi supersymmetry is not restricted to linear systems. Another important example of nonlinear Bose–Fermi supersymmetry is given in chapter 6.

6

Applications involving topology

6.1 Nontrivial configuration spaces

Standard canonical systems

At the beginning of chapter 5 it was remarked that configuration spaces in classical mechanics can be arbitrary differentiable manifolds and that the configuration spaces of superclassical dynamical systems can be arbitrary supermanifolds. In none of the cases studied, however, was the configuration space other than $\mathbf{R}_c^m \times \mathbf{R}_a^n$ for some (m, n). This chapter moves beyond these cases.

In ordinary classical mechanics the action functional is generally chosen to have the form

$$S = \int L(x, \dot{x})\,\mathrm{d}t, \tag{6.1.1}$$

where the *Lagrangian* L is a real-valued scalar function on the *tangent bundle* of the configuration space C. Here the argument x denotes a point of C, and the argument $\dot{x}$ denotes an element of the tangent space at x. In the integral, the point x is understood to be a function of t, tracing out a curve (known as a *trajectory* or *history*) in C, and $\dot{x}$ is the tangent to the curve at t. In a specific chart in C, x is represented by its coordinates x^i, and the components of $\dot{x}$ are given by

$$\dot{x}^i = \mathrm{d}x^i/\mathrm{d}t. \tag{6.1.2}$$

If m is the dimension of C, it is a trivial matter to extend C to a supermanifold of dimension $(m, 0)$, and in this chapter such an extension will be understood. The parameter t, however, will always be restricted to the real line $\mathbf{R}$, and the dynamical trajectory will always be an ordinary curve in C, not a *supercurve* (see chapter 2). Since the x^i are c-numbers, no distinction need be made in this section between left and right functional differentiation.

The dynamical equations of the system (6.1.1) are,

$$0 = S_{,i} \equiv \frac{\partial L}{\partial x^i} - \frac{\mathrm{d}}{\mathrm{d}t}\frac{\partial L}{\partial \dot{x}^i}. \tag{6.1.3}$$

and the second functional derivative of the action is given by

$$S_{,ij'} = \frac{\partial^2 L}{\partial x^i\,\partial x^j}\delta(t,t') + \frac{\partial^2 L}{\partial x^i\,\partial \dot{x}^j}\frac{\partial}{\partial t}\delta(t,t') - \frac{\partial}{\partial t}\left[\frac{\partial^2 L}{\partial \dot{x}^i\,\partial x^j}\delta(t,t') + \frac{\partial^2 L}{\partial \dot{x}^i\,\partial \dot{x}^j}\frac{\partial}{\partial t}\delta(t,t')\right]$$

$$= \left(-\frac{\partial}{\partial t}A_{ij}\frac{\partial}{\partial t} + \frac{1}{2}\left\{B_{ij},\frac{\partial}{\partial t}\right\} - C_{ij}\right)\delta(t,t'), \tag{6.1.4}$$

with

$$A_{ij} = \frac{\partial^2 L}{\partial \dot{x}^i\,\partial \dot{x}^j}, \tag{6.1.5}$$

$$B_{ij} = \frac{\partial^2 L}{\partial x^i\,\partial \dot{x}^j} - \frac{\partial^2 L}{\partial \dot{x}^i\,\partial x^j}, \tag{6.1.6}$$

$$C_{ij} = -\frac{\partial^2 L}{\partial x^i\,\partial x^j} + \frac{1}{2}\frac{\mathrm{d}}{\mathrm{d}t}\left(\frac{\partial^2 L}{\partial x^i\,\partial \dot{x}^j} + \frac{\partial^2 L}{\partial \dot{x}^i\,\partial x^j}\right). \tag{6.1.7}$$

Here $\delta(t,t')$ is the delta function on $\mathbf{R}$, and we have relaxed the condensed notation of sections 5.1 to 5.3 a little by no longer making the indices do double duty as both discrete and continuous labels. The times $t, t', \ldots$ now appear explicitly, and primes are placed on indices to indicate with which of the t's they are associated. It will be noted that $A_{ij} = A_{ji}$, $B_{ij} = -B_{ji}$, and $C_{ij} = C_{ji}$. These symmetries are a reflection of the commutativity of functional differentiation ($S_{,ij'} = S_{,j'i}$) and guarantee that the differential operator appearing in the last line of (6.1.4) is self-adjoint.

Suppose

$$\det(A_{ij}) \neq 0 \quad \text{in all charts in } C. \tag{6.1.8}$$

Then the equations

$$p_i \stackrel{\text{def}}{=} \frac{\partial L}{\partial \dot{x}^i} \tag{6.1.9}$$

can be solved for the $\dot{x}^i$ in terms of the x^i and p_i, and system (6.1.1) is said to be *standard canonical*. Since L is a scalar, the p_i (at t) are the components of a covariant vector p (at $x(t)$) known as the *canonical momentum*. When a system is standard canonical, its energy or *Hamiltonian*,

$$H \stackrel{\text{def}}{=} p_i\dot{x}^i - L, \tag{6.1.10}$$

is a scalar function $H(x,p)$ on the cotangent bundle of C.

Green's functions

The retarded and advanced Green's functions $G^{\pm ij'}$ for small disturbances of the system (6.1.1) satisfy

$$\left(-\frac{\partial}{\partial t}A_{ik}\frac{\partial}{\partial t}+\frac{1}{2}\left\{B_{ik},\frac{\partial}{\partial t}\right\}-C_{ik}\right)G^{\pm kj'}=-\delta_i{}^{j'}\stackrel{\text{def}}{=}\delta_i{}^j\delta(t,t') \quad (6.1.11)$$

(see eq. (5.1.12)). Although one cannot in general write down exact solutions of this equation, it is straightforward in the case of standard canonical systems to obtain solutions in the form of power series in $(t-t')$, valid for t' close to t, that satisfy the boundary conditions appropriate to retarded and advanced Green's functions. With only a little computation one finds

$$G^{+ij'}=-\theta(t',t)\left[(t-t')A^{-1i'j'}+\tfrac{1}{2}(t-t')^2A^{-1i'k'}\left(B_{k'l'}-\frac{\mathrm{d}A_{k'l'}}{\mathrm{d}t'}\right)A^{-1l'j'}+\ldots\right], \quad (6.1.12)$$

$$G^{-ij'}=\theta(t,t')\left[(t-t')A^{-1i'j'}+\tfrac{1}{2}(t-t')^2A^{-1i'k'}\left(B_{k'l'}-\frac{\mathrm{d}A_{k'l'}}{\mathrm{d}t'}\right)A^{-1l'j'}+\ldots\right], \quad (6.1.13)$$

where $\theta(t,t')$ is the step function,

$$\theta(t,t')\stackrel{\text{def}}{=}\begin{cases}1, & t>t',\\ 0, & t<t',\end{cases} \quad (6.1.14)$$

and (A^{-1ij}) is the matrix inverse to (A_{ij}). It is not difficult to verify, up to order $(t-t')^2$, that the series (6.1.12) and (6.1.13) satisfy the reciprocity relation

$$G^{\pm ij'}=G^{\mp j'i} \quad (6.1.15)$$

(cf. eq. (5.1.18)).

With the retarded and advanced Green's functions in hand, one can compute all Peierls brackets (see section 5.1) in terms of the basic Peierls bracket

$$\begin{aligned}(x^i,x^{j'}) &= \tilde{G}^{ij'}\\ &= G^{+ij'}-G^{-ij'}\\ &= -(t-t')A^{-1i'j'}-\tfrac{1}{2}(t-t')^2A^{-1i'k'}\left(B_{k'l'}-\frac{\mathrm{d}A_{k'l'}}{\mathrm{d}t'}\right)A^{-1l'j'}+\ldots.\end{aligned} \quad (6.1.16)$$

For example,

$$(\dot{x}^i, x^{j'}) = \frac{\partial}{\partial t}\tilde{G}^{ij'} = -A^{-1i'j'} - (t-t')A^{-1i'k'}\left(B_{k'l'} - \frac{\mathrm{d}A_{k'l'}}{\mathrm{d}t'}\right)A^{-1l'j'} + \dots, \tag{6.1.17}$$

$$(\dot{x}^i, \dot{x}^{j'}) = \frac{\partial^2}{\partial t\,\partial t'}\tilde{G}^{ij'} = A^{-1i'k'}B_{k'l'}A^{-1l'j'} + \dots, \tag{6.1.18}$$

etc. It should be noted that the A_{ij} are the components of a symmetric covariant tensor (at $x(t)$) and the A^{-1ij} are the components of the reciprocal contravariant tensor.

Equivalence of Peierls and Poisson brackets

The series expansions (6.1.16)–(6.1.18) are particularly useful in computing *equal-time* Peierls brackets. Thus,

$$(x^i, x^j) = (\tilde{G}^{ij'})_{t'=t} = 0, \tag{6.1.19}$$

$$(\dot{x}^i, x^j) = (\dot{x}^i, x^{j'})_{t'=t} = -A^{-1ij}, \tag{6.1.20}$$

$$(\dot{x}^i, \dot{x}^j) = (\dot{x}^i, \dot{x}^{j'})_{t'=t} = A^{-1ik}B_{kl}A^{-1lj}. \tag{6.1.21}$$

From these results it is straightforward to obtain also

$$\begin{aligned}(x^i, p_j) &= (x^i, \partial L/\partial \dot{x}^j)\\ &= (x^i, x^k)\frac{\partial^2 L}{\partial x^k\,\partial \dot{x}^j} + (x^i, \dot{x}^k)\frac{\partial^2 L}{\partial \dot{x}^k\,\partial \dot{x}^j}\\ &= A^{-1ik}A_{kj} = \delta^i{}_j,\end{aligned} \tag{6.1.22}$$

$$\begin{aligned}(p_i, p_j) &= (\partial L/\partial \dot{x}^i, \partial L/\partial \dot{x}^j)\\ &= \frac{\partial^2 L}{\partial \dot{x}^i\,\partial x^k}(x^k, x^l)\frac{\partial^2 L}{\partial x^l\,\partial \dot{x}^j} + \frac{\partial^2 L}{\partial \dot{x}^i\,\partial x^k}(x^k, \dot{x}^l)\frac{\partial^2 L}{\partial \dot{x}^l\,\partial \dot{x}^j}\\ &\quad + \frac{\partial^2 L}{\partial \dot{x}^i\,\partial \dot{x}^k}(\dot{x}^k, x^l)\frac{\partial^2 L}{\partial x^l\,\partial \dot{x}^j} + \frac{\partial^2 L}{\partial \dot{x}^i\,\partial \dot{x}^k}(\dot{x}^k, \dot{x}^l)\frac{\partial^2 L}{\partial \dot{x}^l\,\partial \dot{x}^j}\\ &= \frac{\partial^2 L}{\partial \dot{x}^i\,\partial x^j} - \frac{\partial^2 L}{\partial x^i\,\partial \dot{x}^j} + B_{ij} = 0.\end{aligned} \tag{6.1.23}$$

Equations (6.1.19), (6.1.22) and (6.1.23) show that for standard canonical systems the Peierls bracket is identical to the canonical Poisson bracket. Because use of the Peierls bracket commutes with use of the dynamical equations (see eq. (5.1.26)), this equivalence extends to brackets involving unequal times.

6.2 Quantization

Problems with the naive quantization rule

To quantize a standard canonical system, one tries to impose the quantization rule

$$[x^i, x^{j'}] = i\tilde{G}^{ij'}, \tag{6.2.1}$$

which is the form that eq. (5.3.1) takes in this case. In the quantum theory $\tilde{G}^{ij'}$ is generally an operator. It may be regarded as determined, from a set of equal-time commutators, by the dynamical equations, which themselves depend on a particular factor-ordering prescription for the Hamiltonian. Since the right hand sides of eqs. (6.1.19), (6.1.22) and (6.1.23) do not involve any dynamical variables their quantum versions,

$$[x^i, x^j] = 0, \quad [x^i, p_j] = i\delta^i{}_j, \quad [p_i, p_j] = 0, \tag{6.2.2}$$

appear to constitute an easy set of equal-time commutators (defining the operator superalgebra) on which to base the super Hilbert space of the theory.

But there is a problem. Since, in the classical theory, the x^i and p_i are physical observables, one wants them to be physical observables also in the quantum theory (i.e., physical observables as defined in section 5.2). But this implies that the eigenvalues of the x^i and p_i must range over the whole of $\mathbf{R}_c$. (See the argument immediately following eq. (5.6.8).) The eigenvalues of the x^i are naturally interpreted as possible sets of coordinates in some chart in C with which the x^i must be assumed to be associated. But the range of coordinates in a given chart need *not* be the whole of $\mathbf{R}_c^m$, so there is a contradiction unless one insists on admitting only those charts that *do* cover $\mathbf{R}_c^m$, which seems an unnatural restriction.

There is also a difficulty with the p_i. Since they do not commute with the x^i, their (eigen)values cannot be specified simultaneously with those of the x^i, and hence one does not know to which cotangent space they belong. Unless C is itself $\mathbf{R}_c^m$, therefore, the naive quantization rule embodied in eqs. (6.2.2) raises serious questions. A more careful statement of what it means to quantize a system having a nontrivial configuration space is needed. In particular, the operator superalgebra of the theory needs to be more carefully defined.

It is clear that this superalgebra will depend on the nature of C. It will therefore be denoted by a symbol $\mathscr{A}(C)$, that reflects this dependence. Nothing will be assumed known a priori about $\mathscr{A}(C)$ beyond the fact that it *is* a superalgebra. That is, its elements may be added and multiplied together and be multiplied by supernumbers, with 1 being the identity.

Operator-valued forms. The projection m-form

Denote by A_p^r the set of all completely antisymmetric rank-r contravariant tensors at a point p of C. Let $\mathbf{\Omega}_p$ be a mapping

$$\mathbf{\Omega}_p : A_p^r \to \mathscr{A}(C) \tag{6.2.3}$$

that satisfies

$$\mathbf{\Omega}_p(\mathbf{T}+\mathbf{U}) = \mathbf{\Omega}_p(\mathbf{T}) + \mathbf{\Omega}_p(\mathbf{U}), \tag{6.2.4}$$

$$\mathbf{\Omega}_p(\mathbf{T}\alpha) = \mathbf{\Omega}_p(\mathbf{T})\,\alpha = (-1)^{\alpha\mathbf{T}}\mathbf{\Omega}_p(\alpha\mathbf{T}) \overset{\text{def}}{=} (-1)^{\alpha\mathbf{T}}(\mathbf{\Omega}_p\,\alpha)(\mathbf{T}) \tag{6.2.5}$$

for all $\mathbf{T}$, $\mathbf{U}$ in A_p^r and all α in Λ_∞. $\mathbf{\Omega}_p$ will be called an *operator-valued r-form at p*. By considering similar mappings at every point of C one may extend $\mathbf{\Omega}_p$ to an *operator-valued r-form* $\mathbf{\Omega}$ *on* C, and because $\mathscr{A}(C)$ is a superalgebra one may give the set of all operator-valued r-forms on C an obvious structure as a supervector space. It is then not difficult to see that the notions of Lie differentiation, covariant differentiation, exterior differentiation and integration† can all be extended to operator-valued forms. The only one of these operations that will be of interest here is integration, and that only in connection with a c-type‡ operator-valued m-form known as the *projection m-form*, denoted by $\mathbf{P}$.

Let $\mathscr{U}$ and $\mathscr{V}$ be any two open subsets of C. The projection m-form is defined by

$$P(\mathscr{U})\,P(\mathscr{V}) = P(\mathscr{U} \cap \mathscr{V}), \quad P(C) = 1. \tag{6.2.6}$$

where

$$P(\mathscr{U}) \overset{\text{def}}{=} \int_{\mathscr{U}} \mathbf{P}, \quad P(\mathscr{V}) \overset{\text{def}}{=} \int_{\mathscr{V}} \mathbf{P}, \text{ etc.} \tag{6.2.7}$$

Equations (6.2.6) in fact constitute not so much a definition of $\mathbf{P}$ as a partial construction of $\mathscr{A}(C)$. It will be noted that if C is a *non*orientable supermanifold, the condition $P(C) = 1$ leads to an inconsistency unless a

† For the theory of integration of ordinary forms in ordinary manifolds, see, for example, *Analysis, Manifolds and Physics* (*Revised Edition*) by Choquet-Bruhat and DeWitt-Morette with Dillard-Bleick (North Holland, Amsterdam, 1982).

‡ Because $\mathscr{A}(C)$ has elements that may be identified as c-type, a-type or impure, the notions of c-type, a-type and impure operator-valued forms may be introduced in an obvious way.

special convention is adopted. The transformation law for the components of **P** between two overlapping charts must include an extra minus sign whenever the charts are oppositely oriented. An m-form that transforms in this way is sometimes called a *pseudo*form.

It will be noted that the operators $P(\mathscr{U})$ obey the idempotency relations characteristic of projection operators,

$$P(\mathscr{U})^2 = P(\mathscr{U}) \tag{6.2.8}$$

and that the $P(\mathscr{U})$ for all open $\mathscr{U} \subset C$ generate a supercommuting sub-superalgebra of $\mathscr{A}(C)$. This sub-superalgebra will be assumed to be a *maximal* supercommuting sub-superalgebra.

The position operator

How can one define the notion of position in the quantum theory without introducing something like the 'component operators' x^i of eqs. (6.2.2)? The answer is to introduce not an operator but a type-preserving *mapping*

$$x : \mathfrak{F}(C) \rightarrow \mathscr{A}(C) \tag{6.2.9}$$

from the set $\mathfrak{F}(C)$ of all scalar fields on C to $\mathscr{A}(C)$, which satisfies

$$x(\phi)\, P(\mathscr{U}) = \int_{\mathscr{U}} \phi \mathbf{P} = \int_{\mathscr{U}} \mathbf{P} \phi = P(\mathscr{U})\, x(\phi), \tag{6.2.10}$$

$$x(\phi_1)\, x(\phi_2) = x(\phi_1 \phi_2), \tag{6.2.11}$$

$$x(\alpha\phi_1 + \beta\phi_2) = \alpha x(\phi_1) + \beta x(\phi_2), \tag{6.2.12}$$

for all ϕ, ϕ_1, ϕ_2 in $\mathfrak{F}(C)$, all α, β in Λ_∞, and all open $\mathscr{U} \subset C$. The products $\phi\mathbf{P}$ and $\phi_1 \phi_2$ are here defined in the obvious manner:

$$(\phi\mathbf{P})_p(\mathbf{T}) \overset{\text{def}}{=} \phi(p)\, \mathbf{P}_p(\mathbf{T}), \quad (\phi_1 \phi_2)(p) = \phi_1(p)\, \phi_2(p), \tag{6.2.13}$$

for all p in C and all $\mathbf{T}$ in A^m_p. Note that because **P** is c-type the operators $x(\phi)$ and $P(\mathscr{U})$ commute regardless of the type of ϕ. Note also that

$$\begin{aligned} [x(\phi_1), x(\phi_2)] &= [x(\phi_1), x(\phi_2)]\, P(C) \\ &= \int_{\mathscr{C}} [\phi_1 \phi_2 - (-1)^{\phi_1 \phi_2} \phi_2 \phi_1]\, \mathbf{P} = 0, \end{aligned} \tag{6.2.14}$$

and hence all elements of $\mathscr{A}(C)$ of the form $x(\phi)$ supercommute with one another.

Despite the fact that x itself is not an element of $\mathscr{A}(C)$ but a mapping *to* $\mathscr{A}(C)$, it will be called the *position operator* for reasons that will appear as we develop some of its properties.

It will be convenient to include in $\mathfrak{F}(C)$ not only differentiable scalar fields but also piecewise-differentiable scalar fields. $\mathfrak{F}(C)$ then contains the so-called *characteristic functions*

$$\chi_{\mathscr{U}}(p) \stackrel{\text{def}}{=} \begin{cases} 1 & \text{if} \quad p \in \mathscr{U} \\ 0 & \text{if} \quad p \notin \mathscr{U} \end{cases} \tag{6.2.15}$$

for all open $\mathscr{U} \subset C$. The $\chi_{\mathscr{U}}$ are obviously c-type and satisfy

$$\chi_{\mathscr{U}}\chi_{\mathscr{V}} = \chi_{\mathscr{U} \cap \mathscr{V}} \tag{6.2.16}$$

where the product $\chi_{\mathscr{U}}\chi_{\mathscr{V}}$ is defined as for ϕ_1 and ϕ_2 above (eq. (6.2.13)). From eqs. (6.2.11) and (6.2.16) it follows that

$$x(\chi_{\mathscr{U}})\,x(\chi_{\mathscr{V}}) = x(\chi_{\mathscr{U}}\chi_{\mathscr{V}}) = x(\chi_{\mathscr{U} \cap \mathscr{V}}), \tag{6.2.17}$$

and hence the operators $x(\chi_{\mathscr{U}})$ behave algebraically just like the operators $P(\mathscr{U})$. The $x(\chi_{\mathscr{U}})$, which commute with the $P(\mathscr{U})$, may be adjoined to the $P(\mathscr{U})$ to generate a supercommuting sub-superalgebra of $\mathscr{A}(C)$. But the sub-superalgebra generated by the $P(\mathscr{U})$ alone has been assumed to be a maximal supercommuting sub-superalgebra. Hence the adjunction of the $x(\chi_{\mathscr{U}})$ cannot add anything new. From eqs. (6.2.6) and (6.2.17), together with

$$x(\chi_{\mathscr{U}})\,P(\mathscr{V}) = \int_{\mathscr{V}} \chi_{\mathscr{U}}\mathbf{P} = \int_{\mathscr{U} \cap \mathscr{V}} \mathbf{P} = P(\mathscr{U} \cap \mathscr{V}), \tag{6.2.18}$$

one must, in fact, conclude that $x(\chi_{\mathscr{U}})$ and $P(\mathscr{U})$ are identical:

$$x(\chi_{\mathscr{U}}) \equiv P(\mathscr{U}). \tag{6.2.19}$$

Moreover, since any piecewise differentiable scalar field may be approximated arbitrarily closely by a sequence of linear combinations of characteristic functions, it is clear that the sub-superalgebra generated by the $P(\mathscr{U})$ is just the set of operators $x(\phi)$. Evidently eqs. (6.2.10) to (6.2.13) together suffice uniquely to define the mapping x.

Vector operators

The superalgebra $\mathscr{A}(C)$ is given additional structure through the introduction of *vector operators*. A *contravariant vector operator* is any mapping

$$X : \mathfrak{F}(C) \to \mathscr{A}(C) \tag{6.2.20}$$

that satisfies the chain rule

$$XF = (Xe^A)\frac{\overleftarrow{\partial}}{\partial e^A}F \quad \text{or} \quad FX = F\frac{\overleftarrow{\partial}}{\partial e^A}(e^A X), \tag{6.2.21}$$

analogous to eq. (2.2.8) or eq. (2.2.22), where F and the e^A are as defined in the latter equations. Note that X is a mapping from $\mathfrak{F}(C)$ to $\mathscr{A}(C)$, *not* to $\mathfrak{F}(C)$, and hence is not a vector *field*, which is why a lightface symbol rather than a boldface one is used to denote it.

Let $\mathfrak{X}(C)$ denote the set of all contravariant vector fields on C. A *covariant vector operator* is any mapping

$$\omega: \mathfrak{X}(C) \to \mathscr{A}(C)$$

that satisfies the law†

$$\omega(\phi_1 \mathbf{X}_1 + \phi_2 \mathbf{X}_2) = \tfrac{1}{2}(-1)^{\phi_1 X_1}\{\omega(\mathbf{X}_1), x(\phi_1)\} + \tfrac{1}{2}(-1)^{\phi_2 X_2}\{\omega(\mathbf{X}_2), x(\phi_2)\}, \tag{6.2.22}$$

for all ϕ_1, ϕ_2 in $\mathfrak{F}(C)$ and all $\mathbf{X}_1, \mathbf{X}_2$ in $\mathfrak{X}(C)$, the products $\phi_1 \mathbf{X}_1, \phi_2 \mathbf{X}_2$ being here understood as tensor products. (See the remark at the top of p. 76). The mapping ω, like the mapping X, is not a vector field. Neither is it an operator-valued 1-form on C. Hence, a lightface symbol is again used to denote it. Both contravariant and covariant vector operators can obviously be classified as c-type, a-type or impure. By introducing the standard notation

$$\omega(\mathbf{X}) \stackrel{\text{def}}{=} \omega \cdot \mathbf{X}, \tag{6.2.23}$$

one can also view the mapping ω in the alternative guise

$$\mathbf{X} \mapsto \mathbf{X} \cdot \omega \stackrel{\text{def}}{=} (-1)^{X\omega} \omega \cdot \mathbf{X}. \tag{6.2.24}$$

Denote the set of all contravariant vector operators and the set of all covariant vector operators by $\mathfrak{X}_{\mathscr{A}}(C)$ and $\Omega_{\mathscr{A}}(C)$, respectively. $\mathfrak{X}_{\mathscr{A}}(C)$ and $\Omega_{\mathscr{A}}(C)$ have obvious structures as supervector spaces. It will be convenient to extend the domain and range of the position-operator mapping x to

$$\left.\begin{aligned} x&: \mathfrak{X}(C) \to \mathfrak{X}_A(C) \\ x&: \mathbf{\Omega}(C) \to \Omega_{\mathscr{A}}(C) \end{aligned}\right\} \tag{6.2.25}$$

by defining

$$x(\mathbf{X})\phi = x(\mathbf{X}\phi), \tag{6.2.26}$$

$$x(\omega)\cdot \mathbf{X} = x(\omega \cdot \mathbf{X}), \tag{6.2.27}$$

for all ϕ in $\mathfrak{F}(C)$, all $\mathbf{X}$ in $\mathfrak{X}(C)$ and all ω in $\mathbf{\Omega}(C)$.

† For the role and significance of the antisupercommutators appearing in this law see eq. (6.2.55)ff.

The momentum operator

Since contravariant vector operators are mappings from $\mathfrak{F}(C)$ to $\mathscr{A}(C)$ and not from $\mathfrak{F}(C)$ to itself, they cannot be multiplied together (i.e. applied consecutively), and hence the concept of super Lie bracket does not exist for them. However, the super Lie bracket does make its appearance in connection with a *c*-type covariant vector operator called the *momentum operator*. This operator, denoted here by p, is (partly) defined by the supercommutator law

$$[p\cdot\mathbf{X}, x(\phi)] = -\mathrm{i}x(\mathbf{X})\,\phi = -\mathrm{i}x(\mathbf{X}\phi), \qquad (6.2.28)$$

for all ϕ in $\mathfrak{F}(C)$ and all $\mathbf{X}$ in $\mathfrak{X}(C)$. Using the fact that all operators of the form $x(\phi)$ supercommute with one another, it is not difficult to verify that (6.2.28) is consistent with the law (6.2.22). The operators $p\cdot\mathbf{X}$ are seen *not* to supercommute with the elements of the maximal supercommuting sub-superalgebra. They also do not generally supercommute with each other.

Let $\mathbf{X}$ and $\mathbf{Y}$ be any two elements of $\mathfrak{X}(C)$ and ϕ an arbitrary element of $\mathfrak{F}(C)$. Then

$$\begin{aligned}
[[p\cdot\mathbf{X}, p\cdot\mathbf{Y}], x(\phi)] &= (-1)^{\mathbf{Y}\phi}[[p\cdot\mathbf{X}, x(\phi)], p\cdot\mathbf{Y}] + [p\cdot\mathbf{X}, [p\cdot\mathbf{Y}, x(\phi)]] \\
&= -\mathrm{i}(-1)^{\mathbf{Y}\phi}[x(\mathbf{X}\phi), p\cdot\mathbf{Y}] - \mathrm{i}[p\cdot\mathbf{X}, x(\mathbf{Y}\phi)] \\
&= \mathrm{i}(-1)^{\mathbf{X}\mathbf{Y}}[p\cdot\mathbf{Y}, x(\mathbf{X}\phi)] - \mathrm{i}[p\cdot\mathbf{X}, x(\mathbf{Y}\phi)] \\
&= (-1)^{\mathbf{X}\mathbf{Y}}x(\mathbf{Y}\mathbf{X}\phi) - x(\mathbf{X}\mathbf{Y}\phi) \\
&= -x([\mathbf{X},\mathbf{Y}]\phi) = [-\mathrm{i}p\cdot[\mathbf{X},\mathbf{Y}], x(\phi)]. \qquad (6.2.29)
\end{aligned}$$

The definition of the momentum operator can evidently be made more precise by demanding that it satisfy the additional law

$$[p\cdot\mathbf{X}, p\cdot\mathbf{Y}] = -\mathrm{i}p\cdot[\mathbf{X},\mathbf{Y}]. \qquad (6.2.30)$$

Thus $p\cdot\mathbf{X}$ and $p\cdot\mathbf{Y}$ will supercommute with one another as operators if and only if $\mathbf{X}$ and $\mathbf{Y}$ supercommute (i.e. have vanishing super Lie bracket) as vector fields.

Restriction to a local chart

Let $(\mathscr{U}, \psi)$ be a local chart in C. Denote by x^i_ψ the coordinate functions (see p. 54) defined by the mapping $\psi: \mathscr{U} \to \mathbf{R}^m_c$. That is,

$$\psi(p) = (x^1_\psi(p), \ldots, x^m_\psi(p)) \quad \text{for all } p \text{ in } \mathscr{U}. \qquad (6.2.31)$$

The x^i_ψ are *c*-type scalar fields on $\mathscr{U}$, and one may introduce the corresponding elements of $\mathscr{A}(\mathscr{U})$, namely the c-type operators $x(x^i_\psi)$. These operators, which commute with each other, may be identified with the operators x^i appearing in eqs. (6.2.2):

$$x^i \stackrel{\text{def}}{=} x(x^i_\psi). \tag{6.2.32}$$

Because any scalar field ϕ on $\mathscr{U}$ may be regarded as a function of the x^i_ψ and because the $x(x^i_\psi)$ commute with each other, one may write, by natural extension of the rules (6.2.11) and (6.2.12),

$$\begin{aligned} x(\phi) &= x(\phi(x^1_\psi, \ldots, x^m_\psi)) \\ &\stackrel{\text{def}}{=} \phi(x(x^1_\psi), \ldots, x(x^m_\psi)) = \phi(x^1, \ldots, x^m). \end{aligned} \tag{6.2.33}$$

This is often abbreviated to

$$x(\phi) = \phi(x), \tag{6.2.34}$$

but it must be understood that the symbol on the right strictly has meaning only with respect to a local chart.

Let X^i be the components of a contravariant vector field $\mathbf{X}$ in the local chart. The X^i are scalar fields on $\mathscr{U}$ and define corresponding elements of $\mathscr{A}(\mathscr{U})$:

$$x(X^i) = X^i(x). \tag{6.2.35}$$

This equation suggests that one write

$$x(\mathbf{X}) \stackrel{\text{def}}{=} \mathbf{X}(x), \tag{6.2.36}$$

with the understanding that the notation is again valid only in the local chart.

Let σ be a covariant vector operator. One may define its components in the local chart by

$$\sigma_i \stackrel{\text{def}}{=} \sigma \cdot \frac{\partial}{\partial x^i_\psi}. \tag{6.2.37}$$

In particular, the components of $x(\omega)$, where ω is a covariant vector field on $\mathscr{U}$, may be expressed in the form

$$x(\omega)_i = x(\omega) \cdot \frac{\partial}{\partial x^i_\psi} = x\left(\omega \cdot \frac{\partial}{\partial x^i_\psi}\right) = x(\omega_i) = \omega_i(x), \tag{6.2.38}$$

which suggests the notation

$$x(\omega) \stackrel{\text{def}}{=} \omega(x), \tag{6.2.39}$$

again valid only in the local chart. Moreover, eqs. (6.2.11) and (6.2.12) allow one to write

$$(\omega\cdot\mathbf{X})(x) = x(\omega\cdot\mathbf{X}) = x(\omega_i X^i) = x(\omega_i)\, x(X^i) = \omega_i(x)\, X^i(x), \quad (6.2.40)$$

which can be abbreviated to

$$(\omega\cdot\mathbf{X})(x) = \omega(x)\cdot\mathbf{X}(x), \quad (6.2.41)$$

valid locally.

Since the contravariant vector fields $\partial/\partial x^i_\psi$ on $\mathscr{U}$ commute with each other, so also do the operators

$$p_i = p\cdot\frac{\partial}{\partial x^i_\psi}. \quad (6.2.42)$$

These operators may be identified with the p_i appearing in eqs. (6.2.2). In fact, eqs. (6.2.2) may now be *derived*. The vanishing of $[x^i, x^j]$ and $[p_i, p_j]$ is obvious, while eq. (6.2.28) implies

$$[x^i, p_j] = -\left[p\cdot\frac{\partial}{\partial x^j_\psi}, x(x^i_\psi)\right] = \mathrm{i}x\left(\frac{\partial}{\partial x^j_\psi}x^i_\psi\right)$$

$$= \mathrm{i}x(\delta^i{}_j) = \mathrm{i}\delta^i{}_j.$$

The commutation relations (6.2.2) do imply that the eigenvalues of the x^i range over the whole of $\mathbf{R}_c$. But one now sees that these eigenvalues are conceptually distinct from the values assumed by the scalar fields x^i_ψ and that the corresponding ranges need not coincide.

Lack of uniqueness of the momentum operator

Let ω be a c-type covariant vector field on C and let $\mathbf{X}$ and $\mathbf{Y}$ be arbitrary contravariant vector fields on C. Then

$$[(p+x(\omega))\cdot\mathbf{X}, (p+x(\omega))\cdot\mathbf{Y}]$$

$$= [p\cdot\mathbf{X}, p\cdot\mathbf{Y}] + [p\cdot\mathbf{X}, x(\omega\cdot\mathbf{Y})] + [x(\omega\cdot\mathbf{X}), p\cdot\mathbf{Y}] + [x(\omega\cdot\mathbf{X}), x(\omega\cdot\mathbf{Y})]$$

$$= -\mathrm{i}p\cdot[\mathbf{X},\mathbf{Y}] - \mathrm{i}x(\mathbf{X}(\mathbf{Y}\cdot\omega)) + \mathrm{i}(-1)^{\mathbf{XY}}x(\mathbf{Y}(\mathbf{X}\cdot\omega)). \quad (6.2.43)$$

From eqs. (2.6.26), (2.6.29) and (2.6.30) one obtains

$$\omega\cdot[\mathbf{X},\mathbf{Y}] = [\mathbf{X},\mathbf{Y}]\cdot\omega = I_{[\mathbf{X},\mathbf{Y}]}\,\omega$$

$$= [\mathscr{L}_{\mathbf{X}}, I_{\mathbf{Y}}]\,\omega = [[\mathbf{d}, I_{\mathbf{X}}], I_{\mathbf{Y}}]\,\omega$$

$$= [\mathbf{d}I_{\mathbf{X}}I_{\mathbf{Y}} + I_{\mathbf{X}}\mathbf{d}I_{\mathbf{Y}} - (-1)^{\mathbf{XY}}I_{\mathbf{Y}}\mathbf{d}I_{\mathbf{X}} - (-1)^{\mathbf{XY}}I_{\mathbf{Y}}I_{\mathbf{X}}\mathbf{d}]\,\omega$$

$$= \mathbf{X}(\mathbf{Y}\cdot\omega) - (-1)^{\mathbf{XY}}\mathbf{Y}(\mathbf{X}\cdot\omega) - \mathbf{X}\cdot(\mathbf{d}\omega)\cdot\mathbf{Y}, \quad (6.2.44)$$

whence it follows that

$$[(p+x(\omega))\cdot\mathbf{X},(p+x(\omega))\cdot\mathbf{Y}] = -\mathrm{i}(p+x(\omega))\cdot[\mathbf{X},\mathbf{Y}]-\mathrm{i}x(\mathbf{X}\cdot(\mathbf{d}\omega)\cdot\mathbf{Y}), \tag{6.2.45}$$

Since also

$$\begin{aligned}[(p+x(\omega))\cdot\mathbf{X},x(\phi)] &= [p\cdot\mathbf{X},x(\phi)]+[x(\omega\cdot\mathbf{X}),x(\phi)]\\ &= -\mathrm{i}x(\mathbf{X}\phi),\end{aligned} \tag{6.2.46}$$

it is seen that if $\mathbf{d}\omega = 0$, then $p+x(\omega)$ has all the algebraic properties of p. That is, the *momentum operator is defined only up to a covariant vector operator $x(\omega)$ where ω is an arbitrary closed c-type 1-form.*†

Overlapping charts. Transformation of coordinates

Since the variables x^i, p_i in the classical theory suffice completely to describe a standard canonical system – in particular, the *state* (complete Cauchy data) of the system at any instant of time – all operators appearing in the quantum theory must be expressible in terms of the operators x^i, p_i at a given instant. We therefore have already at hand the complete operator superalgebra needed for constructing the super Hilbert space of the theory. Before undertaking this construction, however, we pause to examine how x^i and p_i, which are always defined relative to a given chart, change under a transformation to another chart.

Let $(\mathscr{U},\psi)$ and $(\bar{\mathscr{U}},\bar{\psi})$ be two overlapping charts, with corresponding coordinate functions x^i_ψ and $x^i_{\bar{\psi}}$ respectively. The corresponding operators in $\mathscr{A}(C)$ are

$$x^i \overset{\text{def}}{=} x(x^i_\psi), \quad \bar{x}^i \overset{\text{def}}{=} x(x^i_{\bar{\psi}}). \tag{6.2.47}$$

All these operators commute with each other, and since the $x^i_{\bar{\psi}}$ are invertible functions of the x^i_ψ, viz.

$$x^i_{\bar{\psi}} = x^i_{\bar{\psi}}(x^1_\psi,\dots,x^m_\psi), \tag{6.2.48}$$

one may write, without ambiguity,

$$\begin{aligned}\bar{x}^i &= x(x^i_{\bar{\psi}}(x^1_\psi,\dots,x^m_\psi)) = x^i_{\bar{\psi}}(x(x^1_\psi),\dots,x(x^m_\psi))\\ &= x^i_{\bar{\psi}}(x^1,\dots,x^m) = x^i_{\bar{\psi}}(x),\end{aligned} \tag{6.2.49}$$

$$\frac{\partial\bar{x}_i}{\partial x^j} = x\left(\frac{\partial x^i_{\bar{\psi}}}{\partial x^j_\psi}\right), \quad \frac{\partial x^i}{\partial\bar{x}^j} = x\left(\frac{\partial x^i_\psi}{\partial x^j_{\bar{\psi}}}\right), \text{ etc.} \tag{6.2.50}$$

† A form is said to be *closed* if its exterior derivative vanishes. It is said to be *exact* it if *is* an exterior derivative. Exact forms are necesarily closed, but not vice versa.

From eq. (6.2.22), rewritten in the form

$$\omega\cdot(\phi_1\mathbf{X}_1+\phi_2\mathbf{X}_2)=\tfrac{1}{2}(-1)^{\phi_1\omega}\{x(\phi_1),\omega\cdot\mathbf{X}_1\}+\tfrac{1}{2}(-1)^{\phi_2\omega}\{x(\phi_2),\omega\cdot\mathbf{X}_2\},\tag{6.2.51}$$

it is now straightforward to obtain the coordinate transformation law for the components of the momentum operator:

$$\bar{p}_i=p\cdot\frac{\partial}{\partial x^i_{\bar\psi}}=p\cdot\left(\frac{\partial x^j_\psi}{\partial x^i_{\bar\psi}}\frac{\partial}{\partial x^j_\psi}\right)$$
$$=\frac{1}{2}\left\{x\left(\frac{\partial x^i_\psi}{\partial x^i_{\bar\psi}}\right),p\cdot\frac{\partial}{\partial x^j_\psi}\right\}=\frac{1}{2}\left\{\frac{\partial x^j}{\partial\bar{x}^i},p_j\right\}.\tag{6.2.52}$$

It will be noted that if the *c-type* operators $x^i, \bar{x}^i, p_i$ are assumed to be physical observables, and hence self-adjoint, then the $\bar{p}_i$ too will be self-adjoint. It is the symmetrization effected by the anticommutator in (6.2.52) that guarantees this. In fact, it was precisely in order to yield this result that the antisupercommutator was introduced into the law (6.2.22) in the first place.

With the aid of the formula

$$\left[\frac{\partial x^i}{\partial\bar{x}^i},p_j\right]=\mathrm{i}\frac{\partial}{\partial x^j}\frac{\partial x^j}{\partial\bar{x}^i}=\mathrm{i}\frac{\partial\bar{x}^k}{\partial x^j}\frac{\partial^2 x^j}{\partial\bar{x}^k\,\partial\bar{x}^i}$$
$$=\mathrm{i}\frac{\partial}{\partial\bar{x}^i}\ln\frac{\partial(x)}{\partial(\bar{x})}=\mathrm{i}\frac{\partial x^j}{\partial\bar{x}^i}\frac{\partial}{\partial x^j}\ln\frac{\partial(x)}{\partial(\bar{x})}\tag{6.2.53}$$

one may rewrite the transformation law (6.2.52) in the alternative forms

$$\bar{p}_i=\frac{\partial x^j}{\partial\bar{x}^i}p_j-\frac{1}{2}\left[\frac{\partial x^j}{\partial\bar{x}^i},p_j\right]=p_j\frac{\partial x^j}{\partial\bar{x}^i}+\frac{1}{2}\left[\frac{\partial x^j}{\partial\bar{x}^i},p_j\right]$$
$$=\frac{\partial x^j}{\partial\bar{x}^i}\left[p_j-\frac{1}{2}\frac{\partial}{\partial x^j}\ln\frac{\partial(x)}{\partial(\bar{x})}\right]=\left[p_j+\frac{\mathrm{i}}{2}\frac{\partial}{\partial x^j}\ln\frac{\partial(x)}{\partial(\bar{x})}\right]\frac{\partial x^j}{\partial\bar{x}^i}.\tag{6.2.54}$$

It is to be noted that the symbol $\partial/\partial x^i(\partial/\partial\bar{x}^i)$ appearing in eqs. (6.2.50) to (6.2.54) indicates that a straightforward differentiation is to be performed with respect to the commuting operators $x^i(\bar{x}^i)$. It does *not* represent $x(\partial/\partial x^i_\psi)\,(x(\partial/\partial x^i_{\bar\psi}))$, which is a different kind of object (see eqs. (6.2.25–26)).

The position representation

The position representation makes use of basis supervectors that specify the position (or configuration) of the dynamical system. Let p be a point† of the configuration space C. To specify a state in which the dynamical system is at p at time t it is natural to introduce a supervector $|p, t\rangle$ that is an eigenvector of all the operators $x^i(t)$ associated with any chart $(\mathscr{U}, \psi)$ containing p:

$$x^i(t)\,|p,t\rangle = x^i_\psi(p)\,|p,t\rangle \quad \text{for all } i. \tag{6.2.55}$$

(Note that we are now making the t dependence of the dynamical variables x^i explicit.) Although the spectrum of each of the $x^i(t)$ is the whole of $\mathbf{R}_c$, only the eigenvalues $x^i_\psi(p)$, with p in $\mathscr{U}$, are relevant. When the system moves out of $\mathscr{U}$ another chart must be used. Equation (6.2.55) is obviously consistent for overlapping charts. Thus, if $(\bar{\mathscr{U}}, \bar{\psi})$ is any other chart containing p, we also have

$$\bar{x}^i(t)\,|p,t\rangle = x^i_{\bar{\psi}}(p)\,|p,t\rangle \quad \text{for all } i. \tag{6.2.56}$$

One obtains a minimal representation of the operator superalgebra of the theory by assuming that for each t the set of supervectors $|p, t\rangle$, for all p in C, contains a complete basis for the super Hilbert space of the theory. There is a difficulty in using the $|p, t\rangle$, however. There is no way of normalizing them without introducing a measure function μ into C (see section 2.9), and such a function, in order to be 'natural', would have to be found by examining the specific structure of the Lagrangian. An alternative convention exists that is independent of the Lagrangian. Instead of introducing a measure function, which would of course be chart dependent, one makes the basis supervectors themselves chart dependent and writes

$$x^i(t)\,|x', t\rangle = x'^i\,|x', t\rangle, \quad \bar{x}^i(t)\,|\bar{x}', t\rangle = \bar{x}'^i\,|\bar{x}', t\rangle, \tag{6.2.57}$$

with the following transformation law between overlapping charts:

$$|\bar{x}', t\rangle = \left|\frac{\partial(x')}{\partial(\bar{x}')}\right|^{\frac{1}{2}} |x', t\rangle. \tag{6.2.58}$$

This allows δ-function normalization to be imposed,

$$\langle x'', t\,|\,x', t\rangle = \delta(x'', x'), \tag{6.2.59}$$

† In this and the immediately following paragraph the reader should not confuse p with the momentum operator.

and the completeness relation to be written in the simple form

$$\int_C |x', t\rangle \langle x', t| \mathrm{d}^m x' = 1. \tag{6.2.60}$$

The left hand side of (6.2.60) must be understood as an abbreviation for the process described in section 2.9 in which a locally finite atlas and an associated partition of unity are introduced. (Such a process would be required even if a measure function were available.) The eigenvalues x'^i in each chart range only over the values appropriate to that chart.

The completeness relation (6.2.60) allows an arbitrary state supervector $|\psi\rangle$ to be decomposed in the form

$$|\psi\rangle = \int_C |x', t\rangle \, \psi(x', t) \, \mathrm{d}^m x' \tag{6.2.61}$$

where

$$\psi(x', t) \stackrel{\text{def}}{=} \langle x', t | \psi\rangle. \tag{6.2.62}$$

$\psi(x', t)$ is known as the *wave function* of the system in the state represented by $|\psi\rangle$. From the transformation law (6.2.58) one infers that it is a scalar *density* of weight $1/2$.† If $|\psi\rangle$ is normalized to unity, then

$$\langle \psi | \psi \rangle = 1 \quad \text{or, equivalently,} \quad \int_C |\psi(x', t)|^2 \mathrm{d}^m x' = 1. \tag{6.2.63}$$

The momentum operator in the position representation

Let $\delta x'^i$, $i = 1, \ldots, m$, be a set of infinitesimal real c-numbers. Since the $\delta x'^i$ are real, it follows that the operator $p_i(t)\,\delta x'^i$ is self-adjoint and the operator

$$U(t) \stackrel{\text{def}}{=} 1 - \mathrm{i} p_i(t) \, \delta x'^i \tag{6.2.64}$$

is unitary. It is easy to see that

$$[x^i(t), U(t)] = \delta x'^i \tag{6.2.65}$$

and hence

$$\begin{aligned} x^i(t)\, U(t)\, |x', t\rangle &= \{[x^i(t), U(t)] + U(t)\, x^i(t)\}\, |x', t\rangle \\ &= [\delta x'^i + x'^i U(t)]\, |x', t\rangle = (x'^i + \delta x'^i)\, U(t)\, |x', t\rangle \end{aligned} \tag{6.2.66}$$

† A *tensor density of weight w and rank* (r, s) transforms like the product $\mu^w \mathrm{T}$ where μ is a measure function and T is a tensor field of rank (r, s).

correct to first infinitesimal order. From this it follows that $U(t)|x',t\rangle$ must be proportional to $|x'+\delta x',t\rangle$, and since a unitary operator preserves normalizations, the proportionality constant must be a phase factor $\exp[\mathrm{i}\omega_i(x',t)\,\delta x'^i]$ where the ω_i are real c-numbers. Thus,

$$[1-\mathrm{i}p_i(t)\,\delta x'^i]\,|x',t\rangle = [1+\mathrm{i}\omega_i(x',t)\,\delta x'^i]\,|x'+\delta x',t\rangle \qquad (6.2.67)$$

or, to first infinitesimal order,

$$-\mathrm{i}p_i(t)\,\delta x'^i\,|x',t\rangle = \delta x'^i\left[\frac{\partial}{\partial x'^i}+\mathrm{i}\omega_i(x(t),t)\right]|x',t\rangle, \qquad (6.2.68)$$

the ω_i here being viewed as the components of a real c-type t-dependent 1-form ω, and the symbol $\partial|x',t\rangle/\partial x'^i$ denoting the supervector that maps the dual super Hilbert space $\mathfrak{H}^*$ to C_c by the rule $\langle\psi|\mapsto\partial\langle\psi|x',t\rangle/\partial x'^i$ for all $\langle\psi|$ in $\mathfrak{H}^*$ for which $\langle\psi|x',t\rangle$ is differentiable.

Since the $\delta x'^i$ are arbitrary, it follows that

$$p_i(t)\,|x',t\rangle = \left[\mathrm{i}\frac{\partial}{\partial x'^i}-\omega_i(x',t)\right]|x',t\rangle. \qquad (6.2.69)$$

For consistency we must have

$$0=[p_i(t),p_j(t)]\,|x',t\rangle = \left[\mathrm{i}\frac{\partial}{\partial x'^i}-\omega_i(x',t),\mathrm{i}\frac{\partial}{\partial x'^j}-\omega_j(x',t)\right]|x',t\rangle$$

$$=\mathrm{i}[\omega_{i,j}(x',t)-\omega_{j,i}(x',t)]\,|x',t\rangle, \qquad (6.2.70)$$

and hence $\omega(t)$ must be a closed 1-form:

$$\mathbf{d}\omega(t)=0. \qquad (6.2.71)$$

Note that although the representation (6.2.69) is chart dependent, the 1-form $\omega(t)$ is global. This follows from the coordinate transformation laws (6.2.54) and (6.2.58), which yield

$$\bar{p}_i(t)\,|\bar{x}',t\rangle = \left[p_j(t)+\frac{\mathrm{i}}{2}\frac{\partial}{\partial x^j(t)}\ln\frac{\partial(x(t))}{\partial(\bar{x}(t))}\right]\frac{\partial x^j(t)}{\partial\bar{x}^i(t)}|x',t\rangle\left|\frac{\partial(x')}{\partial(\bar{x}')}\right|^{\frac{1}{2}}$$

$$=\mathrm{i}\left|\frac{\partial(x')}{\partial(\bar{x}')}\right|^{\frac{1}{2}}\frac{\partial x'^j}{\partial\bar{x}'^i}\left[\frac{\partial}{\partial x'^j}+\mathrm{i}\omega_j(x',t)+\frac{\mathrm{i}}{2}\frac{\partial}{\partial x'^j}\ln\frac{\partial(x')}{\partial(\bar{x}')}\right]|x',t\rangle$$

$$=\left[\mathrm{i}\frac{\partial}{\partial\bar{x}'^i}-\bar{\omega}_i(\bar{x}',t)\right]|\bar{x}',t\rangle, \qquad (6.2.72)$$

showing that the representation (6.2.69) extends throughout the configuration space C.

The presence of the undetermined closed 1-form $\omega(t)$ in the representation (6.2.69) is a consequence of the lack of uniqueness of the momentum operator that has already been noted. One element of arbitrariness, however, may be disposed of. The physics of the quantized system remains unaffected if the position eigenvectors $|x', t\rangle$ are subjected to a phase transformation

$$\overline{|x', t\rangle} = e^{-i\phi(x', t)}|x', t\rangle, \tag{6.2.73}$$

where ϕ is any differentiable t-dependent scalar field on C. This yields

$$\begin{aligned} p_i(t)\overline{|x', t\rangle} &= e^{-i\phi(x', t)}\left[i\frac{\partial}{\partial x'^i} - \omega_i(x', t)\right]|x', t\rangle \\ &= \left[i\frac{\partial}{\partial x'^i} - \omega_i(x', t) - \phi_{,i}(x', t)\right]\overline{|x', t\rangle}, \end{aligned} \tag{6.2.74}$$

showing that the phase transformation adds to $\omega(t)$ an exact 1-form $\mathbf{d}\phi(t)$. This means that $\omega(t)$ may be replaced by a representative of the de Rham cohomology class† to which it belongs. Since the topology of C is assumed not to change with time, the representative may be chosen t-independent, and one may finally write

$$\left.\begin{aligned} p_i(t)|x', t\rangle &= \left[i\frac{\partial}{\partial x'^i} - \omega_i(x')\right]|x', t\rangle, \\ \langle x', t|p_i(t) &= \left[-i\frac{\partial}{\partial x'^i} - \omega_i(x')\right]\langle x', t|, \end{aligned}\right\} \tag{6.2.75}$$

where ω is some linear combination, with real c-number coefficients, of a set of b_1 linearly-independent cohomology-class representatives, where b_1 is the first Betti number of C. If C is compact these representatives may be chosen to be real-valued harmonic 1-forms relative to some positive definite metric tensor field on C.† One cannot be more precise than this.

These results imply that corresponding to each standard canonical classical system there corresponds a b_1-parameter family of distinct quantum systems. That these quantum systems are *physically* distinct is illustrated in exercises **6.5** and **6.6** at the end of the chapter. Only when the first Betti number of C vanishes (e.g., when $C = \mathbf{R}_c^m$ or $C_B = S^m$) is the quantum system unique.

† See Choquet-Bruhat and DeWitt–Morette, with Dillard–Bleick, *loc. cit.*

The Schrödinger equation

In the classical theory the energy (6.1.10) is conserved whenever L has no explicit dependence on t. Even when L and H are explicitly time dependent, expression (6.1.10) is still an important quantity by virtue of the equal-time Peierls brackets.

$$(x^i, H) = (x^i, p_j)\,\dot{x}^j + p_j(x^i, \dot{x}^j) - (x^i, x^j)\frac{\partial L}{\partial x^j} - (x^i, \dot{x}^j)\frac{\partial L}{\partial \dot{x}^j}$$
$$= \dot{x}^i = \mathrm{d}x^i/\mathrm{d}t, \tag{6.2.76}$$

$$(p_i, H) = (p_i, p_j)\,\dot{x}^j + p_j(p_i, \dot{x}^j) - (p_i, x^j)\frac{\partial L}{\partial x^j} - (p_i, \dot{x}^j)\frac{\partial L}{\partial \dot{x}^j}$$
$$= \frac{\partial L}{\partial x^i} = \frac{\mathrm{d}}{\mathrm{d}t}\frac{\partial L}{\partial \dot{x}^i} = \frac{\mathrm{d}p_i}{\mathrm{d}t}. \tag{6.2.77}$$

The final forms for these brackets are consequences of eqs. (6.1.3), (6.1.9), (6.1.20), (6.1.23) and (6.1.24).

The corresponding equal-time brackets in the quantum theory are

$$[x^i(t), H(x(t), p(t), t)] = \mathrm{i}\,\mathrm{d}x^i(t)/\mathrm{d}t, \tag{6.2.78}$$

$$[p_i(t), H(x(t), p(t), t)] = \mathrm{i}\,\mathrm{d}p_i(t)/\mathrm{d}t, \tag{6.2.79}$$

where a specific factor-ordering must be assumed for the dynamical variables $x^i(t)$, $p_i(t)$ out of which the Hamiltonian *operator* $H(x(t), p(t), t)$ is built. This ordering must be chart invariant and must make H self-adjoint.

Let δt be an infinitesimal real number. Since δt is real, the operator $H(x(t), p(t), t)\,\delta t$ is self-adjoint, and the operator

$$V(t) \stackrel{\text{def}}{=} 1 + \mathrm{i}H(x(t), p(t), t)\,\delta t \tag{6.2.80}$$

is unitary. It is easy to see that

$$[x^i(t), V(t)] = -\delta t\,\mathrm{d}x^i(t)/\mathrm{d}t, \tag{6.2.81}$$

$$[p_i(t), V(t)] = -\delta t\,\mathrm{d}p^i(t)/\mathrm{d}t, \tag{6.2.82}$$

and hence, to first infinitesimal order,

$$x^i(t+\delta t)\,V(t)\,|x', t\rangle = [x^i(t)\,V(t) + \delta t\,\mathrm{d}x^i(t)/\mathrm{d}t]\,|x', t\rangle$$
$$= V(t)\,x^i(t)\,|x', t\rangle = x'^i\,V(t)\,|x', t\rangle \tag{6.2.83}$$

$$p_i(t+\delta t)\,V(t)\,|x', t\rangle = [p_i(t)\,V(t) + \delta t\,\mathrm{d}p_i(t)/\mathrm{d}t]\,|x', t\rangle$$
$$= V(t)\,p_i(t)\,|x', t\rangle = \left[\mathrm{i}\frac{\partial}{\partial x'^i} - \omega_i(x')\right]V(t)\,|x', t\rangle. \tag{6.2.84}$$

From this it follows that $V(t)|x', t\rangle$ must be proportional to $|x', t+\delta t\rangle$. Since $V(t)$ is unitary and since we are now assuming, for all t, the special representation (6.2.75) for the momentum operator, the proportionality constant must be an x'-independent phase factor $\exp[\mathrm{i}f(t)\,\delta t]$ where $f(t)$ is some real-valued function of the time. Thus,

$$[1+\mathrm{i}H(x(t), p(t), t)\,\delta t]\,|x', t\rangle = [1+\mathrm{i}f(t)\,\delta t]\,|x', t+\delta t\rangle \tag{6.2.85}$$

or, to first infinitesimal order,

$$\mathrm{i}H(x(t), p(t), t)\,\delta t\,|x', t\rangle = \delta t\left[\frac{\partial}{\partial t}+\mathrm{i}f(t)\right]|x', t\rangle. \tag{6.2.86}$$

Factoring δt out of both sides, we get

$$H(x(t), p(t), t)\,|x', t\rangle = \left[-\mathrm{i}\frac{\partial}{\partial t}+f(t)\right]|x', t\rangle. \tag{6.2.87}$$

Now carry out the phase transformation

$$\overline{|x', t\rangle} = \mathrm{e}^{\mathrm{i}\lambda(t)}\,|x', t\rangle, \quad \lambda(t) = \int^t f(t')\,\mathrm{d}t'. \tag{6.2.88}$$

This yields

$$-\mathrm{i}\frac{\partial}{\partial t}\overline{|x', t\rangle} = \mathrm{e}^{\mathrm{i}\lambda(t)}\left[-\mathrm{i}\frac{\partial}{\partial t}+f(t)\right]|x', t\rangle = H(x(t), p(t), t)\,\overline{|x', t\rangle}. \tag{6.2.89}$$

Without loss of generality one may assume the transformation already to have been carried out. Then

$$-\mathrm{i}\frac{\partial}{\partial t}|x', t\rangle = H(x(t), p(t), t)\,|x', t\rangle \tag{6.2.90}$$

or

$$\begin{aligned}\mathrm{i}\frac{\partial}{\partial t}\langle x', t| &= \langle x', t|\,H(x(t), p(t), t)\\ &= H(x', -\mathrm{i}\partial/\partial x' - \omega(x'), t)\,\langle x', t|.\end{aligned} \tag{6.2.91}$$

Multiplying this last equation on the right by an arbitrary state supervector $|\psi\rangle$, one obtains the *Schrödinger equation*

$$\mathrm{i}\partial\psi(x', t)/\partial t = H(x', -\mathrm{i}\partial/\partial x' - \omega(x'), t)\,\psi(x', t). \tag{6.2.92}$$

The position representation of the projection m-form

The explicit construction of a super Hilbert space, of a representation for the operator algebra $\mathscr{A}(C)$, and of a dynamics (eq. (6.2.90)) for the basis supervectors $|x', t\rangle$ is now complete. It is of interest to return to our starting point and to reexamine the operator-valued projection m-form with which we began. We first note that if ϕ is an arbitrary scalar field on C, then

$$x_t(\phi)\,|x', t\rangle = \phi(x(t))\,|x', t\rangle = \phi(x')\,|x', t\rangle, \tag{6.2.93}$$

where x_t is the position-operator mapping (6.2.9) at time t. Equation (6.2.93), together with (6.2.60), allows one to infer

$$x_t(\phi) = \int_C \phi(x')\,|x', t\rangle\langle x', t|\,\mathrm{d}^m x'. \tag{6.2.94}$$

In particular,

$$P_t(\mathscr{U}) = x_t(\chi_{\mathscr{U}}) = \int_C \chi_{\mathscr{U}}(x')\,|x', t\rangle\langle x', t|\,\mathrm{d}^m x'$$

$$= \int_{\mathscr{U}} |x', t\rangle\langle x', t|\,\mathrm{d}^m x' \tag{6.2.95}$$

(see eq. (6.2.19)). But

$$P_t(\mathscr{U}) = \int_{\mathscr{U}} \mathbf{P}_t \tag{6.2.96}$$

where $\mathbf{P}_t$ is the projection m-form at time t (see eqs. (6.2.7)). From this follows the explicit representation

$$\mathbf{P}_{tx'} = |x', t\rangle\langle x', t|\,\mathbf{d}x'^1 \wedge \ldots \wedge \mathbf{d}x'^m, \tag{6.2.97}$$

which may make $\mathbf{P}_t$ more concrete to some readers.

6.3 Curved configuration spaces

A special class of systems

In this and the next three sections we shall study systems for which the Lagrangian takes, in every chart, the special form

$$L = \tfrac{1}{2} g_{ij}(x)\,\dot{x}^i \dot{x}^j + a_i(x)\,\dot{x}^i - v(x), \tag{6.3.1}$$

with $(g_{ij})_B$ positive definite. This Lagrangian may be regarded as describing a nonrelativistic particle of unit mass and charge moving in an $(m, 0)$-dimensional Riemmanian configuration space of metric g, in a stationary magnetic field described by the 1-form potential $\boldsymbol{a}$, and in a stationary electric field described by the scalar potential v.† If the Riemann tensor field of $\boldsymbol{g}$ is nonvanishing the configuration space is curved.

The coefficients A_{ij}, B_{ij}, C_{ij} of eqs. (6.1.5) to (6.1.7) are readily computed:

$$A_{ij} = g_{ij}, \tag{6.3.2}$$

$$B_{ij} = (g_{jk,i} - g_{ik,j})\dot{x}^k + a_{j,i} - a_{i,j}, \tag{6.3.3}$$

$$C_{ij} = -\tfrac{1}{2}g_{kl,ij}\dot{x}^k\dot{x}^l - a_{k,ij}\dot{x}^k + v_{,ij} + \frac{1}{2}\frac{d}{dt}[(g_{jk,i} + g_{ik,j})\dot{x}^k + a_{j,i} + a_{i,j}]. \tag{6.3.4}$$

However, if these coefficients are substituted into expression (6.1.14) for the second functional derivative of the action, a result is obtained that is far from manifestly covariant. It will be useful to introduce a notation with which covariance (i.e., chart independence of form) may be more directly displayed.

Covariant variation

Let $\mathbf{T}[x, t]$ be a tensor (or tensor density) at $x(t)$, which depends functionally, but locally, on the curve x in C. Examples of such an object are: (1) the Lagrangian, (2) $\boldsymbol{g}(x(t))$, (3) the velocity vector at time t, having components $\dot{x}^i(t)$, etc. 'Locally' means that the object depends on no more than a finite number of t derivatives of the $x^i(t)$. Denote by $T[x, t]$ the column vector obtained by arranging the components of $\mathbf{T}[x, t]$ (in any chart) in a vertical array. From the results of chapters 2 and 4 it is not difficult to show that the change in T resulting from a parallel translation through an infinitesimal displacement δx^i takes the general form

$$\delta_{\parallel} T[x, t] = -G^j{}_k \Gamma^k{}_{ji}[x(t)]\, T[x, t]\, \delta x^i, \tag{6.3.5}$$

where the $\Gamma^k{}_{ji}$ are the components of the Riemannian connection on C and the $G^j{}_k$ are the generators of the matrix representation of the group $GL(m, 0)$ to which T corresponds. The $G^j{}_k$ satisfy the commutation relations

$$[G^i{}_j, G^k{}_l] = \delta^i{}_l G^k{}_j - \delta^k{}_j G^i{}_l. \tag{6.3.6}$$

Let the curve x be modified by an infinitesimal displacement having

† Stationarity is here assumed for simplicity. The reader may easily generalize to the nonstationary case. Even the metric may be allowed to vary with time.

components $\delta x^i(t)$ at time t. Then $T[x, t]$ will suffer a corresponding change $\delta T[x, t]$. The *covariant variation* of $T[x, t]$ under this displacement is defined to be

$$\bar{\delta} T[x, t] \stackrel{\text{def}}{=} \delta T[x, t] - \delta_{\parallel} T[x, t]. \tag{6.3.7}$$

An example of a covariant variation is

$$\bar{\delta}\dot{x}^i(t) = \delta\dot{x}^i(t) + \Gamma^i{}_{jk}[x(t)]\,\dot{x}^j(t)\,\delta x^k(t). \tag{6.3.8}$$

We shall frequently suppress the arguments and write this in the abbreviated form

$$\bar{\delta}\dot{x}^i = \delta\dot{x}^i + \Gamma^i{}_{jk}\dot{x}^j\delta x^k = (\mathrm{d}\delta x^i/\mathrm{d}t) + \Gamma^i{}_{jk}\dot{x}^j\delta x^k. \tag{6.3.9}$$

As another example, suppose $\mathbf{T}[x, t] = \mathbf{T}_{x(t)}$ where $\mathbf{T}$ is an ordinary tensor (or tensor density) field on C. Then

$$\bar{\delta} T = \delta T - \delta_{\parallel} T = T_{,i}\delta x^i + G^j{}_k\Gamma^k{}_{ji}T\delta x^i = T_{;i}\delta x^i. \tag{6.3.10}$$

In particular,

$$\bar{\delta} g = 0. \tag{6.3.11}$$

Covariant differentiation with respect to t

Another useful concept is the *covariant derivative of T* with respect to t, defined by

$$\frac{\mathrm{D}}{\mathrm{d}t}T \stackrel{\text{def}}{=} \frac{\mathrm{d}}{\mathrm{d}t}T + G^j{}_k\Gamma^k{}_{ji}T\dot{x}^i. \tag{6.3.12}$$

Examples of covariant t-derivatives are

$$\frac{\mathrm{D}}{\mathrm{d}t}\dot{x}^i = \frac{\mathrm{d}}{\mathrm{d}t}\dot{x}^i + \Gamma^i{}_{jk}\dot{x}^j\dot{x}^k, \tag{6.3.13}$$

$$\frac{\mathrm{D}}{\mathrm{d}t}\delta x^i = \frac{\mathrm{d}}{\mathrm{d}t}\delta x^i + \Gamma^i{}_{jk}\dot{x}^k\delta x^j = \bar{\delta}\dot{x}^i, \tag{6.3.14}$$

and, if $\mathbf{T}$ is a tensor (density) field on C,

$$\frac{\mathrm{D}}{\mathrm{d}t}T = T_{;i}\dot{x}^i. \tag{6.3.15}$$

We shall often denote the covariant t-derivative simply by a dot:

$$\dot{T} \stackrel{\text{def}}{=} \frac{\mathrm{D}}{\mathrm{d}t}T. \tag{6.3.16}$$

This abbreviation leads us to write

$$\ddot{x}^i = \frac{\mathrm{D}}{\mathrm{d}t}\dot{x}^i = \frac{\mathrm{d}^2 x^i}{\mathrm{d}t^2} + \Gamma^i{}_{jk}\dot{x}^j\dot{x}^k, \tag{6.3.17}$$

$$\dddot{x}^i = \frac{\mathrm{D}}{\mathrm{d}t}\ddot{x}^i = \frac{\mathrm{d}}{\mathrm{d}t}\ddot{x}^i + \Gamma^i{}_{jk}\ddot{x}^j\dot{x}^k, \text{etc.}, \tag{6.3.18}$$

and the reader must then remember that repeated dots denote covariant, *not* ordinary, t-differentiation.

Covariant variation and covariant differentiation with respect to t do not generally commute:

$$\begin{aligned}
\bar{\delta}\dot{T} - \frac{\mathrm{D}}{\mathrm{d}t}\bar{\delta}T &= \delta(\mathrm{d}T/\mathrm{d}t + G^j{}_i\Gamma^i{}_{jk}T\dot{x}^k) + G^j{}_i\Gamma^i{}_{jl}(\mathrm{d}T/\mathrm{d}t + G^m{}_n\Gamma^n{}_{mk}T\dot{x}^k)\delta x^l \\
&\quad - \mathrm{d}(\delta T + G^j{}_i\Gamma^i{}_{jl}T\delta x^l)/\mathrm{d}t - G^j{}_i\Gamma^i{}_{jk}(\delta T + G^m{}_n\Gamma^n{}_{ml}T\delta x^l)\dot{x}^k \\
&= -\{G^j{}_i(\Gamma^i{}_{jl,k} - \Gamma^i{}_{jk,l}) - [G^j{}_i, G^m{}_n]\Gamma^i{}_{jl}\Gamma^n{}_{mk}\}T\dot{x}^k\delta x^l \\
&= -G^j{}_i(\Gamma^i{}_{jl,k} - \Gamma^i{}_{jk,l} + \Gamma^i{}_{km}\Gamma^m{}_{lj} - \Gamma^i{}_{lm}\Gamma^m{}_{kj})T\dot{x}^k\delta x^l \\
&= -G^j{}_i R^i{}_{jkl}T\dot{x}^k\delta x^l.
\end{aligned} \tag{6.3.19}$$

In the special case in which $\mathbf{T}$ is an ordinary tensor (density) field on C, this result follows from the noncommutativity of covariant differentiation and the easily verified fact that the operations $\bar{\delta}$ and $\mathrm{D}/\mathrm{d}t$ both obey Leibniz' rule.

The dynamical equations

The utility of covariant variation and covariant t-differentiation can be seen in the ease with which they permit the dynamical equations for the system (6.3.1) to be obtained. One need keep in mind only three simple facts: (1) When acting on a scalar the operations $\bar{\delta}$ and $\mathrm{D}/\mathrm{d}t$ reduce to δ and $\mathrm{d}/\mathrm{d}t$. (2) These operations give zero when acting on the metric tensor. (3) Integration by parts can be performed with $\mathrm{D}/\mathrm{d}t$ just as with $\mathrm{d}/\mathrm{d}t$.

To obtain the functional derivative of the action, one assumes that the δx^i have compact support in t. Then

$$\begin{aligned}
\delta S &= \int \delta L\,\mathrm{d}t = \int \bar{\delta}L\,\mathrm{d}t \\
&= \int (g_{ij}\dot{x}^j\bar{\delta}\dot{x}^i + a_{j;i}\dot{x}^j\delta x^i + a_i\bar{\delta}\dot{x}^i - v_{;i}\delta x^i)\,\mathrm{d}t
\end{aligned}$$

$$= \iint \left[-\frac{\mathrm{D}}{\mathrm{d}t}(g_{ij}\dot{x}^j) + a_{j;i}\dot{x}^j - \frac{\mathrm{D}}{\mathrm{d}t}a_i - v_{;i} \right] \delta x^i \, \mathrm{d}t$$

$$= \int (-g_{ij}\ddot{x}^j + F_{ij}\dot{x}^j - v_{;i})\, \delta x^i \, \mathrm{d}t \tag{6.3.20}$$

where

$$F_{ij} \stackrel{\text{def}}{=} a_{j;i} - a_{i;j} = a_{j,i} - a_{i,j}. \tag{6.3.21}$$

From this the dynamical equations immediately follow:

$$0 = S_{,i} \equiv -g_{ij}\ddot{x}^j + F_{ij}\dot{x}^j - v_{;i}. \tag{6.3.22}$$

Covariant functional differentiation

There is a natural extension of the covariant derivative from the configuration space C to the full space of histories Φ (see section 5.1). Let $x^i(t)$ be the coordinates of a curve in C and $x^i(t) + \delta x^i(t)$ the coordinates of an infinitesimally displaced curve. One may regard these coordinate functions as the 'coordinates' of two infinitesimally close histories, i.e., *points* of Φ. The infinitesimal distance $\mathrm{d}s$ between these histories is defined, in a natural way, as

$$\mathrm{d}s^2 = \int g_{ij}(x(t))\, \delta x^i(t)\, \delta x^j(t)\, \mathrm{d}t = \int \mathrm{d}t \int \mathrm{d}t' g_{ij'}[x]\, \delta x^i \, \delta x^{j'} \tag{6.3.23}$$

where

$$g_{ij'}[x] \stackrel{\text{def}}{=} g_{ij}(x(t))\, \delta(t, t'). \tag{6.3.24}$$

The $g_{ij'}$ are the components of the metric tensor field on Φ, and the corresponding connection components are given by

$$\Gamma^i{}_{j'k''}[x] = \frac{1}{2} \int g^{il'''}(g_{l'''j',k''} + g_{l'''k'',j'} - g_{j'k'',l'''})\, \mathrm{d}t'''$$

$$= \Gamma^i{}_{jk}(x(t))\, \delta(t, t')\, \delta(t, t''). \tag{6.3.25}$$

An important quantity in which these connection components enter is the second *covariant* functional derivative of the action. Since S is a scalar on Φ, we have

$$S_{;ij'} = S_{,ij'} - \int S_{,k''} \Gamma^{k''}{}_{ij'}\, \mathrm{d}t''. \tag{6.3.26}$$

The explicit computation of $S_{;ij'}$ can be carried out either tediously, with the aid of eqs. (6.1.4), (6.3.2), (6.3.3), (6.3.4) and (6.3.26), or expeditiously, by introducing a coordinate frame at $x(t)$ in which the $\Gamma^i{}_{jk}$ vanish and noting that

$$\begin{aligned}\int S_{;ij'}\,\delta x^{j'}\,\mathrm{d}t' &= \bar{\delta} S_{,i} \\ &= -g_{ij}\,\bar{\delta}\ddot{x}^j + F_{ij;k}\,\dot{x}^j\delta x^k + F_{ij}\,\bar{\delta}\dot{x}^j - v_{;ij}\,\delta x^j \\ &= -g_{ij}\left(\frac{\mathrm{D}}{\mathrm{d}t}\bar{\delta}\dot{x}^j - R^j{}_{klm}\,\dot{x}^k\dot{x}^l\delta x^m\right) + F_{ij;k}\,\dot{x}^j\delta x^k + F_{ij}\,\bar{\delta}\dot{x}^j - v_{;ij}\,\delta x^j \\ &= \left(g_{ij}\frac{\mathrm{D}^2}{\mathrm{d}t^2} + \frac{1}{2}\left\{F_{ij}, \frac{\mathrm{D}}{\mathrm{d}t}\right\} - [R_{ikjl}\,\dot{x}^k\dot{x}^l - \tfrac{1}{2}(F_{ik;j} + F_{jk;i})\,\dot{x}^k + v;_{ij}]\right)\delta x^j.\end{aligned} \tag{6.3.27}$$

Evidently,

$$S_{;ij'} = \left(-g_{ij}\frac{\mathrm{D}^2}{\mathrm{d}t^2} + \frac{1}{2}\left\{F_{ij}, \frac{\mathrm{D}}{\mathrm{d}t}\right\} - [R_{ijkl}\,\dot{x}^k\dot{x}^l - \tfrac{1}{2}(F_{ik;j} + F_{jk;i})\,\dot{x}^k + v_{;ij}]\right)\delta(t, t'). \tag{6.3.28}$$

It is easy to see that $S_{;ij'}$, like $S_{,ij'}$, is self-adjoint as a differential operator.

6.4 The Feynman functional integral and its meaning

Formal computation of $\det G^+[x]$

Instead of imposing on the Hamiltonian operator an *a priori* ordering of factors, we shall, in this and the following section, *derive* the quantum theory of the system (6.3.1) directly from the Feynman functional integral (5.3.46). This is the modern approach to quantization. It requires that one give meaning to the functional integral by taking certain steps that, to date, are valid only formally, depending for their justification on the consistency of the results to which they lead. One of these steps is the computation of the superdeterminant that appears in the integrand of (5.3.46)

In the present problem the superdeterminant becomes an ordinary determinant $\det G^+[x]$. It can be (formally) evaluated by integrating the variational formula

$$\delta \ln \det G^+[x]^{x_2, t_2}_{x_1, t_1} = \int_{t_1+}^{t_2} \mathrm{d}t \int_{t_1}^{t_2} \mathrm{d}t'\,\delta S_{,ij'}\,G^{+j'i} \tag{6.4.1}$$

appropriate to the boundary conditions $|\mathrm{in}\rangle = |x_1, t_1\rangle$, $|\mathrm{out}\rangle = |x_2, t_2\rangle$,

$t_2 > t_1$. The lower limit of the t integration in (6.4.1) is taken infinitesimally greater than t_1 because the integrand vanishes unless $t > t'$.†

The variation δS may be assumed to arise from independent variations in g_{ij}, a_i, and v in the Lagrangian. We then have

$$\int \delta S_{,ij'}\, G^{+j'k''}\,\mathrm{d}t' = \left(-\frac{\partial}{\partial t}\delta g_{ij}\frac{\partial}{\partial t}+\frac{1}{2}\left\{\delta B_{ij}, \frac{\partial}{\partial t}\right\}-\delta C_{ij}\right)G^{+jk''}$$

$$= -\delta(t,t'')\,\delta g_{ij}\,g^{jk}+\theta(t'',t)\left[\delta g_{ij}\left(g^{jl}B_{lm}\,g^{mk}+\frac{\mathrm{d}g^{jk}}{\mathrm{d}t}\right)\right.$$

$$\left.+\frac{\mathrm{d}\delta g_{ij}}{\mathrm{d}t}g^{jk}-\delta B_{ij}\,g^{jk}+\ldots\right] \qquad (6.4.2)$$

in which eqs. (6.1.11), (6.1.12) and (6.3.2) have been used. Setting $k = i$ and $t'' = t$, and noting the antisymmetry of B_{ij}, we find

$$\delta \ln\det G^{+}[x]^{x_2,t_2}_{x_1,t_1} = \int_{t_1+}^{t_2}\left[-\delta(t,t)\,g^{ji}\delta g_{ij}+\theta(t,t)\frac{\mathrm{d}}{\mathrm{d}t}(g^{ji}\delta g_{ij})\right]\mathrm{d}t$$

$$= \delta\int_{t_1+}^{t_2}\left[-\delta(t,t)\ln g+\theta(t,t)\frac{\mathrm{d}}{\mathrm{d}t}\ln g\right]\mathrm{d}t, \qquad (6.4.3)$$

where $g = |\det(g_{ij})|$ (cf. eq. (2.9.8)).

We now make two formal identifications. The first is

$$\theta(t,t) = \tfrac{1}{2}, \qquad (6.4.4)$$

which follows from setting $t' = t$ in the identity

$$\theta(t,t')+\theta(t',t) = 1. \qquad (6.4.5)$$

The second is

$$\delta(t,t)\,\mathrm{d}t = 1, \qquad (6.4.6)$$

which has an obvious meaning (and validity) in a lattice approximation to the functional integral. Recalling that the symbol $\int$ is an old-fashioned summation sign, we see that eqs. (6.4.4) and (6.4.6) allow us to rewrite expression (6.4.3) in the form

$$\delta\Big[-\sum_{t_1<t\leqslant t_2}\ln g(x(t))+\ln g^{\frac{1}{2}}(x_2)-\ln g^{\frac{1}{2}}(x_1)\Big]$$

$$= \delta\ln\Big\{g^{-\frac{1}{2}}(x_2)\Big[\prod_{t_1<t<t_2} g^{-1}(x(t))\Big]g^{-\frac{1}{2}}(x_1)\Big\}, \qquad (6.4.7)$$

† Another reason for applying this rule is to make the determinant transposition invariant, so that $\det G^{+} = \det G^{-}$, and to obtain the symmetric result (6.4.8).

from which it follows that

$$(\det G^{+}[x]^{x_2, t_2}_{x_1, t_1})^{-\frac{1}{2}} = \text{const} \times g^{\frac{1}{4}}(x_2) \Big[\prod_{t_1 < t < t_2} g^{\frac{1}{2}}(x(t)) \Big] g^{\frac{1}{4}}(x_1). \qquad (6.4.8)$$

In obtaining the factors standing outside the square brackets in (6.4.8), we have used the fact that the functional integral, in the present case, is constrained to range over trajectories for which

$$x(t_1) = x_1, \quad x(t_2) = x_2. \qquad (6.4.9)$$

This constraint may be inferred from the following corollary of eqs. (5.3.33) and (5.3.46):

$$\langle \text{out} | \, T(A[\varphi]) \, | \text{in} \rangle = Z \int A[\varphi] \, e^{i(S[\varphi] + J\varphi)} \, (\text{sdet}\, G^{+}[\varphi])^{-\frac{1}{2}} \, d\varphi. \qquad (6.4.10)$$

Simply set $|\text{in}\rangle = |x_1, t_1\rangle$, $|\text{out}\rangle = |x_2, t_2\rangle$, and $A[\varphi] = x^i(t_1)$ or $x^i(t_2)$, and note that $|x_1, t_1\rangle$ and $|x_2, t_2\rangle$ are eigenvectors of $x^i(t_1)$ and $x^i(t_2)$ respectively.

The functional integral

Insertion of (6.4.8) into the functional integral (5.3.46) yields the manifestly covariant formula

$$\langle x_2, t_2 \,|\, x_1, t_1 \rangle = Z g^{\frac{1}{4}}(x_2) \, g^{1/4}(x_1) \int \exp\left[i \int_{x_1, t_1}^{x_2, t_2} (\tfrac{1}{2} g_{ij} \dot{x}^i \dot{x}^j + a_i \dot{x}^i - v) \, dt \right]$$

$$\times \prod_{t_1 < t < t_2} g^{\frac{1}{2}}(x(t)) \prod_{i=1}^{m} dx^i(t). \qquad (6.4.11)$$

Note especially that the coordinate transformation character of $\langle x_2, t_2 \,|\, x_1, t_1 \rangle$, which is a biscalar density of weight $\frac{1}{2}$ at both x_1 and x_2, is given correctly by this formula. This is a surprise, for neither $S_{,ij'}$ nor $G^{+ij'}$ are covariant objects. However, it is easy to see that the same result for the determinant (6.4.8) would have been obtained had we used the Green's function of the covariant operator $S_{;ij'}$. The term involving the connection components in eq. (6.3.26) introduces a correction solely to the quantity C_{ij} appearing in eq. (6.1.4), and the final result (6.4.8) for the determinant depends on neither B_{ij} nor C_{ij}. We note that use of the bitensor $S_{;ij'}$ in place of $S_{,ij}$ allows one immediately to infer the correct coordinate transformation law (infinite product of Jacobians) for the determinant.

What is more remarkable is that the determinant $\det G^+$ *by itself* leads to a unitary quantum theory for the system (6.3.1). It will be recalled that eq. (5.3.46) was obtained by omitting the terms '$+\ldots$' in the infinite series (5.3.39). This omission can be justified for the system (6.3.1) as follows. Note that for this system all the 'vertex functions' $S_{,ijk}[\varphi_c]$, $S_{,ijkl}[\varphi_c]$, etc., in the expansion (5.3.36) are effectively differential operators of the second order in $\partial/\partial t$. When the factors ϕ^j, ϕ^k, etc., in this expansion are rearranged in chronological order, the advanced Green's function *operator* $G^{+ij}[\varphi_c+\phi]$ appears as a factor, and when the ϕs on which this operator itself depends are rearranged in chronological order, further factors $G^{+ij}[\varphi_c+\phi]$ appear. The reordering process can be carried out recursively, and one obtains $T(S_{,i}[\varphi])$ plus a series of terms involving either products of chronologically-ordered operator vertex functions and Green's functions or products of classical vertices ($S_{,ijk}[\varphi_c]$, etc.) and classical Green's functions $G^{+ij}[\varphi_c]$. Each of these terms can be represented by a diagram involving lines (representing the Green's functions) connecting vertices. There is always one vertex that has an 'external' prong to which no line is attached.

Let V be the number of vertices in a given diagram, L the number of lines, and K the number of independent closed loops. These numbers satisfy the topological relation

$$L-V=K-1. \tag{6.4.12}$$

In each diagram there is always at least one closed loop in which all the advanced Green's functions in the loop have the same orientation around the loop. The only way this loop can contribute anything but a vanishing factor to the corresponding term is for all the times in the loop to coincide and for the operators $\partial/\partial t$ in the loop vertices to act twice on each Green's function in the loop, so that delta functions appear (see, for example, eq. (6.4.2)). Since each vertex is of the second differential order and since there are as many lines as vertices in each loop, no operators $\partial/\partial t$ are left over to act on the ends of any other string of alternating lines and vertices that may be attached to the given loop to form a second closed loop.

It is a property of the diagrams generated by the reordering process that the successively attached strings can always be chosen in such a way that the Green's functions in each string all have the same orientation. Since the times in the first loop all coincide (because of the delta functions), the times in the second loop are all forced to coincide as well, and so on. But eq. (6.4.12) shows that whenever $K>1$ there are more lines than vertices and hence not enough operators $\partial/\partial t$ to create enough delta functions to keep the corresponding term from vanishing. Thus only those diagrams that contain only a single closed loop contribute to the series. In fact, the only term that survives is expression (5.3.40), and one is led directly to eq. (5.3.46).

Normalization

One way of defining (and computing) the functional integral (6.4.11) is to divide the time interval $t_2 - t_1$ into small bits and to assign a value to the contribution from each bit, as was done for the examples studied in chapter 5. This is not the method that we shall use in the present chapter, but we *can* use the division into bits to obtain a formal expression for the normalization constant Z.

In chapter 5 the action for short time intervals was evaluated using straight-line paths. In a curved supermanifold the corresponding paths are geodesics. When the time interval is very small the first term in the Lagrangian (6.3.1) dominates the action, and one obtains

$$\langle x'', t'' | x', t' \rangle \underset{t'' \to t'}{\longrightarrow} Z_{t''t'}\, g^{\frac{1}{4}}(x'')\, g^{\frac{1}{4}}(x')\, \mathrm{e}^{\mathrm{i}\sigma(x'', x')/(t''-t')} \tag{6.4.13}$$

for some $Z_{t''t'}$, where $\sigma(x'', x')$ is the so-called *world function*, equal to half the square of the length of the (shortest) geodesic between x' and x''.† By introducing a quasi-Cartesian coordinate system at x' one sees that expression (6.4.13) takes the form of a well-known representation of the delta function (as is required, in the limit $t'' \to t'$, by eq. (6.2.59)) provided

$$Z_{t''t'} = [2\pi\mathrm{i}(t'' - t')]^{-m/2}. \tag{6.4.14}$$

Multiplying together the factors $Z_{t''t'}$ for each of the short intervals into which $t_2 - t_1$ is broken, one obtains the formal result

$$Z = (2\pi\mathrm{i}\, \mathrm{d}t)^{-m(t_2 - t_1)/2\mathrm{d}t}, \tag{6.4.15}$$

which may be compared with eq. (5.6.71).

Ambiguity in the functional integral

Basic to the functional integral (5.3.46) is the idea that the points over which the integral ranges are the 'points' of the space Φ of dynamical histories. Even if one defines the integral by dividing the time interval into

† The world function is well defined in a strict sense only when there are no conjugate points (see the remarks following eq. (6.6.16)) along the geodesic between x' and x''.

small bits (or, in the field theoretic case, by putting the field on a spacetime lattice) and then passing to the continuum limit, some kind of interpolation across each time interval (or between the points of the lattice) must always be assumed in order that the action take on a definite value for each history. The dominant histories in the integral may become nondifferentiable in the continuum limit (as in the case of the integral (6.4.11)) or even *discontinuous* (as in the field theoretic case). Nevertheless, the idea that there *is* a history, and hence a *homotopy class*, associated with each point in the domain of the functional integral must always be maintained.

The need for homotopy considerations is easy to see in the case of the integral (6.4.11). The histories over which this integral ranges are paths, or trajectories, between the points x_1 and x_2. Corresponding to each path is a phase factor given by the exponential in the integrand. For the paths in a given homotopy equivalence class one simply adds the phase factors together. But if the configuration space C is not simply connected there is no *a priori* phase relation between the contributions from paths in different homotopy equivalence classes (although one may assume that the *magnitudes* of these contributions are all equally important). This means that there is an ambiguity in the definition of the functional integral.

There did not *seem* to be an ambiguity when the integral (5.3.46) was 'derived' in section 5.3. But in that derivation a functional Fourier transform was introduced, a device that is valid only in supermanifolds having trivial topology, like $\mathbf{R}_c^m \times \mathbf{R}_a^n$. When the topology is nontrivial one must go beyond what is strictly allowed by the arguments of section 5.3 if the integral (6.4.11) is to be regarded, in its own right, as something on which a quantum theory can be based. The price that must be paid for this is the above-mentioned ambiguity.

Homotopy

Analysis of the ambiguity requires some standard homotopy notions.[†] For later convenience the symbols x_1, t_1, x_2, t_2 in eq. (6.4.11) will be replaced by x', t', x'', t'' respectively. In homotopy theory, a path from x' to x'' is defined as a continuous mapping from the closed interval $[0, 1]$ of $\mathbf{R}$ to C,

[†] A useful reference for the beginner is F. H. Croom, *Basic Concepts of Algebraic Topology*, Springer-Verlag (New York, 1978). The homotopy analysis of the Feynman integral was first carried out by M. G. G. Laidlaw and C. M. DeWitt, *Phys. Rev.*, **D3**, 1375 (1971).

$$\lambda_{x''x'}:[0,1]\to C, \quad s\mapsto \lambda_{x''x'}(s), \tag{6.4.16}$$

satisfying

$$\lambda_{x''x'}(0)=x', \quad \lambda_{x''x'}(1)=x''. \tag{6.4.17}$$

The 'product' of a path $\lambda_{x'''x'}$ from x' to x''' and a path $\mu_{x''x'''}$ from x''' to x'' is also defined:

$$(\mu_{x''x'''}\lambda_{x'''x'})(s) \stackrel{\text{def}}{=} \begin{cases} \lambda_{x'''x'}(2s) & \text{for } 0\leqslant s\leqslant \tfrac{1}{2}, \\ \mu_{x''x'''}(2s-1) & \text{for } \tfrac{1}{2}\leqslant s\leqslant 1. \end{cases} \tag{6.4.18}$$

The class of paths homotopically equivalent to $\lambda_{x''x'}$ (i.e., which can be continuously deformed to $\lambda_{x''x'}$) is denoted by $[\lambda_{x''x'}]$. There is a natural definition of multiplication for homotopy equivalence classes:

$$[\mu_{x''x'''}][\lambda_{x'''x'}] \stackrel{\text{def}}{=} [\mu_{x''x'''}\lambda_{x'''x'}]. \tag{6.4.19}$$

It is not hard to show that this multiplication is associative. The set of all homotopy equivalence classes of paths from x' to x'' will be denoted by $\Pi(C,x'',x')$.

The special case of *closed* paths, in which $x''=x'$, is of importance. Let

$$\lambda_{x'} \stackrel{\text{def}}{=} \lambda_{x'x'}, \quad \Pi(C,x') \stackrel{\text{def}}{=} \Pi(C,x',x'). \tag{6.4.20}$$

$\Pi(C,x')$ is a *group* under the multiplication (6.4.19), its identity element being $[e_{x'}]$ where

$$e_{x'}(s)=x' \quad \text{for all } s \text{ in } [0,1]. \tag{6.4.21}$$

Moreover,

$$[\lambda_{x'}]^{-1}=[\lambda_{x'}^{-1}] \tag{6.4.22}$$

where

$$\lambda_{x''x'}^{-1}(s) \stackrel{\text{def}}{=} \lambda_{x''x'}(1-s) \tag{6.4.23}$$

for all s in $[0,1]$.

C is usually assumed to be a connected supermanifold. In this case, for all x' and x'' in C, the groups $\Pi(C,x')$ and $\Pi(C,x'')$ are isomorphic. For let $\nu_{x''x'}$ be an arbitrary fixed path from x' to x''. Then the mapping $\phi:\Pi(C,x')\to\Pi(C,x'')$ defined by

$$\phi([\lambda_{x'}])=[\nu_{x''x'}][\lambda_{x'}][\nu_{x''x'}^{-1}] \tag{6.4.24}$$

is an isomorphism:

$$\left.\begin{aligned} \phi([e_{x'}])=[e_{x''}], \quad \phi([\lambda_{x'}])^{-1}=\phi([\lambda_{x'}]^{-1}), \\ \phi([\mu_{x'}])\,\phi([\lambda_{x'}])=\phi([\mu_{x'}][\lambda_{x'}]). \end{aligned}\right\} \tag{6.4.25}$$

The abstract group to which $\Pi(C,x')$ and $\Pi(C,x'')$ are both isomorphic is called the *fundamental group* or *first homotopy group* of C and denoted by $\pi_1(C)$.

Homotopy mesh

Let x_0 be a fixed point of C. For every point x' in C choose a unique path $\mu_{x'x_0}$ from x_0 to x'. The set of paths $\mu_{x'x_0}$ is called a *homotopy mesh based on* x_0 and will be assumed to be chosen in a nonpathological way. (For the definition of 'nonpathological' see the paragraphs on the universal covering space below.) Introduction of a homotopy mesh allows one to use the elements of $\Pi(C, x_0)$ as labels for homotopy equivalence classes of *nonclosed* paths in C. For every pair of points x', x'' in C one introduces a one-to-one mapping $\rho_{x''x'}: \Pi(C, x_0) \to \Pi(C, x'', x')$ defined by

$$\rho_{x''x'}(\alpha) = [\mu_{x''x_0}]\,\alpha[\mu^{-1}_{x'x_0}]. \tag{6.4.26}$$

The mappings satisfy the combination law

$$\rho_{x'''x''}(\beta)\,\rho_{x''x'}(\alpha) = \rho_{x'''x'}(\beta\alpha) \tag{6.4.27}$$

for all x', x'', x''' in C and all α, β in $\Pi(C, x_0)$.

Let x be one of the trajectories that contributes, in the functional integral, to the transition amplitude $\langle x'', t'' \mid x', t' \rangle$. Then

$$x(t') = x' \quad x(t'') = x''. \tag{6.4.28}$$

Denote by $\Lambda_{t''t'}$ the one-to-one mapping from trajectories to paths, defined by

$$[\Lambda_{t''t'}(x)](s) = x(st'' + (1-s)\,t') \tag{6.4.29}$$

for all s in $[0, 1]$ and all x in the space $\Phi^{x''t''}_{x't'}$ of trajectories contributing to $\langle x'', t'' \mid x', t' \rangle$. the space $\Phi^{x''t''}_{x't'}$ decomposes into equivalence classes of trajectories:

$$\Phi^{x''t''}_{x't'}(\alpha) \stackrel{\text{def}}{=} \Lambda^{-1}_{t''t'}(\rho_{x''x'}(\alpha)), \quad \alpha \in \Pi(C, x_0). \tag{6.4.30}$$

Corresponding to each of these equivalence classes there is a *partial* amplitude, given by

$$\begin{aligned} K^{\alpha}(x'', t'' \mid x', t') \stackrel{\text{def}}{=}\; & (2\pi \mathrm{i}\, \mathrm{d}t)^{-m(t''-t')/2\mathrm{d}t}\, g^{\frac{1}{4}}(x'')\, g^{\frac{1}{4}}(x')\, \mathrm{e}^{\mathrm{i}[\phi(x'')-\phi(x')]} \\ & \times \int_{\Phi^{x''t''}_{x't'}(\alpha)} \exp\left[\mathrm{i}\int_{x't'}^{x''t''} (\tfrac{1}{2} g_{ij}\, \dot{x}^i \dot{x}^j + a_i \dot{x}^i - v)\, \mathrm{d}t\right] \\ & \times \sum_{t' < t < t''} g^{\frac{1}{2}}(x(t)) \prod_{i=1}^{m} \mathrm{d}x^i(t), \end{aligned} \tag{6.4.31}$$

where the α-independent factor $e^{i[\phi(x'')-\phi(x')]}$ has been inserted to allow for the phase arbitrariness that always exists. (We shall see presently that this factor is needed in order that the total amplitude $\langle x'', t'' | x', t' \rangle$ be independent of the homotopy mesh chosen.) In view of eq. (6.4.27) and the structure of expression (6.4.31) it is easy to see that the partial amplitudes satisfy the composition law

$$K^{\alpha}(x'', t'' | x', t') = \sum_{\beta} \int K^{\alpha\beta^{-1}}(x'', t'' | x''', t''') K^{\beta}(x''', t''' | x', t') \, d^m x''', \quad (6.4.32)$$

where $t' < t''' < t''$ and the summation is over $\Pi(C, x_0)$.

The total amplitude

We have already remarked that there is no *a priori* phase relation between contributions to the functional integral from trajectories in different homotopy equivalence classes. This means that there is no *a priori* phase relation between the contributions that different partial amplitudes make to the total transition amplitude. One *can* say, however, that there must exist a mapping χ from $\Pi(C, x_0)$ to the unit circle in $\mathbf{C}_c$,† such that the total amplitude is given by

$$\langle x'', t'' | x', t' \rangle_{\chi} = \sum_{\alpha} \chi(\alpha) K^{\alpha}(x'', t'' | x', t'). \quad (6.4.33)$$

There is, additionally, a condition that χ must satisfy, stemming from the composition law for total amplitudes:

$$\begin{aligned} \langle x'', t'' | x', t' \rangle_{\chi} &= \int \langle x'', t'' | x''', t''' \rangle_{\chi} \langle x''', t''' | x', t' \rangle_{\chi} \, d^m x''' \\ &= \sum_{\gamma, \beta} \chi(\gamma) \chi(\beta) \int K^{\gamma}(x'', t'' | x''', t''') K^{\beta}(x''', t''' | x', t') \, d^m x''' \\ &= \sum_{\alpha, \beta} \chi(\alpha\beta^{-1}) \chi(\beta) \int K^{\alpha\beta^{-1}}(x'', t'' | x''', t''') K^{\beta}(x''', t''' | x', t') \, d^m x'''. \end{aligned}$$

(6.4.34)

Comparing expressions (6.4.33) and (6.4.34) and taking note of the partial composition law (6.4.32), one sees that a sufficient condition for consistency is $\chi(\alpha\beta^{-1})\chi(\beta) = \chi(\alpha)$, or

$$\chi(\gamma)\chi(\beta) = \chi(\gamma\beta), \quad (6.4.35)$$

† The unit circle in $\mathbf{C}_c$ is the set of complex c-numbers z satisfying $|z| = 1$, where the absolute value is as defined in exercise **1.6**.

for all α, β, γ in $\Pi(C, x_0)$. We shall see presently that this is also a necessary condition and hence that the χs must constitute a 1-dimensional unitary representation of the fundamental group $\pi_1(C)$.

Denote by $[\pi_1(C), \pi_1(C)]$ the *commutator subgroup of* $\pi_1(C)$, i.e. the set of all elements of the form $\alpha_1 \beta_1 \alpha_1^{-1} \beta_1^{-1} \ldots \alpha_n \beta_n \alpha_n^{-1} \beta_n^{-1}$. The following facts are readily verified: (1) The commutator subgroup is a normal subgroup, and the factor group $\pi_1(C)/[\pi_1(C), \pi_1(C)]$ is Abelian. (2) If α is an element of the commutator subgroup, then $\chi(\alpha) = 1$ for all 1-dimensional representations χ. From this it follows that the 1-dimensional representations are in fact representations of the factor group.

Change of homotopy mesh

The equivalence classes $\Phi_{x't'}^{x''t''}(\alpha)$ are determined by the homotopy mesh. However, the total transition amplitude $\langle x'', t'' \,|\, x', t' \rangle_\chi$ must not depend on which homotopy mesh is chosen. Let $\bar{\mu}_{x'x_0}$ be a new homotopy mesh. It leads to a new mapping $\bar{\rho}_{x''x'} : \Pi(C, x_0) \to \Pi(C, x'', x')$, given by

$$\begin{aligned} \bar{\rho}_{x''x'}(\alpha) &= [\bar{\mu}_{x''x_0}]\,\alpha\,[\bar{\mu}_{x'x_0}^{-1}] \\ &= [\mu_{x''x_0}][\mu_{x''x_0}^{-1}][\bar{\mu}_{x''x_0}]\,\alpha[\bar{\mu}_{x'x_0}^{-1}][\mu_{x'x_0}][\mu_{x'x_0}^{-1}] \\ &= \rho_{x''x'}(\gamma(x'')\,\alpha\gamma^{-1}(x')), \end{aligned} \tag{6.4.36}$$

where

$$\gamma(x') = [\mu_{x'x_0}^{-1}][\bar{\mu}_{x'x_0}] \in \Pi(C, x_0) \tag{6.4.37}$$

for all x' in C. This in turn leads to new equivalence classes of trajectories,

$$\bar{\Phi}_{x't'}^{x''t''}(\alpha) = \Lambda_{t''t'}^{-1}(\bar{\rho}_{x''x'}(\alpha)) = \Phi_{x't'}^{x''t''}(\gamma(x'')\,\alpha\gamma^{-1}(x')), \tag{6.4.38}$$

and to new partial amplitudes

$$\begin{aligned} \bar{K}^{\alpha}(x'', t'' \,|\, x', t') &= (2\pi \mathrm{i}\, \mathrm{d}t)^{-m(t''-t')/2\mathrm{d}t} g^{\frac{1}{4}}(x'')\, g^{\frac{1}{4}}(x')\, \mathrm{e}^{\mathrm{i}[\bar{\phi}(x'')-\bar{\phi}(x')]} \\ &\quad \times \int_{\Phi_{x't'}^{x''t''}(\alpha)} \exp\left[\mathrm{i} \int_{x't'}^{x''t''} (\tfrac{1}{2} g_{ij}\, \dot{x}^i \dot{x}^j + a_i \dot{x}^i - v)\, \mathrm{d}t\right] \\ &\quad \times \prod_{t' < t < t''} g^{\frac{1}{2}}(x(t)) \prod_{i=1}^{n} \mathrm{d}x^i(t) \\ &= \mathrm{e}^{\mathrm{i}[\bar{\phi}(x'')-\phi(x'')-\bar{\phi}(x')+\phi(x')]} K^{\gamma(x'')\alpha\gamma^{-1}(x')}(x'', t'' \,|\, x', t'). \end{aligned} \tag{6.4.39}$$

The total amplitude will be homotopy-mesh invariant, i.e.,

$$\langle x'', t'' \,|\, x', t' \rangle_\chi = \sum_{\alpha} \chi(\alpha)\, \bar{K}^{\alpha}(x'', t'' \,|\, x', t') \tag{6.4.40}$$

for all new homotopy meshes, if and only if eq. (6.4.35) is satisfied and the phase factors associated with the old and new meshes are related by

$$e^{i\acute{\phi}(x')} = \chi(\gamma(x'))\, e^{i\phi(x')} \tag{6.4.41}$$

for all x' in C and all new meshes.

The role of homology

We have seen that the amplitudes $\langle x'', t'' | x', t' \rangle_\chi$ are labelled by 1-dimensional representations, or *characters* χ, of the factor group $\pi_1(C)/[\pi_1(C), \pi_1(C)]$. It is not difficult to show that every element of the commutator subgroup $[\pi_1(C), \pi_1(C)]$ is a homotopy equivalence class of 1-boundaries. The converse of this statement is also true, namely every 1-boundary lies in an element of $[\pi_1(C), \pi_1(C)]$. From this it can be shown that the factor group is isomorphic to the first homology group $H_1(C, Z)$ of C over the additive group Z of the integers† and hence that the amplitudes $\langle x'', t'' | x', t' \rangle_\chi$ may equally well be regarded as labeled by characters of $H_1(C, Z)$.

In section 6.2 we have seen that the only ambiguity in the quantization of standard canonical systems is the ambiguity in the momentum operator and that there is precisely a b_1-parameter family of physically distinct quantum systems corresponding to a given classical system, b_1 being the first Betti number of C. From the point of view of homology the first Betti number is determined by the factor group $H_1(C, Z)/\text{Tor}$ where Tor is the *torsion* subgroup of $H_1(C, Z)$. Indeed,

$$H_1(C, Z)/\text{Tor} = \overbrace{Z \times \ldots \times Z}^{b_1 \text{ factors}}. \tag{6.4.42}$$

Although the arguments of this section seem to allow *all* the characters of $H_1(C, Z)$ to occur as labels for physically distinct quantum theories of the system (6.3.1), in fact only the characters of $H_1(C, Z)/\text{Tor}$ can be used, the contribution from the torsion subgroup being always trivial. If one tries to use a nontrivial character of the torsion subgroup one finds that the phase factors $e^{i\phi(x'')}, e^{-i\phi(x')}$ in (6.4.31) cannot then be chosen in such a way as to yield a total amplitude (6.4.33) that is continuous. It is intuitively clear that if C is not simply connected, the element $[\mu_{x'x_0}]$ of $\Pi(C, x', x_0)$ will depend discontinuously on x' at certain places, the function $\gamma: C \to \Pi(C, x_0)$ defined by (6.4.37) will possess corresponding discontinuities, and the phase factors $e^{i\phi(x'')}$, $e^{-i\phi(x')}$ will have to possess

† See F. H. Croom, *loc. cit.*

discontinuities in order that $\langle x'', t'' | x', t' \rangle_\chi$ itself possess none. Illustrative examples of these phenomena will be found in exercises **6.6**, **6.7** and **6.8** at the end of the chapter.

Although we have arrived in the end at the same result regarding the quantization ambiguity as that found in section 6.2 from cohomological considerations, the Feynman functional integral gives us quite a different view of it, namely as a phenomenon of homotopy and homology. It has the added virtue of allowing us to infer, without further analysis, that the b_1 independent real parameters describing a given physical quantum theory do not range over the whole of $\mathbf{R}_c$. This is because the group of which χ is a character is always of the form $Z \times \ldots \times Z$, and hence $\chi : Z \times \ldots \times Z \to \mathbf{C}$ is given by

$$\chi(n_1, \ldots, n_{b_1}) = \exp\left(\mathrm{i} \sum_{r=1}^{b_1} n_r \theta_r\right), \tag{6.4.43}$$

where the θ_r are the independent parameters. To avoid redundancy one must restrict these parameters, for example by requiring

$$0 \leqslant (\theta_r)_B < 2\pi, \quad r = 1, 2, \ldots, b_1, \tag{6.4.44}$$

a result that is not obtainable from cohomological considerations alone. The examples of exercises **6.5** to **6.8** show that the souls of the θ_r must generally vanish in actual cases in order that the energy eigenvalues be physical.

The universal covering space[†]

The above results can be brought into sharper focus by introducing the *universal covering space* of C. This space, denoted here by $\bar{C}$, is most easily defined for our purposes as the set of all homotopy equivalence classes $[\lambda_{x', x_0}]$ with x_0 fixed. $\bar{C}$ can be given the structure of a topological space, indeed of a supermanifold, by defining its open sets to be the unions of arbitrary collections of *basic neighbourhoods*. A basic neighbourhood of a point $[\lambda_{x'x_0}]$ in $\bar{C}$ is the set of all homotopy equivalence classes of the form $[\nu_{x''x'}][\lambda_{x'x_0}]$ where $\nu_{x''x'}$ is a path lying in some fixed open subset of C containing x'. The universal covering space can be shown to be simply connected.

Corresponding to each homotopy mesh of paths $\mu_{x'x_0}$ in C is a subset of homotopy equivalence classes $[\mu_{x'x_0}]$ in $\bar{C}$. A mesh can always be chosen in such a way that this subset is simply connected in $\bar{C}$. The mesh is then said

[†] The author is indebted to Gary Hamrick for help in presenting the following analysis.

to be *nonpathological*, and the subset is called a *fundamental domain*. Since there is in general no canonical way of choosing a homotopy mesh, there is no canonical way of choosing a fundamental domain. A fundamental domain is generally neither closed nor open. It contains some of its boundary points. However, the boundary points that it contains may themselves be chosen to form a simply connected subset of $\bar{C}$.

There is a natural projection $\pi:\bar{C}\to C$ defined by

$$\pi([\lambda_{x'x_0}]) = x' \tag{6.4.45}$$

for all $[\lambda_{x'x_0}]$ in $\bar{C}$. This projection is differentiable, and its inverse π^{-1} is a 'multivalued diffeomorphism'. That is, each open subset of a fundamental domain is a copy, under π^{-1}, of the corresponding open subset in C.

For every α in $\Pi(C, x_0)$ there exists a mapping $f_\alpha:\bar{C}\to\bar{C}$ defined by

$$f_\alpha([\lambda_{x'x_0}]) = [\lambda_{x'x_0}]\,\alpha \tag{6.4.46}$$

for all $[\lambda_{x'x_0}]$ in $\bar{C}$. Each f_α is a diffeomorphism called a *covering translation*. Let F be a fundamental domain. Then $f_\alpha(F)$ is also a fundamental domain, for all α. The $f_\alpha(F)$ are all disjoint, but their union is $\bar{C}$ itself. It is easy to see that

$$\pi\circ f_\alpha = \pi, \tag{6.4.47}$$

$$f_\alpha\circ f_\beta = f_{\beta\alpha}, \quad f_\alpha^{-1} = f_{\alpha^{-1}}, \tag{6.4.48}$$

for all α, β in $\Pi(C, x_0)$ and hence that the covering translations constitute a realization of $\Pi(C, x_0)$ through the correspondence $\alpha\to f_{\alpha^{-1}}$.

Let ω be the closed 1-form appearing in eqs. (6.2.75). The projection π induces a corresponding closed 1-form $\bar{\omega}$ on $\bar{C}$ by the pullback operation[†] or, equivalently (via π^{-1}) by the derivative mapping (see section 2.5). This 1-form repeats itself on every fundamental domain, i.e.,

$$f'_\alpha(\bar{\omega}) = \bar{\omega} \quad \text{for all } \boldsymbol{\alpha} \text{ in } \boldsymbol{\Pi}(C, x_0). \tag{6.4.49}$$

Since $\bar{C}$ is simply connected, $\bar{\omega}$ must be exact, i.e., $\bar{\omega} = \mathbf{d}\bar{\phi}$ for some scalar field $\bar{\phi}$ on $\bar{C}$. $\bar{\phi}$ satisfies the law

$$\mathbf{d}(\bar{\phi}\circ f_\alpha) = \mathbf{d}f'_{\alpha^{-1}}(\bar{\phi}) = f'_{\alpha^{-1}}(\mathbf{d}\bar{\phi}) = f'_{\alpha^{-1}}(\bar{\omega}) = \bar{\omega} = \mathbf{d}\bar{\phi} \tag{6.4.50}$$

for all α (see eqs. (2.5.29) and (6.4.48) and exercise **2.5**).

Let $\bar{x}'$ and $\bar{y}'$ be any two points of $\bar{C}$. Then

$$\bar{\phi}(f_\alpha(\bar{x}')) - \bar{\phi}(f_\alpha(\bar{y}')) = \int_{\bar{y}'}^{\bar{x}'} \mathbf{d}(\bar{\phi}\circ f_\alpha) = \int_{\bar{y}'}^{\bar{x}'} \mathbf{d}\bar{\phi} = \bar{\phi}(\bar{x}') - \bar{\phi}(\bar{y}'), \tag{6.4.51}$$

[†] See Choquet-Bruhat and DeWitt–Morette, with Dillard–Bleick, *loc. cit.*

from which it follows that the difference $\bar{\phi}(f_\alpha(\bar{x}')) - \bar{\phi}(\bar{x}')$ is independent of $\bar{x}'$ and depends only on α. That is, there must exist a real-valued function h on $\Pi(C, x_0)$ such that

$$\bar{\phi}(f_\alpha(\bar{x}')) - \bar{\phi}(\bar{x}') = h(\alpha) \tag{6.4.52}$$

for all $\bar{x}'$ in $\bar{C}$. It is easy to see that this function satisfies

$$h(\alpha\beta) = h(\beta\alpha) = h(\alpha) + h(\beta), \tag{6.4.53}$$

and hence that the $h(\alpha)$ are elements of an Abelian group (under addition) homomorphic to $\Pi(C, x_0)$. This group is turned into a multiplicative group through introduction of the character

$$\chi(\alpha) = e^{ih(\alpha)}. \tag{6.4.54}$$

This is the character of $H_1(C, Z)$ that we have already encountered in eqs. (6.4.33) and (6.4.35).

It is now easy to see why the torsion subgroup of $H_1(C, Z)$ can make no contribution to χ. If ω is a closed 1-form on C then so is $a\omega$ where a is any real c-number. The corresponding character is $e^{iah(\alpha)}$. But the multiplication of the phase $h(\alpha)$ by a continuous parameter a is not compatible with the nontrivial characters of the torsion subgroup, which always have discrete phases.

It is also easy to see that all closed 1-forms in a given cohomology equivalence class yield the same χ. Equations (6.4.43) and (6.4.44) show that the χ's may be regarded as points in the b_1-torus, or rather in its trivial extension to a $(b_1, 0)$-dimensional supermanifold, which we shall denote by $\mathbf{T}_c^{b_1}$. A mapping is thus defined from the space $\mathbf{R}_c^{b_1}$ of equivalence classes of 1-forms to $\mathbf{T}_c^{b_1}$. If $b_1 \neq 0$ the images of a point in $\mathbf{T}_c^{b_1}$ under the inverse of this mapping are infinite in number, and hence for each χ there is an infinity of equivalence classes of closed 1-forms that yield that χ.

The total amplitude revisited

Let $\bar{g}$, $\bar{a}$ and $\bar{v}$ be respectively the pullbacks, under π, of the metric tensor g, the 1-form potential $\boldsymbol{a}$ and the scalar potential v appearing in the Lagrangian (6.3.1). These pullbacks are invariant under covering translations and define a similarly invariant dynamical system having $\bar{C}$ as its configuration space. The amplitude for this system to have the configuration $\bar{x}''$ at time t'' if it had the configuration $\bar{x}'$ at time t' is given by the functional integral

$$\overline{\langle \bar{x}'', t'' | \bar{x}', t' \rangle_\chi} = (2\pi \mathrm{i}\, \mathrm{d}t)^{-m(t''-t')/2\mathrm{d}t} \bar{g}^{\frac{1}{4}}(\bar{x}'')\, \bar{g}^{\frac{1}{4}}(\bar{x}')\, \mathrm{e}^{\mathrm{i}[\bar{\phi}(\bar{x}'')-\bar{\phi}(\bar{x}')]}$$

$$\times \int_{\bar{\Phi}_{\bar{x}'t'}^{\bar{x}''t''}} \exp\left[\mathrm{i}\int_{\bar{x}'t'}^{\bar{x}''t''} (\tfrac{1}{2}\bar{g}_{ij}\dot{\bar{x}}^i\dot{\bar{x}}^j + \bar{a}_i\dot{\bar{x}}^i - \bar{v})\, \mathrm{d}t\right]$$

$$\times \prod_{t'<t<t''} \bar{g}^{\frac{1}{2}}(\bar{x}(t)) \prod_{i=1}^{m} \mathrm{d}\bar{x}^i(t), \tag{6.4.55}$$

where $\bar{\Phi}_{\bar{x}'t'}^{\bar{x}''t''}$ is the space of all trajectories in $\bar{C}$ having the indicated endpoints.

The subscript χ on the transition amplitude (6.4.55) has no physical significance for the system on $\bar{C}$ because $\bar{C}$ is simply connected, and one may carry out a phase transformation like (6.2.73) on the position eigenvectors, which sets the scalar field $\bar{\phi}$ equal to zero. However, χ becomes significant if the transition amplitude is projected onto C.

Let

$$x' = \pi(\bar{x}'), \quad x'' = \pi(\bar{x}''). \tag{6.4.56}$$

Then

$$\bar{g}_{ij}(\bar{x}') = g_{ij}(x'), \quad \bar{a}_i(\bar{x}') = a_i(x'), \quad \bar{v}(\bar{x}') = v(x'), \text{ etc.} \tag{6.4.57}$$

Let F be the fundamental domain defined by the homotopy mesh $\mu_{x'x_0}$, and let $\bar{x}'$ and $\bar{x}''$ be any two points of F. Then, in view of eqs. (6.4.26), (6.4.30), (6.4.31), (6.4.46), (6.4.52), (6.4.54), and (6.4.57), we see that

$$\overline{\langle f_\alpha(\bar{x}''), t'' | \bar{x}', t' \rangle_\chi} = \chi(\alpha)\, K^\alpha(x'', t'' | x', t'), \tag{6.4.58}$$

provided we understand the scalar field ϕ appearing in eq. (6.4.31) to be defined by

$$\phi(x') = \bar{\phi}(\bar{x}') \quad \text{with } \bar{x}' \text{ in } F. \tag{6.4.59}$$

Equation (6.4.33) immediately allows us to write

$$\langle x'', t'' | x', t' \rangle_\chi = \sum_\alpha \overline{\langle f_\alpha(\bar{x}''), t'' | \bar{x}', t' \rangle_\chi}, \tag{6.4.60}$$

which shows plainly that the total transition amplitude in C is generally both continuous and single valued despite the fact that the scalar field ϕ defined by (6.4.59) typically has discontinuities at the projections of the bounding points of F on C.

6.5 The Hamiltonian operator: a nonlattice derivation

Integration over phase space

We have remarked that we shall not compute the Feynman path integral by dividing the time interval into a lattice of small bits and assigning an *a priori* value to the contribution from each bit. We have, however, used the division into small bits to yield the formal expression (6.4.15) for the normalization constant. We may also use it to derive an important alternative expression for the functional integral itself. We first note that the momenta and Hamiltonian of the system (6.3.1) are given by

$$p_i = \partial L/\partial \dot{x}^i = g_{ij}\dot{x}^j + a_i, \tag{6.5.1}$$

$$H = p_i\dot{x}^i - L = \tfrac{1}{2}g_{ij}\dot{x}^i\dot{x}^j + v = \tfrac{1}{2}g^{ij}(p_i - a_i)(p_j - a_j) + v, \tag{6.5.2}$$

and we then invoke the formal identity

$$\begin{aligned}
\int e^{i[p_i\dot{x}^i - H(x,p)]dt}\prod_i dp_i &= \int e^{i[p_i\dot{x}^i - \frac{1}{2}g^{ij}(p_i-a_i)(p_j-a_j)-v]dt}\prod_i dp_i \\
&= \int e^{i[-\frac{1}{2}g^{ij}(p_i-a_i-g_{ik}\dot{x}^k)(p_j-a_j-g_{jl}\dot{x}^l)+\frac{1}{2}g_{ij}\dot{x}^i\dot{x}^j+a_i\dot{x}^i-v]dt}\prod_i dp_i \\
&= \left(\frac{2\pi}{i\,dt}\right)^{m/2} g^{\frac{1}{2}} e^{i[\frac{1}{2}g_{ij}\dot{x}^i\dot{x}^j+a_i\dot{x}^i-v]dt}.
\end{aligned} \tag{6.5.3}$$

If this result is applied to each of the factors in a decomposition of the integrand of (6.4.31) into contributions from short time intervals dt and if the p_i are assumed to be associated with the midpoints of these intervals, then the metric appearing in (6.5.3) is effectively the average of the metric at the endpoints, the $g^{\frac{1}{2}}$ that arises from the integration is effectively the product of $g^{\frac{1}{4}}$'s coming from each endpoint, and expression (6.4.31) may be replaced by

$$\begin{aligned}
K^\alpha(x'', t''\,|\,x', t') = (2\pi)^{-m(t''-t')/dt}\, e^{i[\phi(x'')-\phi(x')]} \\
\times \int_{\Phi^{x''t''}_{x't'}(\alpha)} \exp\left\{i\int_{x't'}^{x''t''}[p_i\dot{x}^i - H(x,p)]\,dt\right\} \\
\times \left(\prod_{t'<t<t''}^{+}\prod_{i=1}^{m} dp_i(t)\right)\left(\prod_{t'<t<t''}\prod_{i=1}^{m} dx^i(t)\right).
\end{aligned} \tag{6.5.4}$$

A plus sign is attached to the first $\prod$ to signal the fact that the product of the dp's in the functional volume element is over *one more instant of time* than is the product of the dx's. Despite this fact, the functional integral

(6.5.4) is often loosely regarded as an integral over an infinite product of copies of *phase space* (the cotangent bundle of C).

It is worth stressing that expression (6.5.4) is of *wider applicability* than (6.4.31), for it is valid for systems having Lagrangians more general than (6.3.1) (e.g., for relativistic single particles). It is again a special case of (5.3.46), in which the action is now expressed in terms of the canonical variables x^i, p_i. It is not difficult to verify in this case that the $(\det G^+)^{-\frac{1}{2}}$ in (5.3.46) is a simple constant, independent of the x's and p's, and that the terms represented by '...' in eq. (5.3.39) again contribute nothing.

The Schrödinger equation

Use of expression (6.5.4) makes it easy to determine what Schrödinger equation the amplitude $\langle x'', t'' | x', t' \rangle_\chi$ satisfies and hence what the quantum Hamiltonian operator $H^Q(x, p)$ is for the system defined by the functional integral. Let us abbreviate by $H^Q_{x'}$ the position representation of this operator:

$$H^Q_{x'} \equiv H^Q\left(x', -\mathrm{i}\frac{\partial}{\partial x'} - \omega(x')\right), \tag{6.5.5}$$

$$H^Q_{x'}\langle x', t'| = \langle x', t'| H^Q(x(t'), p(t')), \tag{6.5.6}$$

where ω is any of the 1-forms that yields the character χ. We may then write the Schrödinger equation in the form

$$\left(\mathrm{i}\frac{\partial}{\partial t''} - H^Q_{x''}\right)\langle x'', t'' | x', t' \rangle_\chi = 0 \tag{6.5.7}$$

(see eqs. (6.2.91) and (6.2.92)). By virtue of the necessary boundary condition

$$\langle x'', t' | x', t' \rangle_\chi = \delta(x'', x') \tag{6.5.8}$$

(cf. eq. (6.2.59)), it is easy to verify that the expression

$$G^-(x'', t'' | x', t') \overset{\text{def}}{=} \mathrm{i}\theta(t'', t')\langle x'', t'' | x', t' \rangle_\chi \tag{6.5.9}$$

is the retarded Green's function of the operator $\mathrm{i}\partial/\partial t'' - H^Q_{x''}$:

$$\left(\mathrm{i}\frac{\partial}{\partial t''} - H^Q_{x''}\right)G^-(x'', t'' | x', t') = -\delta(t'', t')\,\delta(x'', x'). \tag{6.5.10}$$

Now let the Hamiltonian operator suffer an infinitesimal variation produced by infinitesimal changes in the functions g_{ij}, a_i, v having supports contained in the interval (t'', t'). Equation (6.5.7) suffers a corresponding variation

$$\left(i\frac{\partial}{\partial t''} - H^Q_{x''}\right)\delta\langle x'', t''|x', t'\rangle_\chi = \delta H^Q_{x'',t''}\langle x'', t''|x', t'\rangle, \qquad (6.5.11)$$

of which the solution consistent with the support conditions on δg_{ij}, δa_i and δv is

$$\delta\langle x'', t''|x', t'\rangle_\chi = -\int_{t'}^{t''} dt \int_C d^m x''' G^-(x'', t''|x''', t)\delta H^Q_{x''',t}\langle x''', t|x', t'\rangle_\chi$$

$$= -i\int_{t'}^{t''} dt \int_C d^m x'''\langle x'', t''|x''', t\rangle_\chi \delta H^Q_{x''',t}\langle x''', t|x', t'\rangle_\chi$$

$$= -i\left\langle x'', t''\left|\delta\int_{t'}^{t''} H^Q((x(t), p(t)))\,dt\right|x', t'\right\rangle_\chi. \qquad (6.5.12)$$

An alternative expression for the variation in the transition amplitude follows from eqs. (6.4.10), (6.4.33) and (6.5.4), namely,

$$\delta\langle x'', t''|x', t'\rangle_\chi = -i\left\langle x'', t''\left|\delta\int_{t'}^{t''} T(H(x(t), p(t)))\,dt\right|x', t'\right\rangle_\chi. \qquad (6.5.13)$$

Comparison of (6.5.12) and (6.5.13) shows that, up to an ignorable additive constant, the following identity must hold:

$$H^Q(x, p) \equiv T(H(x, p)). \qquad (6.5.14)$$

Evaluation of the chronologically ordered Hamiltonian

The function H standing on the right of eq. (6.5.14) is the classical Hamiltonian (6.5.2), which is built out of factors all evaluated at the same instant of time. The chronologically ordered form of the corresponding quantum operator must be understood as the limit, as the times coincide, of the chronologically ordered form of the operator obtained by evaluating the noncommuting factors at different times. Thus,

$$T(H(x, p)) = \lim_{t', t'' \to t} \tfrac{1}{2}[\theta(t, t')\theta(t', t'')p_i g^{i'j'}p_{j'} + \theta(t', t'')\theta(t'', t)g^{i'j'}p_{j'}p_i$$

$$+ \theta(t'', t)\theta(t, t')p_{j'}p_i g^{i'j'} + \theta(t, t'')\theta(t'', t')p_i p_{j'} g^{i'j'}$$

$$+ \theta(t'', t')\theta(t', t)p_{j'}g^{i'j'}p_i + \theta(t', t)\theta(t, t'')g^{i'j'}p_i p_{j'}]$$

$$- \lim_{t' \to t}[\theta(t, t')a^i p_{i'} + \theta(t', t)p_{i'}a^i] + \tfrac{1}{2}a^i a_i + v, \qquad (6.5.15)$$

where

$$a^i \overset{\text{def}}{=} g^{ij} a_j \tag{6.5.16}$$

and where the primes on indices indicate the times at which the operators to which the indices are affixed are to be evaluated.

There is an immediate difficulty with formula (6.5.15): Momentum-component operators at different times (and hence associated with different tangent spaces) cannot strictly be compared, and therefore a danger exists that the formula will yield an operator that is not coordinate invariant in the limit. This difficulty will be dealt with by applying the operator to a position eigenvector $|x', t\rangle$ and choosing the coordinate system to be quasi-Cartesian at x', i.e., such that first derivatives of the metric components vanish there and such that infinitesimal parallel displacement leaves components unchanged.

It is not difficult to display an operator that is both Hermitian and coordinate invariant and reduces to (6.5.2) in the classical limit, namely, the *naive* Hamiltonian operator:

$$\begin{aligned} H^N(x,p) &\overset{\text{def}}{=} \tfrac{1}{2} g^{-\frac{1}{4}}(p_i - a_i)\, g^{\frac{1}{2}} g^{ij} (p_j - a_j)\, g^{-\frac{1}{4}} + v \\ &= \tfrac{1}{2} g^{-\frac{1}{4}} p_i g^{\frac{1}{2}} g^{ij} p_j g^{-\frac{1}{4}} - \tfrac{1}{2}\{a^i, p_i\} + \tfrac{1}{2} a^i a_i + v. \end{aligned} \tag{6.5.17}$$

Use of eqs. (6.2.54) together with the coordinate transformation law of g yields

$$\bar{g}^{-\frac{1}{4}} \bar{p}_i \bar{g}^{\frac{1}{4}} = g^{-\frac{1}{4}} p_j g^{\frac{1}{4}} \frac{\partial x^j}{\partial \bar{x}^i}, \quad \bar{g}^{\frac{1}{4}} \bar{p}_i \bar{g}^{-\frac{1}{4}} = \frac{\partial x^j}{\partial \bar{x}^i} g^{\frac{1}{4}} p_j g^{-\frac{1}{4}}, \tag{6.5.18}$$

from which the coordinate invariance of H^N immediately follows. H^Q will now be compared to H^N.

H^N is first rewritten in the form

$$H^N(x,p) = \tfrac{1}{2} p_i g^{ij} p_j - \tfrac{1}{2}\{a^i, p_i\} + \tfrac{1}{2} a^i a_i + v + \tfrac{1}{8} g^{ij} g^{kl} g_{ij,kl} + \ldots, \tag{6.5.19}$$

which is easily obtained by reordering factors and using the basic commutation relations. The unwritten terms, denoted by '...', all involve products of first derivatives of the metric components. Next, the factors in the first two terms are evaluated at different times, and the terms themselves are multiplied by step-function decompositions of unity:

$$\begin{aligned} H^N(x,p) = {} & \lim_{t',t''\to t} \tfrac{1}{2}[\theta(t,t')\,\theta(t',t'') + \theta(t',t'')\,\theta(t'',t) + \theta(t'',t)\,\theta(t,t') \\ & + \theta(t,t'')\,\theta(t'',t') + \theta(t'',t')\,\theta(t',t) + \theta(t',t)\,\theta(t,t'')]\, p_i g^{i'j'} p_{j''} \\ & - \lim_{t'\to t} \tfrac{1}{2}[\theta(t,t') + \theta(t',t)]\,\{a^i, p_{i'}\} + \tfrac{1}{2} a^i a_i + v + \tfrac{1}{8} g^{ij} g^{kl} g_{ij,kl} + \ldots. \end{aligned} \tag{6.5.20}$$

Subtracting (6.5.20) from (6.5.15) one gets

$$H^Q(x,p) - H^N(x,p) = T(H(x,p)) - H^N(x,p)$$

$$= \lim_{t',t''\to t} \tfrac{1}{4}\{\theta(t',t'')\,\theta(t'',t)\,[g^{i'j'}p_{j''},p_i] + \theta(t'',t)\,\theta(t,t')\,[p_{j''},p_i g^{i'j'}]$$

$$+\theta(t,t'')\,\theta(t'',t')\,p_i[p_{j''},g^{i'j'}] + \theta(t',t)\,\theta(t,t'')\,[g^{i'j'},p_i]\,p_{j''}$$

$$+\theta(t'',t')\,\theta(t',t)\,([p_{j''},g^{i'j'}]\,p_i + [g^{i'j'}p_{j''},p_i])\}$$

$$-\lim_{t'\to t}\tfrac{1}{2}\{\theta(t,t')\,[a^i,p_{i'}] + \theta(t',t)\,[p_{i'},a^i]\} - \tfrac{1}{8}g^{ij}g^{kl}g_{ij,kl} + \ldots. \qquad (6.5.21)$$

After discarding commutators that vanish when the times coincide, one easily sees that the first, fourth and sixth commutators in (6.5.21) together yield $\theta(t',t)\,[g^{i'j'},p_i]\,p_{j''}$, while the second and third commutators together yield $\theta(t'',t')\,\theta(t,t')\,p_i[p_{j''},g^{i'j'}]$. Moreover, the seventh and eighth commutators cancel in view of (6.4.4), leaving one with

$$H^Q(x,p) - H^N(x,p) = \lim_{t',t''\to t}\tfrac{1}{2}\{\theta(t',t)\,[g^{i'j'},p_i]\,p_{j''} + \theta(t'',t')\,\theta(t,t')\,p_i[p_{j''},g^{i'j'}]$$

$$+\theta(t'',t')\,\theta(t',t)\,[p_{j''},g^{i'j'}]\,p_i\} - \tfrac{1}{8}g^{ij}g^{kl}g_{ij,kl} + \ldots$$

$$= \lim_{t',t''\to t}\tfrac{1}{2}\{\theta(t',t)\,[g^{i'j'},p_i]\,p_{j''} + \theta(t'',t')\,[p_{j''},g^{i'j'}]\,p_i$$

$$+\theta(t'',t')\,\theta(t,t')\,[p_i,[p_{j''},g^{i'j'}]]\} - \tfrac{1}{8}g^{ij}g^{kl}g_{ij,kl} + \ldots. \qquad (6.5.22)$$

Taking the limit of coincident times and once again using (6.4.4), one arrives at the Hermitian result

$$H^Q(x,p) - H^N(x,p) = -\tfrac{1}{8}(g^{ij}{}_{,ij} + g^{ij}g^{kl}g_{ij,kl}) + \ldots. \qquad (6.5.23)$$

Expression (6.5.23) is now applied to the position eigenvector $|x',t\rangle$. The terms represented by '...' contribute nothing, while the term involving second derivatives of the metric components yields $\tfrac{1}{8}$ times the curvature scalar R evaluated at x'. Invoking coordinate invariance one may therefore infer that

$$H^Q(x,p) - H^N(x,p) = \tfrac{1}{8}R(x), \qquad (6.5.24)$$

whence

$$H^Q(x,p) = \tfrac{1}{2}g^{-\frac{1}{4}}(p_i - a_i)\,g^{\frac{1}{2}}g^{ij}(p_j - a_j)\,g^{-\frac{1}{4}} + v + \tfrac{1}{8}R. \qquad (6.5.25)$$

6.6 Approximate evaluation of the path integral

Brief review of Hamilton–Jacobi theory

The action functional appearing in the functional integrals (6.4.31) is the one appropriate to the boundary conditions specified by the variables x'', t'', x', t':

$$S_{x't'}^{x''t''}[x] = \int_{x't'}^{x''t''} L(x, \dot{x})\,\mathrm{d}t. \tag{6.6.1}$$

These variables determine one or more (super)classical trajectories x_c satisfying

$$S_{,i}[x_c] = 0, \quad x_c^i(t'') = x''^i, \ x_c^i(t') = x'^i. \tag{6.6.2}$$

Different (super)classical trajectories corresponding to the same boundary conditions may or may not belong to different homotopy equivalence classes, but each of them, by the principle of stationary phase, is a point in the space Φ of dynamical histories around which is clustered a neighbourhood of histories whose contributions to the functional integral reinforce each other by constructive interference.

Depending on the length of the time interval $t'' - t'$, or on the locations of x' and x'' in the configuration space C, or on yet other factors, one of these (super)classical trajectories may dominate the functional integral. Or we may have a special interest in a particular trajectory x_c. Whatever the reason, we pick that trajectory and define the so-called action *function* corresponding to it:

$$S(x'', t'' \mid x', t') \stackrel{\text{def}}{=} S_{x't'}^{x''t''}[x_c]. \tag{6.6.3}$$

Because dynamical systems can be highly nontrivial, exhibiting simultaneously both coherent and chaotic behaviours, the analytic structure of the action function can be exceedingly complex, possessing branch points and other singularities of bewildering intricacy. We stress the analytic structure here because, as we have already seen in chapter 5 (sections 5.5 and 5.6), important stationary points of the action functional may correspond to trajectories lying outside the (super) classical configuration space, in its complex extension.

Because of the richness and variety of the analytic structures associated with dynamical systems, the standard theorems involving the action function constitute not so much a theory as a framework for a theory. The most familiar theorem states that the action function satisfies a pair of Hamilton–Jacobi equations:

Let the variables x''^i, t'' suffer infinitesimal variations for $\delta x''^i$, $\delta t''$

respectively. The trajectory x_c suffers a corresponding variation δx_c^i satisfying

$$0 = \int S_{,ij''}[x_c]\,\delta x_c^{j'''}\,dt''' = \left(-\frac{\partial}{\partial t}A_{ij}\frac{\partial}{\partial t} + \frac{1}{2}\left\{B_{ij}, \frac{\partial}{\partial t}\right\} - C_{ij}\right)\delta x_c^j \tag{6.6.4}$$

(see eqs. (6.1.4) and (6.6.2)) together with the boundary conditions

$$\delta x_c^i(t') = 0, \quad \delta x_c^i(t'') = \delta x''^i - \dot{x}_c^i(t'')\,\delta t''. \tag{6.6.5}$$

These variations in turn induce a change in the action function given by

$$\delta S(x'', t'' \,|\, x', t') = L_c''\,\delta t'' + \int_{x't'}^{x''t''} \left(\frac{\partial L_c}{\partial x_c^i}\delta x_c^i + \frac{\partial L_c}{\partial \dot{x}_c^i}\delta \dot{x}_c^i\right) dt$$
$$= L_c''\,\delta t'' + p_{ci}''(\delta x''^i - \dot{x}_c''^i\,\delta t'') \tag{6.6.6}$$

in which we have used eqs. (6.1.3) (or (6.6.2)) and (6.6.5) in carrying out the integration by parts leading to the last line. The abbreviations

$$L_c = L(x_c, \dot{x}_c), \qquad p_{ci} = \partial L(x_c, \dot{x}_c)/\partial \dot{x}_c^i, \tag{6.6.7}$$

$$L_c'' = L(x_c(t''), \dot{x}_c(t'')), \quad p_{ci}'' = p_{ci}(t''), \text{ etc.}, \tag{6.6.8}$$

are also used here.

Referring to expression (6.1.10) for the Hamiltonian one sees that eq. (6.6.6) may be rewritten in the form

$$\delta S(x'', t'' \,|\, x', t') = p_{ci}''\,\delta x''^i - H_c''\,\delta t'', \tag{6.6.9}$$

$$H_c = H(x_c, p_c), \text{ etc.}, \tag{6.6.10}$$

whence (suppressing the arguments of the action function)

$$p_{ci}'' = \partial S/\partial x^i, \tag{6.6.11}$$

$$\partial S/\partial t'' + H(x'', \partial S/\partial x'') = 0. \tag{6.6.12}$$

By making infinitesimal changes in the variables x'^i, t' one can show in a similar way that

$$p_{ci}' = -\partial S/\partial x'^i, \tag{6.6.13}$$

$$-\partial S/\partial t' + H(x', -\partial S/\partial x') = 0. \tag{6.6.14}$$

Equations (6.6.12) and (6.6.14) are the Hamilton–Jacobi equations. It is an interesting fact that, even for simple systems, these equations often cannot be solved by finite-difference methods on a computer. In such cases every differencing scheme leads to an unconditionally unstable iteration. This is a reflection of the complexity of the analytic structure of S and of the fact that the long-time behaviour of the corresponding dynamical system is extremely sensitive to small perturbations (in this case round-off and differencing errors).

The Van Vleck–Morette determinant

The (super)classical trajectory appearing in (6.6.3) is specified by endpoint data: x''^i, x'^i with the times t'' and t' held fixed. Another specification is in terms of initial (Cauchy) data: $x'^i, \dot{x}_c'^i$ or, what is the same thing, x'^i, p'_{ci} (for standard canonical systems). The Jacobian of the transformation from one set of data to the other is

$$\frac{\partial(p'_c, x')}{\partial(x'', x')} = \frac{\partial(p'_c)}{\partial(x'')} = D \overset{\text{def}}{=} \det(D_{ij}) \tag{6.6.15}$$

$$D_{ij} \overset{\text{def}}{=} \frac{\partial p'_{cj}}{\partial x''^i} = -\frac{\partial^2 S}{\partial x''^i \partial x'^j}. \tag{6.6.16}$$

D is known as the *Van Vleck–Morette determinant.*

It can happen that an infinitesimal change in the p'_{ci} leaves the x''^i unaffected. It is easy to see that D^{-1} must then vanish and that D itself diverges. The points x' and x'' are said to be *conjugate* to each other, and x'' typically lies on a *caustic* (envelope) of the family of superclassical trajectories emanating from x'.† This family of trajectories satisfies a conservation law in which the determinant D appears:

$$\frac{\partial D}{\partial t''} + \frac{\partial}{\partial x''^i}(D\dot{x}_c''^i) = 0, \quad \dot{x}_c''^i = \frac{\partial H_c''}{\partial p_{ci}''}. \tag{6.6.17}$$

A proof of this law is outlined in exercise **6.9.** It is evident that the trajectories emanating from x' crowd together with infinite density at points conjugate to x'.

The Van Vleck–Morette matrix (D_{ij}) satisfies some technically useful identities stemming from the additivity of the action (6.6.1) and its stationarity at the point x_c in the space Φ of trajectories. Additivity implies

$$S(x'', t'' \mid x', t') = S(x'', t'' \mid x_c, t) + S(x_c, t \mid x', t') \tag{6.6.18}$$

where x_c is to be understood as an abbreviation for $x_c(t)$ with t lying between t' and t'', and the two action functions on the right side of (6.6.18)

† Although other types of conjugate points can and do occur.

are evaluated for different portions of the same trajectory x_c. Stationarity implies

$$\partial S(x'', t'' \mid x_c, t)/\partial x_c^i + \partial S(x_c, t \mid x', t')/\partial x_c^i = 0, \tag{6.6.19}$$

whence (making the arguments of D_{ij} explicit)

$$D_{ij}(x'', t'' \mid x', t') = -\frac{\partial^2}{\partial x''^i \partial x'^j}[S(x'', t'' \mid x_c, t) + S(x_c, t \mid x', t')]$$

$$= \begin{cases} -\dfrac{\partial}{\partial x''^i}\left\{\left[\dfrac{\partial S(x'', t'' \mid x_c, t)}{\partial x_c^k} + \dfrac{\partial S(x_c, t \mid x', t')}{\partial x_c^k}\right]\dfrac{\partial x_c^k}{\partial x'^j} + \dfrac{\partial S(x_c, t \mid x', t')}{\partial x'^j}\right\} \\ \text{or} \\ -\dfrac{\partial}{\partial x'^j}\left\{\dfrac{\partial S(x'', t'' \mid x_c, t)}{\partial x''^i} + \left[\dfrac{\partial S(x'', t'' \mid x_c, t)}{\partial x_c^k} + \dfrac{\partial S(x_c, t \mid x', t')}{\partial x_c^k}\right]\dfrac{\partial x_c^k}{\partial x''^i}\right\}. \end{cases} \tag{6.6.20}$$

Depending on whether eq. (6.6.19) is used before or after the differentiations, this expression can be cast in various forms:

$$D_{ij}(x'', t'' \mid x', t') = \frac{\partial x_c^k}{\partial x''^i} D_{kj}(x_c, t \mid x', t') = D_{ik}(x'', t'' \mid x_c, t)\frac{\partial x_c^k}{\partial x'^j}$$

$$= D_{ik}(x'', t'' \mid x_c, t)\frac{\partial x_c^k}{\partial x'^j} - \frac{\partial x_c^k}{\partial x''^i}\frac{\partial x_c^l}{\partial x'^j} M_{kl} + \frac{\partial x_c^k}{\partial x''^i} D_{kj}(x_c, t \mid x', t')$$

$$= \frac{\partial x_c^k}{\partial x''^i}\frac{\partial x_c^l}{\partial x'^j} M_{kl}, \tag{6.6.21}$$

where

$$M_{kl} = \partial^2[S(x'', t'' \mid x_c, t) + S(x_c, t \mid x', t')]/\partial x_c^k \partial x_c^l. \tag{6.6.22}$$

In the generic case none of the factors appearing in the above terms is a singular matrix, and one may write also

$$\frac{\partial x_c^k}{\partial x''^i} = D_{il}(x'', t'' \mid x', t')\, D^{-1lk}(x_c, t \mid x', t') = D_{il}(x'', t'' \mid x_c, t)\, M^{-1lk}, \tag{6.6.23}$$

$$\frac{\partial x_c^k}{\partial x'^j} = D^{-1kl}(x'', t'' \mid x_c, t)\, D_{lk}(x'', t'' \mid x', t') = M^{-1kl} D_{lj}(x_c, t \mid x', t'), \tag{6.6.24}$$

where the indicated inverses are matrix inverses.

Jacobi fields and the Green's function for the trajectory x_c

The variation δx_c^i appearing in eq. (6.6.4) satisfies the homogeneous equation of infinitesimal disturbances and is called a *Jacobi field* along the trajectory x_c. Since this variation is produced by variations in the endpoint data it is evident that the derivatives $\partial x_c^k/\partial x''^i$ and $\partial x_c^k/\partial x'^j$ are themselves Jacobi fields:

$$\left.\begin{aligned} \left(-\frac{\partial}{\partial t}A_{kl}\frac{\partial}{\partial t}+\frac{1}{2}\left\{B_{kl},\frac{\partial}{\partial t}\right\}-C_{kl}\right)\frac{\partial x_c^l}{\partial x''^i}=0,\\ \left(-\frac{\partial}{\partial t}A_{kl}\frac{\partial}{\partial t}+\frac{1}{2}\left\{B_{kl},\frac{\partial}{\partial t}\right\}-C_{kl}\right)\frac{\partial x_c^l}{\partial x'^j}=0,\end{aligned}\right\} \tag{6.6.25}$$

the coefficients A_{kl}, B_{kl}, C_{kl} being those appropriate to the trajectory x_c. The Jacobi field $\partial x_c^k/\partial x''^i$ vanishes when $t = t'$ and equals $\delta^k{}_i$ when $t = t''$, whereas $\partial x_c^k/\partial x'^j$ vanishes when $t = t''$ and equals $\delta^k{}_j$ when $t = t'$.

Let δS be a variation in the action functional such that the support of $\delta S_{,i}$ lies between t' and t''. This variation will induce a change in the trajectory x_c, which satisfies the inhomogeneous equation of infinitesimal disturbances (cf. eq. (5.1.11)).

$$\int S_{,ij'''}[x_c]\,\delta x_c^{j'''}\,dt''' = -\delta S_{,i}. \tag{6.6.26}$$

If the endpoints of the trajectory are held fixed then the solution of this equation is

$$\delta x_c^i = \int G^{ij'''}\delta S_{,j'''}\,dt''' \tag{6.6.27}$$

where $G^{ij'''}$ is the Green's function appropriate to these boundary conditions:†

$$\begin{aligned} G^{ij'''} = &-\theta(t,t''')\frac{\partial x_c^i}{\partial x'^k}D^{-1kl}\frac{\partial x_c^{j'''}}{\partial x''^l}\\ &-\theta(t''',t)\frac{\partial x_c^{j'''}}{\partial x'^k}D^{-1kl}\frac{\partial x_c^i}{\partial x''^l}. \end{aligned} \tag{6.6.28}$$

The symbols x_c^i and $x_c^{j'''}$ are here abbreviations for $x_c^i(t)$ and $x_c^j(t''')$ respectively, and the arguments of D^{-1kl} are x'', t'', x', t'. It is easy to see that $G^{ij'''}$ vanishes when t is equal to t' or t'' with $t' \leqslant t''' \leqslant t''$, or when t'''

† An expression equivalent to (6.6.28) was first obtained by C. DeWitt-Morette (*Annals of Physics*, N.Y., **97**, 367 (1976)).

is equal to t' or t'' with $t' \leqslant t \leqslant t'''$. That it is indeed a Green's function (of the small-disturbance operator $S_{,ij''}$) is readily proved:

Taking note of eqs. (6.6.25) we have

$$\left(-\frac{\partial}{\partial t}A_{im}\frac{\partial}{\partial t}+\frac{1}{2}\left\{B_{im},\frac{\partial}{\partial t}\right\}-C_{im}\right)G^{mj'''}$$

$$=-\left(\frac{\partial}{\partial t}A_{im}+B_{im}\right)\delta(t,t''')\left(\frac{\partial x_c^m}{\partial x'^k}D^{-1kl}\frac{\partial x_c^j}{\partial x''^l}-\frac{\partial x_c^j}{\partial x'^k}D^{-1kl}\frac{\partial x_c^m}{\partial x''^l}\right)$$

$$+A_{im}\,\delta(t,t''')\left(\frac{\partial^2 x_c^m}{\partial t\,\partial x'^k}D^{-1kl}\frac{\partial x_c^j}{\partial x''^l}-\frac{\partial x_c^j}{\partial x'^k}D^{-1kl}\frac{\partial^2 x_c^m}{\partial t\,\partial x''^l}\right). \qquad (6.6.29)$$

Making use of eqs. (6.6.23) and (6.6.24) one easily verifies that the final factor in the first term on the right above is equal to $M^{-1mj}-M^{-1jm}$ and hence vanishes by the symmetry of M_{ij}. The final factor in the second term may be evaluated through use of the Hamilton equation

$$\partial x_c^m/\partial t = \dot{x}_c{}^m = \partial H_c/\partial p_{cm} \qquad (6.6.30)$$

with

$$p_{cm} = -\partial S(x'',t''\,|\,x_c,t)/\partial x_c^m = \partial S(x_c,t\,|\,x',t')/\partial x_c^m. \qquad (6.6.31)$$

Thus

$$\frac{\partial^2 x_c^m}{\partial t\,\partial x'^k}=\frac{\partial^2 H_c}{\partial p_{cm}\,\partial x_c^n}\frac{\partial x_c^n}{\partial x'^k}+\frac{\partial^2 H_c}{\partial p_{cm}\,\partial p_{cr}}\frac{\partial p_{cr}}{\partial x'^k}$$

$$=\left[\frac{\partial^2 H_c}{\partial p_{cm}\,\partial x_c^n}-\frac{\partial^2 H_c}{\partial p_{cm}\,\partial p_{cr}}\frac{\partial^2 S(x'',t''\,|\,x_c,t)}{\partial x_c^r\,\partial x_c^n}\right]\frac{\partial x_c^n}{\partial x'^k} \qquad (6.2.32)$$

and, similarly,

$$\frac{\partial^2 x_c^m}{\partial t\,\partial x''^l}=\left[\frac{\partial^2 H_c}{\partial p_{cm}\,\partial x_c^n}+\frac{\partial^2 H_c}{\partial p_{cm}\,\partial p_{cr}}\frac{\partial^2 S(x_c,t\,|\,x',t')}{\partial x_c^r\,\partial x_c^n}\right]\frac{\partial x_c^n}{\partial x''^l}. \qquad (6.6.33)$$

The final factor in (6.6.29) is therefore

$$\frac{\partial^2 H_c}{\partial p_{cm}\,\partial p_{cr}}\left[-\frac{\partial^2 S(x'',t''\,|\,x_c,t)}{\partial x_c^r\,\partial x_c^n}M^{nj}-M^{jn}\frac{\partial^2 S(x_c,t\,|\,x',t')}{\partial x_c^n\,\partial x_c^r}\right]$$

$$=-A^{-1mr}M_{rn}M^{nj}=-A^{-1mj} \qquad (6.6.34)$$

where we have used eq. (6.6.22) and

$$\frac{\partial^2 H_c}{\partial p_{cm}\,\partial p_{cr}}=\frac{\partial \dot{x}_c^m}{\partial p_{cr}}=A^{-1mr} \quad \text{(evaluated for trajectory } x_c\text{)}, \qquad (6.6.35)$$

which is the equation inverse to $\partial p_i/\partial \dot{x}^j = A_{ij}$ (see eqs. (6.1.5) and (6.1.9)). The right hand side of eq. (6.6.29) is seen to reduce to $-\delta_i^{\,j}\,\delta(t,t''')$, q.e.d.

We finally note that $G^{ij'''}$ may be expressed in the alternative forms

$$G^{ij'''} = -\theta(t, t''')\left(\frac{\partial x_c^i}{\partial x'^k} D^{-1kl} \frac{\partial x_c^{j'''}}{\partial x''^l} - \frac{\partial x_c^{j'''}}{\partial x'^k} D^{-1kl} \frac{\partial x_c^i}{\partial x''^l}\right) - \frac{\partial x_c^{j'''}}{\partial x'^k} D^{-1kl} \frac{\partial x_c^i}{\partial x''^l} \tag{6.6.36a}$$

$$= \theta(t''', t)\left(\frac{\partial x_c^i}{\partial x'^k} D^{-1kl} \frac{\partial x_c^{j'''}}{\partial x''^l} - \frac{\partial x_c^{j'''}}{\partial x'^k} D^{-1kl} \frac{\partial x_c^i}{\partial x''^l}\right) - \frac{\partial x_c^i}{\partial x'^k} D^{-1kl} \frac{\partial x_c^{j'''}}{\partial x''^l}, \tag{6.6.36b}$$

which allows one to infer the following forms for the retarded and advanced green's functions:

$$G^{-ij'''} = -\theta(t, t''')\left(\frac{\partial x_c^i}{\partial x'^k} D^{-1kl} \frac{\partial x_c^{j'''}}{\partial x''^l} - \frac{\partial x_c^{j'''}}{\partial x'^k} D^{-1kl} \frac{\partial x_c^i}{\partial x''^l}\right), \tag{6.6.37}$$

$$G^{+ij'''} = \theta(t''', t)\left(\frac{\partial x_c^i}{\partial x'^k} D^{-1kl} \frac{\partial x_c^{j'''}}{\partial x''^l} - \frac{\partial x_c^{j'''}}{\partial x'^k} D^{-1kl} \frac{\partial x_c^i}{\partial x''^l}\right). \tag{6.6.38}$$

Determinantal relations

Suppose the functional form of the action $S[x]$ suffers an infinitesimal change δS such that the support of $\delta S_{,i}$ lies between t' and t''. Suppose that the endpoint conditions that fix the trajectory remain unchanged. Then the operator $S_{,ij'''}[x_c]$ suffers the change

$$\delta(S_{,ij'''}[x_c]) = \delta S_{,ij'''}[x_c] + \int S_{,ij'''k''''}[x_c]\, \delta x_c^{k''''}\, \mathrm{d}t''''$$

$$= \delta S_{,ij'''}[x_c] + \int \mathrm{d}t'''' \int \mathrm{d}t'''''\, S_{,ij'''k''''}[x_c]\, G^{k''''l'''''}\, \delta S_{,l'''''}[x_c], \tag{6.6.39}$$

δx_c^i being given by eq. (6.6.27). This change induces a corresponding change in the formal determinants $\det G$ and $\det G^+$ given by

$$\delta \ln \det G = \int_{t'}^{t''} \mathrm{d}t \int_{t'}^{t''} \mathrm{d}t'''\, G^{j'''i}\, \delta(S_{,ij'''}[x_c]), \tag{6.6.40}$$

$$\delta \ln \det G^+ = \int_{t}^{t''} \mathrm{d}t \int_{t'}^{t''} \mathrm{d}t'''\, G^{+j'''i}\, \delta(S_{,ij'''}[x_c]), \tag{6.6.41}$$

where the limits of integration are determined by the range of the parameter t labelling the integration variables in the functional integral (where these results will be used). Subtracting eq. (6.6.41) from eq. (6.6.40) and making use of eqs. (6.6.36b) and (6.6.38), we find

$$\delta \ln \frac{\det G}{\det G^{+}} = -\int_{t'}^{t''} \mathrm{d}t \int_{t'}^{t''} \mathrm{d}t''' \frac{\partial x_c^{j'''}}{\partial x'^k} D^{-1kl} \frac{\partial x_c^{i}}{\partial x''^l} \delta(S_{,ij'''}[x_c]). \qquad (6.6.42)$$

To simplify this expression with the aid of eq. (6.6.39) we differentiate the Jacobi field equation

$$\int S_{,il'''}[x_c] \frac{\partial x_c^{l'''}}{\partial x''^j} \mathrm{d}t''' = 0 \qquad (6.6.43)$$

with respect to the endpoint variables x'^k, obtaining

$$\int S_{,il'''}[x_c] \frac{\partial^2 x_c^{l'''}}{\partial x''^j \partial x'^k} \mathrm{d}t''' + \int \mathrm{d}t''' \int \mathrm{d}t'''' S_{,il'''m''''}[x_c] \frac{\partial x_c^{l'''}}{\partial x''^j} \frac{\partial x_c^{m''''}}{\partial x'^k} = 0. \qquad (6.6.44)$$

If t lies between t' and t'' the integrations in the second term on the left may be restricted to this interval. In view of the boundary conditions on $\partial x_c^{l'''}/\partial x''^j$ it is easy to see that $\partial^2 x_c^{l'''}/\partial x''^j \partial x'^k$ must vanish when t''' equals t' or t'' and hence that eq. (6.6.44) is solved by

$$\frac{\partial^2 x_c^{i}}{\partial x''^j \partial x'^k} = \int_{t'}^{t''} \mathrm{d}t''' \int_{t'}^{t''} \mathrm{d}t'''' \int_{t'}^{t''} \mathrm{d}t''''' G^{il'''} S_{,l'''m''''n'''''}[x_c] \frac{\partial x_c^{m''''}}{\partial x''^j} \frac{\partial x_c^{n'''''}}{\partial x'^k}. \qquad (6.6.45)$$

Now substituting (6.6.39) into (6.6.42) and using (6.6.45), we find

$$\begin{aligned}
\delta \ln \frac{\det G}{\det G^{+}} &= -D^{-1kl}\left(\int_{t'}^{t''} \mathrm{d}t \int_{t'}^{t''} \mathrm{d}t''' \frac{\partial x_c^{i}}{\partial x''^l} \delta S_{,ij'''}[x_c] \frac{\partial x_c^{j'''}}{\partial x'^k} \right. \\
&\qquad \left. + \int_{t'}^{t''} \delta S_{,i}[x_c] \frac{\partial^2 x_c^{i}}{\partial x''^l \partial x'^k} \mathrm{d}t \right) \\
&= -D^{-1kl} \frac{\partial^2 \delta S[x_c]}{\partial x''^l \partial x'^k} = D^{-1kl} \delta D_{lk} \\
&= \delta \ln D, \qquad (6.6.46)
\end{aligned}$$

whence

$$\frac{\det G}{\det G^{+}} = \text{const} \times D. \qquad (6.6.47)$$

This remarkable result establishes an important relation between the finite Van Vleck–Morette determinant and the two formal infinite determinants that appear in the loop expansion of the functional integral (6.4.31) (see below).

The loop expansion

The *loop expansion* is a method of computing the functional integral (6.4.31) (or the more general integral (5.3.46)) that is fundamentally naive but often yields useful results. One begins by replacing the integration variables $x^i(t)$ by $\xi^i(t)$ where

$$x^i(t) = x^i_c(t) + \xi^i(t), \tag{6.6.48}$$

$x_c(t)$ being one of the (super)classical trajectories defined by the endpoint data x'', t'', x', t'. One then rewrites (6.4.31) in the form

$$K^\alpha(x'', t'' \mid x', t')$$

$$= \text{const} \times e^{i[\phi(x'')-\phi(x')]} \int \exp(iS^{x''t''}_{x't'}[x_c+\xi])(\det G^+[x_c+\xi])^{-\frac{1}{2}}\, d\xi$$

$$= \text{const} \times e^{i[\phi(x'')-\phi(x')]} \int \exp(iS^{x''t''}_{x',t'}[x_c+\xi] - \tfrac{1}{2}\ln\det G^+[x_c+\xi])\, d\xi, \tag{6.6.49}$$

expands the exponent in the final integrand as a functional Taylor expansion in ξ and integrates over the ξ as if they were vector components in the tangent space $T_{x_c}(\Phi)$, ignoring the topology of the space Φ of trajectories and hence of the configuration space C. Thus

$$K^\alpha(x'', t'' \mid x', t') = \text{const} \times e^{i[\phi(x'')-\phi(x')]+iS(x'',t''|x',t')}$$

$$\times \int \exp\Big(\frac{i}{2} S_{,ij}\xi^i\xi^j + \frac{i}{6} S_{,ijk}\xi^i\xi^j\xi^k + \frac{i}{24} S_{,ijkl}\xi^i\xi^j\xi^k\xi^l + \dots$$

$$- \tfrac{1}{2}\ln\det G^+ - \tfrac{1}{2} G^{+ij} S_{,ijk}\xi^k + \Big)\, d\xi, \tag{6.6.50}$$

in which use is made of eq. (6.6.3), the coefficients of the ξ's are evaluated at the (super)classical trajectory x_c, the term in $S_{,i}[x_c]\xi^i$ is dropped in view of (6.6.2), and the integrations implied by the dummy indices on the ξ's are understood to range from t' to t''. One then expands the exponential itself in the form

$$K^\alpha(x'', t''|x', t') = \text{const} \times (\det G^+)^{-\frac{1}{2}} e^{i[S(x'',t''|x',t')+\phi(x'')-\phi(x')]}$$

$$\times \int e^{(i/2)S_{,ab}\xi^a\xi^b} \Big(1 + \frac{i}{24} S_{,ijkl}\xi^i\xi^j\xi^k\xi^l - \tfrac{1}{72} S_{,ijk} S_{,lmn}\xi^i\xi^j\xi^k\xi^k\xi^l\xi^m$$

$$- \frac{i}{12} S_{,ijk} S_{,lmn} G^{+mn}\xi^i\xi^j\xi^k\xi^l - \tfrac{1}{8} G^{+ij} S_{,ijk} S_{,lmn} G^{+mn}\xi^k\xi^l + \dots\Big)\, d\xi \tag{6.6.51}$$

and carries out a series of Gaussian integrations. Note that only even powers of the ξ's need be retained in the expansion.

The reason the expansion is called the *loop* expansion is that each term can be represented graphically, with the functions $S_{,ijk}\ldots$ being represented by vertices and the Green's functions G^{+ij}, as well as those arising from the integrals

$$\int e^{(i/2)S_{,ab}\xi^a\xi^b}\,d\xi = \text{const}\times(\det G)^{\frac{1}{2}}, \tag{6.6.52a}$$

$$\int e^{(i/2)S_{,ab}\xi^a\xi^b}\,\xi^i\xi^j\,d\xi = (-i)\,\text{const}\times(\det G)^{\frac{1}{2}}G^{ij}, \tag{6.6.52b}$$

$$\int e^{(i/2)S_{,ab}\xi^a\xi^b}\,\xi^i\xi^j\xi^k\xi^l\,d\xi = (-i)^2\,\text{const}\times(\det G^{\frac{1}{2}})$$
$$\times(G^{ij}G^{kl}+G^{ik}G^{jl}+G^{il}G^{jk}),\text{etc.,} \tag{6.6.52c}$$

being represented by lines joining the vertices to one another. The Green's-function lines all join up into closed loops (because all the indices are dummies), and the order of approximation that one reaches in truncating the expansion at a given point is determined by the highest number of closed loops reached.

All graphs having the same number of closed loops are of the same approximation order in the expansion. This is easily proved by making Planck's constant explicit rather than choosing units in which $\hbar = 1$. The exponent e^{iS} is replaced by $e^{iS/\hbar}$, and then the action functional, the small disturbance operator $S_{,ij}$, and the vertex functions $S_{,ijk\ldots}$, each carry a factor $\hbar^{-1}$, while every Green's function carries a factor $\hbar$. The topological identity (6.4.12) implies that every graph with K closed loops is proportional to $\hbar^{K-1}$.

The WKB approximation

Making use of eqs. (6.6.47), (6.6.52a) and the normalization conditions (6.4.13)–(6.4.14) one easily obtains, from eq. (6.6.51),

$$K^{\alpha}(x'',t''\,|\,x',t') = (2\pi i)^{-m/2}D^{\frac{1}{2}}(x'',t''\,|\,x',t')\,e^{i[S(x'',t''|x',t')+\phi(x'')-\phi(x')]}$$
$$\times[1+(\text{2-loop and higher order terms})]. \tag{6.6.53}$$

With the two-loop and higher order terms omitted this is just the well-known WKB approximation to the transition amplitude. Note that it is essential to retain the factor $(\det G^{+})^{-\frac{1}{2}}$ in the functional integrand in order to get, via eq. (6.6.47), the Van Vleck–Morette determinant that stands in the WKB approximation. If one actually carries out the integrations involved in the two-loop and higher order terms of the expansion one finds that the factor $(\det G^+)^{-\frac{1}{2}}$ is also precisely what is needed to cancel terms in which the delta function with coincident arguments $(\delta(t',t'))$ appears, so that no formal divergences survive. The factor $(\det G^+)^{\frac{1}{2}}$ does *not* cancel the divergences that arise in quantum field theory, but even there it has a role to play, as is illustrated in exercise **6.11** at the end of the chapter.

We should remark that the determinantal factors, particularly when they are elevated into the exponent with logarithm symbols attached, are often referred to as *one-loop terms*. Thus the WKB approximation is also known as the *one-loop approximation*, a terminology common in quantum field theory.

For systems that obey linear dynamics (e.g., free particle, Bose oscillator, charged particle in a constant magnetic field) the WKB approximation to the transition amplitude is exact. In general, however, it is an asymptotic approximation, and the loop expansion is an asymptotic series. It is unfortunately not possible to give a general statement about the nature of the asymptotics; this depends on the details of the system in question.

We have remarked that when there are several (super)classical trajectories between the endpoints x', t' and x'', t'', one of them may dominate in the functional integral. This is usually the shortest trajectory and is the one chosen for the WKB approximation. When there are several trajectories the WKB approximation can often be improved by adding similar WKB contributions from the other trajectories, and this improvement is often better than that obtained by including higher order loop terms. The WKB contributions must be added with appropriate phase factors. When the configuration space is simply connected the $D^{\frac{1}{2}}(x'',t''\,|\,x',t')$ appearing in eq. (6.6.53) must be understood as standing for $i^\nu\,|D^{\frac{1}{2}}(x'',t''\,|\,x',t')|$ where ν is the *Morse index* of the trajectory, namely, the number of conjugate points along the trajectory from x' to x'', each counted with its *multiplicity*. The multiplicity of a conjugate point $x_c(t_0)$ is defined as the number of zero eigenvalues of the inverse Van Vleck–Morette matrix $D^{-1ij}(x_c(t_0),t_0\,|\,x',t')$.

When the configuration space is not simply connected then, in addition to taking the Morse index into account, one must also multiply the WKB approximations to the partial amplitudes $K^\alpha(x'',t''\,|\,x',t')$ by the corresponding phase factors $\chi(\alpha)$ (see eq. (6.4.33)) before adding them together to obtain the WKB approximation to the total amplitude $\langle x'',t''\,|\,x',t'\rangle_\chi$.

The WKB approximation often fails near conjugate points, but not always. In the case of a particle moving geodesically in a *group manifold* (carrying a group invariant metric) one arrives at an exact total amplitude by adding partial WKB amplitudes together according to the above rules.

We have not space here to go further into the fascinating subject of the WKB approximation, but we draw the reader's attention to exercise **6.12** which, by making use of some of the identities proved earlier in this section, provides an independent check on the correctness of the normalization factor $(2\pi i)^{-m/2}$ appearing in expression (6.6.53).

The heat kernel expansion

In topological studies of manifolds it has been found convenient to introduce an alternative to the loop expansion, based on the Schrödinger equation satisfied by the transition amplitude $\langle x'', t'' | x', t' \rangle_\chi$. The functional form of the Hamiltonian operator appearing in this equation is identical to that of the Hamiltonian operator appearing in the Schrödinger equation for the amplitude (6.4.55) in the universal covering space. In view of eqs. (6.4.57) and (6.4.58) it follows that the partial amplitudes $K^\alpha(x'', t'' | x', t')$ too satisfy the same Schrödinger equation:

$$i \frac{\partial}{\partial t''} K^\alpha(x'', t'' | x', t') = H^{\mathrm{Q}}_{x''} K^\alpha(x'', t'' | x', t'), \tag{6.6.54}$$

$$H^{\mathrm{Q}}_{x''} = H^{\mathrm{Q}}(x'', -i\partial/\partial x'' - \omega(x'')) \tag{6.6.55}$$

(cf. eqs. (6.2.92), (6.5.5) and (6.5.6)). The explicit form of H^{Q} is given by eq. (6.5.25), which also specifies the order of factors.

It is not difficult to see that $H^{\mathrm{Q}}_{x''}$ includes a covariant Laplacian operator, and if the variable t'' is rotated through 90° in the complex plane eq. (6.6.54) looks like a diffusion equation. Historically, one of the earliest quantities to be studied as a 'diffusing substance' was heat, and therefore the diffusion equation is often called the *heat equation* and $K^\alpha(x'', t'' | x', t')$ is known as a *heat kernel*.

The *heat kernel expansion*, like the loop expansion, is asymptotic. It differs from the latter, however, in that it takes as its lowest approximation (starting term) not the WKB approximation, but what the WKB approximation would become if the 1-form potential $\boldsymbol{a}$ and the scalar potential v were to vanish. This has the consequence that the expansion becomes a power series in the time interval $t'' - t'$ rather than in $\hbar$:

$$K^{\alpha}(x'',t''\,|\,x',t') = [2\pi\mathrm{i}(t''-t')]^{-m/2}\hat{D}^{\frac{1}{2}}(x''\,|\,x')\,\mathrm{e}^{\mathrm{i}[\sigma(x'',x')/(t''-t')+\phi(x'')-\phi(x')]}$$

$$\times\sum_{n=0}^{\infty}\mathrm{i}^n\mathfrak{a}_n(x'',x')(t''-t')^n, \qquad (6.6.56)$$

$$\mathfrak{a}_0(x',x') = 1. \qquad (6.6.57)$$

Here $\sigma(x'',x')$ is the world function (see eq. (6.4.13)) corresponding to a geodesic between x' and x'' that lies in the equivalence class of trajectories $\Phi^{x''t''}_{x't'}(\alpha)$ (see eq. (6.4.30)). $\hat{D}$ is $(t''-t')^m$ times what the Van Vleck–Morette determinant for that equivalence class would be if $\boldsymbol{a}=0$, $v=0$. Using the fact that the action function for the system with $\boldsymbol{a}=0$, $v=0$ is simply $\sigma(x'',x')/(t''-t')$, one easily sees that

$$\hat{D} = \det(\hat{D}_{ij}), \quad \hat{D}_{ij} = -\partial^2\sigma/\partial x''^i\partial x'^j, \qquad (6.6.58)$$

in which arguments have been suppressed.

The world function σ satisfies an important identity that is an immediate corollary of the Hamilton–Jacobi equations (6.6.12) and (6.6.14) for the system with $\boldsymbol{a}=0$, $v=0$:

$$\sigma = \tfrac{1}{2}\sigma_{;i''}\sigma_{;}^{\ i''} = \tfrac{1}{2}\sigma_{;i'}\sigma_{;}^{\ i'}. \qquad (6.6.59)$$

(Here all derivatives are expressed as covariant derivatives, and primes on indices indicate which of the suppressed arguments is involved.) Similarly, eqs. (6.6.17) yield, for this system, the following important corollaries:

$$\dot{x}_c''^i = \sigma_{;}^{\ i''}/(t''-t'), \qquad (6.6.60)$$

$$\hat{D}^{-1}(\hat{D}\sigma_{;}^{\ i''})_{;i''} = m. \qquad (6.6.61)$$

Note that the covariant divergence in the last equation is just an ordinary divergence because $\hat{D}$ is a biscalar density.

Inserting the expansion (6.6.56) into eq. (6.6.54) and making use of (6.5.25), (6.6.59) and (6.6.61), one obtains the following recursion relations for the coefficients $\mathfrak{a}_n$:

$$\sigma_{;}^{\ i''}(\mathfrak{a}_{0;i''} - ia_{i''}\mathfrak{a}_0) = 0, \qquad (6.6.62)$$

$$\sigma_{;}^{\ i''}(\mathfrak{a}_{n;i''} - ia_{i''}\mathfrak{a}_n) + n\mathfrak{a}_n = \hat{D}^{-\frac{1}{2}}[\tfrac{1}{2}(\hat{D}^{\frac{1}{2}}\mathfrak{a}_{n-1})_{;i''}{}^{i''} - ia^{i''}(\hat{D}^{\frac{1}{2}}\mathfrak{a}_{n-1})_{;i''}]$$

$$-(\tfrac{1}{2}ia^{i''}{}_{;i''} + \tfrac{1}{2}a_{i''}a^{i''} + v'' + \tfrac{1}{8}R'')\,\mathfrak{a}_{n-1}, \quad n = 1,2,3,\ldots \qquad (6.6.63)$$

It will be noted that the $\mathfrak{a}$'s are independent of the 1-form ω and the character χ. This follows from the fact that $\omega = \mathbf{d}\phi$ (at points where ϕ is not discontinuous, cf. eq. (6.4.50)) and hence that the $\mathfrak{a}$'s are unchanged if the ϕ's in eq. (6.6.56) and the ω's in eq. (6.6.55) are set equal to zero.

Equations (6.6.62) and (6.6.63) can in principle be solved successively by integrating along each geodesic emanating from x'. This procedure fails, however, as soon as one arrives at a point geodesically conjugate to x', where $\hat{D}$ diverges. This means that as an approximation to the partial amplitude K^α, the heat kernel expansion is generally poorer than the loop expansion. In principle the two expansions become identical when $\boldsymbol{a} = 0$, $v = 0$, and hence in this case there must exist a way of extending the $\mathfrak{a}$-coefficients (by analytic continuation) beyond the conjugate points. In practice the utility of the heat kernel expansion is limited to the case in which $t'' - t'$ is small, x'' is close to x', and the geodesic to which $\sigma(x'', x')$ refers is the shortest geodesic between x' and x''.

In a number of applications, notably in index theory in topology and in renormalization theory for quantum fields in curved spacetime, it is precisely this case which is of greatest interest. In these applications, in fact, one is interested in the coincidence limit $x'' \to x'$. There is a straightforward procedure for computing the $\mathfrak{a}$-coefficients in this limit. One starts from the basic limits

$$\left.\begin{aligned} [\sigma_{;i'}] &= 0, \quad [\sigma_{;i'j'}] = [\sigma_{;i'j'}] = g_{i'j'}, \\ [\hat{D}_{i'j'}] &= g_{i'j'}, \quad [\hat{D}] = g', \end{aligned}\right\} \tag{6.6.64}$$

where the brackets indicate that x'' is to be set equal to x'. By taking repeated covariant derivatives of eq. (6.6.59), commuting indices, making use of the laws for interchanging the order of covariant derivatives, and then passing to the limit, one obtains an infinite sequence of further limits:

$$[\sigma_{;i'j'k'}] = 0, \quad [\sigma_{;i'j'k'l'}] = -\tfrac{1}{3}(R_{i'k'j'l'} + R_{i'l'j'k'}), \quad \ldots. \tag{6.6.65}$$

By performing similar operations on eqs. (6.6.61), (6.6.62) and (6.6.63) respectively, one obtains also

$$[\hat{D}^{\frac{1}{2}}{}_{;i'}] = 0, \quad [\hat{D}^{\frac{1}{2}}{}_{;i'j'}] = \tfrac{1}{6} g'^{\frac{1}{2}} R_{i'j'}, \quad \ldots, \tag{6.6.66}$$

$$[\mathfrak{a}_{0;i'}] = \mathrm{i} a_{i'}, \quad [\mathfrak{a}_{0;i'j'}] = \tfrac{1}{2}\mathrm{i}(a_{i';j'} + a_{j';i'}) + a^{i'} a_{i'}, \quad \ldots, \tag{6.6.67}$$

$$[\mathfrak{a}_1] = -(v' + \tfrac{1}{24} R'), \quad \ldots. \tag{6.6.68}$$

Coincidence limits of $\mathfrak{a}_2$ and $\mathfrak{a}_3$ have been computed by various authors, but $[\mathfrak{a}_1]$ will suffice for our needs here.

Role of the two-loop term in the independent verification of (6.5.25)

When $\boldsymbol{a} = 0$, $v = 0$ the two-loop term in the expansion (6.6.53) should be precisely equal to $\mathrm{i}\mathfrak{a}_1(x'', x')(t'' - t')$. In order to verify this, and hence to have *an independent check that expression (6.5.25) is indeed the Hamiltonian operator that the Feynman functional integral defines*, one must compute the two-loop term directly. The easiest way to do this is to start from the functional integral in the Hamiltonian form (6.5.4) and to make use of the expression (6.6.51) with the understanding that the ξ's include components representing the departure of the p's as well as the x's from the (super)classical trajectory, and that the coefficients $S_{,ijk...}$ include functional derivatives with respect to the p's as well as the x's. This allows one to avoid dealing with the advanced Green's functions and with the formal terms involving $\delta(t', t')$ that appear (but finally cancel) when (6.4.31) is used. It should be stressed that the final result is the same with either starting point,† but the computation is considerably longer when (6.4.31) is used.

New variables

It is convenient to choose a set of variables for the functional integration that makes the computation manifestly covariant. Instead of the variables $x^i(t)$, which refer to arbitrary charts, we shall use

$$\xi_a(t) \overset{\text{def}}{=} -\sigma_{;a}(x_c(t), x(t)), \qquad (6.6.69)$$

which refer to tangent spaces along the (super)classical trajectory x_c between x', t' and x'', t'', and which, for each t, relate the arbitrary point $x(t)$ to the point $x_c(t)$. For clarity, indices from the first part of the alphabet will here be used to refer to coordinates of points on x_c, whereas indices from the middle of the alphabet will refer to coordinates of the arbitrary point $x(t)$. Note that the Jacobian matrix of the transformation from the $x^i(t)$ to the $\xi_a(t)$ is given by

$$\frac{\partial \xi_a(t)}{\partial x^i(t)} = \hat{D}_{ai}(x_c(t), x(t)), \quad \frac{\partial x^i(t)}{\partial \xi_a(t)} = \hat{D}^{-1ia}(x_c(t), x(t)). \qquad (6.6.70)$$

The coordinate transformation (6.6.70) depends on the time t through $x_c(t)$, and this has the consequence that the time derivatives of the ξ_a become

$$\dot{\xi}_a = \hat{D}_{ai}\dot{x}^i - \sigma_{;ab}\dot{x}^b_c \quad \text{(arguments suppressed)}. \qquad (6.6.71)$$

(These are actually covariant time derivatives, but one may easily verify that the Lagrangian and Hamiltonian formalisms, in which they appear in

† This has been verified directly by several different methods (unpublished).

what follows, carry through as usual.) By solving eqs. (6.6.71) for the $\dot{x}^i$ one finds that the Lagrangian (with $\boldsymbol{a} = 0$, $v = 0$) now takes the form

$$L = \tfrac{1}{2} g_{ij} \dot{x}^i \dot{x}^j = \tfrac{1}{2} g_{ij} \hat{D}^{-1ia} \hat{D}^{-1jb} (\dot{\xi}_a + \sigma_{;ac} \dot{x}^c_c)(\dot{\xi}_b + \sigma_{;bd} \dot{x}^d_c). \quad (6.6.72)$$

Because of the t-dependence of the transformation (6.6.70), this form for the Lagrangian, regarded as a function of the ξ_a and $\dot{\xi}_a$, has also an explicit dependence on t. The corresponding Hamiltonian therefore differs from the usual $\tfrac{1}{2} g^{ij} p_i p_j$ by a term of the form $\partial s/\partial t$ where '$\partial/\partial t$' refers to the explicit t-dependence of s. The new Hamiltonian is

$$\tilde{H} = \pi^a \dot{\xi}_a - L = \tfrac{1}{2} g^{ij} \hat{D}_{ai} \hat{D}_{bj} \pi^a \pi^b - \sigma_{;ab} \pi^a \dot{x}^b_c, \quad (6.6.73)$$

the π^a being the momentum variables conjugate to the ξ_a:

$$\pi^a = \frac{\partial L}{\partial \dot{\xi}_a} = g_{ij} \hat{D}^{-1ia} \hat{D}^{-1jb} (\dot{\xi}_b + \sigma_{;bd} \dot{x}^d_c)$$

$$= g_{ij} \hat{D}^{-1ia} \dot{x}^j = \hat{D}^{-1ia} p_i. \quad (6.6.74)$$

The function s is evidently $-\sigma_{;a} \pi^a$.

The action, now regarded as a functional of the functions $\xi_a(t)$ and $\pi^a(t)$, is simply

$$S[\xi, \pi] = \int_{t'}^{t''} (\pi^a \dot{\xi}_a - \tilde{H}) \, dt, \quad (6.6.75)$$

with the boundary conditions

$$\xi_a(t') = \xi_a(t'') = 0. \quad (6.6.76)$$

Note that there are no boundary constraints on the $\pi^a(t)$ and that the boundary data x', t', x'', t'', which appear on the left in eq. (6.6.1), are now subsumed in the choice of the trajectory x_c. Making use of eqs. (6.6.70) and (6.6.73), as well as the technique of covariant variation introduced in section 6.3, one easily finds that the action (6.6.75) yields the dynamical equations in the form

$$0 = \frac{\delta S}{\delta \xi_a} = -\dot{\pi}^a - (g^{ij} \hat{D}_{bi} \hat{D}_{cj;k} \pi^b \pi^c - \sigma_{;bck} \pi^b \dot{x}^c_c) \hat{D}^{-1ka}, \quad (6.6.77)$$

$$0 = \frac{\delta S}{\delta \pi^a} = \dot{\xi}_a - g^{ij} \hat{D}_{ai} \hat{D}_{bj} \pi^b + \sigma_{;ab} \dot{x}^b_c, \quad (6.6.78)$$

the dots denoting covariant time derivatives.

Computation of the two-loop term

To verify the correctness of the Hamiltonian operator (6.6.25) it suffices to examine the partial amplitude K^α in the limit when $t''-t'$ becomes infinitesimally small, and hence to retain only the terms in $\mathfrak{a}_0$ and $\mathfrak{a}_1$ in expression (6.6.56). Because the prefactor in this expression tends toward a δ-function in this limit it also suffices to examine $\mathfrak{a}_1$, and hence the two-loop term itself, in the coincidence limit $x''\to x'$.

In the coincidence limit we have $x_c^a(t'')=x_c^a(t')$ and $\dot{x}_c^a=0$. The (super)classical trajectory becomes $\xi_a=0$, $\pi^a=0$, and the second functional derivatives of the action, evaluated at this trajectory, reduce to the simple form

$$\left.\begin{aligned}
\left[\frac{\delta^2 S}{\delta\xi_a\,\delta\xi_{b'''}}\right]&=0,\\
\left[\frac{\delta^2 S}{\delta\xi_a\,\delta\pi^{b'''}}\right]&=-\frac{\partial}{\partial t}\delta^a{}_{b'''},\\
\left[\frac{\delta^2 S}{\delta\pi^a\delta\xi_{b'''}}\right]&=\frac{\partial}{\partial t}\delta_a{}^{b'''},\\
\left[\frac{\delta^2 S}{\delta\pi^a\delta\pi^{b'''}}\right]&=-[g^{ij}\hat{D}_{ai}\,\hat{D}_{bj}]\,\delta(t,t''')=-g_{ab}\,\delta(t,t''').
\end{aligned}\right\}\qquad(6.6.79)$$

Moreover, the 3-pronged vertex functions all vanish, while the only nonvanishing 4-pronged vertex function is

$$\left[\frac{\delta^4 S}{\delta\xi_a\,\delta\xi_{b'''}\,\delta\pi^{c''''}\,\delta\pi^{d'''''}}\right]=-[g^{ij}\{(\hat{D}_{ci}\,\hat{D}_{dj})_{;k}\,\hat{D}^{-1ka}\}_{;l}\,\hat{D}^{-1lb}]$$
$$\times\,\delta(t,t''')\,\delta(t,t'''')\,\delta(t,t''''').\qquad(6.6.80)$$

The coincidence limits in (6.6.80) can be computed by the same methods as those used in obtaining the limits (6.6.65). One finds

$$[\hat{D}_{ai}]=g_{ai},\quad [\hat{D}_{ai;j}]=0,\quad [\hat{D}_{ai;jk}]=-\tfrac{1}{3}(R_{aikj}+R_{ajki}),\qquad(6.6.81)$$

whence

$$\left[\frac{\delta^4 S}{\delta\xi_a\,\delta\xi_{b'''}\,\delta\pi^{c''''}\,\delta\pi^{d'''''}}\right]=\tfrac{1}{3}(R_{cd}{}^{ba}+R_c{}^{ab}{}_d+R_{dc}{}^{ba}+R_d{}^{ab}{}_c)\,\delta(t,t''')\,\delta(t,t'''')\,\delta(t,t''''')$$
$$=-\tfrac{1}{3}(R^a{}_c{}^b{}_d+R^a{}_d{}^b{}_c)\,\delta(t,t''')\,\delta(t,t'''')\,\delta(t,t''''').\qquad(6.6.82)$$

The only term in expression (6.6.51) that will contribute, in the present case, to the coincidence limit of the two-loop portion of the functional integral is the term in $(i/24)S_{,ijkl}\xi^i\xi^j\xi^k\xi^l$. Making use of eq. (6.6.52c) and noting that there are six distinct permutations of the symbols ξ, ξ, π, π, we get for this contribution

$$-\frac{i}{4}\int_{t'}^{t''} dt \int_{t'}^{t''} dt''' \int_{t'}^{t''} dt'''' \int_{t'}^{t''} dt''''' \left[\frac{\delta^4 S}{\delta\xi_a \delta\xi_{b'''} \delta\pi^{c''''} \delta\pi^{d'''''}}\right]$$
$$\times (G_{ab'''}G^{c''''d'''''} + G_a{}^{c''''}G_{b'''}{}^{d'''''} + G_a{}^{d'''''}G_{b'''}{}^{c''''}), \quad (6.6.83)$$

where the G's are Green's functions, appropriate to the boundary conditions (6.6.76), satisfying

$$\begin{pmatrix} 0 & -\delta^a{}_c\partial/\partial t \\ \delta_a{}^c\partial/\partial t & -g_{ac} \end{pmatrix}\begin{pmatrix} G_{cb'''} & G_c{}^{b'''} \\ G^c{}_{b'''} & G^{cb'''} \end{pmatrix} = -\begin{pmatrix} \delta^a{}_b & 0 \\ 0 & \delta_a{}^b \end{pmatrix}\delta(t, t'''). \quad (6.6.84)$$

It is easy to verify that the following are solutions of (6.6.84):

$$\left.\begin{aligned} G_{ab'''} &= -\frac{g_{ab}}{t''-t'}[\theta(t,t''')(t''-t)(t'''-t') + \theta(t''',t)(t''-t''')(t-t')], \\ G_a{}^{b'''} &= -\frac{\delta_a{}^b}{t''-t'}[\theta(t,t''')(t''-t) - \theta(t''',t)(t-t')] = G^{b'''}{}_a, \\ G^a{}_{b'''} &= \frac{\delta^a{}_b}{t''-t'}[\theta(t,t''')(t'''-t') - \theta(t''',t)(t''-t''')] = G_{b'''}{}^a, \\ G^{ab'''} &= \frac{g^{ab}}{t''-t'}. \end{aligned}\right\} \quad (6.6.85)$$

Since $G_{ab'''}$ and $G_a{}^{b'''}$ vanish when t is equal to t' or t'', provided t''' lies in the interval (t', t''), and since $G_{ab'''}$ and $G^a{}_{b'''}$ vanish when t''' is equal to t' or t'', provided t lies in the interval (t', t''), these are in fact the solutions appropriate to the conditions (6.6.76).

Inserting expressions (6.6.82) and (6.6.85) into expression (6.6.83), and making use of (6.4.4), we finally get

$$\frac{i}{12}\int_{t'}^{t''}(R^a{}_c{}^b{}_d + R^a{}_d{}^b{}_c)(G_{ab}G^{cd} + G_a{}^cG_b{}^d + G_a{}^dG_b{}^c)\,dt$$
$$= \frac{i}{12}(t''-t')^{-2}\int_{t'}^{t''}(R^a{}_c{}^b{}_d + R^a{}_d{}^b{}_c)[-g_{ab}g^{cd}(t''-t)(t-t')$$
$$+\tfrac{1}{4}(\delta_a{}^c\delta_b{}^d + \delta_a{}^d\delta_b{}^c)(t''+t'-2t)^2]\,dt$$
$$= -\frac{i}{24}\int_t^{t''} R\,dt = -\frac{i}{24}R'(t''-t'), \quad (6.6.86)$$

in which we have used the fact that R is constant and equal to R' in the coincidence limit. Referring to eq. (6.6.68) we see that when $v = 0$ this is just $\mathrm{i}[\mathfrak{a}_1](t'' - t')$, q.e.d.

6.7 Supersymmetry and the Euler–Poincaré characteristic

Inclusion of a-type dynamical variables

In the preceding sections we have seen how a knowledge of the topology of the configuration space C of a classical system is necessary for the proper definition of the quantum theory of the system or, equivalently, for the proper construction of the Feynman functional integral that defines the quantum theory. In particular we have seen the importance of the fundamental group, the first homology group and the first Betti number b_1. In this section we shall see how a set of a-type dynamical variables can be added to the system of section 6.3 in such a way that the fully superclassical augmented system possesses a supersymmetry group, and how the quantum theory of this system in turn gives information on the de Rham cohomology of C, and hence on all the other Betti numbers. Moreover, the Feynman version of the quantum theory will lead naturally to a derivation, via a functional integral with respect to the a-number variables, of the Chern–Gauss–Bonnet formula for the Euler–Poincaré characteristic of C.

The extended system is obtained by setting $\boldsymbol{a}$ and v equal to zero in (6.3.1), adding $2m$ new variables y^i_α ($i = 1, \ldots, m$; $\alpha = 1, 2$), and taking the action functional in the form

$$S = \int (\tfrac{1}{2} g_{ij} \dot{x}^i \dot{x}^j + \tfrac{1}{2}\mathrm{i} g_{ij} y^i_\alpha \dot{y}^j_\alpha + \tfrac{1}{8} R_{ijkl} y^i_\alpha y^j_\alpha y^k_\beta y^l_\beta)\, \mathrm{d}t. \qquad (6.7.1)$$

For each point x of C, and for each value of the index α, the y^i_α are to be understood as the a-number-valued components of a real a-type contravariant vector at x. The configuration space of this augmented system may be viewed as an $(m, 2m)$-dimensional supermanifold, which will be denoted by $\hat{C}$.

Remembering that the dot on $\dot{y}^i_\alpha$ denotes the covariant time derivative (see eq. (6.3.12)), and taking note of eq. (6.3.19), one easily obtains the variational formula

$$\delta S = \int [\delta x^i(-g_{ij} \ddot{x}^j + \tfrac{1}{2}\mathrm{i} R_{ijkl} \dot{x}^j y^k_\alpha y^l_\alpha + \tfrac{1}{8} R_{jklm;i} y^j_\alpha y^k_\alpha y^l_\beta y^m_\beta)$$

$$+ \bar{\delta} y^i_\alpha(\mathrm{i} g_{ij} \dot{y}^j_\alpha + \tfrac{1}{2} R_{ijkl} y^j_\alpha y^k_\beta y^l_\beta)]\, \mathrm{d}t, \qquad (6.7.2)$$

which yields the dynamical equations

$$\ddot{x}^i = \tfrac{1}{2}\mathrm{i}R^i{}_{jkl}\,\dot{x}^j y^k_\alpha y^l_\alpha + \tfrac{1}{8}R_{jklm;}{}^i y^j_\alpha y^k_\alpha y^l_\beta y^m_\beta, \qquad (6.7.3)$$

$$\dot{y}^i_\alpha = \tfrac{1}{2}\mathrm{i}R^i{}_{jkl}\,y^j_\alpha y^k_\beta y^l_\beta. \qquad (6.7.4)$$

Green's functions and Peierls brackets

To obtain the equations for the Green's functions that define the Peierls brackets of the theory, one needs the ordinary functional derivatives of the action. Remembering that the coefficient $\bar{\delta}y^i_\alpha$ in the integrand of (6.7.2) depends on δx^i one easily computes

$$\frac{\vec{\delta}}{\delta x^i}S = -g_{ij}\ddot{x}^j + \tfrac{1}{2}\mathrm{i}R_{ijkl}\,\dot{x}^j y^k_\alpha y^l_\alpha + \tfrac{1}{8}R_{jklm;i}\,y^j_\alpha y^k_\alpha y^l_\beta y^m_\beta$$

$$+\Gamma^j{}_{ki}\,y^k_\alpha(\mathrm{i}g_{jl}\,\dot{y}^l_\alpha + \tfrac{1}{2}R_{jlmn}\,y^l_\alpha y^m_\beta y^n_\beta), \qquad (6.7.5)$$

$$\frac{\vec{\delta}}{\delta y^i_\alpha}S = \mathrm{i}g_{ij}\,\dot{y}^j_\alpha + \tfrac{1}{2}R_{ijkl}\,y^j_\alpha y^k_\beta y^l_\beta, \qquad (6.7.6)$$

and hence

$$\frac{\vec{\delta}}{\delta x^i}S\frac{\overleftarrow{\delta}}{\delta x^{j'}} = \left(-g_{ij}\frac{\partial^2}{\partial t^2} - 2\Gamma_{ijk}\,\dot{x}^k\frac{\partial}{\partial t} + \tfrac{1}{2}\mathrm{i}R_{ijkl}\,y^k_\alpha y^l_\alpha\frac{\partial}{\partial t}\right.$$

$$\left. + \mathrm{i}\Gamma^k{}_{il}\,y^l_\alpha\Gamma_{kjm}\,y^m_\alpha\frac{\partial}{\partial t} + \ldots\right)\delta(t, t'), \qquad (6.7.7)$$

$$\frac{\vec{\delta}}{\delta x^i}S\frac{\overleftarrow{\delta}}{\delta y^{j'}_{\beta'}} = \left(\mathrm{i}\Gamma_{jik}\,y^k_\beta\frac{\partial}{\partial t} + \ldots\right)\delta(t, t'), \qquad (6.7.8)$$

$$\frac{\vec{\delta}}{\delta y^i_\alpha}S\frac{\overleftarrow{\delta}}{\delta x^{j'}} = \left(\mathrm{i}\Gamma_{ijk}\,y^k_\alpha\frac{\partial}{\partial t} + \ldots\right)\delta(t, t'), \qquad (6.7.9)$$

$$\frac{\vec{\delta}}{\delta y^i_\alpha}S\frac{\overleftarrow{\delta}}{\delta y^{j'}_{\beta'}} = \left(\mathrm{i}g_{ij}\,\delta_{\alpha\beta}\frac{\partial}{\partial t} + \ldots\right)\delta(t, t'), \qquad (6.7.10)$$

where '...' denotes terms that do not contain the operator $\partial/\partial t$.

It is straightforward to check that the series solutions of the Green's-function equation

$$\int\begin{pmatrix} \frac{\vec{\delta}}{\delta x^i}S\frac{\overleftarrow{\delta}}{\delta x^{k''}} & \frac{\vec{\delta}}{\delta x^i}S\frac{\overleftarrow{\delta}}{\delta y^{k''}_{\gamma''}} \\ \frac{\vec{\delta}}{\delta y^i_\alpha}S\frac{\overleftarrow{\delta}}{\delta x^{k''}} & \frac{\vec{\delta}}{\delta y^i_\alpha}S\frac{\overleftarrow{\delta}}{\delta y^{k''}_{\gamma''}} \end{pmatrix}\begin{pmatrix} G^{\pm k''j'} & G^{\pm k''j'}{}_{\beta'} \\ G^{\pm k''j'}{}_{\gamma''} & G^{\pm k''j'}{}_{\gamma''\beta'} \end{pmatrix}\mathrm{d}t'' = \begin{pmatrix} \delta_i^{\,j} & 0 \\ 0 & \delta_i^{\,j}\delta^\alpha{}_\beta \end{pmatrix}\delta(t, t')$$

(6.7.11

are given by

$$G^{-ij'} = G^{+j'i} = \theta(t,t')[(t-t')\,g^{i'j'} + \tfrac{1}{2}(t-t')^2 g^{i'k'}(B_{k'l'} - \mathrm{d}g_{k'l'}/\mathrm{d}t')\,g^{l'j'} + O(t-t')^3], \tag{6.7.12}$$

$$G^{-ij'}{}_{\beta'} = G^{+j'i}{}_{\beta'} = -\theta(t,t')\,[(t-t')\,g^{i'k'}\Gamma^{j'}{}_{k'l'}\,y^{l'}_{\beta'} + O(t-t')^2], \tag{6.7.13}$$

$$G^{-ij'}{}_{\alpha} = G^{+j'i}{}_{\alpha} = -\theta(t,t')\,[(t-t')\,g^{j'k'}\Gamma^{i'}{}_{k'l'}\,y^{l'}_{\alpha'} + O(t-t')^2], \tag{6.7.14}$$

$$G^{-ij'}_{\alpha\beta'} = -G^{+j'i}_{\beta'\alpha} = \mathrm{i}\theta(t,t')\,[g^{i'j'}\delta_{\alpha'\beta'} + O(t-t')], \tag{6.7.15}$$

where

$$B_{ij} = (g_{jk,i} - g_{ik,j})\,\dot{x}^k + \tfrac{1}{2}\mathrm{i}R_{ijkl}\,y^k_\alpha\,y^l_\alpha. \tag{6.7.16}$$

The series expansion of the supercommutator function is therefore

$$\tilde{G}^{ij'} = -(t-t')\,g^{i'j'} - \tfrac{1}{2}(t-t')^2 g^{i'k'}(B_{k'l'} - \mathrm{d}g_{k'l'}/\mathrm{d}t')\,g^{l'j'} + O(t-t')^3, \tag{6.7.17}$$

$$\tilde{G}^{ij'}_{\beta'} = (t-t')\,g^{i'k'}\Gamma^{j'}{}_{k'l'}\,y^{l'}_{\beta'} + O(t-t')^2, \tag{6.7.18}$$

$$\tilde{G}^{ij'}_{\alpha} = (t-t')\,g^{j'k'}\Gamma^{i'}{}_{k'l'}\,y^{l'}_{\alpha'} + O(t-t')^2, \tag{6.7.19}$$

$$\tilde{G}^{ij'}_{\alpha\beta'} = -\mathrm{i}g^{i'j'}\delta_{\alpha'\beta'} + O(t-t'), \tag{6.7.20}$$

from which the following equal-time Peierls brackets may be immediately obtained:

$$(x^i, x^j) = 0, \quad (x^i, \dot{x}^j) = g^{ij}, \quad (\dot{x}^i, \dot{x}^j) = g^{ik}B_{kl}\,g^{lj}, \tag{6.7.21}$$

$$(x^i, y^j_\alpha) = 0, \quad (\dot{x}^i, y^j_\alpha) = g^{ik}\Gamma^j{}_{kl}\,y^l_\alpha, \tag{6.7.22}$$

$$(y^i_\alpha, y^j_\beta) = -\mathrm{i}g^{ij}\delta_{\alpha\beta}. \tag{6.7.23}$$

One also finds

$$\begin{aligned}(x^i, \dot{y}^j_\alpha) &= (x^i, \mathrm{d}y^j_\alpha/\mathrm{d}t + \Gamma^j{}_{kl}\,\dot{x}^k y^l_\alpha) \\ &= -g^{ik}\Gamma^j{}_{kl}\,y^l_\alpha + g^{ik}\Gamma^j{}_{kl}\,y^l_\alpha = 0,\end{aligned} \tag{6.7.24}$$

which is consistent with the dynamical equation (6.7.4) and the other Peierls brackets.

It will be convenient to introduce two additional quantities,

$$y_{\alpha i} \stackrel{\text{def}}{=} g_{ij}\,y^j_\alpha, \tag{6.7.25}$$

$$p_i \stackrel{\text{def}}{=} g_{ij}\,\dot{x}^j, \tag{6.7.26}$$

which satisfy the bracket relations

$$(x^i, y_{\alpha j}) = 0, \quad (\dot{x}^i, y_{\alpha j}) = -g^{ik}\Gamma^l{}_{kj}\, y_{\alpha l}, \tag{6.7.27}$$

$$(y_{\alpha i}, y_{\beta j}) = -\mathrm{i} g_{ij}\, \delta_{\alpha\beta}, \tag{6.7.28}$$

$$(x^i, p_j) = \delta^i{}_j, \quad (p_i, p_j) = \tfrac{1}{2}\mathrm{i} R_{ijkl}\, y^k_\alpha y^l_\alpha, \tag{6.7.29}$$

$$(p_i, y^j_\alpha) = \Gamma^j{}_{ik}\, y^k_\alpha, \quad (p_i, y_{\alpha j}) = -\Gamma^k{}_{ij}\, y_{\alpha k}. \tag{6.7.30}$$

Energy and supersymmetry group

Denoting by L (for Lagrangian) the integrand of expression (6.7.1), we find for the energy or Hamiltonian of the system,

$$H = L\frac{\overleftarrow{\partial}}{\partial \dot{x}^i}\dot{x}^i + L\frac{\overleftarrow{\partial}}{\partial \dot{y}^i_\alpha}\dot{y}^i_\alpha - L = \tfrac{1}{2}g_{ij}\,\dot{x}^i\dot{x}^j - \tfrac{1}{8}R_{ijkl}\, y^i_\alpha y^j_\alpha y^k_\beta y^l_\beta \tag{6.7.31a}$$

$$= \tfrac{1}{2}g^{ij}p_i p_j - \tfrac{1}{8}R_{ijkl}\, y^i_\alpha y^j_\alpha y^k_\beta y^l_\beta. \tag{6.7.31b}$$

Using the dynamical equations (6.7.3) and (6.7.4) and the anticommutativity of the y's, one easily verifies that the energy is conserved:

$$\dot{H} = 0. \tag{6.7.32}$$

Let δt, $\delta\xi_\alpha$ ($\alpha = 1, 2$) be infinitesimal functions of t of compact support, δt being real-c-number-valued and the $\delta\xi_\alpha$ real-a-number-valued. Let the dynamical variables suffer the changes

$$\left.\begin{aligned} \delta x^i &= \dot{x}^i\delta t + \mathrm{i} y^i_\alpha\,\delta\xi_\alpha, \\ \delta y^i_\alpha &= (\mathrm{d}y^i_\alpha/\mathrm{d}t)\,\delta t + \dot{x}^i\delta\xi_\alpha + \mathrm{i}\Gamma^i{}_{jk}\, y^j_\alpha y^k_\beta\,\delta\xi_\beta, \\ \bar{\delta} y^i_\alpha &= \delta y^i_\alpha + \Gamma^i{}_{jk}\, y^j_\alpha\,\delta x^k = \dot{y}^i_\alpha\delta t + \dot{x}^i\delta\xi_\alpha. \end{aligned}\right\} \tag{6.7.33}$$

When the dynamical equations are satisfied infinitesimal changes must leave the action functional unaffected, and hence (see eq. (6.7.2))

$$\begin{aligned} 0 = \int &[(\dot{x}^i\delta t + \mathrm{i}y^i_\alpha\delta\xi_\alpha)(-g_{ij}\ddot{x}^j + \tfrac{1}{2}\mathrm{i}R_{ijkl}\,\dot{x}^j y^k_\beta y^l_\beta + \tfrac{1}{8}R_{jklm;i}\, y^j_\beta y^k_\beta y^l_\gamma y^m_\gamma) \\ &+ (\dot{y}^i_\alpha\delta t + \dot{x}^i\delta\xi_\alpha)(\mathrm{i}g_{ij}\,\dot{y}^j_\alpha + \tfrac{1}{2}R_{ijkl}\, y^j_\alpha y^k_\beta y^l_\beta)]\,\mathrm{d}t \\ = \int &[(-g_{ij}\,\dot{x}^i\ddot{x}^j + \tfrac{1}{8}R_{jklm;i}\,\dot{x}^i y^j_\beta y^k_\beta y^l_\gamma y^m_\gamma + \tfrac{1}{2}R_{ijkl}\,\dot{y}^i_\alpha y^j_\alpha y^k_\beta y^l_\beta)\,\delta t \\ &- \mathrm{i}(y_{\alpha i}\ddot{x}^i + \dot{y}_{\alpha i}\dot{x}^i)\,\delta\xi^\alpha]\,\mathrm{d}t \end{aligned} \tag{6.7.34}$$

in which certain terms that vanish by symmetry have been dropped and others have disappeared through cancellation. Note that the term involving five y's and the covariant derivative of the Riemann tensor vanishes by virtue of the Bianchi identity (2.8.41) and the fact that the indices α, β, γ can assume only the values 1 and 2.

Equation (6.7.34) can be rewritten in the form

$$0 = \int (-\dot{H}\delta t - \mathrm{i}\dot{Q}_\alpha \delta\xi^\alpha)\,\mathrm{d}t, \tag{6.7.35}$$

where

$$Q_\alpha = y_{\alpha i}\dot{x}^i = y^i_\alpha p_i, \tag{6.7.36}$$

and because of the arbitrariness of δt and the $\delta\xi_\alpha$ we infer, in addition to the conservation law (6.7.32), also

$$\dot{Q}_\alpha = 0. \tag{6.7.37}$$

As in the case of the new conserved quantities found in section 5.7, it is straightforward to verify that H and the Q_α together generate, *modulo* the dynamical equations, the very changes that give rise to the conservation laws:

$$\left.\begin{aligned} (x^i, H\delta t + \mathrm{i}Q_\alpha\delta\xi_\alpha) &= \delta x^i, \\ (y^i_\alpha, H\delta t + \mathrm{i}Q_\beta\delta\xi_\beta) &= \delta y^i_\alpha. \end{aligned}\right\} \tag{6.7.38}$$

If δt and the $\delta\xi_\alpha$ are chosen to be time-*independent* then eqs. (6.7.33) describe the infinitesimal actions of an invariance group possessed by the system, for, as is easily checked, these equations then change the Lagrangian by a total time derivative and hence leave the dynamical equations unaffected. The invariance group, as an abstract group, is identical to the supersymmetry group of section 5.7, for if we make the identification

$$\delta\alpha = \begin{pmatrix} \delta\xi_1 \\ \delta\xi_2 \end{pmatrix}, \tag{6.7.39}$$

we find that the commutator of two group actions takes exactly the form (5.7.31). In the present case the closure property of the group actions holds independently of the dynamical equations and does not require the introduction of an auxiliary variable.

Quantization

The quantum theory of the system (6.7.1) is obtained by converting Peierls brackets to supercommutator brackets:

$$\left.\begin{aligned} &[x^i, x^j] = 0, \quad [x^i, p_j] = \mathrm{i}\delta^i{}_j, \quad [p_i, p_j] = -\tfrac{1}{2}R_{ijkl}\, y^k_\alpha y^l_\alpha, \\ &[x^i, y^j_\alpha] = 0, \quad [x^i, y_{\alpha j}] = 0, \quad [y^i_\alpha, y^j_\beta] = g^{ij}\delta_{\alpha\beta}, \\ &[p_i, y^j_\alpha] = \mathrm{i}\Gamma^j{}_{ik}\, y^k_\alpha, \quad [p_i, y_{\alpha j}] = -\mathrm{i}\Gamma^k{}_{ij}\, y_{\alpha k}. \end{aligned}\right\} \tag{6.7.40}$$

Because of the commutativity of the x's with the y's there is no factor-ordering ambiguity in the expressions standing on the right of these equations. Nor is there any factor-ordering-ambiguity in eq. (6.7.25) or in the first two of eqs. (6.7.33). Equation (6.7.26) must, however, be modified to

$$p_i = \tfrac{1}{2}\{g_{ij}, \dot{x}^j\}, \quad \dot{x}^i = \tfrac{1}{2}\{g^{ij}, p_j\}, \tag{6.7.41}$$

and special factor orderings must be adopted for the operators Q_α and H.

The proper ordering for Q_α may be deduced from the requirement that Q_α be a scalar. Combining the coordinate transformation laws (6.5.18) with

$$\bar{y}^i_\alpha = \frac{\partial \bar{x}^i}{\partial x^j} y^j_\alpha = y^j_\alpha \frac{\partial \bar{x}^i}{\partial x^j}, \tag{6.7.42}$$

one sees that Q_α may be taken in either of two forms:

$$Q_\alpha = y^i_\alpha g^{\frac{1}{4}} p_i g^{-\frac{1}{4}} \quad \text{or} \quad Q_\alpha = g^{-\frac{1}{4}} p_i g^{\frac{1}{4}} y^i_\alpha. \tag{6.7.43}$$

These two forms are, in fact, identical, as follows from the easily verified commutator

$$[g^{\frac{1}{4}} y^i_\alpha, p_i] = 0. \tag{6.7.44}$$

The identity of the two forms also implies that Q_α is self-adjoint (assuming the x's, p's and y's to be self-adjoint).

The factor ordering for the Hamiltonian follows from the factor ordering for the Q's by the requirement that H and the Q's continue to be the generators of the supersymmetry group, in the quantum theory as in the superclassical theory. This implies (cf. eq. (5.7.22))

$$\begin{aligned} 2\delta_{\alpha\beta} H &= [Q_\alpha, Q_\beta] = Q_\alpha Q_\beta + Q_\beta Q_\alpha = g^{\frac{1}{4}}[y^i_\alpha p_i, y^j_\beta p_j]\, g^{-\frac{1}{4}} \\ &= g^{\frac{1}{4}}(\delta_{\alpha\beta} g^{ij} p_j p_i + \mathrm{i}[y^i_\alpha, y^j_\beta]\,\Gamma^k{}_{ij} p_k - \tfrac{1}{2}R_{ijkl}\, y^i_\alpha y^j_\beta y^k_\gamma y^l_\gamma)\, g^{-\frac{1}{4}} \\ &= \delta_{\alpha\beta} g^{\frac{1}{4}}[g^{ij} p_i p_j - \mathrm{i} g^{-\frac{1}{2}}(g^{\frac{1}{2}} g^{ij})_{,j} p_i - \tfrac{1}{4}R_{ijkl}\, y^i_\gamma y^j_\gamma y^k_\delta y^l_\delta]\, g^{-\frac{1}{4}} \\ &= \delta_{\alpha\beta}(g^{-\frac{1}{4}} p_i g^{\frac{1}{2}} g^{ij} p_j g^{-\frac{1}{4}} - \tfrac{1}{4}R_{ijkl}\, y^i_\gamma y^j_\gamma y^k_\delta y^l_\delta), \end{aligned} \tag{6.7.45}$$

whence

$$H = \tfrac{1}{2}g^{-\frac{1}{4}}p_i g^{\frac{1}{2}} g^{ij} p_j g^{-\frac{1}{4}} - \tfrac{1}{8}R_{ijkl}\, y^i_\alpha y^j_\alpha y^k_\beta y^l_\beta. \tag{6.7.46}$$

In passing from the second to the third line of (6.7.45) one makes use of the algebraic symmetries of the Riemann and Ricci tensors together with the anticommutation relations satisfied by the y's and the fact that the Greek indices on the y's range over only two values. By taking the commutator of the equation $[Q_\alpha, Q_\beta] = 2\delta_{\alpha\beta} H$ with Q_γ, using the super Jacobi identity, and setting $\beta = \gamma$, one obtains also (cf. eq. (5.7.23))

$$[Q_\alpha, H] = 0. \tag{6.7.47}$$

It is often convenient to replace the y's by the equivalent non-self-adjoint variables

$$\left.\begin{aligned} z^i &= \frac{1}{\sqrt{2}}(y^i_1 - \mathrm{i}y^i_2), \quad z^{i*} = \frac{1}{\sqrt{2}}(y^i_1 + \mathrm{i}y^i_2), \\ z_i &= g_{ij} z^j = \frac{1}{\sqrt{2}}(y_{1i} - \mathrm{i}y_{2i}), \quad z_i{}^* = g_{ij} z^{j*} = \frac{1}{\sqrt{2}}(y_{1i} + \mathrm{i}y_{2i}), \end{aligned}\right\} \tag{6.7.48}$$

which satisfy the supercommutation relations

$$\left.\begin{aligned} &[x^i, z_j] = 0, \quad [x^i, z_j{}^*] = 0, \\ &[z_i, z_j{}^*] = g_{ij}, \quad [z_i z_j] = 0, \quad [z_i{}^*, z_j{}^*] = 0, \\ &[z_i, p_j] = \mathrm{i}\Gamma^k{}_{ij} z_k, \quad [z_i{}^*, p_j] = \mathrm{i}\Gamma^k{}_{ij} z_k{}^*, \\ &[z^i, p_j] = -\mathrm{i}\Gamma^i{}_{jk} z^k, \quad [z^{i*}, p_j] = -\mathrm{i}\Gamma^i{}_{jk} z^{k*}. \end{aligned}\right\} \tag{6.7.49}$$

It is easy to check that

$$y^i_\alpha y^j_\alpha = z^{i*} z^j + z^i z^{j*} = z^{i*} z^j - z^{j*} z^i + g^{ij}, \tag{6.7.50}$$

which permits the Hamiltonian to be recast in the form

$$H = \tfrac{1}{2}g^{-\frac{1}{4}}p_i g^{\frac{1}{2}} g^{ij} p_j g^{-\frac{1}{4}} - \tfrac{1}{2}R_{ijkl}\, z^{i*} z^j z^{k*} z^l. \tag{6.7.51}$$

We note that the super-Jacobian of the transformation (6.7.48) is

$$\frac{\partial(z^*, z)}{\partial(y_1, y_2)} = \left[\mathrm{sdet}\begin{pmatrix} 1/\sqrt{2} & i/\sqrt{2} \\ 1/\sqrt{2} & -i/\sqrt{2} \end{pmatrix}\right]^m = \mathrm{i}^m. \tag{6.7.52}$$

When the z's and z^*'s are used it is convenient also to introduce the operators

$$\left.\begin{aligned} \mathfrak{Q} &= \frac{1}{\sqrt{2}}(Q_1 - \mathrm{i}Q_2) = z^i g^{\frac{1}{4}} p_i g^{-\frac{1}{4}} = g^{-\frac{1}{4}} p_i g^{\frac{1}{4}} z^i, \\ \mathfrak{Q}^* &= \frac{1}{\sqrt{2}}(Q_1 + \mathrm{i}Q_2) = z^{i*} g^{\frac{1}{4}} p_i g^{-\frac{1}{4}} = g^{-\frac{1}{4}} p_i g^{\frac{1}{4}} z^{i*}, \end{aligned}\right\} \tag{6.7.53}$$

(cf. eqs. (5.7.73)) which satisfy the supercommutation relations

$$\left.\begin{aligned} [\mathfrak{Q}, \mathfrak{Q}] = 2\mathfrak{Q}^2 = 0, \quad [\mathfrak{Q}^*, \mathfrak{Q}^*] = 2\mathfrak{Q}^{*2} = 0, \\ [\mathfrak{Q}, \mathfrak{Q}^*] = 2H, \quad [\mathfrak{Q}, H] = 0, \quad [\mathfrak{Q}^*, H] = 0. \end{aligned}\right\} \tag{6.7.54}$$

Basis supervectors

Equations (6.7.40) and (6.7.49) determine the operator superalgebra of the quantum theory. By extending the analysis of sections 5.5 and 6.2 one easily sees that the following equations define (up to a phase transformation) a complete set of basis supervectors $|x', i_1, \ldots, i_r, t\rangle$, $r = 0, 1, 2, \ldots$:

$$x^i(t)\,|x', t\rangle = x'^i\,|x', t\rangle, \quad z_i(t)\,|x', t\rangle = 0, \tag{6.7.55}$$

$$\left.\begin{aligned} |x', i_1, \ldots, i_r, t\rangle &\overset{\text{def}}{=} z_{i_1}{}^*(t) \ldots z_{i_r}{}^*(t)\,|x', t\rangle, \\ \langle x', i_r, \ldots, i_1, t| &\overset{\text{def}}{=} |x', i_1, \ldots, i_r, t\rangle^* = \langle x', t|\, z_{i_r}(t) \ldots z_{i_1}(t). \end{aligned}\right\} \tag{6.7.56}$$

If the normalization condition

$$\langle x'', t\,|\,x', t\rangle = \delta(x'', x') \tag{6.7.57}$$

is imposed on $|x', t\rangle$ then the supercommutation relations (6.7.49) imply

$$\langle x'', j_s, \ldots, j_1, t\,|\,x', i_1, \ldots, i_r, t\rangle = \delta_{sr} \sum_{\text{perm}} (\pm) g_{j_1 i_1}(x') \ldots g_{j_r i_r}(x')\, \delta(x'', x'), \tag{6.7.58}$$

where the sum is over all permutations of the indices $i_1, \ldots, i_r$, a plus sign being taken for the even permutations and a minus sign for the odd ones. The requirement that the basis vectors form a complete set is equivalent to the statement

$$\sum_{r=0}^{m} \frac{1}{r!} \int_C |x', i_r, \ldots, i_1, t\rangle\, g^{i_1 j_1}(x') \ldots g^{i_r j_r}(x')\, \langle x', j_1, \ldots, j_r, t|\, \mathrm{d}^m x' = 1, \tag{6.7.59}$$

in which summation over repeated indices is understood. The sum over r in this equation terminates at $r = m$ because of the antisymmetry of the basis vectors in their indices, which follows from the anticommutativity of the z's.

We shall choose to call the supervectors $|x', t\rangle$ c-type.† The supervectors $|x', i_1, \ldots, i_r, t\rangle$ are then a-type or c-type according as r is odd or even. The integer r is *counted* by the self-adjoint c-type operator

$$N \stackrel{\text{def}}{=} g_{ij} z^{i*} z^j = g^{ij} z_i{}^* z_j = N^*, \tag{6.7.60}$$

which satisfies the following commutation relations:

$$\left.\begin{aligned} &[N, z_i] = -z_i, \quad [N, z_i{}^*] = z_i{}^*, \quad [N, x^i] = 0, \\ &[N, p_i] = [g^{jk} z_j{}^* z_k, p_i] = \mathrm{i}(-g_{jk,i} + \Gamma_{jki} + \Gamma_{kji}) z^{j*} z^k = 0, \\ &[N, z^{i*} z^j] = z^{i*} z^j - z^{i*} z^j = 0. \end{aligned}\right\} \tag{6.7.61}$$

Using these relations one easily finds

$$\dot{N} = -\mathrm{i}[N, H] = 0 \tag{6.7.62}$$

$$N |x', i_1, \ldots, i_r, t\rangle = r |x', i_1, \ldots, i_r, t\rangle, \tag{6.7.63}$$

$$(-1)^N z_i = -z_i(-1)^N, \quad (-1)^N z_i{}^* = -z_i{}^*(-1)^N, \tag{6.7.64}$$

where $(-1)^N = \mathrm{e}^{\pi \mathrm{i} N}$.

If $|\psi\rangle$ is the state supervector of the quantum system then the basis vectors above yield a description of the system in terms of 2^m wave functions

$$\psi_{i_1 \ldots i_r}(x', t) \stackrel{\text{def}}{=} \langle x, i_1, \ldots, i_r, t | \psi\rangle. \tag{6.7.65}$$

The transformation character of these wave functions is completely fixed by the coordinate transformation law (6.7.42) together with eqs. (6.7.55) and the normalization condition (6.7.57). For each r the $\psi_{i_1 \ldots i_r}$ are the components of an r-form density of weight $\frac{1}{2}$. Since nothing prevents ψ's of more than one rank from appearing as coefficients in the expansion of $|\psi\rangle$ in terms of the basis supervectors, it will be convenient to introduce a symbol ψ to represent the *formal sum*, over r, of these r-form densities. One may regard ψ as $g^{\frac{1}{4}}$ times an element of the direct sum $\Omega(C) \stackrel{\text{def}}{=} \oplus_r \Omega_r(C)$, where $\Omega_r(C)$ is the supervector space of r-forms on C.

† There is always an arbitrariness about type assignment when a-type dynamical variables are present (see sections 5.4 and 5.5).

Differential representation of operators

Consider the operator superalgebra generated by the first three of the supercommutation relations (6.7.40). Note that with use of eq. (6.7.50) the third of these relations may be rewritten in the form

$$[p_i, p_j] = -R_{ijkl}z^{k*}z^l, \tag{6.7.66}$$

which together with the adjoints of eqs. (6.7.55), implies

$$\langle x', t|[p_i(t), p_j(t)] = -R_{ijkl}(x')\langle x', t|z^{k*}(t)z^l(t) = 0. \tag{6.7.67}$$

From the results of the preceding sections of this chapter it is clear that the operator $p_i(t)$, when bearing *directly* on the supervector $\langle x', t|$, may be regarded as having the effect of a differential operator, namely

$$\langle x', t|p_i(t) = [-\mathrm{i}\partial/\partial x'^i - \omega_i(x')]\langle x', t| \tag{6.7.68}$$

(cf. eq. (6.2.75)), where ω is an arbitrary harmonic 1-form.

In this section we shall choose

$$\omega = 0 \tag{6.7.69}$$

and hence confine our attention, in the Feynman formulation of the theory, to the trivial representation of $\pi_1(C)$. This choice yields

$$\begin{aligned}\langle x', i_1, \ldots, i_r, t|\mathfrak{Q}^*(t) &= \langle x', t|z_{i_1}(t)\ldots z_{i_r}(t)z^{j*}(t)g^{\frac{1}{4}}(x(t))p_j(t)g^{-\frac{1}{4}}(x(t))\\ &= \sum_{s=1}^{r}(-1)^{r-s}\langle x', t|z_{i_1}(t)\ldots z_{i_{s-1}}(t)z_{i_{s+1}}(t)\ldots z_{i_r}(t)\\ &\qquad\times g^{\frac{1}{4}}(x(t))p_{i_s}(t)g^{-\frac{1}{4}}(x(t))\\ &= \frac{1}{(r-1)!}\sum_{\text{perm}}(\pm)\langle x', t|z_{i_1}(t)\ldots z_{i_{r-1}}(t)g^{\frac{1}{4}}(x(t))p_{i_r}(t)\\ &\qquad\times g^{-\frac{1}{4}}(x(t))\\ &= -\frac{\mathrm{i}}{(r-1)!}g'^{\frac{1}{4}}\sum_{\text{perm}}(\pm)\frac{\partial}{\partial x'^{i_r}}(g'^{-\frac{1}{4}}\langle x', i_1, \ldots, i_{r-1}, t|),\end{aligned} \tag{6.7.70}$$

where the sum is over all permutations of the indices $i_1, \ldots, i_r$, the plus or minus sign being taken according as the permutation is even or odd. When commuting the operator p_{i_r} through the z's to arrive at the last line of (6.7.70), one finds that the connection components that are generated (see eqs. (6.7.49)) all cancel because of the antisymmetrization effected by the permutation sum.

Comparison of eq. (6.7.70) with eq. (2.6.21) shows that the operator $\mathfrak{Q}^*$ has, up to a factor $\pm i$, the effect of exterior differentiation. Indeed, generalizing the definition of exterior differentiation in Riemannian manifolds, to r-form densities $\boldsymbol{\Omega}$ of weight w, by the rule

$$(\mathbf{d}\boldsymbol{\Omega})_{i_1 \dots i_{r+1}} \overset{\text{def}}{=} \frac{(-1)^r}{r!} g^{\frac{1}{2}w} \sum_{\text{perm}} (\pm)(g^{-\frac{1}{2}w} \Omega_{i_1 \dots i_r})_{,i_{r+1}}, \qquad (6.7.71)$$

and invoking the formal symbol ψ, we may write

$$i\langle x', i_1, \dots, i_r, t | \mathfrak{Q}^*(t)(-1)^N | \psi \rangle = (\mathbf{d}\psi)_{i_1 \dots i_r}(x', t). \qquad (6.7.72)$$

We have therefore the formal correspondence

$$\mathbf{d} \to i\mathfrak{Q}^*(-1)^N = -i(-1)^N \mathfrak{Q}^*. \qquad (6.7.73)$$

The adjoint of this correspondence is

$$\mathbf{d}^* \to i\mathfrak{Q}(-1)^N = -i(-1)^N \mathfrak{Q}, \qquad (6.7.74)$$

where the symbol $\mathbf{d}^*$, known as the *Hodge operator*† and often written $\boldsymbol{\delta}$, is the adjoint of $\mathbf{d}$ with respect to the inner product defined by (6.7.59).

Equations (6.7.73) and (6.7.74) yield immediately a correspondence between the Laplacian operator on forms† and the Hamiltonian (6.7.51):

$$\Delta \overset{\text{def}}{=} \mathbf{d}\mathbf{d}^* + \mathbf{d}^*\mathbf{d} \to \mathfrak{Q}^*\mathfrak{Q} + \mathfrak{Q}\mathfrak{Q}^* = 2H. \qquad (6.7.75)$$

By straightforward computation one may show that the term in the Riemann tensor in expression (6.7.51) yields precisely the terms in the Riemann tensor that appear when the Laplacian of a form is written out explicitly in terms of components. The Schrödinger equations satisfied by the wave functions (6.7.65), namely

$$i\partial\psi_{i \dots i_r}(x', t)/\partial t = \langle x', i_1, \dots, i_r, t | H | \Psi \rangle \qquad (6.7.76)$$

are equivalent to the single equation

$$i\partial\psi/\partial t = \tfrac{1}{2}\Delta\psi. \qquad (6.7.77)$$

† See Choquet-Bruhat and DeWitt-Morette, with Dillard-Bleick, *loc. cit.*

Coherent states

Let z'^i $(i = 1, \ldots, m)$ be arbitrary complex *a*-numbers. Introduce the supervectors

$$\left.\begin{aligned} |x', z', t\rangle &\stackrel{\text{def}}{=} \mathrm{e}^{-\frac{1}{2}z_i'^* z'^i + z_i^*(t)\, z'^i}\, |x', t\rangle, \\ \langle x', z'^*, t| &\stackrel{\text{def}}{=} |x', z', t\rangle^* = \langle x', t|\, \mathrm{e}^{-\frac{1}{2}z_i'^* z'^i + z_i'^* z^i(t)}, \end{aligned}\right\} \tag{6.7.78}$$

where

$$z_i' \stackrel{\text{def}}{=} g_{ij}(x')\, z'^j. \tag{6.7.79}$$

It is easy to verify that

$$\mathrm{e}^{-z_j^*(t)\, z'^j}\, z^i(t)\, \mathrm{e}^{z_k^*(t)\, z'^k} = z^i(t) + z'^i \tag{6.7.80}$$

and hence that the $|x', z', t\rangle$ are right eigenvectors of the $z^i(t)$ while the $\langle x', z'^*, t|$ are left eigenvectors of the $z^{i*}(t)$:

$$\left.\begin{aligned} z^i(t)\, |x', z', t\rangle &= z'^i |x', z', t\rangle, \\ \langle x', z'^*, t|\, z^{i*}(t) &= \langle x', z', t|\, z'^{i*}. \end{aligned}\right\} \tag{6.7.81}$$

In analogy with the terminology introduced for the supervectors $|a', t\rangle$ in section 5.5, the $|x', z', t\rangle$ will be called supervectors corresponding to *coherent states*.

Although the coherent-state supervectors do not constitute a complete set (they are all *c*-type) they nevertheless satisfy an integral identity that has the formal appearance of a completeness relation. This is most easily shown through the introduction of orthonormal frame fields $\{e_a\}$ having components, in coordinate frames, satisfying

$$e_a{}^i e_a{}^j = g^{ij}, \quad g_{ij}\, e_a{}^i e_b{}^j = \delta_{ab}. \tag{6.7.82}$$

Defining

$$\bar{z}_a(t) = e_a{}^i(x(t))\, z_i(t), \quad \bar{z}_a' = e_a{}^i(x')\, z_i', \tag{6.7.83}$$

one may write

$$|x', z', t\rangle = \mathrm{e}^{-\frac{1}{2}\bar{z}_a'^* \bar{z}_a' + \bar{z}_a^*(t)\, \bar{z}_a'}\, |x', t\rangle. \tag{6.7.84}$$

Introducing the 'volume elements'

$$\mathrm{d}^m \bar{z}' = \mathrm{d}\bar{z}_1' \ldots \mathrm{d}\bar{z}_m', \quad d^m \bar{z}'^* = \mathrm{d}\bar{z}_m'^* \ldots \mathrm{d}\bar{z}_1'^*,$$

taking note of the algebra of *a*-numbers, and making use of the rules for integration with respect to a-numbers, one may also write

$$\frac{1}{(2\pi \mathrm{i})^m}\int_{\hat{C}} |x', z', t\rangle\langle x', z'^*, t|\, \mathrm{d}^m x'\, \mathrm{d}^m \bar{z}'^*\, \mathrm{d}^m \bar{z}'$$

$$= \frac{1}{(2\pi \mathrm{i})^m}\int_{\hat{C}} \mathrm{e}^{-\bar{z}'_a{}^*\bar{z}'_a}\,\mathrm{e}^{\bar{z}_b{}^*(t)\,\bar{z}'_b}|x', t\rangle\langle x', t|\,\mathrm{e}^{\bar{z}'_c{}^*\bar{z}_c(t)}\, \mathrm{d}^m x'\, \mathrm{d}^m \bar{z}'^*\, \mathrm{d}^m \bar{z}'$$

$$= \frac{1}{(2\pi \mathrm{i})^m}\sum_{r=0}^{m}\int_{\hat{C}} \frac{1}{(m-r)!\,(r!)^2}(\bar{z}'_a\bar{z}'_a{}^*)^{m-r}(\bar{z}_b{}^*(t)\,\bar{z}'_b)^r\,|x', t\rangle\langle x', t|(\bar{z}'_c{}^*\bar{z}_c(t))^r$$
$$\times\, \mathrm{d}^m x'\, \mathrm{d}^m \bar{z}'^*\, \mathrm{d}^m \bar{z}'$$

$$= \sum_{r=0}^{m}\frac{1}{r!}\int_{C} |x', i_r, \ldots, i_1, t\rangle\, g^{i_1 j_1}(x') \ldots g^{i_r j_r}(x')\,\langle x', j_1, \ldots, j_r, t|\, \mathrm{d}^m x'$$

$$= 1. \tag{6.7.85}$$

Another important relation may be derived by similar analysis. Let A be any c-type operator. Then

$$\frac{1}{(2\pi \mathrm{i})^m}\int_{\hat{C}} \langle x', z'^*, t|\, A\, |x', z', t\rangle\, \mathrm{d}^m x'\, \mathrm{d}^m \bar{z}'^*\, \mathrm{d}^m \bar{z}'$$

$$= \frac{1}{(2\pi \mathrm{i})^m}\int_{\hat{C}} \mathrm{e}^{-\bar{z}'_a{}^*\bar{z}'_a}\,\langle x', t|\,\mathrm{e}^{\bar{z}'_c{}^*\bar{z}_c(t)}\, A\,\mathrm{e}^{\bar{z}_b{}^*(t)\,\bar{z}'_b}|x', t\rangle\, \mathrm{d}^m x'\, \mathrm{d}^m \bar{z}'^*\, \mathrm{d}^m \bar{z}'$$

$$= \frac{1}{(2\pi \mathrm{i})^m}\sum_{r=0}^{m}\int_{\hat{C}} \frac{1}{(m-r)!\,(r!)^2}(\bar{z}'_a\bar{z}'_a{}^*)m-r\langle x', t|\,(\bar{z}'_c{}^*\bar{z}_c(t))^r A(\bar{z}_b{}^*(t)\,\bar{z}'_b)^r\,|x', t\rangle$$
$$\times\, \mathrm{d}^m x'\, \mathrm{d}^m \bar{z}'^*\, \mathrm{d}^m \bar{z}'$$

$$= \sum_{r=0}^{m}\frac{(-1)^r}{r!}\int_{C} g^{i_1 j_1}(x') \ldots g^{i_r j_r}(x')\,\langle x', i_1, \ldots, i_r, t|\, A\, |x', j_r, \ldots, j_r, t\rangle\, \mathrm{d}^m x'$$

$$= \operatorname{str} A, \tag{6.7.86}$$

the final summation-integration being the obvious expression for the supertrace (cf. exercise 5.10).

It will be convenient to replace the integration variables $\bar{z}'_a, \bar{z}'_a{}^*$ in (6.7.85) and (6.7.86) by the original variables z'^i, z'^{i*}. It is easy to see that the super Jacobian for this change of variables is $g^{-1}(x')$, and hence

$$\frac{1}{(2\pi \mathrm{i})^m}\int_{\hat{C}} |x', z', t\rangle\langle x', z'^*, t|\, g^{-1}(x')\, \mathrm{d}^m x'\, \mathrm{d}^m z'^*\, \mathrm{d}^m z' = 1, \tag{6.7.85$'$}$$

$$\frac{1}{(2\pi i)^m}\int_{\bar{C}} \langle x', z'^*, t|\, A\, |x', z', t\rangle\, g^{-1}(x')\, \mathrm{d}^m x'\, \mathrm{d}^m z'^*\, \mathrm{d}^m z' = \mathrm{str}\, A \tag{6.7.86$'$}$$

Energy eigenfunctions

The supercommutation relations (6.7.54) allow the Hamiltonian to be expressed in the form

$$H = \tfrac{1}{2}(\mathfrak{Q} + \mathfrak{Q}^*)^2. \tag{6.7.87}$$

Since $\mathfrak{Q} + \mathfrak{Q}^*$ is Hermitian its eigenvalues are real and the eigenvalues of the Hamiltonian are nonnegative. The latter eigenvalues are conveniently grouped into two sets: those that are positive and those that vanish. In the case of the nonlinear supersymmetric system studied in section 5.7 the positive eigenvalues were doubly degenerate, each being associated with a *c*-type eigenvector and an *a*-type eigenvector. When C is compact the energy spectrum of the present system is similar. Corresponding to each positive eigenvalue there is a certain number of linearly independent *c*-type eigenvectors and an equal number of linearly independent *a*-type eigenvectors. This is, in fact, a classic result, deriving, via the correspondence (6.7.75), from the theory of the Laplacian operator on forms. We give a brief resumé of the classic proof, using the language forms.†

One begins by splitting the supervector space $\Omega(C)$ of all forms on C into the direct sum of the supervector space $\Omega_+(C)$ of forms of even degree and the supervector space $\Omega_-(C)$ of forms of odd degree. By restricting the exterior derivative and its adjoint to these spaces, respectively, one makes a corresponding splitting of the operators $\mathbf{d}$ and $\mathbf{d}^*$:

$$\left.\begin{aligned} \mathbf{d} = \mathbf{d}_+ + \mathbf{d}_-, \quad \mathbf{d}^* = \mathbf{d}_+{}^* + \mathbf{d}_-{}^*, \\ \mathbf{d}_\pm = \mathbf{d}|_{\Omega_\pm(C)}, \quad \mathbf{d}_\pm{}^* = \mathbf{d}^*|_{\Omega_\mp(C)}. \end{aligned}\right\} \tag{6.7.88}$$

The operators $\mathbf{d}_\pm, \mathbf{d}_\pm{}^*$ are mappings of the following types:

$$\left.\begin{aligned} \mathbf{d}_+ : \Omega_+(C) \to \Omega_-(C), \quad \mathbf{d}_- : \Omega_-(C) \to \Omega_+(C), \\ \mathbf{d}_+{}^* : \Omega_-(C) \to \Omega_+(C), \quad \mathbf{d}_-{}^* : \Omega_+(C) \to \Omega_-(C). \end{aligned}\right\} \tag{6.7.89}$$

They may be extended to the full space $\Omega(C)$ by defining

$$\mathbf{d}_+\Omega_-(C) = 0, \quad \mathbf{d}_-\Omega_+(C) = 0, \quad \mathbf{d}_+{}^*\Omega_+(C) = 0, \quad \mathbf{d}_-{}^*\Omega_-(C) = 0, \tag{6.7.90}$$

† A generalization of the following analysis to the noncompact case is possible but will not be included here.

It is then easy to see that the Laplacian operator can be decomposed in the form

$$\begin{aligned}\Delta &= \mathbf{dd}^* + \mathbf{d}^*\mathbf{d} \\ &= (\mathbf{d}_+ + \mathbf{d}_-)(\mathbf{d}_+^* + \mathbf{d}_-^*) + (\mathbf{d}_+^* + \mathbf{d}_-^*)(\mathbf{d}_+ + \mathbf{d}_-) \\ &= \Delta_+ + \Delta_-, \end{aligned} \tag{6.7.91}$$

where

$$\left.\begin{aligned}\Delta_+ &= \mathbf{d}_+^*\mathbf{d}_+ + \mathbf{d}_-\mathbf{d}_-^*, \\ \Delta_- &= \mathbf{d}_-^*\mathbf{d}_- + \mathbf{d}_+\mathbf{d}_+^*.\end{aligned}\right\} \tag{6.7.92}$$

Using the corollaries $\mathbf{d}_+\mathbf{d}_- = 0$, $\mathbf{d}_-\mathbf{d}_+ = 0$ of $\mathbf{d}^2 = 0$, as well as the adjoints of these corollaries, one may also write

$$\Delta_+ = P^*P, \quad \Delta_- = PP^*, \tag{6.7.93}$$

where

$$P = \mathbf{d}_+ + \mathbf{d}_-^*, \quad P^* = \mathbf{d}_+^* + \mathbf{d}_-. \tag{6.7.94}$$

Denote by $\Omega_{\lambda\pm}$ the subspace of $\Omega_\pm(C)$ corresponding to the eigenvalue λ of Δ. Let ψ be $g^{\frac{1}{4}}$ times an element of $\Omega_{\lambda+}$. Then $(P+P^*)\,\psi$ will be $g^{\frac{1}{4}}$ times an element of $\Omega_-(C)$. Since $P+P^* = \mathbf{d}+\mathbf{d}^*$ and $\Delta = (\mathbf{d}+\mathbf{d}^*)^2 = (P+P^*)^2$, it follows that

$$\Delta\psi = \lambda\psi \quad \text{and} \quad \Delta(P+P^*)\,\psi = (P+P^*)\,\Delta\psi = \lambda(P+P^*)\,\psi. \tag{6.7.95}$$

Moreover, using the notation $\langle\ ,\ \rangle$ to abbreviate the inner product defined by (6.7.59), one may write

$$\begin{aligned}\langle (P+P^*)\,\psi, (P+P^*)\,\psi\rangle &= \langle \psi, (P+P^*)^2\,\psi\rangle \\ &= \langle \psi, \Delta\psi\rangle = \lambda\langle\psi,\psi\rangle,\end{aligned} \tag{6.7.96}$$

which is necessarily positive if $\lambda > 0$. In this case $(P+P^*)\,\psi$ is $g^{\frac{1}{4}}$ times a nonvanishing element of $\Omega_{\lambda-}$. It follows that for every eigenfunction of Δ of even degree there is a corresponding eigenfunction of odd degree, provided the eigenvalue λ is positive. The converse statement is proved similarly. This implies

$$\dim\Omega_{\lambda+} = \dim\Omega_{\lambda-} \quad \text{when} \quad \lambda > 0, \tag{6.7.97}$$

where 'dim' denotes 'c-type dimension of'. (The a-type dimension is zero.)

Equation (6.7.97) cannot strictly be inferred until one proves that $\dim\Omega_{\lambda+}$ and $\dim\Omega_{\lambda-}$ are finite. It is a curious fact that the techniques needed for the proof lie far outside the standard techniques of de Rham cohomology theory where (6.7.97) is used. Physicists, used to thinking of bound (compact) systems as having discrete finitely-degenerate energy spectra, will probably accept the finiteness of $\dim \Omega_{\lambda-}$ and $\dim \Omega_{\lambda-}$ without proof. The proof will certainly not be given here!

The Euler–Poincaré characteristic

A zero-eigenvalue eigenfunction of the Laplacian operator, of rank r, is called a *harmonic* r-form. Denote by Ω_{0r} the supervector space of harmonic r-forms. Then

$$\Omega_{0+} = \bigoplus_{k=0}^{\infty} \Omega_{0\,2k}, \quad \Omega_{0-} = \bigoplus_{k=0}^{\infty} \Omega_{0\,2k+1}. \tag{6.7.98}$$

When C is compact its rth Betti number is given by

$$b_r = \dim \Omega_{0r}, \tag{6.7.99}$$

and the *Euler–Poincaré characteristic* is

$$\chi(C) \overset{\text{def}}{=} \sum_{r=0}^{m} (-1)^r b_r = \dim \Omega_{0+} - \dim \Omega_{0-}. \tag{6.7.100}$$

The existence of the identity (6.7.97) makes it possible to express the Euler–Poincaré characteristic very simply as a supertrace. Let z be a complex number lying in the upper half plane. Then

$$\begin{aligned} \operatorname{str} e^{iz\Delta} &= \sum_{\lambda} e^{iz\lambda}(\dim \Omega_{\lambda+} - \dim \Omega_{\lambda-}) \\ &= \dim \Omega_{0+} - \dim \Omega_{0-} = \chi(C). \end{aligned} \tag{6.7.101}$$

Two facts should be noted. This formula is independent of z and holds no matter how close z is to the origin as long as its imaginary part is positive. The exponential damping produced by the imaginary part serves to render the sums $\Sigma e^{iz\lambda} \dim \Omega_{\lambda+}$ and $\Sigma e^{iz\lambda} \dim \Omega_{\lambda-}$ separately convergent despite the typical tendency of $\dim \Omega_{\lambda\pm}$ to grow with λ.

The separate convergence of each sum allows one to give a heuristic proof of the topological invariance of the Euler–Poincaré characteristic. Let the metric tensor field on C suffer an infinitesimal change, and let δP, δP^* be the corresponding changes induced in the operators P, P^* defined by eqs. (6.7.94). Invoking the cyclic invariance of the trace we then have

$$\delta \sum_{\lambda} e^{iz\lambda} \dim \Omega_{\lambda+} = \delta \operatorname{tr} e^{iz\Delta_+} = \delta \operatorname{tr} e^{izP^*P}$$

$$= iz \operatorname{tr} [(\delta P^* P + P^* \delta P) e^{izP^*P}]$$
$$= iz \operatorname{tr} [(P \delta P^* + \delta P P^*) e^{izPP^*}]$$

$$= \delta \operatorname{tr} e^{izPP^*} = \delta \operatorname{tr} e^{iz\Delta_-} = \delta \sum_{\lambda} e^{iz\lambda} \dim \Omega_{\lambda-}, \quad (6.7.102)$$

whence $\delta\chi(C) = 0$.

The connection (6.7.75) between the Laplacian operator Δ and the Hamiltonian operator H of the system (6.7.1) allows one to express the Euler–Poincaré characteristic in terms of the coherent-state transition amplitude for the system. Setting $z = -\frac{1}{2}t$ and making use of eq. (6.7.86), we may write

$$\chi(C) = \operatorname{str} e^{-\frac{1}{2}it\Delta} = \operatorname{str} e^{-iHt}$$

$$= \frac{1}{(2\pi i)^m} \int \langle x', z'^*, t' | e^{-iHt} | x', z', t' \rangle g^{-1}(x') \, d^m x' \, d^m z'^* \, d^m z'$$

$$= \frac{1}{(2\pi i)^m} \int \langle x', z'^*, t' + t | x', z', t' \rangle g^{-1}(x') \, d^m x' \, d^m z'^* \, d^m z'. \quad (6.7.103)$$

In this equation t must be understood as having a small negative imaginary part, but because the transition amplitude is analytic in t this causes no problem.

Functional integral for the coherent-state transition amplitude

Breaking the time interval $t'' - t'$ into small bits, recalling the derivation of the amplitude (5.5.51) for the Fermi oscillator, and using either the completeness relation (6.7.85′) or the results of exercise **6.14**, one may infer that the coherent-state transition amplitude for the present system is given by the functional integral

$$\langle x'', z''^*, t'' | x', z', t' \rangle = Z g^{\frac{1}{4}}(x'') g^{\frac{1}{4}}(x') \int_{\Phi^{x''z''^*t''}_{x'z't'}}$$

$$\times \exp\left\{ i \int_{x'z't'}^{x''z''^*t''} [\tfrac{1}{2} g_{ij} (\dot{x}^i \dot{x}^j + i z^{i*} \dot{z}^j - i \dot{z}^{i*} z^j) + \tfrac{1}{2} R_{ijkl} z^{i*} z^j z^{k*} z^l] \, dt \right\}$$

$$\times \prod_{t' < t < t''} g^{-\frac{1}{2}}(x(t)) \prod_{i=1}^{m} dx^i(t) \, dz^{i*}(t) \, dz^i(t), \quad (6.7.104)$$

with

$$Z = (2\pi \mathrm{i}\, \mathrm{d}t)^{-m(t''-t')/2\mathrm{d}t}\, (2\pi \mathrm{i})^{-m(t''-t')/\mathrm{d}t+m}. \tag{6.7.105}$$

The functional integration is here taken over all trajectories in the space $\Phi^{x''z''*t''}_{x'z't'}$ of dynamical histories having the indicated endpoints. Note that no extra phase factors are inserted here for trajectories in differing homotopy equivalence classes.

The only possible uncertainty in the choice of integrand in expression (6.7.104) concerns the term $\frac{1}{8}R$ that appears in the Hamiltonian (6.5.25) for the system when the dynamical variables z^i, z^{i*} are absent. One must ask whether $\frac{1}{8}R$ should be added to the integrand in the exponent in (6.7.104) to cancel the $\frac{1}{8}R$ in the Hamiltonian, in view of the fact that the Laplacian operator on forms (which appears in the Schrödinger equation (6.7.77)) never contains a term involving the curvature scalar. It turns out that such a cancellation is already effected by the term in $\frac{1}{2}R_{ijkl}\, z^{i*}z^j z^{k*}z^l$. (Indeed, a result like this is to be expected because the direct addition of a term $\frac{1}{8}R$ would break the supersymmetry invariance of the exponent.)

To show this it suffices to work to two-loop order in the loop expansion. In this order the term in the Riemann tensor, which may be rewritten in the form appearing in eq. (6.7.1), contributes (in part)

$$-\frac{\mathrm{i}}{8}\int_{t'}^{t''} R_{ijkl}(G^{ij}_{\alpha\alpha}\, G^{kl}_{\beta\beta} - G^{ik}_{\alpha\beta}\, G^{jl}_{\alpha\beta} + G^{il}_{\alpha\beta}\, G^{jk}_{\alpha\beta})\, \mathrm{d}t. \tag{6.7.106}$$

Here use has been made of eq. (6.6.52c) modified, by appropriate insertion of a minus sign, for the case in which the ξ's are *a*-type. The *G*'s are propagators appropriate to boundary conditions that fix $y'^i_1 - iy'^i_2$ and $y''^i_1 + iy''^i_2$ (cf. eqs. (5.5.70) and (5.5.71)), namely

$$G^{ij'''}_{\alpha\beta'''} = \tfrac{1}{2}\mathrm{i}[\theta(t, t''') - \theta(t''', t)]\, g^{ij}\, \delta_{\alpha\beta} + \tfrac{1}{2}g^{ij}\varepsilon_{\alpha\beta} + O(t - t'''), \tag{6.7.107}$$

where $(\varepsilon_{\alpha\beta})$ is the matrix M of eq. (5.5.2). Setting $t''' = t$ and making use of (6.4.4) we find

$$G^{ij}_{\alpha\beta} = \tfrac{1}{2}g^{ij}\varepsilon_{\alpha\beta}, \tag{6.7.108}$$

so that expression (6.7.106) reduces to $\frac{1}{8}\mathrm{i}R'(t'' - t')$, which is just what a term in $\frac{1}{8}R$ would contribute to order $t'' - t'$ in the heat-kernel expansion, q.e.d.

The Chern–Gauss–Bonnet formula

Inserting expression (6.7.104) into (6.7.103) one gets

$$\chi(C) = \bar{Z}\int_{\Phi_{x'z't'}^{x'z'^{*}t'+t}} \times \exp\left\{\mathrm{i}\int_{x'z't'}^{z'z'^{*}t'+t}[\tfrac{1}{2}g_{ij}(\dot{x}^i\dot{x}^j + \mathrm{i}z^{i*}\dot{z}^j - \mathrm{i}\dot{z}^{i*}z^j) + \tfrac{1}{2}R_{ijkl}z^{i*}z^jz^{k*}z^l]\,\mathrm{d}t\right\} \times \prod_{t'\leqslant t<t'+t} g^{-\frac{1}{2}}(x(t))\prod_{i=1}^{m}\mathrm{d}x^i(t)\,\mathrm{d}z^{i*}(t)\,\mathrm{d}z^i(t) \qquad (6.7.109)$$

with

$$\bar{Z} = (2\pi\mathrm{i}\,\mathrm{d}t)^{-mt/2\mathrm{d}t}(2\pi\mathrm{i})^{-mt/\mathrm{d}t}, \qquad (6.7.110)$$

where it is understood that $x^i(t') = x'^i$, $z^i(t') = z'^i$, $z^{i*}(t') = z'^{i*}$. Now recall that the supertrace formula for $\chi(C)$ is independent of t. This allows one to shrink the time interval to a single bit: $t = \mathrm{d}t$, and to reduce the functional integral to an ordinary integral. Taking note of eqs. (5.5.48) and (6.4.13), which hold when the time interval is infinitesimal, and observing that the endpoint data are the same at both ends of the interval, one sees that the terms in parentheses in the integrand of the exponent in (6.7.109) contribute nothing when $t \to \mathrm{d}t$. One is therefore left with

$$\chi(C) = (2\pi\mathrm{i}\,\mathrm{d}t)^{-\frac{1}{2}m}(2\pi\mathrm{i})^{-m}\int_{\hat{C}} \exp[\tfrac{1}{2}\mathrm{i}R_{ijkl}(x')z'^{i*}z'^jz'^{k*}z'^l\,\mathrm{d}t] \times g^{-\frac{1}{2}}(x')\prod_{i=1}^{m}\mathrm{d}x'^i\,\mathrm{d}z'^{i*}\,\mathrm{d}z'^i. \qquad (6.7.111)$$

To evaluate this integral it is convenient to reintroduce the variables y^i_α. Taking note of the super Jacobian (6.7.52) and dropping the primes on the integration variables, one gets

$$\chi(C) = (2\pi\mathrm{i}\,\mathrm{d}t)^{-\frac{1}{2}m}(2\pi)^{-m}\int \exp(\tfrac{1}{4}\mathrm{i}R_{ijkl}y^i_1y^j_1y^k_2y^l_2\,\mathrm{d}t)\,g^{-\frac{1}{2}}\mathrm{d}^mx\prod_{i=1}^{m}\mathrm{d}y^i_1\,\mathrm{d}y^i_2. \qquad (6.7.112)$$

Expanding the exponent and remembering the rules for integrating over $\mathbf{R}_a^{2m}$, one sees immediately that $\chi(C)$ vanishes when m is odd. When m is even one has

$$\chi(C) = \frac{(2\pi)^{-3m/2}}{4^{m/2}(m/2)!}\int (R_{ijkl}y^i_1y^j_1y^k_2y^l_2)^{m/2}g^{-\frac{1}{2}}\mathrm{d}^mx\,\mathrm{d}y^1_1\,\mathrm{d}y^1_2\ldots\mathrm{d}y^m_1\,\mathrm{d}y^m_2$$

$$
\begin{aligned}
&= (-1)^{\frac{1}{2}m(m-1)} \frac{(2\pi)^{-3m/2}}{4^{m/2}(m/2)!} \int g^{-\frac{1}{2}} \varepsilon^{i_1 \dots i_m} \varepsilon^{j_1 \dots j_m} R_{i_1 i_2 j_1 j_2} \dots R_{i_{m-1} i_m j_{m-1} j_m} \\
&\quad \times y_1^1 \dots y_1^m \, y_2^1 \dots y_2^m \, \mathrm{d}y_2^m \dots \mathrm{d}y_2^1 \, \mathrm{d}y_1^m \dots \mathrm{d}y_1^1 \\
&= \frac{1}{(8\pi)^{m/2}(m/2)!} \int_C g^{-\frac{1}{2}} \varepsilon^{i_1 \dots i_m} \varepsilon^{j_1 \dots j_m} R_{i_1 i_2 j_1 j_2} \dots R_{i_{m-1} i_m j_{m-1} j_m} \, \mathrm{d}^m x,
\end{aligned}
\tag{6.7.113}
$$

in which the antisymmetric permutation symbol $\varepsilon^{i_1 \dots i_m}$ has been introduced and, in passing to the last line, use has been made of the fact that $\mathrm{i}^m(-1)^{\frac{1}{2}m(m-1)} = 1$ when m is even.

Expression (6.7.113) is the well-known Chern–Gauss–Bonnet formula for the Euler–Poincaré characteristic. Its derivation via a Feynman functional integral involving integration over both c- and a-type variables shows unequivocally that the functional integral is of more than purely formal validity. Note should be taken too of the fact that the normalization constant in front of the final integral derives ultimately from the 'formal' normalization constant (6.7.110) and its antecedents.

Exercises

6.1 Give an axiomatic construction of the set of all operator-valued r-forms on C as a supervector space. By taking the direct sum (over r) of the sets of operator-valued r-forms on C and by defining the exterior product of operator-valued forms in a natural way, given an axiomatic construction of this sum as a superalgebra over Λ_∞.

6.2 Verify the consistency of the laws (6.2.22) and (6.2.28).

6.3 Show that in a local chart one may write

$$p \cdot \mathbf{X} = \tfrac{1}{2}\{X^i(x), p_i\} \quad \text{for all } \mathbf{X} \text{ in } \mathfrak{X}(C),$$

where p is the momentum operator and the p_i are its components in the chart.

6.4 The exterior derivative $\mathbf{d}$ converts a scalar field ϕ into a 1-form $\mathbf{d}\phi$ on C. Introduce a symbol d that converts ϕ into a covariant vector *operator*, by the definition

$$\mathrm{d}\phi \overset{\text{def}}{=} x(\mathbf{d}\phi) = (\mathbf{d}\phi)(x) \quad \text{(see eq. (6.2.39))}.$$

Show that

$$\mathbf{X} \cdot \mathrm{d}\phi = x(\mathbf{X})\phi = \mathbf{X}(x)\phi \quad \text{(see eq. (6.2.36))}.$$

and verify that $\mathrm{d}\phi$ satisfies the law (6.2.22) for all ϕ in $\mathfrak{F}(C)$.

6.5 Consider a $(1, 0)$-dimensional system having Lagrangian $\frac{1}{2}\dot{x}^2$ and Hamiltonian $\frac{1}{2}p^2$. Suppose the body of the configuration space of the system is a circle of circumference a_B. Then in the quantum theory the chart with coordinate x may be used globally provided one makes, for the position eigenvectors, the identification

$$|x'+a, t\rangle = |x', t\rangle$$

where a is a real c-number with body a_B. This corresponds to imposing the periodicity condition

$$\psi(x'+a, t) = \psi(x', t)$$

on the wave function. Show that eq. (6.2.75) takes the form

$$\langle x', t|p(t) = \left(-\mathrm{i}\frac{\partial}{\partial x'} - \omega\right)\langle x', t|$$

where ω, without loss of generality, can be chosen to be an arbitrary constant real c-number. Show that the spectrum of the momentum operator is restricted to the discrete values

$$p' = -\omega + 2\pi n/a, \quad n = ..., -2, -1, 0, 1, 2, ...$$

corresponding to the energy eigenvalues $\frac{1}{2}(-\omega + 2\pi n/a)^2$. Conclude that the souls of both ω and a must vanish if the momentum and energy are to be physical observables and that, corresponding to the single classical system, there is an infinity of physically distinct quantum systems, one for each value of ω in the range $0 \leqslant \omega < 2\pi/a$.

6.6 Suppose the body of the configuration space is a flat Klein bottle and the system is a free particle. The universal covering space of the Klein bottle is $\mathbf{R}^2$. To describe the system one may use Cartesian coordinates x, y in $\mathbf{R}_c{}^2$ provided the following identifications are made for the position eigenvectors:

$$|(-1)^n x' + ma, y' + nb, t\rangle = |x', y', t\rangle$$

for all $m, n \in Z$. Here a and b are positive real numbers, a being the circumference of the Klein bottle in the x direction and b the circumference in the y direction. The Lagrangian of the system is $L = \frac{1}{2}(\dot{x}^2 + \dot{y}^2)$, and the Hamiltonian is $H = \frac{1}{2}(p_x^2 + p_y^2)$. The first Betti number of the Klein bottle is $b_1 = 1$, and the representatives of the first cohomology classes may be chosen to be the 1-parameter family of 1-forms ω with

$$\omega_x = 0, \quad \omega_y = \text{const.}$$

Equation (6.2.75) then takes the form

$$\langle x', y', t | p_x(t) = -\mathrm{i}\frac{\partial}{\partial x'}\langle x', y', t|.$$

$$\langle x', y', t | p_y(t) = \left(-\mathrm{i}\frac{\partial}{\partial y'} - \omega_y\right)\langle x', y', t|.$$

Show that the operator $p_x(t)$ has no eigenfunctions satisfying the periodicity conditions of the Klein bottle but that its square does. Show that the eigenvalues of the energy are

$$\frac{1}{2}\left[\left(\frac{2\pi k}{a}\right)^2 + \left(-\omega_y + \frac{\pi l}{b}\right)^2\right], \quad k = 0, 1, 2, \ldots, \; l = \ldots -2, -1, 0, 1, 2, \ldots,$$

and that ω_y may be restricted to the range $0 \leqslant \omega_y < \pi/b$ on the real line. Find the corresponding normalized energy eigenfunctions (include the time dependent Schrödinger phase factor) and show that they form a complete set on the Klein bottle.

6.7 The configuration space of the system considered in exercise **6.5** is the manifold of the group $U(1)$, and the Lagrangian is that of a free particle. The WKB approximation to the partial amplitude (6.4.31) for this system is therefore exact. In carrying out the homotopy analysis of trajectories one may limit oneself to the chart $0 \leqslant x'_B < a$. Let the origin be the fixed point x_0 of the homotopy mesh, and let the path $\mu_{x'x_0}$ consist simply of the points sx' with $0 \leqslant s \leqslant 1$. The first homotopy group of the configuration space is $\pi_1(S^1) = Z$. Let the closed path labelled by the integer $m \in Z$ be the path that passes around the circle S^1 m times in the positive x direction. Let θ be the parameter that labels the 1-dimensional representation χ of $\pi_1(S^1)$ according to eqs. (6.4.43) and (6.4.44). Then $\chi(m) = \mathrm{e}^{\mathrm{i}m\theta}$. Let $\omega = (\theta + 2n\pi)/a$ where n is any integer, and let the phase $\phi(x')$ appearing in eq. (6.4.31) be chosen to be

$$\phi(x') = \omega x'.$$

(Note that this phase has a discontinuity of magnitude ωa at $x' = 0$ (or $x' = a$).) Then the partial amplitude labelled by the element m of $\pi_1(S^1)$ is given by

$$K^m(x'', t'' | x', t') = \frac{1}{\sqrt{[2\pi\mathrm{i}(t'' - t')]}}\exp\left[\frac{\mathrm{i}}{2}\frac{(x'' - x' + ma)^2}{t'' - t'} + \mathrm{i}\omega(x'' - x')\right].$$

Show that the total amplitude

$$\langle x,t\,|\,x',t'\rangle_\chi = \sum_{m=-\infty}^{\infty} \chi(m)\, K^m(x'',t''\,|\,x',t')$$

is everywhere continuous on the configuration space (and hence satisfies the necessary periodicity condition for the circle) and is also given by

$$\langle x'',t''\,|\,x',t'\rangle_\chi = \sum_{E'} \langle x'',t''\,|\,E'\rangle\langle E'\,|\,x',t'\rangle$$

where $\langle x'',t''\,|\,E'\rangle$ is the energy eigenfunction for the system (with the time dependent Schrödinger phase factor included) the E' being the energy eigenvalues found in exercise **6.5**.

6.8 Another system for which the WKB approximation to the partial amplitudes is exact is the free particle on the flat Klein bottle, studied in exercise **6.6**. The first homotopy group of the Klein bottle is non-Abelian, but its 1-dimensional characters are functions only of the numbers, m and n, of times a closed path circles around the Klein bottle, in the positive x and y directions respectively, regardless of the order of circling. In fact, since the cofactor of the commutator subgroup is $Z \times Z_2$ the dependence on m is only mod 2. With the torsion subgroup Z_2 eliminated, the 1-dimensional characters depend only on n and are given by

$$\chi(m,n) = \mathrm{e}^{\mathrm{i}n\theta_y}, \quad 0 \leqslant (\theta_y)_B < 2\pi.$$

Confine attention to the chart $0 \leqslant x'_B < a$, $0 \leqslant y'_B < b$, and introduce an obvious homotopy mesh based on the origin of (x',y') coordinates. If the phase $\phi(x',y')$ is chosen to be

$$\phi(x',y') = \omega_y y', \quad \omega_y = (\theta_y + 2r\pi)/b, \quad r \text{ an integer},$$

show that the partial amplitudes for the system are given by

$$K^{m,n}(x'',y'',t''\,|\,x',y',t') = \frac{1}{2\pi\mathrm{i}(t''-t')}$$

$$\times \exp\left[\frac{\mathrm{i}}{2}\frac{((-1)^n x'' - x' + ma)^2 + (y'' - y' + nb)^2}{t''-t'} + \mathrm{i}\omega_y(y''-y')\right].$$

Show that with this choice of phase the total amplitude

$$\langle x'',y'',t''\,|\,x',y',t'\rangle_\chi = \sum_{m,n=-\infty}^{\infty} \chi(m,n)\, K^{m,n}(x'',y'',t''\,|\,x',y',t')$$

is continuous on the Klein bottle and is equally well given by

$$\langle x'', y'', t'' | x', y', t' \rangle_\chi = \sum_{E'} \langle x'', y'', t'' | E' \rangle \langle E' | x', y', t' \rangle,$$

where the E' are the energy eigenvalues found in exercise **6.6** and the $\langle x'', y'', t'' | E' \rangle$ are the corresponding energy eigenfunctions. Show that if one tries to make non-trivial use of the torsion subgroup, by setting

$$\chi(m, n) = (-1)^m \, \mathrm{e}^{\mathrm{i}n\theta_y},$$

no choice for the phase $\phi(x', y')$ can then be found that will make $\langle x'', y'', t'' | x', y', t' \rangle_\chi$ continuous.

6.9 Prove the conservation law (6.6.17). *Hint*: Write

$$\frac{\partial D}{\partial t''} = D D^{-1ji} \frac{\partial D_{ij}}{\partial t''} = -D D^{-1ji} \frac{\partial^3 S}{\partial t'' \, \partial x''^i \, \partial x'^j}$$

where (D^{-1ij}) is the matrix inverse to (D_{ij}). Then use the result of differentiating the Hamilton–Jacobi equation (6.6.12) with respect to x''^i and x'^j.

6.10 Equations (6.6.37) and (6.6.38) yield the following expression for the (super)commutator function (5.1.22):

$$\tilde{G}^{ij'''} = \frac{\partial x_c^i}{\partial x'^k} D^{-1kl} \frac{\partial x_c^{j'''}}{\partial x''^l} - \frac{\partial x_c^{j'''}}{\partial x'^k} D^{-1kl} \frac{\partial x_c^i}{\partial x''^l}.$$

Show that this is equal to the Peierls bracket $(x_c^i, x_c^{j'''})$. *Hint*: Show that it reduces to

$$\frac{\partial x_c^i}{\partial x'^k} \frac{\partial x_c^{j'''}}{\partial p_k'} - \frac{\partial x_c^{j'''}}{\partial x'^k} \frac{\partial x_c^i}{\partial p_k'},$$

and use the equivalence of the Peierls and Poisson brackets.

6.11 Show that the Feynman propagator (5.6.85) for the Bose oscillator may be expressed as an integral:

$$G(s, t) = \frac{1}{2\pi} \int \frac{\mathrm{e}^{-\mathrm{i}p^0(s-t)}}{-(p^0)^2 + \omega^2 - \mathrm{i}\varepsilon} \mathrm{d}p^0,$$

where ε is an infinitesimal positive real number and the integration is over the real axis. Show that the advanced Green's function (5.6.5) for the Bose oscillator may be similarly represented:

$$G^+(s, t) = \frac{1}{2\pi} \int \frac{\mathrm{e}^{-\mathrm{i}p^0(s-t)}}{-(p^0 - \mathrm{i}\varepsilon)^2 + \omega^2} \mathrm{d}p^0.$$

In the remainder of this exercise *assume* that the corresponding Green's functions for a relativistic free scalar field ϕ of mass m are given by analogous integrals

$$G(x, x') = \frac{1}{(2\pi)^4} \int \frac{e^{ip\cdot(x-x')}}{p^2+m^2-i\varepsilon} d^4p,$$

$$G^{+}(x, x') = \frac{1}{(2\pi)^4} \int \frac{e^{ip\cdot(x-x')}}{-(p^0-i\varepsilon)^2+\mathbf{p}^2+m^2} d^4p,$$

reflecting the fact that a relativistic bosonic field may be viewed as an infinite collection of Bose oscillators. Here $\mathbf{p}$ is a 3-vector, and

$$p^2 = -(p^0)^2+\mathbf{p}^2, \quad p\cdot(x-x') = -p^0(x^0-x'^0)+\mathbf{p}\cdot(\mathbf{x}-\mathbf{x}'),$$

$$d^4p = dp^0\, dp^1\, dp^2\, dp^3, \quad \mathbf{x} = (x^1, x^2, x^3), \text{ etc.}$$

Suppose now that terms of cubic and (possibly) higher order in ϕ are added to the field Lagrangian. Then use of the loop expansion (around the classical vacuum $\phi = 0$) becomes a standard method for dealing with the resulting quantum system. One of the first and simplest integrals encountered in the loop expansion for, say, the so-called *2-point function* $\langle T(\phi(x)\,\phi(x'))\rangle$ (the average being the vacuum average) is

$$S_{,ijkl}\, G^{ln} G^{mk} S_{,mnj}.$$

The Fourier transform of this leads to 4-momentum integrals of the form

$$\int \frac{d^4k}{(k^2+m_1^2-i\varepsilon)\,[(p+k)^2+m_2^2-i\varepsilon]}$$

$$= \int_0^1 dx \int d^4k[k^2+x(1-x)\,p^2+xm_1^2+(1-x)\,m_2^2-i\varepsilon]^{-2}, \quad \text{(I)}$$

in which the possible presence of fields of different mass has been allowed for. The determinant $(\det G^{+})^{-\frac{1}{2}}$, on the other hand, contributes to the loop expansion a term of the form

$$S_{,ikl}\, G^{+ln} G^{+mk} S_{,mnj}$$

in which the two advanced Green's functions have the same orientation around the loop. The Fourier transform of this leads to 4-momentum integrals of the form

$$\int_0^1 dx \int d^4k[-(k^0-i\varepsilon)^2+\mathbf{k}^2+x(1-x)\,p^2+xm_1^2+(1-x)\,m_2^2]^{-2}. \quad \text{(II)}$$

The conventional way of dealing with these integrals is to ignore (II) and to evaluate (I) after rotating the k^0 contour through 90° in the complex plane, which is equivalent to working from the beginning in Euclidean rather than Minkowski space. The rotation, however, is permissible only if the induced arcs at infinity contribute nothing. (I) is, in fact, a logarithmically divergent integral, and the contributions from the arcs at infinity do *not* vanish.

Calculate both (I) and (II), evaluating (I) by rotating the k^0 contour in such a way as to avoid the poles. Show that the contributions from the arcs at infinity are precisely cancelled by (II) and hence that the conventional procedure after all leads to correct results. (This is true quite generally.) In evaluating the arc contributions and the integral (II), carry out the 3-momentum integrations (over $\mathbf{k}$) first, reserving the k^0 integrations to the last. (The integration over x need not actually be performed.)

6.12 Suppose the points x' and x'' and the times t' and t'' are close enough together that there is essentially only one trajectory between them that dominates the functional integral for the transition amplitude. Suppose also that there are no conjugate points between x', t' and x'', t''. Then we may ignore homotopy equivalence classes and write the WKB approximation simply in the form

$$\langle x'', t'' | x', t' \rangle \approx (2\pi \mathrm{i})^{-m/2} D^{\frac{1}{2}}(x'', t'' | x', t')\, \mathrm{e}^{\mathrm{i}[S(x'', t'' | x', t') + \phi(x'') - \phi(x')]}.$$

Let t lie between t' and t''. Show that the WKB approximation approximately satisfies the combination law

$$\langle x'', t'' | x', t' \rangle = \int \langle x'', t'' | x''', t \rangle \langle x''', t | x', t' \rangle\, \mathrm{d}^m x''$$

$$\approx (2\pi \mathrm{i})^{-m} \int D^{\frac{1}{2}}(x'', t'' | x''', t)\, D^{\frac{1}{2}}(x''', t | x', t')$$

$$\times\, \mathrm{e}^{\mathrm{i}[S(x'', t'' | x''', t) + S(x''', t | x', t') + \phi(x'') - \phi(x')]}\, \mathrm{d}^m x'''.$$

Hint: Use the method of stationary phase to approximate the final integral by a Gaussian integral, and make use of eqs. (6.6.18), (6.6.19), (6.6.22) and the determinant of the following corollary of eqs. (6.6.23) and (6.6.24):

$$D_{ij}(x'', t'' | x', t') = D_{il}(x'', t'' | x_c, t)\, M^{-1lk} D_{kj}(x_c, t | x', t').$$

Note that since there are no conjugate points between x', t' and x'', t'' all the matrices in this equation are positive definite.

6.13 For the system of section 6.7 show that the operator $\mathfrak{Q}(t)$, when viewed with respect to the wave functions $\psi_{i_1 \ldots i_r}(x, t)$, is the same as $-\mathrm{i}$ times the covariant divergence operator. Specifically, show that

$$\langle x', i_1, \ldots, i_r, t | \mathfrak{Q}(t) | \psi \rangle = -\mathrm{i}\psi_{i_1 \ldots i_r j;}{}^{j}(x', t')$$

or alternatively,

$$(\mathbf{d}^*\psi)_{i_1 \ldots i_r} = -(-1)^r \psi_{i_1 \ldots i_r j;}{}^{j}.$$

6.14 Compute the terms of order $(t-t')^2$ in eqs. (6.7.13) and (6.7.14) and the term of order $(t-t')$ in eq. (6.7.15). Use these results to obtain a formal derivation of the superdeterminant of G^+ for the system of section 6.7. Show that

$$(\operatorname{sdet} G^+[x, z^*, z]_{x'z't'}^{x''z''^*t''})^{-\frac{1}{2}} = \text{const} \times g^{\frac{1}{4}}(x'')[\prod_{t'<t<t''} g^{-\frac{1}{2}}(x(t))]\, g^{\frac{1}{4}}(x').$$

Comments on Chapter 6

The main lesson to be learned from this chapter is that considerable care must be taken in defining quantum mechanics on nontrivial supermanifolds, and even more care must be taken in giving meaning to the Feynman functional integrals for the associated transition amplitudes of the theory. But if this is done, one finds that the functional integral can be manipulated in a way that is not only internally consistent but also useful in computations that are by no means trivial. One of the most beautiful illustrations of this utility, described in section 6.7, stems from the close connection between supersymmetry, cohomology theory, Morse theory and the Atiyah–Singer index theorem, which was brought to light by work of E. Witten and L. Alvarez-Gaumé (see references at the end of the book).

The applications of the functional integral that appear here are (apart from exercise **6.11**) confined to superclassical systems having finite numbers of degrees of freedom. The utility of the functional integral in quantum field theory has long been known, from its beginnings in perturbation theory to recent investigations of lattice quantum field theories on supercomputers. If an attempt were made to include field theoretical applications in this book it would become a book on functional integration, not on supermanifolds. Such a book should be written. But nowadays, in addition to applications to quantum field theory, it would have to include an account of such astonishing developments as the recent work of Witten (see references) on the use of functional integration to compute topological invariants of 3-manifolds.

References

Superanalysis

Berezin, F. A., *The Method of Second Quantization*, Academic Press, New York (1966).

– 'Automorphisms of a Grassmann algbebra', *Akademia Nauk SSSR, Math. Notes*, **1**, 180 (1967).

– *Introduction to Superanalysis*, Reidel, Dordrecht (1987).

Khrennikov, A. Yu., 'Functional superanalysis', *Russ. Math. Surv.*, **43:2**, 103 (1988).

Rogers, A., 'Realizing the Berezin integral as a superspace contour integral', *J. Math. Phys.*, **27**, 710 (1986).

Supermanifolds

Arnowitt, R. & P. Nath, 'Riemannian geometry in spaces with Grassmann coordinates', *Gen. Rel. Grav.*, **7**, 89 (1976).

Bartocci, C. & U. Bruzzo, 'Cohomological methods in supermanifold theory', in *Group Theoretical Methods in Physics, Lecture Notes in Physics*, **313**, Springer, Berlin–New York (1988).

Batchelor, M., 'Graded manifolds and supermanifolds', in *Mathematical Aspects of Superspace*, edited by C. J. S. Clarke, A. Rosenblum and H. J. Siefert, Reidel, Dordrecht (1984).

Berezin, F. A., 'Differential forms on supermanifolds', *Sov. J. Nucl. Phys.*, **30**, 605 (1979).

Berezin, F. A. & D. A. Leites, 'Supermanifolds', *Soviet Math. Dokl.*, **16**, 1218 (1975).

Bernstein, I. N. & D. A. Leites, 'Integrable forms and the Stokes formula on supermanifolds', *Funct. Anal. Appl.*, **11**, 45 (1977).

– 'Integration of differential forms on supermanifolds', *Funct. Anal. Appl.*, **11**, 219 (1977).

Bruzzo, U., 'Supermanifolds, supermanifold cohomology and super vector bundles', in *Differential Geometrical Methods in Theoretical Physics (Como, 1987)*, Kluwer Academic Publishers, Dordrecht (1988).

Choquet-Bruhat, Y., *Graded Bundles and Supermanifolds*, Bibliopolis, Naples (1989).

Kostant, B., 'Graded manifolds, graded Lie theory and prequantization', in *Differential Geometric Methods in Mathematical Physics, Lecture Notes in Mathematics, 570*, edited by K. Bleuler and A. Reetz, Springer, Berlin–New York (1977).

LeBrun, C., 'Moduli of super Riemann surfaces', *Comm. Math. Phys.*, **117**, 159 (1988).

Leites, D. A., 'Introduction to the theory of supermanifolds', *Russian Math. Surveys*, **35**, 1 (1980).

– 'Selected problems of supermanifold theory', *Duke Math. Journal*, **54**, 649 (1987).

Rabin, J. M., 'How different are the supermanifolds of Rogers and DeWitt?' *Comm. Math. Phys.*, **102**, 123 (1985).

– 'Berezin integration on general fermionic supermanifolds', *Comm. Math. Phys.*, **103**, 431 (1986).

– 'Supermanifold cohomology and the Wess–Zumino term of the covariant superstring action', *Comm. Math. Phys.*, **108**, 375 (1987).

Rabin, J. M. & L. Crane, 'Global properties of supermanifolds', *Comm. Math. Phys.*, **100**, 141 (1985).

Rogers, A., 'A global theory of supermanifolds', *J. Math. Phys.*, **21**, 1352 (1980).

– 'Some examples of compact supermanifolds with non-Abelian fundamental group', *J. Math. Phys.*, **22**, 443 (1981).

– 'Consistent superspace integration', *J. Math. Phys.*, **26**, 385 (1985).

– 'On the existence of global integral forms on supermanifolds', *J. Math. Phys.*, **26**, 2749 (1985).

– 'Graded manifolds, supermanifolds and infinite-dimensional Grassmann algebras', *Comm. Math. Phys.*, **105**, 375 (1986).

Rothstein, M. J., 'Deformations of complex supermanifolds', *Proc. Amer. Math. Soc.*, **95**, 255 (1985).

– 'The axioms of supermanifolds and a new structure arising from them', *Trans. Amer. Math. Soc.*, **297**, 159 (1986).

– 'Integration on noncompact supermanifolds', *Trans. Amer. Math. Soc.*, **299**, 387 (1987).

Pestov, V. G., 'Interpreting superanalyticity in terms of convergent series', *Class. Quant. Grav.*, **6**, L145 (1989).

Shander, V. N., 'Vector fields and differential equations on supermanifolds', *Funct. Anal. Appl.* **14**, 160 (1980).

– 'Complete integrability of ordinary differential equations on supermanifolds', *Funct. Anal. Appl.*, **17**, 74 (1983).

Uhlman, A., 'The Cartan algebra of exterior differential forms as a supermanifold: morphisms and manifolds associated with them', *J. Geom. and Phys.*, **1**, 25 (1984).

Voronov, F. F. & A. V. Zorich, 'Complex of forms on a supermanifold', *Akademia Nauk SSSR, Funkt. Anal. Pril.*, **20**, 58 (1986).

Super Lie groups and super Lie algebras

Balantekin, A. B. 'Anomalies and eigenvalues of Casimir operators for Lie groups and supergroups', *J. Math. Phys.* **23**, 486 (1982).

Bars, I., B. Morel & H. Ruegg, 'Kac–Dynkin diagrams and supertableaux', CERN preprint TH.3333 (1982).

Berezin, F. A. & G. I. Kac, 'Lie groups with commuting and anticommuting parameters', *Math. USSR Sbornik*, **11**, 311 (1970).

Bernstein, J. N. & D. A. Leites, 'Irreducible representations of Lie superalgebra of vector fields and invariant differential operators', *Serdika* (*Bulgarian Math. J.*), **7**, 320 (1981) (in Russian).

Corwin, L., Y. Ne'eman & S. Sternberg, 'Graded Lie algebras in mathematics and physics (Bose–Fermi symmetry)', *Rev. Mod. Phys.*, **47**, 573 (1975).

Delduc, F. & M. Gourdin, 'Mixed supertableaux of the superunitary groups'. CERN preprint TH.3568 (1983).

Freund, P. G. O. & I. Kaplansky, 'Simple supersymmetries', *J. Math. Phys.*, **17**, 228 (1976).

Hurni, J. P. & B. Morel, 'Irreducible representations of $SU(m/n)$', *J. Math. Phys.*, **24**, 157 (1983).

Kac, V. C., 'Classification of simple Lie superalgebras', *Funct. Anal. Appl.*, **9**, 263 (1975).

– 'A sketch of Lie superalgebra theory', *Comm. Math. Phys.* **53**, 31 (1977).

– 'Characters of typical representations of classical Lie superalgebras', *Comm. in Algebra*, **5** 889 (1977).

– 'Classification of simple Z-graded Lie superalgebras and simple Jordan superalgebras', *Comm. in Algebra*, **5**, 1375 (1977).
– 'Lie superalgebras', *Advances in Math.*, **26**, 8 (1977).
– 'Classification of simple algebraic supergroups', *Uspekhi Math. Nauk.*, **32**, No. 3, 214 (1977).
– 'Infinite dimensional algebras, Dedekind's η-function, classical Möbius function and the very strange formula', *Advances in Math.*, **30**, 85 (1978).
– 'Representations of classical Lie superalgebras', in *Lecture Notes in Mathematics 676*, Springer-Verlag (1978).
Leites, D. A., 'Irreducible representations of Lie superalgebras of vector fields and invariant differential operators', *Funct. Anal. Appl.*, **16**, 62 (1982).
– 'Representations of Lie superalgebras', *Theor. Math. Phys.* (1982).
Marcu, M., 'The representations of *Spl*(2, 1) – an example of representations of basic superalgebras', *J. Math. Phys.*, **21**, 1277 (1980).
Pestov, V. G., 'On a 'super' version of Lie's Third Fundamental Theorem', *Lett. Math. Phys.*, **18**, 27 (1989).
Picken, R. F., 'Supergroups have no extra topology', *J. Phys.*, **A16**, 3457 (1983).
Rogers, A., 'Super Lie groups: global topology and local structure', *J. Math. Phys.* **22**, 939 (1981).
Scheunert, M., *The Theory of Lie Supergroups. Lecture Notes in Mathematics No. 716*, Springer Verlag (1979).
Scheunert, M., W. Nahm & V. Rittenberg, 'Irreducible representations of the *Osp*(2, 1) and *Spl*(2, 1) graded Lie Algebras', *J. Math. Phys.*, **18**, 155 (1977).
Srivastava, P. P., 'On a generalized supergauge transformation and superfields'. *Lett. Nuovo Cimeto*, **12**, 161 (1975).
– 'Algebra of general covariance group in superspace as the closure of algebras of superconformal, supergauge and linear groups', *Lett. Nuovo Cimento*, **14**, 471 (1975).
– 'On pseudo-Lie algebra in superspace', *Lett. Nuovo Cimento*, **15**, 588 (1976).
– 'Superconformal group in flat superspace', *Lett. Nuovo Cimento*, **17**, 357 (1976).
Stawraki, G. L., 'A nonlocal model of field interactions and the algebra of field operators', in *High Energy Physics and the Theory of Elementary Particles*, edited by V. P. Shelest, Naukova Dumka, Kiev (1967) in Russian.
– 'On a generalization of Lie Algebra', Kiev Institute for Theoretical Physics preprint ITF-67-21 (1967) in Russian.

Index theorems

Alvarez-Gaumé, L., 'Supersymmetry and the Atiyah–Singer index theorem', *Comm. Math. Phys.*, **90**, 161 (1983).
Arai, A., 'Path integral representation of the index of Kähler–Dirac operators on an infinite dimensional manifold', *J. Funct. Anal.*, **82**, 330 (1989).
Borisov, N. V., W. Müller & R. Schrader, 'Relative index theorems and super-symmetric scattering theory', *Comm. Math. Phys.*, **14**, 475 (1988).
Friedan, D. & P. Windey, 'Supersymmetric derivation of the Atiyah–Singer index and the chiral anomaly', *Nucl. Phys.*, **B235**, 395 (1984).
Getzler, E., 'Pseudodifferential operators on supermanifolds and the Atiyah–Singer index theorem', *Comm. Math. Phys.*, **92**, 163 (1983).
Lott, J., 'Supersymmetric path integrals', *Comm. Math. Phys.*, **108**, 605 (1987).
Mavromatos, N. E., ' A note on the Atiyah–Singer index theorem for manifolds with totally antisymmetric *H* torsion', *J. Phys. A: Math. Gen.*, **21**, 2279 (1988).
Perelomov, A. M., 'Supersymmetric chiral models: Geometrical aspects', *Physics Reports*, **174**, No. 4, 229 (1989).

Rempel, S. & T. Schmitt, 'Pseudodifferential operators and the index theorem on supermanifolds', *Math. Nachr.*, **111**, 153 (1983).
Rogers, A., 'A superspace path integral proof of the Gauss–Bonnet–Chern theorem,' *J. Geom. and Phys.* **4**, 417 (1987).
Serganova, V. & A. Weintrob, 'Simple singularities of functions on supermanifolds', *Math. Scand.*, **64**, 251 (1989).
Windey, P., 'Supersymmetric quantum mechanics and the Atiyah–singer index theorem', *Acta Phys. Polon.*, **B15**, 434 (1984).
Witten, E., 'Constraints on supersymmetry breaking', *Nucl. Phys.*, **B202**, 253 (1980).
– 'Supersymmetry and More Theory', *J. Diff. Geom.*, **17**, 661 (1982).
– 'Global aspects of current algebra', *Nucl. Phys.*, **B223**, 422 (1983).

Elementary examples of supersymmetry

Barducci, A., F. Bordi & R. Casalbuoni, 'Path-integral quantization of spinning particles interacting with crossed external electromagnetic fields', CERN preprint TH.2971 (1980).
Salam, A. & J. Strathdee, 'Supergauge transformations', *Nucl. Phys.*, **76B**, 477 (1974).
Salomonson, P. & J. W. van Holten, 'Fermionic coordinates and sypersymmetry in quantum mechanics', CERN preprint TH.3148 (1981).
van Holten, J. W., 'Instantons in supersymmetric quantum mechanics', CERN preprint TH.3191 (1981).
van Hove, L., 'Supersymmetry and positive temperature for simple systems', CERN preprint TH.3280 (1982).
Wess, J. & B. Zumino, 'Supergauge transformations in four dimensions', *Nucl. Phys.*, **70B**, 39 (1974).
– 'A Lagrangian model invariant under supergauge transformations', *Phys. Lett.*, **49B**, 52 (1974).

Topological quantum field theory

Birmingham, D., M. Rakowski & G. Thompson, 'BRST quantization of topological field theories', *Nucl. Phys.*, **B315**, 577 (1989).
Witten, E., 'Topological quantum field theory', *Comm. Math. Phys.*, **117**, 353 (1988).
– 'Quantum field theory and the Jones polynomial', *Comm. Math. Phys.*, **121**, 351 (1989).

Index

Printed in the United States
By Bookmasters